Biological Importance of Annexins

Biological Importance of Annexins

Edited by Sam Williams

SYRAWOOD
PUBLISHING HOUSE
New York

Published by Syrawood Publishing House,
750 Third Avenue, 9th Floor,
New York, NY 10017, USA
www.syrawoodpublishinghouse.com

Biological Importance of Annexins
Edited by Sam Williams

International Standard Book Number: 978-1-64740-396-6 (Hardback)

Cataloging-in-publication Data

Biological importance of annexins / edited by Sam Williams.
p. cm.
Includes bibliographical references and index.
ISBN 978-1-64740-396-6
1. Lipocortins. 2. Calcium-binding proteins. 3. Molecular biology.
4. Biochemistry. I. Williams, Sam.
QP552.L48 B56 2023
570--dc23

TABLE OF CONTENTS

Preface VII

Chapter 1 **Structure of the ALS Mutation Target Annexin A11 Reveals a Stabilising N-Terminal Segment** 1
Peder A. G. Lillebostad, Arne Raasakka, Silje J. Hjellbrekke, Sudarshan Patil, Trude Røstbø, Hanne Hollås, Siri A. Sakya, Peter D. Szigetvari, Anni Vedeler and Petri Kursula

Chapter 2 **Localization of Annexin A6 in Matrix Vesicles during Physiological Mineralization** 18
Ekeveliny Amabile Veschi, Maytê Bolean, Agnieszka Strzelecka-Kiliszek, Joanna Bandorowicz-Pikula, Slawomir Pikula, Thierry Granjon, Saida Mebarek, David Magne, Ana Paula Ramos, Nicola Rosato, José Luis Millán, Rene Buchet, Massimo Bottini and Pietro Ciancaglini

Chapter 3 **(20S)G-Rh2 Inhibits NF-κB Regulated Epithelial-Mesenchymal Transition by Targeting Annexin A2** 34
Yu-Shi Wang, He Li, Yang Li, Shiyin Zhang and Ying-Hua Jin

Chapter 4 **Exploring Biased Agonism at FPR1 as a Means to Encode Danger Sensing** 49
Jieny Gröper, Gabriele M. König, Evi Kostenis, Volker Gerke, Carsten A. Raabe and Ursula Rescher

Chapter 5 **Annexin A1/Formyl Peptide Receptor Pathway Controls Uterine Receptivity to the Blastocyst** 67
Cristina B. Hebeda, Silvana Sandri, Cláudia M. Benis, Marina de Paula-Silva, Rodrigo A. Loiola, Chris Reutelingsperger, Mauro Perretti and Sandra H. P. Farsky

Chapter 6 **Expression of Annexin A2 Promotes Cancer Progression in Estrogen Receptor Negative Breast Cancers** 84
Amira F. Mahdi, Beatrice Malacrida, Joanne Nolan, Mary E. McCumiskey, Anne B. Merrigan, Ashish Lal, Shona Tormey, Aoife J. Lowery, Kieran McGourty and Patrick A. Kiely

Chapter 7 **Annexin A2 Egress during Calcium-Regulated Exocytosis in Neuroendocrine Cells** 108
Marion Gabel, Cathy Royer, Tamou Thahouly, Valérie Calco, Stéphane Gasman, Marie-France Bader, Nicolas Vitale and Sylvette Chasserot-Golaz

Chapter 8 **Annexin A2 in Inflammation and Host Defense** 122
Valentina Dallacasagrande and Katherine A. Hajjar

Chapter 9 **Diverse Roles of Annexin A6 in Triple-Negative Breast Cancer Diagnosis, Prognosis and EGFR-Targeted Therapies** 136
Olga Y. Korolkova, Sarrah E. Widatalla, Stephen D.Williams, Diva S. Whalen, Heather K. Beasley, Josiah Ochieng, Thomas Grewal and Amos M. Sakwe

Chapter 10 **Annexin A1 Regulates NLRP3 Inflammasome Activation and Modifies Lipid Release Profile in Isolated Peritoneal Macrophages** 157
José Marcos Sanches, Laura Migliari Branco, Gustavo Henrique Bueno Duarte, Sonia Maria Oliani, Karina Ramalho Bortoluci, Vanessa Moreira and Cristiane Damas Gil

Chapter 11 **Membrane Binding Promotes Annexin A2 Oligomerization** 172
Anna Lívia Linard Matos, Sergej Kudruk, Johanna Moratz, Milena Heflik, David Grill, Bart Jan Ravoo and Volker Gerke

Chapter 12 **Targeting AnxA1/Formyl Peptide Receptor 2 Pathway Affords Protection against Pathological Thrombo-Inflammation** 183
Shantel A. Vital, Elena Y. Senchenkova, Junaid Ansari and Felicity N. E. Gavins

Chapter 13 **Annexin-A6 in Membrane Repair of Human Skeletal Muscle Cell: A Role in the Cap Subdomain** 201
Coralie Croissant, Céline Gounou, Flora Bouvet, Sisareuth Tan and Anthony Bouter

Chapter 14 **Selective Degradation Permits a Feedback Loop Controlling Annexin A6 and Cholesterol Levels in Endolysosomes of NPC1 Mutant Cells** 221
Elsa Meneses-Salas, Ana García-Melero, Patricia Blanco-Muñoz, Jaimy Jose, Marie-Sophie Brenner, Albert Lu, Francesc Tebar, Thomas Grewal, Carles Rentero and Carlos Enrich

Permissions

List of Contributors

Index

PREFACE

Every book is a source of knowledge and this one is no exception. The idea that led to the conceptualization of this book was the fact that the world is advancing rapidly; which makes it crucial to document the progress in every field. I am aware that a lot of data is already available, yet, there is a lot more to learn. Hence, I accepted the responsibility of editing this book and contributing my knowledge to the community.

Annexins (ANX) are a family of calcium-dependent membrane-binding proteins. They are distributed widely in eukaryotes, but yeasts and prokaryotes do not contain these proteins. ANX are cytosolic proteins, which have pools of a soluble form and a form stably or reversibly linked to the parts of the proteins or cytoskeleton that facilitate interactions among the extracellular matrix and the cell. They interact with a number of cell-membrane elements, involved in growth regulation, organization of cell structure, as well as intracellular signaling done by ion fluxes and enzyme modulation. Furthermore, they can also function as a typical calcium channel. The functions of these proteins are now being translated from biomedical research into the clinical practice. This book explores all the important aspects of annexin biology in the present day scenario. The various advancements in the research on their biological importance are glanced at and their applications as well as ramifications are looked at in detail. The book will serve as a valuable source of reference for graduate and postgraduate students.

While editing this book, I had multiple visions for it. Then I finally narrowed down to make every chapter a sole standing text explaining a particular topic, so that they can be used independently. However, the umbrella subject sinews them into a common theme. This makes the book a unique platform of knowledge.

I would like to give the major credit of this book to the experts from every corner of the world, who took the time to share their expertise with us. Also, I owe the completion of this book to the never-ending support of my family, who supported me throughout the project.

Editor

1

Structure of the ALS Mutation Target Annexin A11 Reveals a Stabilising N-Terminal Segment

Peder A. G. Lillebostad [1,†], Arne Raasakka [1,†], Silje J. Hjellbrekke [1], Sudarshan Patil [1], Trude Røstbø [1], Hanne Hollås [1], Siri A. Sakya [1], Peter D. Szigetvari [1], Anni Vedeler [1,*] and Petri Kursula [1,2,*]

1 Department of Biomedicine, University of Bergen, Jonas Lies vei 91, 5009 Bergen, Norway; plillebostad@gmail.com (P.A.G.L.); arne.raasakka@uib.no (A.R.); silje_johj@hotmail.com (S.J.H.); sudarshan.patil@uib.no (S.P.); trude.kr@gmail.com (T.R.); hanhol1404@gmail.com (H.H.); siriasakya@gmail.com (S.A.S.); peter.szigetvari@uib.no (P.D.S.)

2 Faculty of Biochemistry and Molecular Medicine & Biocenter Oulu, University of Oulu, Aapistie 7, 90220 Oulu, Finland

* Correspondence: anni.vedeler@uib.no (A.V.); petri.kursula@uib.no (P.K.)

† These authors contributed equally.

Abstract: The functions of the annexin family of proteins involve binding to Ca^{2+}, lipid membranes, other proteins, and RNA, and the annexins share a common folded core structure at the C terminus. Annexin A11 (AnxA11) has a long N-terminal region, which is predicted to be disordered, binds RNA, and forms membraneless organelles involved in neuronal transport. Mutations in AnxA11 have been linked to amyotrophic lateral sclerosis (ALS). We studied the structure and stability of AnxA11 and identified a short stabilising segment in the N-terminal end of the folded core, which links domains I and IV. The crystal structure of the AnxA11 core highlights main-chain hydrogen bonding interactions formed through this bridging segment, which are likely conserved in most annexins. The structure was also used to study the currently known ALS mutations in AnxA11. Three of these mutations correspond to buried Arg residues highly conserved in the annexin family, indicating central roles in annexin folding. The structural data provide starting points for detailed structure–function studies of both full-length AnxA11 and the disease variants being identified in ALS.

Keywords: annexin A11; crystal structure; solution structure; calcium binding; membrane binding; protein stability; folding; amyotrophic lateral sclerosis

1. Introduction

The annexin (Anx) family of Ca^{2+}-binding proteins is widely distributed in eukaryotes, with twelve Anxs found in vertebrates [1–3]. All Anxs share multiple structural elements, most notably an evolutionarily conserved C-terminal Anx core. This core consists of four similar domains, each comprised of ~70 amino acids arranged into five α helices, termed A–E [1,4]. The only exception is AnxA6, which contains eight domains, most likely due to a fusion of duplicated *anx*A5 and *anx*A10 genes [5]. The five helices within each domain are connected via short loops, forming a right-handed superhelix. Helices A and B as well as helices D and E are oriented pairwise in an antiparallel fashion, while helix C is positioned as a bridge orthogonally to the others [4]. The loops connecting helix A to helix B and helix D to helix E contain the Ca^{2+} binding sites [1,6].

The Anxs are structurally distinct from the EF-hand and C2-domain Ca^{2+}-binding proteins [2]. They reversibly bind negatively charged lipids in a Ca^{2+}-dependent manner, which is a defining feature of this protein family [1]. The binding is mediated through the convex side of the core structure, coupled to a conformational change induced by Ca^{2+} binding [1,2]. Through lipid binding, Anxs can affect important proteolipid membrane properties, such as curvature and phase separation [7,8].

While the folded C-terminal core structure of the vertebrate Anxs is conserved, the structure of the N terminus varies greatly between different Anxs and determines their individual properties [1]. This region typically contains 10–30 residues, but AnxA7 and AnxA11 harbour unusually long N termini. AnxA11 has the longest N-terminal domain of all Anxs, with ~200 residues [9]. The N termini of AnxA7 and AnxA11 resemble each other, both being hydrophobic, enriched in Gly, Tyr, and Pro [10–12]. The Ca^{2+}-dependent interaction of AnxA11 with phosphatidylserine changes the conformation of its N terminus [13]. Furthermore, the AnxA11 N terminus interacts with the apoptosis-linked gene-2 protein (ALG-2) [14] and S100A6 (calcyclin) [15].

Rabbit AnxA11 appears to exist as two isoforms; AnxA11A and AnxA11B [16]. Only the former binds S100A6 in the presence of Ca^{2+} via the region Gln49-Thr62. This region is absent in AnxA11B [16–18], suggesting isoform-specific functions. S100A6 binds to AnxA1 and AnxA2 in addition to AnxA11 [19,20], indicating functional overlap between these three proteins. According to the UniProt database [21], human AnxA11 is predicted to have two isoforms due to alternative splicing (UniProt ID P50995), leading to a truncation of 33 residues in the N terminus of the shorter form, while rat AnxA11 only exists as a single isoform corresponding to the longer human isoform (UniProt ID Q5XI77).

In certain cell types, AnxA11 is predominantly localised in the cytoplasm, while in others, it is mainly present in the nucleus [15,22]. It is likely that the A isoform is enriched in the nucleus, whereas the B isoform is specifically localised to the cytoplasm [16]. The N terminus was suggested to be responsible for the nuclear localisation [18,23]. In the nucleus, AnxA11 appears to be involved in events related to the cell cycle. It participates in cytokinesis, playing an essential role in the formation of the midbody. Cells lacking AnxA11 cannot complete cytokinesis and undergo apoptosis [24].

Expression levels of AnxA11 have been linked to various cancers [25–30]. Recent studies have revealed that mutations in AnxA11 play an important role in the development of the neurodegenerative disease amyotrophic lateral sclerosis (ALS) [8,31–37]. The molecular basis of these mutations is unclear, as an experimentally determined 3D structure of AnxA11 has been lacking.

Since the long N terminus of AnxA11 renders recombinant production of the protein challenging, N-terminally truncated recombinant forms were generated, and the crystal structure of one form, Δ188AnxA11, was determined. We show that a 4-residue sequence motif in the N terminus of the core structure, conserved in most Anxs, functions as a Velcro-like bridging segment and is important for the thermal stability of the Anx core structure. The currently known ALS mutations are analysed based on the crystal structure, providing a view into conserved features of the Anx core fold.

2. Materials and Methods

2.1. Cloning of Full-Length and N-Terminally Truncated Forms of AnxA11

In order to obtain full-length rat AnxA11, reverse transcriptase-polymerase chain reaction (RT-PCR) was performed using total RNA isolated from PC12 cells by the Trizol method [38]. The cDNA of AnxA11 was obtained using the iScript kit (Bio-Rad, Hercules, CA, USA) according to the manufacturer's instructions and the forward primer 5′-ATCCGG CCATGGGTATGAGCTATCCAGGCTATCCAC-3′ (*Nco*I restriction site is underlined) and the reverse primer 5′-ATCCGGGGTACCTCAGTCGTTGCCACCACAGATC-3′ (*Acc*65I restriction site is underlined). The *anx*A11 cDNA was digested with *Nco*I and *Acc*65I, purified after separation in a 1% agarose gel, and ligated into the same restriction sites of the pETM10 vector (a kind gift from Gunter Stier, Heidelberg). Two truncated versions of AnxA11 were generated by PCR using the pETM10 plasmid harbouring rat full-length *anx*A11 cDNA as a template and using the same reverse primer as for full-length AnxA11. The forward primers were 5′-ATCCGGCCATGGGTAGAGGCACCATCA-3′ and 5′-ATCCGGCCATGGGTACCGATGCTTCTG-3′ (the *Nco*I restriction site is underlined), resulting in Δ188AnxA11 (starting with 189-RGTI) and Δ192AnxA11 (starting with 193-TDAS), respectively. Both constructs code for the tag sequence MLHHHHHHPMG at the N terminus. The cDNAs were ligated into the pETM10 vector after digestion of PCR fragments with *Acc*65I and *Nco*I. Constructs were verified by DNA sequencing.

2.2. Expression and Purification of Full-Length and N-Terminally Truncated Forms of AnxA11

E. coli BL21(DE3) cells were transformed with AnxA11 expression plasmids. The bacteria were grown in 5 mL lysogeny broth (LB) medium with 50 μg/mL kanamycin until an OD_{600} of 0.6, after which 1 mM isopropyl β-D-1-thiogalactopyranoside (IPTG) was added to induce protein expression at +25 °C for 4 h or at +15 °C overnight (ON), to determine optimal expression conditions. At the end of protein expression, 1 mL of the culture was withdrawn and centrifuged at 2000× *g* for 10 min at +4 °C. The supernatant was discarded, and the pellet was resuspended in 500 μL lysis buffer (50 mM Na_2HPO_4, 500 mM NaCl, 5% (*v*/*v*) glycerol, 10 mM imidazole) containing 0.5 μg/mL DNase I (Sigma, St. Louis, MO, USA), 0.25 μg/mL RNase A (Sigma, St. Louis, MO, USA), and cOmplete EDTA-free protease inhibitor cocktail (Roche, Basel, Switzerland). The samples were sonicated on ice, whereafter 100-μL aliquots (total lysate) were withdrawn. Additional 100-μL aliquots of the lysates were transferred into new tubes and centrifuged at 21,000× *g* for 15 min at +4 °C. The supernatant was withdrawn. The pellet was resuspended in 100 μL lysis buffer. After addition of denaturation buffer, 20 μL of the protein samples were separated on 4–20% SDS-PAGE (Mini-PROTEAN TGX Precast Protein Gels, Bio-Rad, Hercules, CA, USA) at 20 mA and 250 V as the limiting voltage.

For large-scale purification, the above procedure was repeated using 300 mL LB medium. The full-length (FL) and N-terminally truncated AnxA11 were induced at +25 °C for 4 h, while FL-AnxA11 was also expressed ON at +15 °C. The cells were collected by centrifugation at 4000× *g* for 15 min. The pellets were frozen at −80 °C in lysis buffer before the addition of 0.5 μg/mL DNase I, 0.25 μg/mL RNase A, and cOmplete EDTA-free protease inhibitor cocktail (Roche, Basel, Switzerland). Subsequently, cells were broken by sonication, and the lysates were centrifuged at 16,000× *g* for 30 min at +4 °C. All purification steps were performed at +4 °C.

We took advantage of an N-terminal His-tag to purify AnxA11, which reduces degradation at the N terminus [13]. Purification using Co^{2+}-NTA gave higher yields than Ni^{2+}-NTA resin. The lysate supernatant was loaded onto a Co^{2+}-NTA agarose column and incubated for 30 min with gentle rotation. Subsequently, the proteins on the Co^{2+} resin were washed with equilibration buffer (50 mM Na_2HPO_4, 0.3 M NaCl, 10 mM imidazole; pH 8) and high-salt buffer (50 mM Na_2HPO_4, 1 M NaCl, 10 mM imidazole; pH 8), both buffers containing EDTA-free protease inhibitors. Elution buffer (50 mM Na_2HPO_4, 0.3 M NaCl, 250 mM imidazole; pH 8) containing EDTA-free protease inhibitors was used to elute the proteins. After adding EGTA to a final concentration of 2 mM, the eluates were quickly transferred onto PD-10 columns for a buffer exchange to 20 mM Tris (pH 8). All recombinant forms of AnxA11 were subjected to size exclusion chromatography using a Superdex 75 or 200 Increase 10/300 GL column (GE Healthcare, Chicago, IL, USA). Protein concentration was determined by absorbance at 280 nm (using molecular masses of 55513, 36828, and 36401 Da and sequence-based extinction coefficients of 42750, 13410, and 13410 M^{-1} cm^{-1} for full-length AnxA11, Δ188AnxA11, and Δ192AnxA11, respectively). The protein size and purity were confirmed by SDS-PAGE and Coomassie Brilliant Blue staining.

2.3. Circular Dichroism Spectroscopy

A Jasco J-810 spectropolarimeter (Jasco, Hampshire, UK) with a Peltier temperature control unit was used for far-UV circular dichroism (CD) spectroscopy. Melting curves were recorded from +25 to +75 °C at 222 nm, at a heating rate of 40 °C/h, to determine the thermal transition temperature (T_m). T_m was estimated in GraphPad Prism 7 (GraphPad, San Diego, CA, USA) using four-parameter logistic regression.

Synchrotron radiation CD (SRCD) data were acquired from 0.5 mg/mL Δ188AnxA11 and Δ192AnxA11 in 20 mM Tris-HCl, pH 8.0, on the AU-CD synchrotron beamline at ASTRID2 (ISA, Aarhus, Denmark). Samples containing 1 mM $CaCl_2$ were included in addition to $CaCl_2$-free samples to detect Ca^{2+}-induced changes in the secondary structure content. Closed 100-μM circular cells were used (Suprasil, Hellma Analytics, Müllheim, Germany). Spectra were recorded from 170 to 280 nm at +10 °C. Baselines were subtracted, and CD units were converted to $\Delta\varepsilon$ (M^{-1} cm^{-1}) in CDtoolX 1.06 [39]. The spectra were truncated based on the HT voltage and the CD spectrum noise levels.

2.4. Small-Angle X-Ray Scattering

SAXS data were collected from 0.41–2.96 mg/mL Δ188AnxA11 and Δ192AnxA11 in 20 mM Tris-HCl on the EMBL/DESY P12 beamline, PETRA III (Hamburg, Germany) [40] (Figure S1). Protein samples at 0.7–1.0 mg/mL were then studied with 0–1 mM $CaCl_2$ to follow potential oligomerisation. Monomeric bovine serum albumin (66.5 kDa) was used as a molecular weight standard. Data were processed and analysed using the ATSAS package 2.8.4 [41], and GNOM 5.0 was used to calculate distance distribution functions [42]. Theoretical scattering of the crystal structure was calculated and fitted against the SAXS data using CRYSOL 2.8.3 [43]. Ab initio modelling was performed using GASBOR 2.3i [44]. SUPCOMB 2.3 [45] was used for crystal structure and ab initio model superposition. SAXS data collection, processing, and fitting parameters are presented in Table 1.

Table 1. Small-angle X-ray scattering parameters and analysis.

Data Collection Parameters		
Instrument	P12, PETRAIII, DESY	
Wavelength (nm)	0.124	
Angular range (nm^{-1})	0.022–7.33	
Exposure time (s)	0.045	
Exposure temperature (°C)	10	
Protein	AnxA11Δ188	AnxA11Δ192
Concentration range (mg/mL)	0.77–2.96	0.41–1.18
Structural Parameters		
I_0 (relative) [from Guinier]	0.02759	0.0274
R_g (nm) [from Guinier]	2.44 ± 0.13	2.43 ± 0.20
I_0 (relative) [from P(r)]	0.02692	0.02748
R_g (nm) [from P(r)]	2.500 ± 0.016	2.503 ± 0.019
D_{max} (nm) [from GNOM]	10.00	9.97
V_{Porod} (nm^3) [from GNOM]	58.03	59.53
Molecular Mass Determination		
Molecular mass M_r (kDa) [from I_0 using Guinier] [1]	34.3	34.1
Molecular mass M_r (kDa) [from I_0 using P(r)] [1]	33.5	34.2
Molecular mass M_r (kDa) [from V_{Porod}]	34.1	35.0
Molecular mass M_r (kDa) [from absolute scale]	34.8	34.8
Theoretical M_r from sequence (kDa)	36.8	36.4
Software		
Primary data reduction & processing	PRIMUS	
Ab initio modelling	GASBOR	
χ^2	0.770	0.629
Fitting of theoretical scattering curve to data	CRYSOL	
χ^2	0.888	0.749

[1] Monomeric BSA standard was used as reference (M_r = 66.5 kDa; I_0 = 0.0534).

2.5. Protein Crystallisation, Data Collection, and Structure Determination

Sitting-drop vapour diffusion crystallisation experiments were set up in 96-well format. X-ray diffraction data were collected on the EMBL P13 beamline, PETRA III (Hamburg, Germany) [46] from a crystal grown in 100 mM Bis-tris (pH 5.5), 3 M NaCl at +20 °C; the drop contained 150 nL of 12.7 mg/mL Δ188AnxA11 and 150 nL of reservoir solution. The crystal was cryoprotected by adding a solution containing 25% glycerol and 75% reservoir solution directly onto the crystallisation drop. Upon mounting, the crystal was flash-frozen in liquid N_2, and it was kept at 100 K during X-ray diffraction data collection.

Diffraction data were processed with XDS [47], and the structure was solved by molecular replacement in Phaser 2.8.3 [48] with PDB entry 1ANN (bovine AnxA4) as a search model [49]. Refinement was carried out using phenix.refine 1.18 [50], and model building with coot 0.8.9.1 [51]. The structure quality was assessed with MolProbity [52]. The data processing and refinement statistics are presented in Table 2.

Table 2. Crystallographic data collection and refinement.

Data Collection	
Construct	Δ188AnxA11
Wavelength (Å)	0.9763
Space group	P1
Unit cell parameters	a = 39.1 Å, b = 86.7 Å, c = 87.3 Å α = 114.0°, β = 102.0°, γ = 97.2°
Resolution range (Å)	50–2.30 (2.36–2.30) [1]
Completeness (%)	95.4 (95.0)
Reflections total/unique	141110/42245 (9818/3046)
⟨I/σI⟩	5.2 (1.1)
R_{sym} (%)	17.5 (164.0)
R_{meas} (%)	20.9 (196.6)
$CC_{1/2}$ (%)	99.2 (47.6)
redundancy	3.3 (3.2)
Data Processing	
R_{cryst}/R_{free} (%)	22.8 (26.6)
Rmsd bond length (Å)	0.002
Rmsd bond angle (°)	0.4
MolProbity score (percentile)	1.58 (98th)
Ramachandran favoured/disallowed (%)	97.2/0.3

[1] Numbers in parentheses refer to the highest-resolution shell.

For structure analysis, the programs PyMOL 2.3.2 [53] and its APBS [54] plugin were used. The refined AnxA11 crystal structure, along with the experimental structure factors, was deposited at the PDB with the entry code 6TU2.

3. Results and Discussion

3.1. Solubility Assays, Large-Scale Expression, and Purification of the N-Terminally Truncated Forms of AnxA11

We studied recombinant rat AnxA11, which is nearly identical to the human AnxA11 (Figure 1A). Induction was performed at +25 °C for 4 h and at +15 °C ON to determine optimal expression conditions. There was little or no difference in yield between the two modes of expression. FL-AnxA11 was consistently more challenging to express than its truncated forms, yielding remarkably low concentrations, and it was notoriously prone to aggregation, as reported previously [13]. The solubility of the truncated forms of AnxA11 in bacterial lysates was assessed after centrifugation at 21,000× *g* by comparing samples from the total bacterial lysate with the corresponding supernatant and pellet by SDS-PAGE (Figure 1B). Two N-terminally truncated versions of AnxA11 were designed, cloned, expressed, and purified, namely, Δ188AnxA11 and Δ192AnxA11. Previously, we noticed by sequence alignment that a segment of ~10 residues in the N terminus preceding the first α helix of domain I is highly conserved in Anxs [55]. In addition, the N-terminal residues that immediately precede the core domain have been suggested to stabilise the overall structure of AnxA2 [55,56].

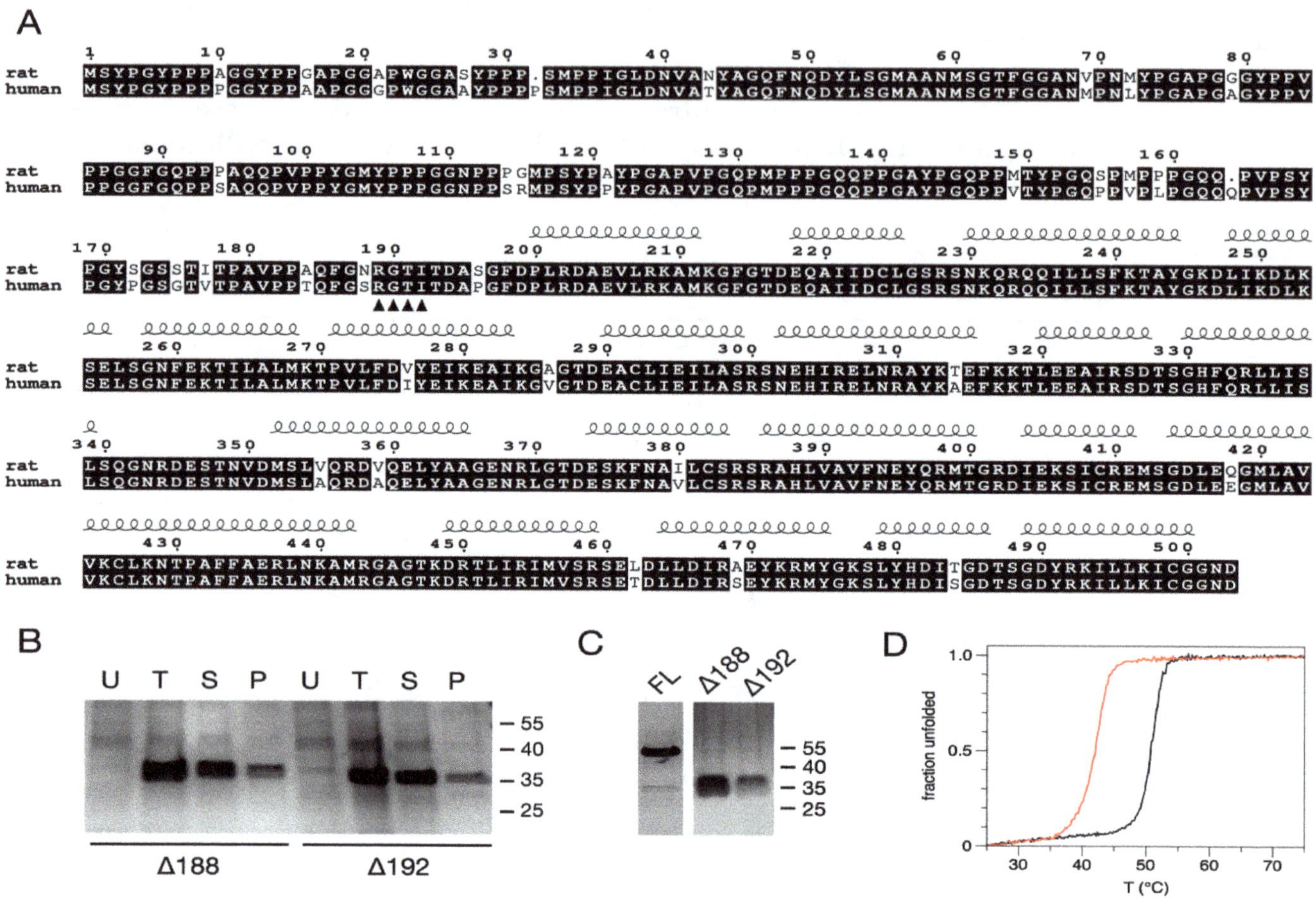

Figure 1. The solubility, purification, and stability of N-terminally truncated forms of rat AnxA11. (**A**) Sequence alignment between rat and human AnxA11. (**B**) Uninduced (U), total (T), supernatant (S), and pellet (P) fractions of bacterial lysates containing Δ188AnxA11 or Δ192AnxA11 as indicated. (**C**) 10 μL of FL-AnxA11 and the N-terminally truncated forms Δ188AnxA11 and Δ192AnxA11, purified using Co^{2+}-resin. Molecular mass standards are indicated to the right. Proteins were separated in 10% SDS-PAGE gels and Coomassie-stained. (**D**) CD-monitored thermal disruption of α-helicity of 20 μM Δ188AnxA11 (black) and Δ192AnxA11 (red).

In contrast to FL-AnxA11, both Δ188AnxA11 and Δ192AnxA11 were mainly soluble (Figure 1B). FL-AnxA11 as well as the N-terminally truncated Δ188AnxA11 and Δ192AnxA11 were purified by affinity chromatography on Co^{2+}-resin, which greatly improved the purity of the recombinant proteins compared to Ni^{2+}-based chromatography. The N-terminal truncations resulted in a marked enhancement in expression, giving a five-to-tenfold increase in yield compared to FL-AnxA11 (Figure 1C). This suggests that the N terminus is responsible for the aggregation of the full-length protein. This may be related to the observed phase separation phenomena linked to the N-terminal segment [8] or misfolding of the protein in bacterial cells.

3.2. *Thermal Stability of the N-Terminally Truncated Δ188AnxA11 and Δ192AnxA11*

To address thermal stability, ellipticity at 222 nm was monitored as a function of temperature for Δ188AnxA11 and Δ192AnxA11. The thermal transition temperatures were estimated from the inflection points of the melting curves (Figure 1D). The T_m of FL-AnxA11 was earlier reported to be +45 °C [13]. The T_m of Δ188AnxA11 showed a T_m of +50.76 ± 0.02 °C, while that of Δ192AnxA11 was substantially lower (+41.87 ± 0.02 °C). This difference highlights the importance of the four amino acid residues 189-RGTI-192 in thermally stabilising the core structure of AnxA11. Cooperative unfolding of the two truncated forms of AnxA11 remained largely unchanged but decreased slightly in Δ192AnxA11. This increased thermal stability of Δ188AnxA11 compared to Δ192AnxA11 may explain why we obtained crystals of Δ188AnxA11, but not of the other N-terminally truncated form. The dramatic difference in T_m is necessarily accounted for by at least one of the four additional residues present in Δ188AnxA11 being involved in the stabilising of the core structure.

3.3. *The Crystal Structure of AnxA11*

Using Δ188AnxA11, we solved the crystal structure of the rat AnxA11 core domain. The asymmetric unit contains three molecules of AnxA11, and the structure presents the canonical Anx core fold. The Ca^{2+} binding sites are partially occupied, whereby two monomers have three Ca^{2+} ions bound, and one monomer has four (Figure 2). As is typical for the Anxs, all Ca^{2+} binding sites are on one face of the molecule, promoting calcium-dependent lipid membrane binding (Figure 2).

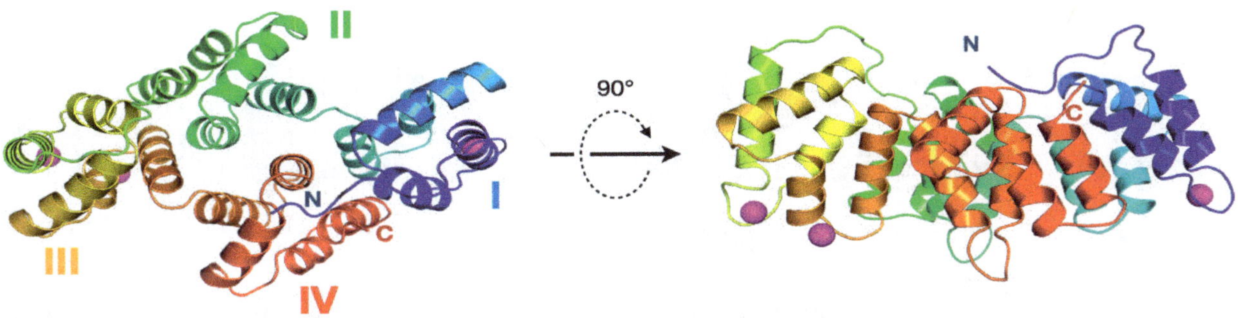

Figure 2. Crystal structure of the AnxA11 core domain. The four subdomains are numbered and coloured from blue to red. The N and C termini are indicated, and bound Ca^{2+} ions are shown in magenta. Shown is one monomer of the three in the asymmetric unit.

The substantial difference in thermal stability induced by the four N-terminal residues of the Δ188AnxA11 construct implies an important structural element at this position. This segment, ~10 residues before the AnxA11 domain I helix A, is central in the folded structure, making interactions with the termini of several helices of the Anx fold, and effectively enabling a tight contact between the N terminus of domain I and the C terminus of domain IV (Figure 3A,B).

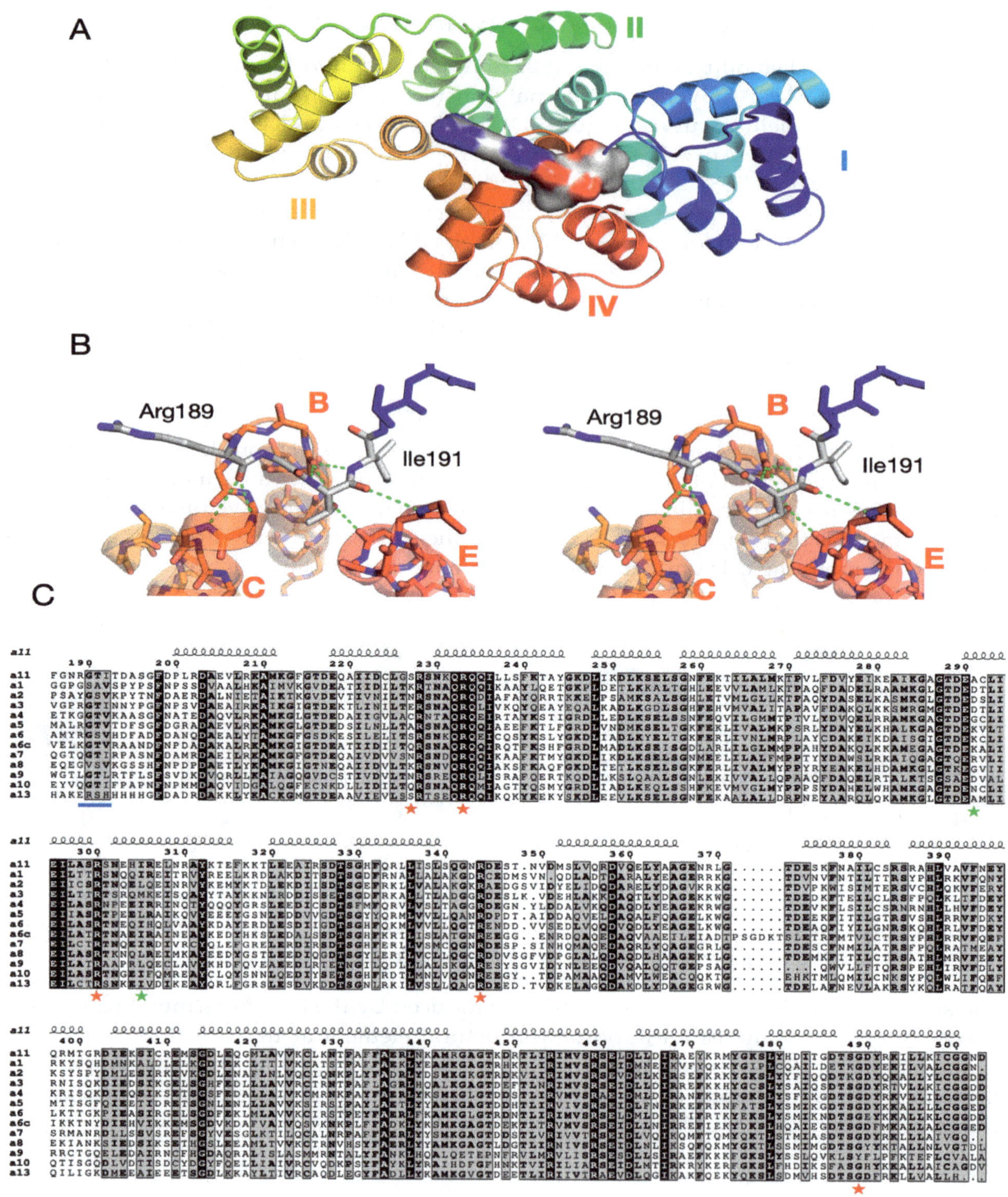

Figure 3. Rat annexins and conservation of the RGTI motif. (**A**) The N-terminal Velcro-like bridging segment is shown in surface mode, bound to domain IV. (**B**) Stereo view of the hydrogen bonding interactions of the motif with three helices in domain IV. (**C**) Sequence alignment of all rat Anxs. The N-terminal motif is underlined in blue, and the mutations (red) and variants (green) linked to ALS are shown with asterisks. A6 and A6c correspond to the N- and C-terminal halves of AnxA6.

Comparisons with other Anxs show that the segment 189–192 is a conserved motif in the Anx family. The only Anx that does not contain the motif at the sequence level is AnxA13; conservation is weak for AnxA1 and AnxA8. The 4-residue segment forms conserved interactions with the ends of helices in domain IV (Figure 3B). These interactions are mainly comprised of main-chain hydrogen bonds, which explains the mode of conservation and the importance of the Gly residue in the motif. Interestingly, it has been proposed that AnxA11 is the common ancestor of all mammalian Anxs [57], except for AnxA7 and AnxA13—the two phylogenetically oldest Anxs [58]. Thus, the AnxA11 core domain has the highest similarity to the other mammalian Anxs, acting as a sort of prototype [59].

The PhosphoSite database lists a number of phosphorylation sites for Anxs, most of which come from high-throughput assays [60]. AnxA11 is phosphorylated, but the exact sites remain unknown [15]. Interestingly, the well-characterised Tyr23 and Ser25 phosphorylation sites of AnxA2 lie within the 4-residue bridging region (Figure 3C). Phosphorylation at Tyr23 and Ser25 results in a closed or open conformation, respectively, of AnxA2 [56]. Similarly, a phosphomimicking mutation in Thr6 of AnxA4 results in an opening of the structure [61], hinting at a general mechanism of Anx conformation regulation through phosphorylation of the RGTI motif. A comparison of the structures of AnxA11, AnxA2, and AnxA4 is presented in Figure 4. Of other Anxs, AnxA6 has a phosphorylation site in the bridging segment [62]. These findings can be understood based on the conserved structure of the observed bridging segment. The regulation of this segment through post-translational modification may be a general means to regulate overall conformation and dynamics of Anxs, possibly resulting in altered binding to functional ligands, such as RNA or membranes.

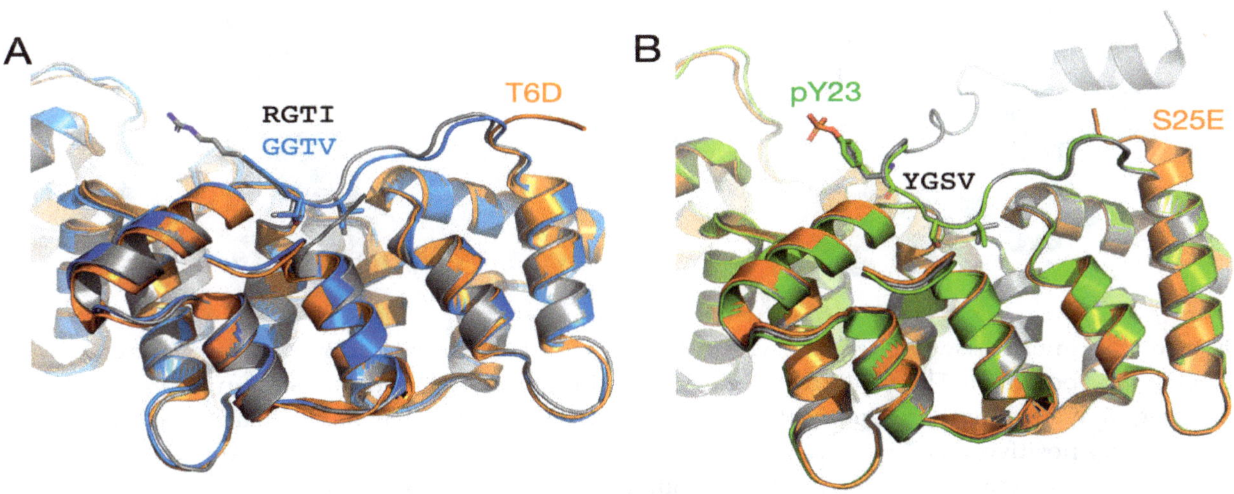

Figure 4. Regulation of the RGTI motif via phosphorylation in other annexins. (**A**) Comparison of AnxA11 (grey) with AnxA4 (blue) [49]. The structure of the phosphomimicking AnxA4 T6D mutant [61] is shown in orange and has an open conformation. (**B**) Three states of AnxA2 [56]. The nonphosphorylated protein is in grey, AnxA2 phosphorylated on Tyr23 in green, and the open conformation of S25E in orange.

The crystal structure of AnxA11 presents trimers very similar to those formed by AnxA5 (Figure 5), and hence, could be expected to interact with membranes in a similar manner. In the crystal, the trimers are arranged in layers, with similar contacts as observed for AnxA4 and AnxA5 between the trimers. Similarly to other Anx assemblies, the Ca^{2+} binding sites of the AnxA11 trimer are all on the same face, which would be in direct contact with the membrane (Figure 5A). The membrane-facing surface in the trimer gains a positive charge potential upon Ca^{2+} binding, while the opposite side is relatively featureless (Figure 5B). This observation further highlights the common function of the Anx core structure in binding membrane surfaces, while the N-terminal domain is responsible for most functional differences between the Anxs [1]. AnxA5 has been the most studied Anx on membrane surfaces, on which it forms a crystal-like lattice of trimers [63–66]. The trimer arrangement of AnxA11 in the crystal closely resembles that of AnxA5 in crystals and on membranes (Figure 5C,D) [63–66]. Notably, similarly to AnxA11, AnxA4 and AnxA5 are monomeric in solution [67–69].

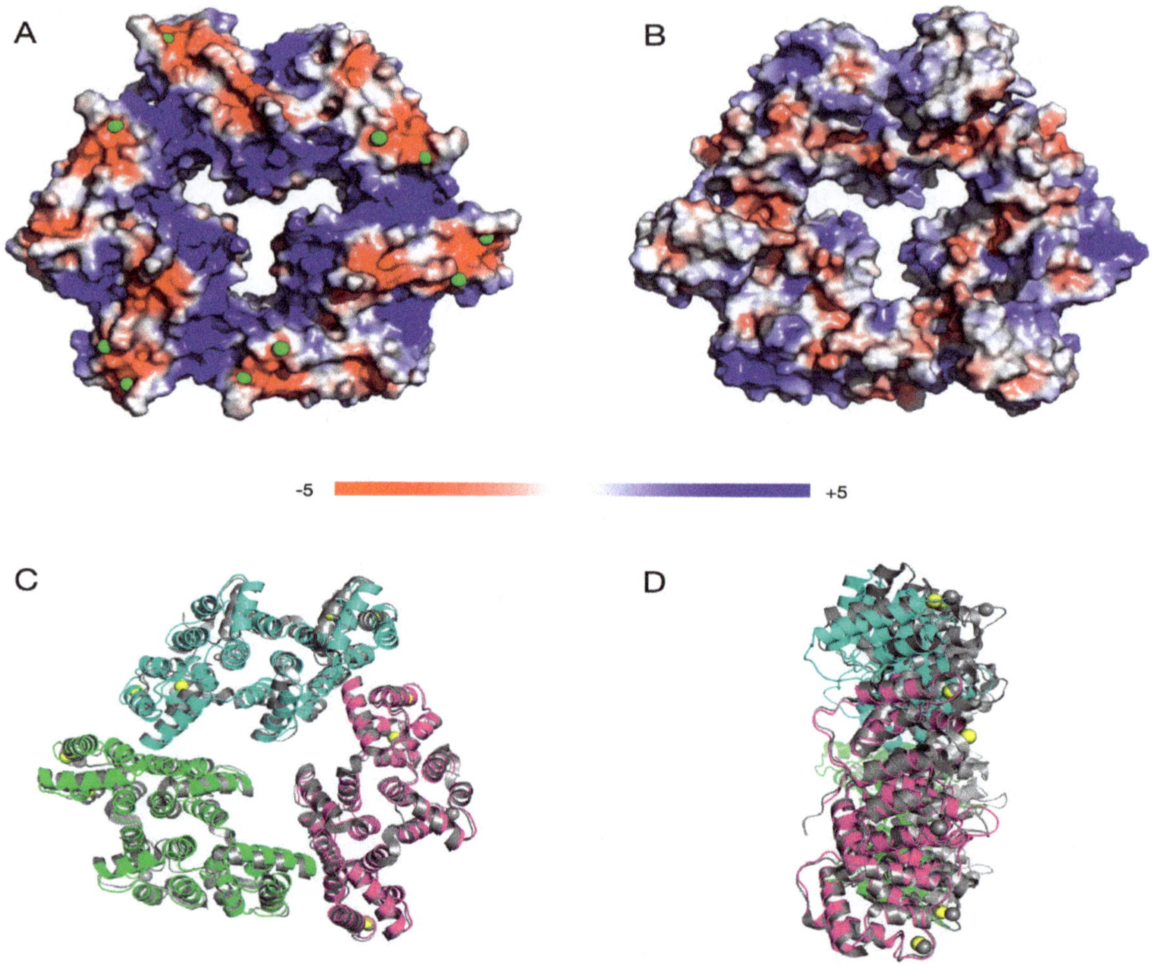

Figure 5. Trimeric assembly of AnxA11 in the crystal. (**A**) AnxA11 trimer surface coloured with electrostatic potential. The view from the Ca^{2+}-binding face. Ca^{2+} ions are shown as green spheres. The negative sites (red) are the locations for Ca^{2+} binding, and their neutralization will make the surface very positive, enabling lipid membrane binding. (**B**) The opposite face of AnxA11 is relatively featureless. (**C**) Superposition of AnxA11 (coloured) with the AnxA5 trimer (gray) [64]. Superposition was done based on one monomer (bottom right) to better highlight possible differences in assembly. The view is the same as in (**B**). (**D**) Side view of the superimposed AnxA11 and AnxA5 trimers.

One might assume that AnxA11 would arrange on a membrane much like it does in the crystal; the N-terminal segments would then point away from the membrane. This is in line with the observations that AnxA11 induces membrane curvature [7], as well as the suggestion that the N terminus of AnxA11 binds to RNA, while the core structure binds to lipids in lysosomes for long-distance co-transport in neurons [8]. The RNA binding site in AnxA2 has been shown to reside within domain IV [70]. The binding dynamics were recently investigated; AnxA2 binds strongly to an RNA 8-mer (5′-GGGGAUUG-3′) with a K_d of 30 nM [71]. The RNA binding sites in AnxA11 have not been determined.

3.4. Structure-Based Analysis of ALS Disease Mutations

The crystal structure of rat AnxA11 allows one to analyse the structural basis of AnxA11 mutations linked to amyotrophic lateral sclerosis (ALS) [35] in humans. The sequence identity between full-length rat and human AnxA11 is 93.3%, being 96.5% for the core structure. The identification of AnxA11 as an ALS gene is rather recent, and mutations are being continuously identified. Thus, it is safe to assume that more ALS mutations will be identified in the future. We studied the mutations thus far linked to ALS with respect to potential structural implications. Many of the mutations, including the best-characterised D40G [35], lie in the N-terminal extension, which is not present in the crystallised construct. The mutations in this region could affect protein–protein interactions, protein localisation, phase separation, or binding to mRNA [8]. More interesting for the current structural study, however, are the mutations identified within the Anx core domain.

Five mutations have thus far been described in ALS patients within the AnxA11 core: S229R, R235Q, R302C, R346C, and G491R; in addition, G189E is located two residues before the start of the crystallised rat AnxA11 construct [35,37]. In the 3D structure, the AnxA11 mutations can be observed clustered at two locations (Figure 6A). Interestingly, three of these mutations involve a buried Arg residue, and a close look at the alignment reveals that these Arg residues are fully conserved in all rat Anxs. The remaining two mutations replace an exposed, non-conserved, small residue with an Arg. Details of the environment of each of the three conserved Arg residues are shown in Figure 6B–D. In brief, two of them coordinate carbonyl groups at C termini of helices, and one is involved in a conserved extended buried network of salt bridges between domains II and IV. These locations suggest that the ALS mutations involving these conserved Arg residues will have effects on AnxA11 folding and/or stability.

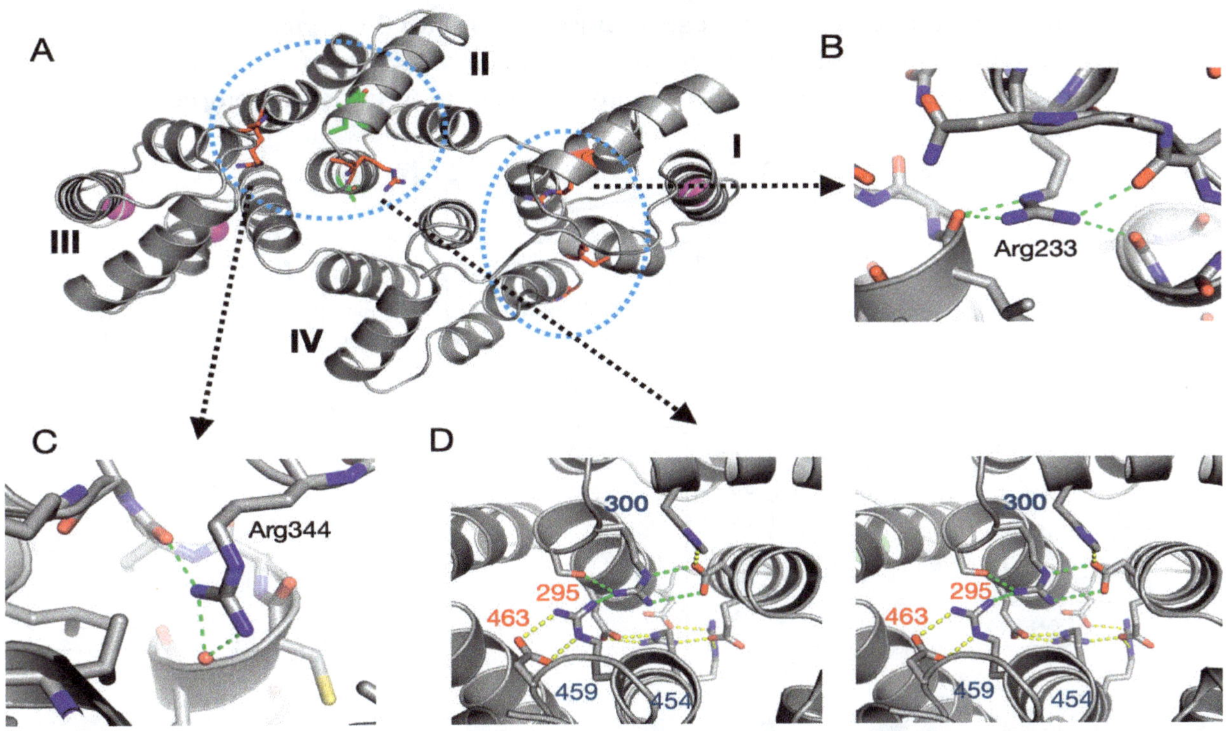

Figure 6. Location of ALS mutations and variants in the AnxA11 structure. (**A**) The five ALS mutations (see text) are highlighted in red and the two rare variants in green. The two 3D clusters of mutations/variants are shown with blue dashed circles. One cluster lies in domain II, possibly affecting folding and/or interactions with domain III. The other cluster is at the interface between domains I and IV, i.e., the N and C terminus of the AnxA11 core structure. (**B**) Arg233 (corresponding to the human R235Q mutation) makes four contacts to carbonyl groups at the C-terminal end of domain I helices B and E. With its buried positive charge, Arg233 interacts with the negative helix dipole from both helices. (**C**) Arg344 (corresponding to R346C) similarly caps helix B of domain III. (**D**) An extended salt bridge network exists between domains II (top) and IV (bottom) (stereo view). Arg300 (corresponds to R302C) is central in this network. The charged residues with numbers indicated are fully conserved in all rat Anxs (Arg, blue; acidic, red).

ΔΔG predictions for the five mutations on the SDM server [72] indicate decreased stability for all mutations except S229R. Hence, the affected residues may be important for correct folding and/or stability of AnxA11. Recent functional characterisation of the AnxA11 ALS mutation R235Q shows that the mutant protein is prone to aggregation and sequesters FL-AnxA11. This in turn abolishes the binding to calcyclin [35]. In addition, two rare missense variants, A293V and I307M, were recently discovered in the Chinese Han population [34]. These residues are located in domain II (Figure 6).

The N-terminal tail of AnxA11 has a unique amino acid composition, reminiscent of disordered proteins able to undergo liquid–liquid phase separation (LLPS) [73]. It has the hallmarks of an extended segment poised for π-π interactions, which are central in LLPS [74]. Indeed, the AnxA11 N-terminal region was recently shown to promote LLPS; its N terminus binds to RNA granules, which do not contain membranes, while the core structure binds to lipids in membranes [8]. The effects of this phase separation on the stability of the AnxA11 core domain are currently unknown. Furthermore, ALS-associated mutations in both the N terminus (D40G) and C terminus (R346C) of AnxA11 were shown to promote phase transition from liquid–liquid droplets to more stable gel-like states within AnxA11 droplets and in addition impair the reversal of the states. As a consequence, these mutations disrupt the function of AnxA11 to tether RNA granules with lysosomes [8].

3.5. The Solution Behaviour of the AnxA11 Core

The thermal stability difference between Δ188AnxA11 and Δ192AnxA11 prompted us to study their conformation in solution using SAXS and SRCD. In SAXS, both variants were monomeric with identical distance distributions and globular folds that superposed well with the crystal structure (Figure 7A–E), indicating that the absence of the RGTI stretch does not lead to large-scale conformational changes. Despite the trimeric packing in the crystal, AnxA11 is monomeric in solution under the employed conditions.

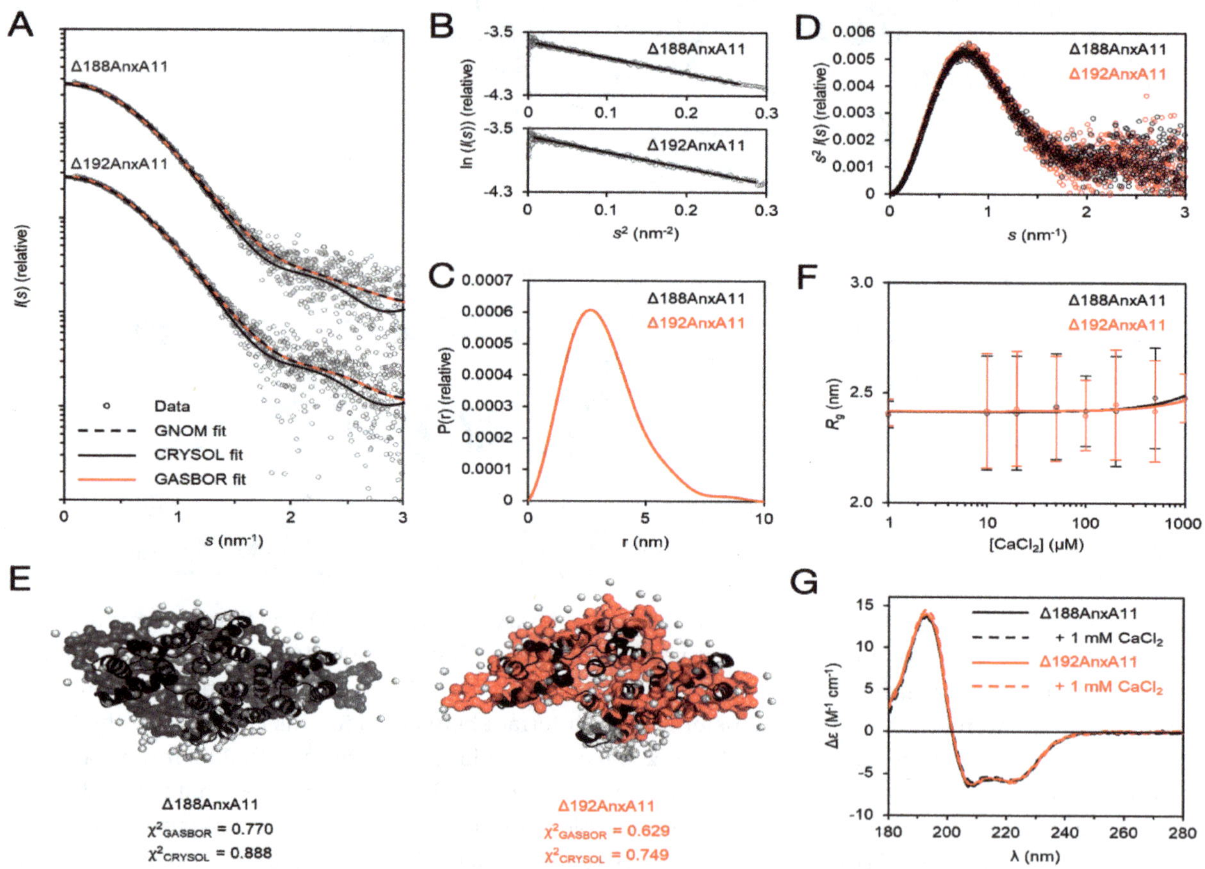

Figure 7. SAXS and SRCD analysis of Δ188AnxA11 and Δ192AnxA11. (**A**) SAXS data for Δ188AnxA11 and Δ192AnxA11 have been offset for clarity. GNOM, GASBOR, and CRYSOL fits are plotted over the measured data to denote the accuracy of distance distribution analysis, ab initio modeling, and theoretical scattering calculated from the crystal structure. (**B**) Guinier analysis. (**C**) Distance distribution diagram from GNOM. (**D**) Kratky plot of the SAXS data. (**E**) GASBOR ab initio models based on the SAXS data. The crystal structure (black) has been fitted inside. χ^2 values of the GASBOR and CRYSOL fits are indicated. (**F**) R_g as a function of $CaCl_2$ concentration. The error bars represent the fitting error from Guinier analysis. (**G**) SRCD spectra of Δ188AnxA11 and Δ192AnxA11 in the presence and absence of 1 mM $CaCl_2$.

Similarly to most Anxs, AnxA11 has been implicated in Ca^{2+} binding [1,13]. Structural studies of many Anxs have revealed Ca^{2+} binding to the convex side of the core structure, essentially through coordination by acidic residues. To test for effect of Ca^{2+} on the folding of the core structure of AnxA11, we titrated Δ188AnxA11 and Δ192AnxA11 with $CaCl_2$ in SAXS and followed the radius of gyration (R_g; Figure 7F). The R_g of both truncated proteins remained the same up to 1 mM $CaCl_2$. Additionally, we carried out SRCD spectroscopy of Δ188AnxA11 and Δ192AnxA11 in the absence and presence of 1 mM $CaCl_2$ (Figure 7G). The spectral quality was excellent, with the data being truncated at 178 nm. The presence of Ca^{2+} did not alter the spectra, indicating that Ca^{2+} does not influence the secondary structure content of the core structure. This result confirms that Δ192AnxA11 is folded despite its dramatic decrease in thermal stability, and that the degree of secondary structure content between the two variants is nearly identical. Ca^{2+} induced a large thermal stabilisation of FL-AnxA11, indicated by a rise in T_m of ~14 °C [13]. The presence of up to 75 mM Ca^{2+} only induced small increases in AnxA11 α-helical content [13]. Thus, Ca^{2+} binding results in only local conformational changes, which primes AnxA11 for lipid membrane binding.

4. Conclusions

As the diversity of the vertebrate Anx family is mainly realised through the uniqueness of the N-terminal tail [1,16,75,76], many functional features are lost by truncation. However, crucial functional properties of interest, such as mRNA binding, Ca^{2+}-dependent binding to phospholipids, as well as Ca^{2+} binding itself, may be retained. Thus, truncated Anxs may serve as a foundation for functional studies, dodging practical issues often encountered due to their long N termini.

The crystal structure of AnxA11, although similar to known Anx core domain structures, provides important starting points to further understand the structure–function relationships in AnxA11 and other Anx family members. The identification of a strongly stabilizing Velcro-like bridging segment at the N terminus of the Anx core, as well as the fact that many known ALS mutations correspond to highly conserved buried Arg residues in the Anx fold, provide new insights into Anx folding. AnxA11, like many other ALS targets, is part of a membraneless organelle, a ribonucleoprotein complex, involved in neuronal transport of mRNA granules [8,77–79]. It is evident that this process can be affected by mutations in either the N-terminal region or the folded core domain of AnxA11, which in turn can affect binding to RNA, lipid membranes, and Ca^{2+}. The fact that the bridging segment is a target for regulatory phosphorylation in AnxA2 shows that in addition to simply stabilizing the core structure, this segment may be a conformational switch common to most Anxs.

Author Contributions: Conceptualization, A.V. and P.K.; methodology, P.A.G.L., A.R., and A.V.; validation, A.R. and P.K.; formal analysis, A.R. and P.K.; investigation, P.A.G.L., A.R., S.J.H., S.P., T.R., H.H., S.A.S., P.D.S., A.V., and P.K.; resources, P.A.G.L., S.J.H., S.P., T.R., H.H., S.A.S., P.D.S., and A.V.; data curation, A.R. and P.K.; writing—original draft preparation, P.A.G.L., A.R., and A.V.; writing—review and editing, A.R., A.V., and P.K.; visualization, A.R. and P.K.; supervision, A.V. and P.K.; project administration, A.V. and P.K.; funding acquisition, A.V. and P.K. All authors have read and agreed to the published version of the manuscript.

Acknowledgments: We thank Marianne Goris for valuable assistance in the laboratory. Crystallographic and SAXS data were collected on EMBL Hamburg beamlines P13 and P12, respectively, at the PETRA III storage ring (DESY, Hamburg, Germany). We are also grateful for the SRCD beamline support at the ASTRID2 synchrotron facility (Aarhus, Denmark). We thank the BiSS facility at the University of Bergen for providing vital instrumentation for this study.

References

1. Gerke, V.; Moss, S.E. Annexins: From structure to function. *Physiol. Rev.* **2002**, *82*, 331–371. [CrossRef] [PubMed]
2. Gerke, V.; Creutz, C.E.; Moss, S.E. Annexins: Linking Ca2+ signalling to membrane dynamics. *Nat. Rev. Mol. Cell Biol.* **2005**, *6*, 449–461. [CrossRef] [PubMed]
3. Moss, S.E.; Morgan, R.O. The annexins. *Genome Biol.* **2004**, *5*, 219. [CrossRef]

4. Liemann, S.; Huber, R. Three-dimensional structure of annexins. *Cell Mol. Life Sci.* **1997**, *53*, 516–521. [CrossRef] [PubMed]
5. Morgan, R.O.; Jenkins, N.A.; Gilbert, D.J.; Copeland, N.G.; Balsara, B.R.; Testa, J.R.; Fernandez, M.P. Novel human and mouse annexin A10 are linked to the genome duplications during early chordate evolution. *Genomics* **1999**, *60*, 40–49. [CrossRef] [PubMed]
6. Rosengarth, A.; Luecke, H. A calcium-driven conformational switch of the N-terminal and core domains of annexin A1. *J. Mol. Biol.* **2003**, *326*, 1317–1325. [CrossRef]
7. Boye, T.L.; Jeppesen, J.C.; Maeda, K.; Pezeshkian, W.; Solovyeva, V.; Nylandsted, J.; Simonsen, A.C. Annexins induce curvature on free-edge membranes displaying distinct morphologies. *Sci. Rep.* **2018**, *8*, 10309. [CrossRef]
8. Liao, Y.C.; Fernandopulle, M.S.; Wang, G.; Choi, H.; Hao, L.; Drerup, C.M.; Patel, R.; Qamar, S.; Nixon-Abell, J.; Shen, Y.; et al. RNA Granules Hitchhike on Lysosomes for Long-Distance Transport, Using Annexin A11 as a Molecular Tether. *Cell* **2019**, *179*, 147–164.e20. [CrossRef]
9. Tokumitsu, H.; Mizutani, A.; Muramatsu, M.; Yokota, T.; Arai, K.; Hidaka, H. Molecular cloning of rabbit CAP-50, a calcyclin-associated annexin protein. *Biochem. Biophys. Res. Commun.* **1992**, *186*, 1227–1235. [CrossRef]
10. Burns, A.L.; Magendzo, K.; Shirvan, A.; Srivastava, M.; Rojas, E.; Alijani, M.R.; Pollard, H.B. Calcium channel activity of purified human synexin and structure of the human synexin gene. *Proc. Natl. Acad. Sci. USA* **1989**, *86*, 3798–3802. [CrossRef]
11. Creutz, C.E.; Snyder, S.L.; Husted, L.D.; Beggerly, L.K.; Fox, J.W. Pattern of repeating aromatic residues in synexin. Similarity to the cytoplasmic domain of synaptophysin. *Biochem. Biophys. Res. Commun.* **1988**, *152*, 1298–1303. [CrossRef]
12. Creutz, C.E. Novel protein ligands of the annexin A7 N-terminal region suggest pro-beta helices engage one another with high specificity. *Gen. Physiol. Biophys.* **2009**, *28*, F7–F13.
13. Lecona, E.; Turnay, J.; Olmo, N.; Guzman-Aranguez, A.; Morgan, R.O.; Fernandez, M.P.; Lizarbe, M.A. Structural and functional characterization of recombinant mouse annexin A11: Influence of calcium binding. *Biochem. J.* **2003**, *373*, 437–449. [CrossRef] [PubMed]
14. Satoh, H.; Shibata, H.; Nakano, Y.; Kitaura, Y.; Maki, M. ALG-2 interacts with the amino-terminal domain of annexin XI in a Ca(2+)-dependent manner. *Biochem. Biophys. Res. Commun.* **2002**, *291*, 1166–1172. [CrossRef] [PubMed]
15. Mizutani, A.; Tokumitsu, H.; Kobayashi, R.; Hidaka, H. Phosphorylation of annexin XI (CAP-50) in SR-3Y1 cells. *J. Biol. Chem.* **1993**, *268*, 15517–15522.
16. Sudo, T.; Hidaka, H. Regulation of calcyclin (S100A6) binding by alternative splicing in the N-terminal regulatory domain of annexin XI isoforms. *J. Biol. Chem.* **1998**, *273*, 6351–6357. [CrossRef]
17. Tokumitsu, H.; Mizutani, A.; Hidaka, H. Calcyclin-binding site located on the NH2-terminal domain of rabbit CAP-50 (annexin XI): Functional expression of CAP-50 in Escherichia coli. *Arch. Biochem. Biophys.* **1993**, *303*, 302–306. [CrossRef]
18. Tomas, A.; Moss, S.E. Calcium- and cell cycle-dependent association of annexin 11 with the nuclear envelope. *J. Biol. Chem.* **2003**, *278*, 20210–20216. [CrossRef]
19. Naka, M.; Qing, Z.X.; Sasaki, T.; Kise, H.; Tawara, I.; Hamaguchi, S.; Tanaka, T. Purification and characterization of a novel calcium-binding protein, S100C, from porcine heart. *Biochim. Biophys. Acta* **1994**, *1223*, 348–353. [CrossRef]
20. Powell, M.A.; Glenney, J.R. Regulation of calpactin I phospholipid binding by calpactin I light-chain binding and phosphorylation by p60v-src. *Biochem. J.* **1987**, *247*, 321–328. [CrossRef]
21. Consortium, T.U. UniProt: A worldwide hub of protein knowledge. *Nucleic Acids Res.* **2019**, *47*, D506–D515. [CrossRef] [PubMed]
22. Mizutani, A.; Usuda, N.; Tokumitsu, H.; Minami, H.; Yasui, K.; Kobayashi, R.; Hidaka, H. CAP-50, a newly identified annexin, localizes in nuclei of cultured fibroblast 3Y1 cells. *J. Biol. Chem.* **1992**, *267*, 13498–13504. [PubMed]
23. Mizutani, A.; Watanabe, N.; Kitao, T.; Tokumitsu, H.; Hidaka, H. The long amino-terminal tail domain of annexin XI is necessary for its nuclear localization. *Arch. Biochem. Biophys.* **1995**, *318*, 157–165. [CrossRef] [PubMed]
24. Tomas, A.; Futter, C.; Moss, S.E. Annexin 11 is required for midbody formation and completion of the terminal phase of cytokinesis. *J. Cell Biol.* **2004**, *165*, 813–822. [CrossRef] [PubMed]

25. Duncan, R.; Carpenter, B.; Main, L.C.; Telfer, C.; Murray, G.I. Characterisation and protein expression profiling of annexins in colorectal cancer. *Br. J. Cancer* **2008**, *98*, 426–433. [CrossRef]
26. Fernandez-Madrid, F.; Tang, N.; Alansari, H.; Granda, J.L.; Tait, L.; Amirikia, K.C.; Moroianu, M.; Wang, X.; Karvonen, R.L. Autoantibodies to Annexin XI-A and Other Autoantigens in the Diagnosis of Breast Cancer. *Cancer Res.* **2004**, *64*, 5089–5096. [CrossRef]
27. Hua, K.; Li, Y.; Zhao, Q.; Fan, L.; Tan, B.; Gu, J. Downregulation of Annexin A11 (ANXA11) Inhibits Cell Proliferation, Invasion, and Migration via the AKT/GSK-3beta Pathway in Gastric Cancer. *Med. Sci. Monit.* **2018**, *24*, 149–160. [CrossRef]
28. Song, J.; Shih Ie, M.; Chan, D.W.; Zhang, Z. Suppression of annexin A11 in ovarian cancer: Implications in chemoresistance. *Neoplasia* **2009**, *11*, 605–614, 1 p following 614. [CrossRef]
29. Wang, J.; Guo, C.; Liu, S.; Qi, H.; Yin, Y.; Liang, R.; Sun, M.Z.; Greenaway, F.T. Annexin A11 in disease. *Clin. Chim. Acta* **2014**, *431*, 164–168. [CrossRef]
30. Xin, W.; Rhodes, D.R.; Ingold, C.; Chinnaiyan, A.M.; Rubin, M.A. Dysregulation of the annexin family protein family is associated with prostate cancer progression. *Am. J. Pathol.* **2003**, *162*, 255–261. [CrossRef]
31. Brenner, D.; Weishaupt, J.H. Update on amyotrophic lateral sclerosis genetics. *Curr. Opin. Neurol.* **2019**, *32*, 735–739. [CrossRef]
32. Fernandopulle, M.; Wang, G.; Nixon-Abell, J.; Qamar, S.; Balaji, V.; Morihara, R.; St George-Hyslop, P.H. Inherited and Sporadic Amyotrophic Lateral Sclerosis and Fronto-Temporal Lobar Degenerations arising from Pathological Condensates of Phase Separating Proteins. *Hum. Mol. Genet.* **2019**, *28*, R187–R196. [CrossRef] [PubMed]
33. Fratta, P.; Birsa, N.; Tosolini, A.P.; Schiavo, G. Travelling Together: A Unifying Pathomechanism for ALS. *Trends Neurosci.* **2020**, *43*, 1–2. [CrossRef] [PubMed]
34. Liu, X.; Wu, C.; He, J.; Zhang, N.; Fan, D. Two rare variants of the ANXA11 gene identified in Chinese patients with amyotrophic lateral sclerosis. *Neurobiol. Aging* **2019**, *74*, 235.e9–235.e12. [CrossRef] [PubMed]
35. Smith, B.N.; Topp, S.D.; Fallini, C.; Shibata, H.; Chen, H.J.; Troakes, C.; King, A.; Ticozzi, N.; Kenna, K.P.; Soragia-Gkazi, A.; et al. Mutations in the vesicular trafficking protein annexin A11 are associated with amyotrophic lateral sclerosis. *Sci. Transl. Med.* **2017**, *9*. [CrossRef] [PubMed]
36. Tsai, P.C.; Liao, Y.C.; Jih, K.Y.; Soong, B.W.; Lin, K.P.; Lee, Y.C. Genetic analysis of ANXA11 variants in a Han Chinese cohort with amyotrophic lateral sclerosis in Taiwan. *Neurobiol. Aging* **2018**, *72*, 188.e1–188.e2. [CrossRef] [PubMed]
37. Zhang, K.; Liu, Q.; Liu, K.; Shen, D.; Tai, H.; Shu, S.; Ding, Q.; Fu, H.; Liu, S.; Wang, Z.; et al. ANXA11 mutations prevail in Chinese ALS patients with and without cognitive dementia. *Neurol. Genet.* **2018**, *4*, e237. [CrossRef]
38. Chomczynski, P.; Sacchi, N. Single-step method of RNA isolation by acid guanidinium thiocyanate-phenol-chloroform extraction. *Anal. Biochem.* **1987**, *162*, 156–159. [CrossRef]
39. Miles, A.J.; Wallace, B.A. CDtoolX, a downloadable software package for processing and analyses of circular dichroism spectroscopic data. *Protein Sci.* **2018**, *27*, 1717–1722. [CrossRef]
40. Blanchet, C.E.; Spilotros, A.; Schwemmer, F.; Graewert, M.A.; Kikhney, A.; Jeffries, C.M.; Franke, D.; Mark, D.; Zengerle, R.; Cipriani, F.; et al. Versatile sample environments and automation for biological solution X-ray scattering experiments at the P12 beamline (PETRA III, DESY). *J. Appl. Crystallogr.* **2015**, *48*, 431–443. [CrossRef]
41. Franke, D.; Petoukhov, M.V.; Konarev, P.V.; Panjkovich, A.; Tuukkanen, A.; Mertens, H.D.T.; Kikhney, A.G.; Hajizadeh, N.R.; Franklin, J.M.; Jeffries, C.M.; et al. ATSAS 2.8: A comprehensive data analysis suite for small-angle scattering from macromolecular solutions. *J. Appl. Crystallogr.* **2017**, *50*, 1212–1225. [CrossRef] [PubMed]
42. Svergun, D.I. Determination of the regularization parameter in indirect-transform methods using perceptual criteria. *J. Appl. Crystallogr.* **1992**, *25*, 495–503. [CrossRef]
43. Svergun, D.; Barberato, C.; Koch, M.H.J. CRYSOL—A Program to Evaluate X-ray Solution Scattering of Biological Macromolecules from Atomic Coordinates. *J. Appl. Crystallogr.* **1995**, *28*, 768–773. [CrossRef]
44. Svergun, D.I.; Petoukhov, M.V.; Koch, M.H. Determination of domain structure of proteins from X-ray solution scattering. *Biophys. J.* **2001**, *80*, 2946–2953. [CrossRef]
45. Kozin, M.B.; Svergun, D.I. Automated matching of high-and low-resolution structural models. *J. Appl. Crystallogr.* **2001**, *34*, 33–41. [CrossRef]

46. Cianci, M.; Bourenkov, G.; Pompidor, G.; Karpics, I.; Kallio, J.; Bento, I.; Roessle, M.; Cipriani, F.; Fiedler, S.; Schneider, T.R. P13, the EMBL macromolecular crystallography beamline at the low-emittance PETRA III ring for high- and low-energy phasing with variable beam focusing. *J. Synchrotron Radiat.* **2017**, *24*, 323–332. [CrossRef] [PubMed]
47. Kabsch, W. XDS. *Acta Crystallogr. D Biol. Crystallogr.* **2010**, *66*, 125–132. [CrossRef]
48. McCoy, A.J.; Grosse-Kunstleve, R.W.; Adams, P.D.; Winn, M.D.; Storoni, L.C.; Read, R.J. Phaser crystallographic software. *J. Appl. Crystallogr.* **2007**, *40*, 658–674. [CrossRef]
49. Sutton, R.B.; Sprang, S.R. Three dimensional structure of annexin IV. In *Annexins: Molecular Structure to Cellular Function*; Seaton, B.A., Ed.; Landes Biosciences: Georgetown, TX, USA, 1996.
50. Afonine, P.V.; Grosse-Kunstleve, R.W.; Echols, N.; Headd, J.J.; Moriarty, N.W.; Mustyakimov, M.; Terwilliger, T.C.; Urzhumtsev, A.; Zwart, P.H.; Adams, P.D. Towards automated crystallographic structure refinement with phenix.refine. *Acta Crystallogr. D Biol. Crystallogr.* **2012**, *68*, 352–367. [CrossRef]
51. Emsley, P.; Lohkamp, B.; Scott, W.G.; Cowtan, K. Features and development of Coot. *Acta Crystallogr. D Biol. Crystallogr.* **2010**, *66*, 486–501. [CrossRef]
52. Davis, I.W.; Murray, L.W.; Richardson, J.S.; Richardson, D.C. MOLPROBITY: Structure validation and all-atom contact analysis for nucleic acids and their complexes. *Nucleic Acids Res.* **2004**, *32*, W615–W619. [CrossRef] [PubMed]
53. DeLano, W.L. Pymol: An open-source molecular graphics tool. *CCP4 Newsl. Protein Crystallogr.* **2002**, *40*, 82–92.
54. Unni, S.; Huang, Y.; Hanson, R.M.; Tobias, M.; Krishnan, S.; Li, W.W.; Nielsen, J.E.; Baker, N.A. Web servers and services for electrostatics calculations with APBS and PDB2PQR. *J. Comput. Chem.* **2011**, *32*, 1488–1491. [CrossRef] [PubMed]
55. Aukrust, I.; Evensen, L.; Hollås, H.; Berven, F.; Atkinson, R.A.; Travé, G.; Flatmark, T.; Vedeler, A. Engineering, biophysical characterisation and binding properties of a soluble mutant form of annexin A2 domain IV that adopts a partially folded conformation. *J. Mol. Biol.* **2006**, *363*, 469–481. [CrossRef]
56. Ecsédi, P.; Kiss, B.; Gógl, G.; Radnai, L.; Buday, L.; Koprivanacz, K.; Liliom, K.; Leveles, I.; Vértessy, B.; Jeszenői, N.; et al. Regulation of the Equilibrium between Closed and Open Conformations of Annexin A2 by N-Terminal Phosphorylation and S100A4-Binding. *Structure* **2017**, *25*, 1195–1207.e5. [CrossRef]
57. Bances, P.; Fernandez, M.R.; Rodriguez-Garcia, M.I.; Morgan, R.O.; Fernandez, M.P. Annexin A11 (ANXA11) gene structure as the progenitor of paralogous annexins and source of orthologous cDNA isoforms. *Genomics* **2000**, *69*, 95–103. [CrossRef]
58. Iglesias, J.M.; Morgan, R.O.; Jenkins, N.A.; Copeland, N.G.; Gilbert, D.J.; Fernandez, M.P. Comparative genetics and evolution of annexin A13 as the founder gene of vertebrate annexins. *Mol. Biol. Evol.* **2002**, *19*, 608–618. [CrossRef]
59. Fernandez, M.P.; Jenkins, N.A.; Gilbert, D.J.; Copeland, N.G.; Morgan, R.O. Sequence and Chromosomal Localization of Mouse Annexin XI. *Genomics* **1996**, *37*, 366–374. [CrossRef]
60. Hornbeck, P.V.; Chabra, I.; Kornhauser, J.M.; Skrzypek, E.; Zhang, B. PhosphoSite: A bioinformatics resource dedicated to physiological protein phosphorylation. *Proteomics* **2004**, *4*, 1551–1561. [CrossRef]
61. Kaetzel, M.A.; Mo, Y.D.; Mealy, T.R.; Campos, B.; Bergsma-Schutter, W.; Brisson, A.; Dedman, J.R.; Seaton, B.A. Phosphorylation mutants elucidate the mechanism of annexin IV-mediated membrane aggregation. *Biochemistry* **2001**, *40*, 4192–4199. [CrossRef]
62. Freye-Minks, C.; Kretsinger, R.H.; Creutz, C.E. Structural and dynamic changes in human annexin VI induced by a phosphorylation-mimicking mutation, T356D. *Biochemistry* **2003**, *42*, 620–630. [CrossRef] [PubMed]
63. Chen, Z.; Mao, Y.; Yang, J.; Zhang, T.; Zhao, L.; Yu, K.; Zheng, M.; Jiang, H.; Yang, H. Characterizing the binding of annexin V to a lipid bilayer using molecular dynamics simulations. *Proteins* **2014**, *82*, 312–322. [CrossRef] [PubMed]
64. Hong, S.; Na, S.; Kim, O.H.; Jeong, S.; Oh, B.C.; Ha, N.C. High-resolution structures of annexin A5 in a two-dimensional array. *J. Struct. Biol.* **2020**, *209*, 107401. [CrossRef] [PubMed]
65. Lin, Y.C.; Chipot, C.; Scheuring, S. Annexin-V stabilizes membrane defects by inducing lipid phase transition. *Nat. Commun.* **2020**, *11*, 230. [CrossRef]
66. Mo, Y.; Campos, B.; Mealy, T.R.; Commodore, L.; Head, J.F.; Dedman, J.R.; Seaton, B.A. Interfacial basic cluster in annexin V couples phospholipid binding and trimer formation on membrane surfaces. *J. Biol. Chem.* **2003**, *278*, 2437–2443. [CrossRef]

67. Concha, N.O.; Head, J.F.; Kaetzel, M.A.; Dedman, J.R.; Seaton, B.A. Annexin V forms calcium-dependent trimeric units on phospholipid vesicles. *FEBS Lett.* **1992**, *314*, 159–162. [CrossRef]
68. Crosby, K.C.; Postma, M.; Hink, M.A.; Zeelenberg, C.H.; Adjobo-Hermans, M.J.; Gadella, T.W. Quantitative analysis of self-association and mobility of annexin A4 at the plasma membrane. *Biophys. J.* **2013**, *104*, 1875–1885. [CrossRef]
69. Ravanat, C.; Torbet, J.; Freyssinet, J.M. A neutron solution scattering study of the structure of annexin-V and its binding to lipid vesicles. *J. Mol. Biol.* **1992**, *226*, 1271–1278. [CrossRef]
70. Aukrust, I.; Hollås, H.; Strand, E.; Evensen, L.; Travé, G.; Flatmark, T.; Vedeler, A. The mRNA-binding site of annexin A2 resides in helices C-D of its domain IV. *J. Mol. Biol.* **2007**, *368*, 1367–1378. [CrossRef]
71. Solbak, S.M.Ø.; Abdurakhmanov, E.; Vedeler, A.; Danielson, U.H. Characterization of interactions between hepatitis C virus NS5B polymerase, annexin A2 and RNA—Effects on NS5B catalysis and allosteric inhibition. *Virol. J.* **2017**, *14*, 236. [CrossRef]
72. Pandurangan, A.P.; Ochoa-Montaño, B.; Ascher, D.B.; Blundell, T.L. SDM: A server for predicting effects of mutations on protein stability. *Nucleic Acids Res.* **2017**, *45*, W229–W235. [CrossRef] [PubMed]
73. Gomes, E.; Shorter, J. The molecular language of membraneless organelles. *J. Biol. Chem.* **2019**, *294*, 7115–7127. [CrossRef]
74. Vernon, R.M.; Chong, P.A.; Tsang, B.; Kim, T.H.; Bah, A.; Farber, P.; Lin, H.; Forman-Kay, J.D. Pi-Pi contacts are an overlooked protein feature relevant to phase separation. *Elife* **2018**, *7*. [CrossRef] [PubMed]
75. Misaki, Y.; Van Venrooij, W.J.; Pruijn, G.J. Prevalence and characteristics of anti-56K/annexin XI autoantibodies in systemic autoimmune diseases. *J. Rheumatol.* **1995**, *22*, 97–102. [PubMed]
76. Sjolin, C.; Movitz, C.; Lundqvist, H.; Dahlgren, C. Translocation of annexin XI to neutrophil subcellular organelles. *Biochim. Biophys. Acta* **1997**, *1326*, 149–156. [CrossRef]
77. Li, H.R.; Chiang, W.C.; Chou, P.C.; Wang, W.J.; Huang, J.R. TAR DNA-binding protein 43 (TDP-43) liquid-liquid phase separation is mediated by just a few aromatic residues. *J. Biol. Chem.* **2018**, *293*, 6090–6098. [CrossRef]
78. Nguyen, H.P.; Van Broeckhoven, C.; van der Zee, J. ALS Genes in the Genomic Era and their Implications for FTD. *Trends Genet.* **2018**, *34*, 404–423. [CrossRef]
79. Wei, Q.; Chen, X.; Chen, Y.; Ou, R.; Cao, B.; Hou, Y.; Zhang, L.; Shang, H.F. Unique characteristics of the genetics epidemiology of amyotrophic lateral sclerosis in China. *Sci. China Life Sci.* **2019**, *62*, 517–525. [CrossRef]

2

Localization of Annexin A6 in Matrix Vesicles during Physiological Mineralization

Ekeveliny Amabile Veschi [1,†], Maytê Bolean [1,†], Agnieszka Strzelecka-Kiliszek [2], Joanna Bandorowicz-Pikula [2], Slawomir Pikula [2], Thierry Granjon [3,4,5,6,7], Saida Mebarek [3,4,5,6,7], David Magne [3,4,5,6,7], Ana Paula Ramos [1], Nicola Rosato [8], José Luis Millán [9], Rene Buchet [3,4,5,6,7], Massimo Bottini [8,9,*] and Pietro Ciancaglini [1,*]

1 Departamento de Química, Faculdade de Filosofia, Ciências e Letras de Ribeirão Preto da Universidade de São Paulo (FFCLRP-USP), Ribeirão Preto, São Paulo 14040-900, Brazil; ekeveliny.veschi@usp.br (E.A.V); maytebolean@usp.br (M.B.); anapr@ffclrp.usp.br (A.P.R.)
2 Nencki Institute of Experimental Biology, 3 Pasteur Street, 02-093 Warsaw, Poland; a.strzelecka-kiliszek@nencki.edu.pl (A.S.K.); j.bandorowicz-pikula@nencki.edu.pl (J.B.-P.); s.pikula@nencki.edu.pl (S.P.)
3 Universite Lyon 1, UFR Chimie Biochimie, CEDEX, 69 622 Villeurbanne, France; thierry.granjon@univ-lyon1.fr (T.G.); saida.mebarek@univ-lyon1.fr (S.M.); david.magne@univ-lyon1.fr (D.M.); rene.buchet@univ-lyon1.fr (R.B.)
4 ICBMS UMR 5246 CNRS, CEDEX, 69 622 Villeurbanne, France
5 INSA, Lyon, CEDEX, 69 622 Villeurbanne, France
6 CPE, Lyon, CEDEX, 69 622 Villeurbanne, France
7 Université de Lyon, CEDEX, 69 622 Villeurbanne, France
8 Department of Experimental Medicine, University of Rome Tor Vergata, 00133 Rome, Italy; nicola.rosato@uniroma2.it
9 Sanford Burnham Prebys Medical Discovery Institute, La Jolla, CA 92037, USA; millan@sbpdiscovery.org
* Correspondence: pietro@ffclrp.usp.br (P.C.); massimo.bottini@uniroma2.it (M.B.)
† These authors contributed equally to this paper.

Abstract: Annexin A6 (AnxA6) is the largest member of the annexin family of proteins present in matrix vesicles (MVs). MVs are a special class of extracellular vesicles that serve as a nucleation site during cartilage, bone, and mantle dentin mineralization. In this study, we assessed the localization of AnxA6 in the MV membrane bilayer using native MVs and MV biomimetics. Biochemical analyses revealed that AnxA6 in MVs can be divided into three distinct groups. The first group corresponds to Ca^{2+}-bound AnxA6 interacting with the inner leaflet of the MV membrane. The second group corresponds to AnxA6 localized on the surface of the outer leaflet. The third group corresponds to AnxA6 inserted in the membrane's hydrophobic bilayer and co-localized with cholesterol (Chol). Using monolayers and proteoliposomes composed of either dipalmitoylphosphatidylcholine (DPPC) to mimic the outer leaflet of the MV membrane bilayer or a 9:1 DPPC:dipalmitoylphosphatidylserine (DPPS) mixture to mimic the inner leaflet, with and without Ca^{2+}, we confirmed that, in agreement with the biochemical data, AnxA6 interacted differently with the MV membrane. Thermodynamic analyses based on the measurement of surface pressure exclusion (π_{exc}), enthalpy (ΔH), and phase transition cooperativity ($\Delta t_{1/2}$) showed that AnxA6 interacted with DPPC and 9:1 DPPC:DPPS systems and that this interaction increased in the presence of Chol. The selective recruitment of AnxA6 by Chol was observed in MVs as probed by the addition of methyl-β-cyclodextrin (MβCD). AnxA6-lipid interaction was also Ca^{2+}-dependent, as evidenced by the increase in π_{exc} in negatively charged 9:1 DPPC:DPPS monolayers and the decrease in ΔH in 9:1 DPPC:DPPS proteoliposomes caused by the addition of AnxA6 in the presence of Ca^{2+} compared to DPPC zwitterionic bilayers. The interaction of AnxA6 with DPPC and 9:1 DPPC:DPPS systems was distinct even in the absence of Ca^{2+} as observed by the larger change in $\Delta t_{1/2}$ in 9:1 DPPC:DPPS vesicles as compared to DPPC vesicles. Protrusions on the surface of DPPC proteoliposomes observed by atomic force microscopy suggested that oligomeric AnxA6 interacted with the vesicle membrane. Further work is needed to delineate possible functions of AnxA6 at its different localizations and ways of interaction with lipids.

Keywords: Annexin A6; matrix vesicles; Langmuir monolayers; proteoliposomes; biomineralization; differential scanning calorimetry.

---

1. Introduction

Bone mineralization is a tightly regulated process that involves mesenchymal cells condensing into tissue elements followed by their differentiation into cartilage (chondrocytes) or bone (osteoblasts) cells [1]. During long bone formation, occurring throughout endochondral ossification, chondrocytes mineralize their extracellular matrix (ECM) by forming apatite. Osteoblasts are responsible for the formation of bone in both endochondral and endomembranous ossification acting in concert with osteoclasts to maintain the integrity of bone tissues. Both chondrocytes and osteoblasts produce a special class of extracellular vesicles, named matrix vesicles (MVs), that provide a suitable microenvironment for initiating apatite formation during bone mineralization [2–8]. MV-mediated mineralization is orchestrated by a constellation of proteins. Among them, the most abundant proteins in MVs isolated from hypertrophic chondrocytes [9] and human osteoblast-like cells Saos2 [10] are the annexins, AnxA1, AnxA2, AnxA5, AnxA6, and AnxA7.

Annexins belong to a family of structurally related proteins which bind to negatively charged phospholipids in a Ca^{2+}-dependent manner. However, their association with membranes can be also Ca^{2+}-insensitive [11,12]. The interaction between annexins and negatively charged phospholipids is also dependent on the presence of cholesterol (Chol) [11] and pH [13,14]. Annexins also interact with collagen: AnxA5- and AnxA6-modified proteoliposomes bind to type II collagen [15–17], while AnxA6 in MVs binds to type I collagen [18]. AnxA6 is the largest member of the annexin family present in MVs, with a molecular weight (MW) of approximately 68 kDa [19]. Unlike other members of the annexin family, which are structurally characterized by the presence of a highly conserved core composed of four homologous domains, AnxA6 has eight domains [19]. The differences between AnxA6 and the other members of the annexin family are also functional, especially in regard to membrane folding [20]. So far, little is known about the localization of annexins in MVs and their interaction with the MV membrane. Herein, we investigated the localization of AnxA6 in chicken embryo cartilage MVs, using a freeze-thaw procedure with and without ethylene glycol tetraacetic acid (EGTA). The co-localization of AnxA6 with Chol domains in MVs was probed by the addition of methyl-β-cyclodextrin (MβCD). To determine the nature of the interaction between AnxA6 and lipids, monolayers and bilayers of different lipid compositions containing AnxA6 were analyzed by means of exclusion pressure analysis, differential scanning calorimetry (DSC), dynamic light scattering (DLS), and atomic force microscopy (AFM). Our findings provide insights into a possible process of AnxA6 translocation across the MV membrane during biomineralization and we propose that AnxA6 has distinct functions in each translocation step from the lumen (interior) to the exterior surface of the MV membrane during biomineralization.

2. Results and Discussion

2.1. AnxA6 Associates With the MV Membrane as Both an Integral and a Peripheric Protein

First, we assessed the association of AnxA6 with the MV membrane by using a freeze-thaw procedure [21]. After 4 freeze-thaw cycles of MV solutions, the presence of full-length transmembrane AnxA6 embedded within the MV bilayer was evidenced by the presence of two isoforms of AnxA6 with a MW of approximately 67 and 60 kDa, respectively, into the pellet. These isoforms were insensitive to trypsin treatment and were not released into the supernatant by addition of EGTA (Figure 1A, lanes 1 and 2). Another population of full-length AnxA6 interacted strongly with the membrane in a Ca^{2+}-dependent manner as indicated by the appearance of an additional small AnxA6 fragment with a MW of approximately 35 kDa. The presence of a 35 kDa-MW AnxA6 fragment in MVs has been previously reported [22] and is due to the collagenase treatment and/or action of endogenous proteases [23,24]. It indicates that a population of AnxA6 associates with the inner leaflet of the MV membrane, accessible to endogenous proteases, as confirmed by the treatment with (Figure 1A, lane 3) or without (Figure 1A, lane 4) trypsin that did not affect the intensity of the bands.

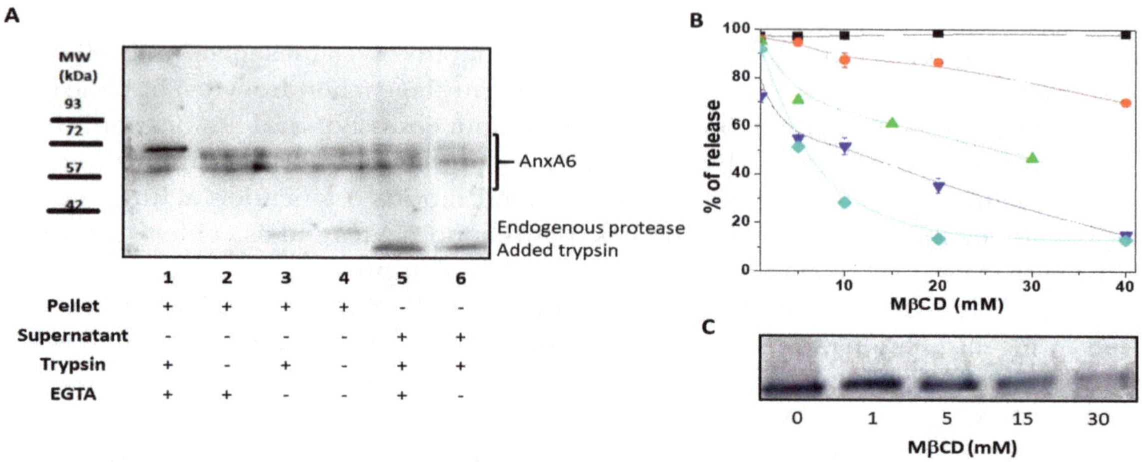

Figure 1. Annexin A6 in matrix vesicles. (**A**) Western blotting performed using mouse anti-AnxA6 monoclonal antibody to detect two isoforms of annexin A6 (AnxA6) in matrix vesicles (MVs) in the presence or absence of 0.1% trypsin after four freeze-thaw procedures followed by centrifugation. Lanes 1, 2, 3, and 4 corresponded to pellets, while lanes 5 and 6 corresponded to supernatants. MVs were treated with or without trypsin, with 2 mM Ca^{2+} or with 2 mM ethylene glycol tetraacetic acid (EGTA), as indicated. (**B**) AnxA6 co-localizes with cholesterol. MVs were incubated in synthetic cartilage lymph (SCL) medium containing 10 mM EGTA and an increasing concentration of methyl-β-cyclodextrin (MβCD) for 30 min in room temperature. Then, the samples were centrifuged. Lactate dehydrogenase (▼) and tissue non-specific alkaline phosphatase (TNAP) (•) activities were tested in the pellets and in the supernatants, as described in Materials and Methods. In the pellets, only the content of cholesterol (♦) and protein (■) was measured. MβCD-treated pellets were analyzed to determine AnxA6 (▲) content by Western blotting using a rabbit polyclonal anti-AnxA6 antibody that recognizes a single protein band. (**C**) Increasing MβCD concentrations led to a decrease of AnxA6 content. The results are expressed as a percentage of release of MV components ± SD.

A part of AnxA6 was digested by trypsin as indicated by the appearance of a 30 kDa-MW band (Figure 1A, lanes 5 and 6). Addition of EGTA (Figure 1A, lane 5) or 2 mM Ca^{2+} (Figure 1A, lane 6) did not eliminate the 30 kDa-MW band. The fragment was released by the freeze-thaw procedure in the supernatant, suggesting that it associates with the outer leaflet of the MV membrane due to its accessibility to trypsin digestion. The two full-length isoforms that were released in the supernatant could originate from the soluble AnxA6 inside MVs (Figure 1A, lane 5) or EGTA-released AnxA6 interacting with the inner leaflet (Figure 1A, lane 6) since both of them were not subjected to trypsin digestion.

2.2. AnxA6 Co-Localizes in Lipid Domains Enriched in Cholesterol

The MV membrane is enriched in Chol [25], therefore, we assessed if AnxA6 co-localizes in Chol-enriched lipid domains by examining the effect of the Chol-depleting agent methyl-β-cyclodextrin (MβCD) on AnxA6 release from the MV membrane. MβCD released Chol from the MV membrane in a concentration-dependent manner, reaching a maximal efficiency of release at a concentration of 15–20 mM (Figure 1B). The remaining 15% percent of membrane Chol, which was resistant to MβCD treatment, represented Chol esters. The calculated molar ratio between free Chol and Chol esters was 7.4 ± 1.0 mol/mol. Approximately 20% of glycosylphosphatidylinositol (GPI)-linked tissue non-specific alkaline phosphatase (TNAP), as probed by its activity, was released from the MV membrane at the maximal MβCD concentration. In addition, the activity of lactate dehydrogenase (LDH), a cytosolic marker, served to monitor the integrity of the MV membrane. Addition of MβCD affected significantly the integrity of the MV membrane at concentrations greater than 1 mM (Figure 1B) as indicated by the decreasing LDH activity in the supernatant. However, the total protein content remained almost stable (it decreased only slightly) in the pellet during the treatment with MβCD at concentrations ranging from 0 mM to 40 mM. Conversely, a significant release of AnxA6 from MV pellets into the supernatant was observed already when 15 mM MβCD was added to the medium (Figure 1B,C). This finding suggested that AnxA6 co-localizes in lipid domains enriched in Chol and its release is selective since the total protein concentration in the pellet did not change significantly (Figure 1B).

2.3. *Interaction of AnxA6 With Lipid Monolayers*

In order to assess how AnxA6 interacts with the leaflets of the MV membrane bilayer, we used Langmuir monolayers formed on the surface of a pendant drop. A monolayer made of 1,2-dipalmitoyl-*sn*-glycero-3-phosphocholine (DPPC) was used to mimic the outer leaflet, whereas the inner leaflet was modeled with a 9:1 DPPC:1,2-dipalmitoyl-*sn*-glycero-3-phospho-L-serine (DPPS) monolayer. Saturated lipids were chosen due to their relative high amount in MVs (approximately 48% are saturated fatty acids) [26]. The effect of Chol was monitored by comparing these monolayers with a 5:4:1 DPPC:Chol:DPPS monolayer. The interaction of AnxA6 with the monolayers was assessed by measuring the value of exclusion pressure (π_{exc}). The linear dependency between $\Delta\pi$ and π_0 for all the monolayers, both in presence and absence of 2 mM Ca^{2+}, enabled us to obtain the abscissa intercept corresponding to π_{exc} (arrows in Figure 2).

Higher π_{exc} values indicate a higher interaction or penetration of the protein into the monolayer even at higher packing conditions. Addition of Chol induced an increase in π_{exc} for all the lipid monolayers tested, indicating a strong interaction between the protein and the mixed monolayers in the presence of this sterol. This behavior was well observed in Chol-containing lipid monolayers, both in the absence (Figure 2A) and presence (Figure 2B) of Ca^{2+}. The presence of Ca^{2+} significantly contributed to the increased interaction of AnxA6 with the DPPS-containing monolayer, 9:1 DPPC:DPPS (molar ratio), as revealed by the increase in π_{exc} values from 23.7 mN/m in the absence of Ca^{2+} to 32.9 mN/m in the presence of Ca^{2+} (Table 1).

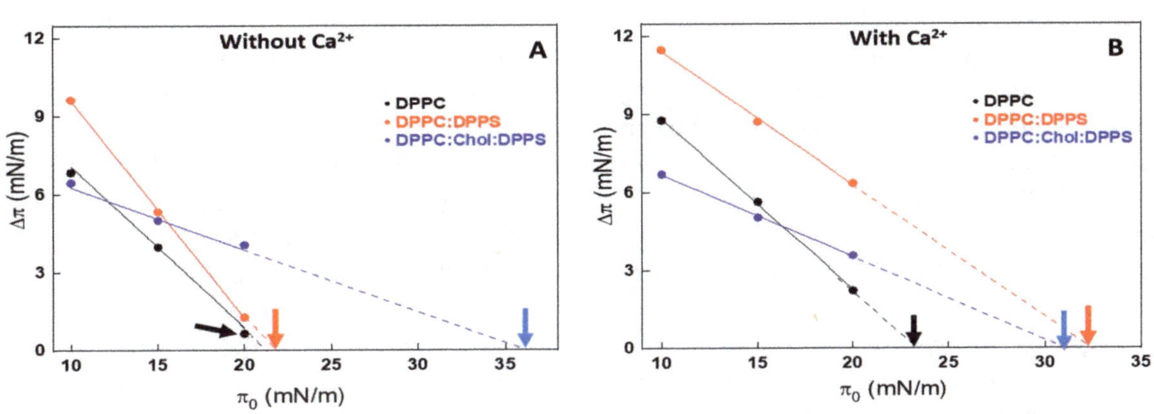

Figure 2. Changes in the surface pressure ($\Delta\pi$) due to the injection of AnxA6 as a function of the initial surface pressure (π_0) of Langmuir monolayers in the absence (**A**) and presence of 2 mM Ca^{2+} (**B**) for 1,2-dipalmitoyl-*sn*-glycero-3-phosphocholine (DPPC) (black dots), 9:1 DPPC:1,2-dipalmitoyl-*sn*-glycero-3-phospho-L-serine (DPPS) (red dots) and 5:4:1 DPPC:cholesterol (Chol):DPPS (blue dots) monolayers as a consequence of interaction with AnxA6. The intercept with abscissa is the exclusion pressure (π_{exc}) and is indicated by the arrows. The analyses were adjusted to π_0 close to 10, 15, and 20 mN/m.

Table 1. Values of exclusion pressure (π_{exc}) for lipid monolayers and concentrations of protein incorporated into lipid bilayers.

Lipid Composition (Molar Ratio)	Ca^{2+} (2mM)	π_{exc} in Monolayers (mN/m)	[Protein] Incorporated in Liposomes (μg/mL)
DPPC	−	22.2 ± 2.5	18.0 ± 0.4
	+	25.1 ± 2.7	59.0 ± 1.1
DPPC:DPPS (9:1)	−	23.7 ± 3.4	17.0 ± 0.3
	+	32.9 ± 2.0	43.0 ± 0.9
DPPC:Chol (6:4)	−	ND*	0.60 ± 0.01
	+	ND*	13.0 ± 0.3
DPPC:Chol:DPPS (5:4:1)	−	36.7 ± 1.0	19.0 ± 0.4
	+	34.3 ± 1.3	54.0 ± 1.0

*Did not form stable monolayer.

2.4. *Interaction of AnxA6 With Lipid Bilayers*

The interaction of AnxA6 with liposomes was assessed using the same lipid compositions of the Langmuir monolayers. In general, the addition of Ca^{2+} led to a significant increase in protein association with all the lipid bilayers (Table 1). It is worth noting that addition of Ca^{2+} led to a significant increase in the amount of AnxA6 incorporated into 5:4:1 DPPC:Chol:DPPS liposomes. However, it did not lead to a significant change in π_{exc}. This apparent controversial result can be assigned to the protein penetration into the monolayers in the presence and absence of Ca^{2+}. However, the stronger interaction with the lipids in the presence of Ca^{2+} translated into a greater incorporation into the liposomes.

DSC analysis showed the effect of AnxA6 addition on the thermodynamic behavior of liposomes (Figures 3 and 4, and Table 2). Addition of AnxA6 decreased the value of enthalpy (ΔH) for all lipid membranes irrespective of the presence of Ca^{2+} (Table 2). For DPPC vesicles, there were also no significant changes in the values of phase transition cooperativity ($\Delta t_{1/2}$) and transition phase temperature (T_m) (Figure 3a,A).

The addition of AnxA6 led to a decrease in the $\Delta t_{1/2}$ values in the presence of Ca^{2+} from 2.80 to 2.29 for 9:1 DPPC:DPPS proteoliposomes containing negative charges (Figure 3b,B) and from 19.37 to 13.89 for 5:4:1 DPPC:Chol:DPPS proteoliposomes (Table 2). However, the presence of Chol increased strongly the $\Delta t_{1/2}$ values irrespective of the presence of Ca^{2+}, resulting in much more cooperative phase transitions and displaying the synergistic effect of protein, Ca^{2+}, negative charges, and Chol, compared to Chol-devoid liposomes (Figure 4a,A,b,B).

The changes in the lipid bilayers' thermodynamic parameters caused by Chol were concentration dependent [27–29]. Addition of 40% molar percentage of Chol to the phospholipids eliminated the pretransition phase of DPPC bilayers (±35 °C) and also resulted in a broadened main lipid phase transition peak, leading to a decrease of ΔH and increase of $\Delta t_{1/2}$ values (Table 2). The ternary 5:4:1 DPPC:Chol:DPPS composition, as well as the binary 6:4 DPPC:Chol composition, showed a significant decrease in ΔH in the presence and absence of Ca^{2+} (Figure 4) as compared with Chol-devoid liposomes (Figure 3). The addition of AnxA6 decreased the value of ΔH of the lipid phase transition in both the ternary 5:4:1 DPPC:Chol:DPPS and binary 6:4 DPPC:Chol compositions. This effect was Ca^{2+}-dependent for both the systems.

The largest ΔH decrease upon addition of AnxA6 was observed for the binary 9:1 DPPC:DPPS composition in the presence of Ca^{2+} (from 8.72 to 4.96 Kcal/mol) and for the ternary 5:4:1 DPPC:Chol:DPPS composition in the absence of Ca^{2+} (from 1.14 to 0.43 Kcal/mol) (Table 2, Figures 3B and 4a). Such a decrease is consistent with AnxA6 interacting with the surfaces of the bilayer or being incorporated into the bilayer, preventing lipid molecules from participating in the phase transition. Such changes in the thermodynamic parameters should be taken into account and correlated to the interactions of the protein with the phospholipids and, consequently, with the physical stability of the liposome formulation.

Table 2. Differential scanning calorimetry (DSC)-calculated thermodynamic parameters [enthalpy (ΔH), transition phase temperature (T_m) and phase transition cooperativity ($\Delta t_{1/2}$)] for liposomes (10 mg/mL) and proteoliposomes with different lipid compositions, harboring or not AnxA6, and in the presence and absence of Ca^{2+} (2 mM).

		Without Ca^{2+}			With Ca^{2+}		
Lipid Composition (Molar Ratio)	**AnxA6**	**ΔH (Kcal/mol)**	**T_m (°C)**	**$\Delta t_{1/2}$**	**ΔH (Kcal/mol)**	**T_m (°C)**	**$\Delta t_{1/2}$**
DPPC	−	9.25	40.8	1.00	8.08	40.8	1.02
	+	6.75	41.2	1.07	6.08	41.0	1.09
DPPC:DPPS (9:1)	−	6.56	41.6	1.42	8.72	42.3	2.80
	+	5.29	41.7	1.50	4.96	42.7	2.29
DPPC:Chol (6:4)	−	1.24	46.7	29.59	0.57	44.3	22.86
	+	0.70	46.2	26.97	0.36	49.5	24.66
DPPC:Chol:DPPS (5:4:1)	−	5.36	43.6	14.54	1.14	46.3	19.37
	+	0.06	42.4	1.63	0.43	46.2	13.89

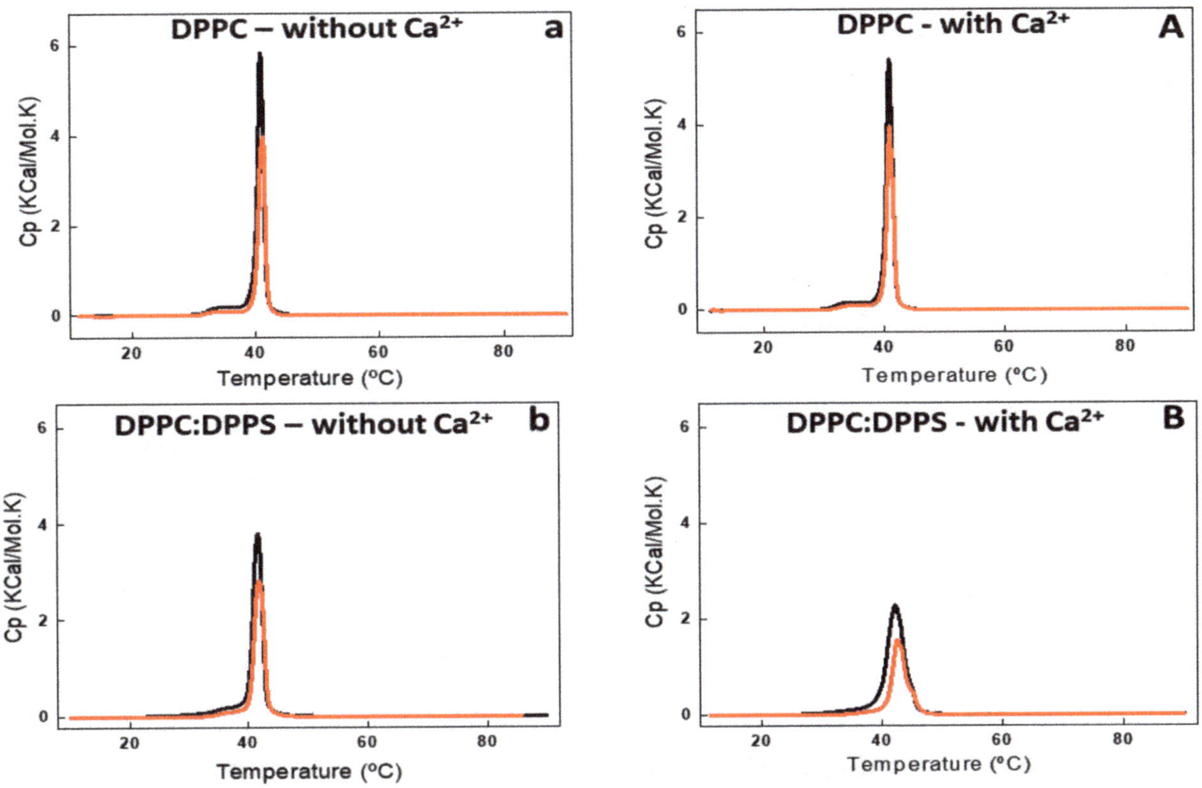

Figure 3. Differential scanning calorimetry (DSC) thermograms of DPPC (**a**,**A**) and 9:1 DPPC:DPPS (**b**,**B**) liposomes (10 mg/mL, total lipid concentration) in the absence (lower case letter) and in the presence (capital letter) of 2 mM Ca^{2+}. DSC thermograms were processed in excess heat capacity (C_p) (kcal/K.mol) as function of temperature (°C) of liposomes (black line) and proteoliposome (red line) harboring AnxA6.

Taken together, the calorimetry studies strongly indicate that AnxA6 can also associate with the outer leaflet of the MV membrane regardless of the presence of phosphatidylserine (PS), sterol, or calcium. However, the best results for protein interaction with liposomes were achieved in more fluid vesicles containing negative charges, Chol, and calcium. It is possible that AnxA6 exhibits different types of membrane interaction depending on the lipid composition, since different changes in the thermodynamic parameters were obtained depending on the proteoliposomes' lipid composition. These results corroborate the hypothesis that AnxA6 can be found in distinct regions of the MV membrane.

Finally, DLS confirmed that the diameter of the liposomes was smaller than 200 nm in absence of Ca^{2+} and did not show significant changes upon addition of Ca^{2+} to vesicle solution. The polydispersity index (PI) was smaller than 0.02 for all the formulations. The vesicles were stable in size (within +/-10 nm) for 28 days as probed by DLS.

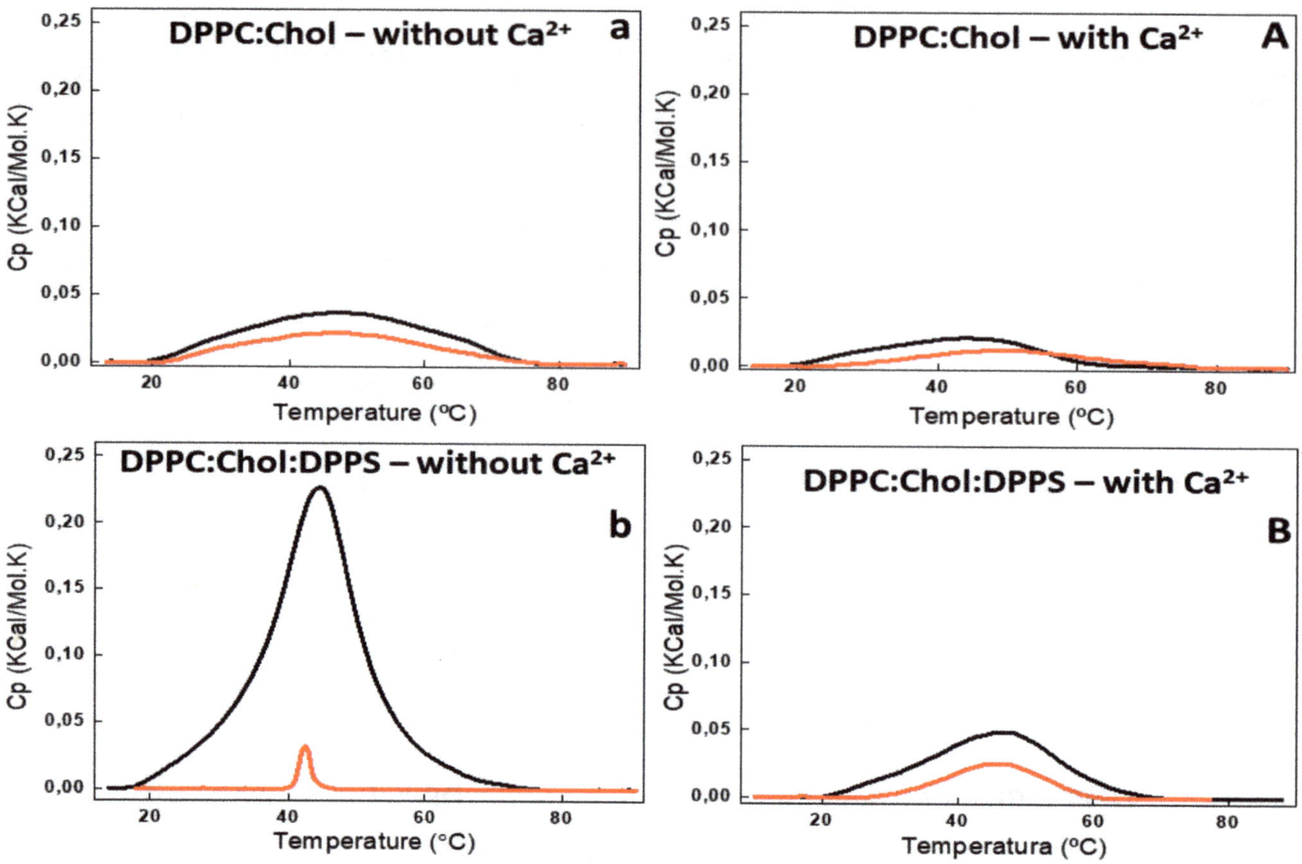

Figure 4. DSC thermograms of 6:4 DPPC:Chol (**a**,**A**) and 5:4:1 DPPC:Chol:DPPS (**b**,**B**) liposomes (10 mg/mL, total lipid concentration) in the absence (lower case) and in the presence (capital letter) of 2 mM Ca^{2+}. DSC thermograms were processed in excess heat capacity (C_P) (kcal/K.mol) as function of temperature (°C) of liposomes (black) and AnxA6-harboring proteoliposome (red).

2.5. Atomic Force Microscopy Analysis

The insertion of AnxA6 into lipid bilayers was further validated by atomic force microscopy (AFM) topographic and phase imaging. Liposomes composed of DPPC (Figure 5A–C) and DPPC:Chol:DPPS (6:4:1) (Figure 5D–F) showed spherical particles with a uniform distribution of size and a smooth and homogeneous surface without significant changes in the phase (Figure 5A,D). Conversely, AnxA6-harboring proteoliposomes showed surface irregularities formed by protrusions that appear to agglutinate at various sites (Figures 6 and 7). By recording the AFM topographic cross-sections, we found that these protrusions were 21.37 ± 4.83 nm wide and 1.15 ± 0.41 nm tall (N = 20) for DPPC proteoliposomes (Figure 6), whereas they were 26.97 ± 7.62 nm wide and 2.01 ± 0.89 nm tall (N = 22) for 6:4:1 DPPC:Chol:DPPS proteoliposomes (Figure 7). The protrusions on the surface of vesicles observed by AFM imaging suggested that oligomeric AnxA6 was on the surface of DPPC proteoliposomes (Figure 6). AnxA6 has two flexible lobes [30], thus it may adopt several conformations [31] with the lobes binding the monolayer in either a parallel orientation or an antiparallel orientation or with one of the lobes protruding out from the monolayer [32].

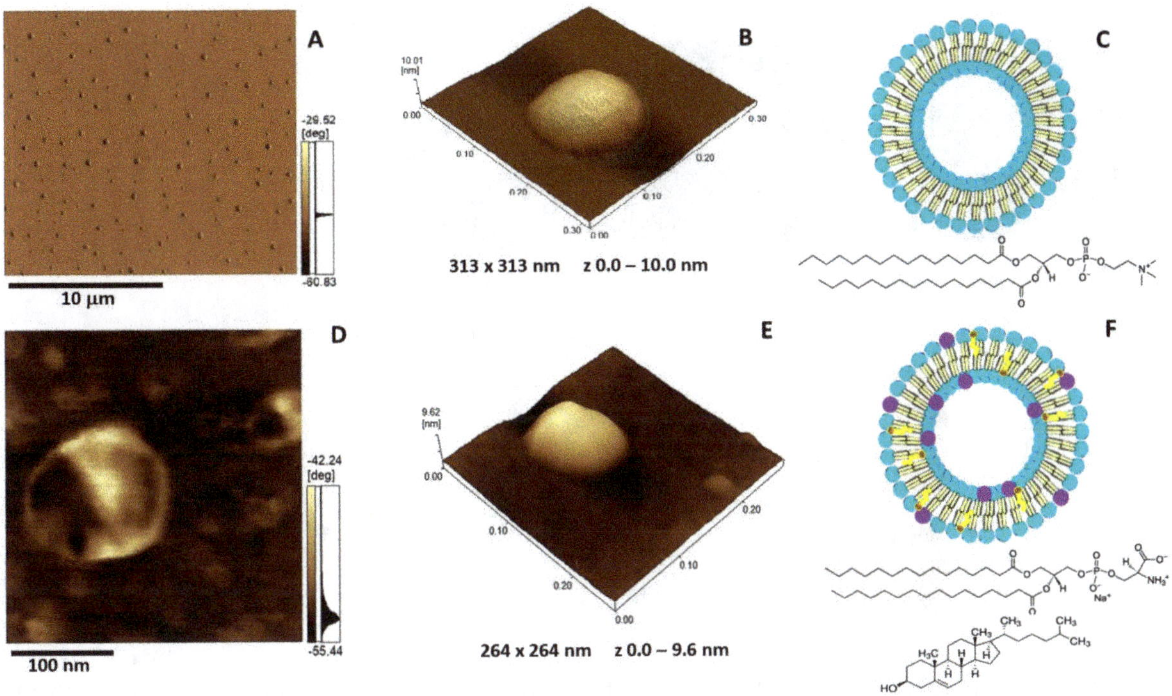

Figure 5. Atomic force microscopy (AFM) images of DPPC (**A–C**) and 5:4:1 DPPC:Chol:DPPS (**D–E**) liposomes (1.5 mg/mL): (**A**,**D**) Phase image, (**B**,**E**) 3D topography, and (**C**,**F**) liposome schematic representation. The blue and purple circles in **C** and **F** are assigned to the polar heads of DPPC and DPPS, respectively. The brown and yellow colors are assigned to Chol. The chemical structure of DPPC is depicted in **C**, and the chemical structures of DPPS and Chol are depicted in **F**.

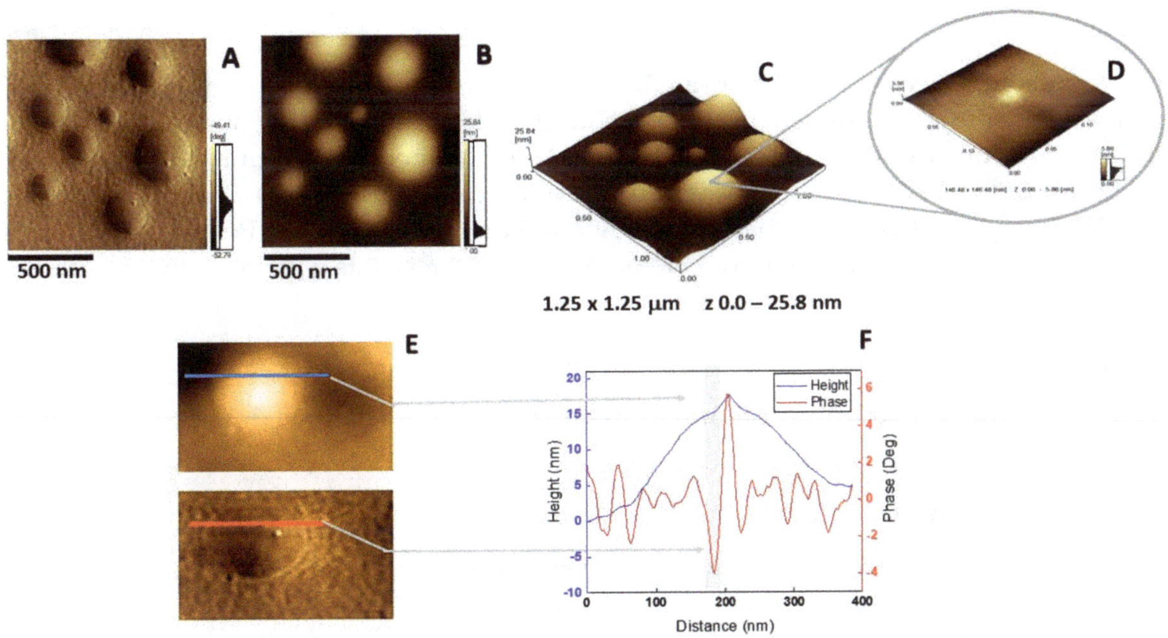

Figure 6. AFM analysis of DPPC proteoliposomes harboring AnxA6: (**A**) Phase image, (**B**) height image, (**C**) 3D topography, and (**D**) magnification of a 146.48 × 146.48 nm^2 surface area showing a protein domain; (**E** and **F**) phase and height line analysis (blue and red lines) of an AnxA6 domain inserted in DPPC proteoliposomes (the gray shadow highlights the protrusion). (**E**) was prepared using WSxM 4.0 Beta 9.1 software developed by Horcas et al. [33].

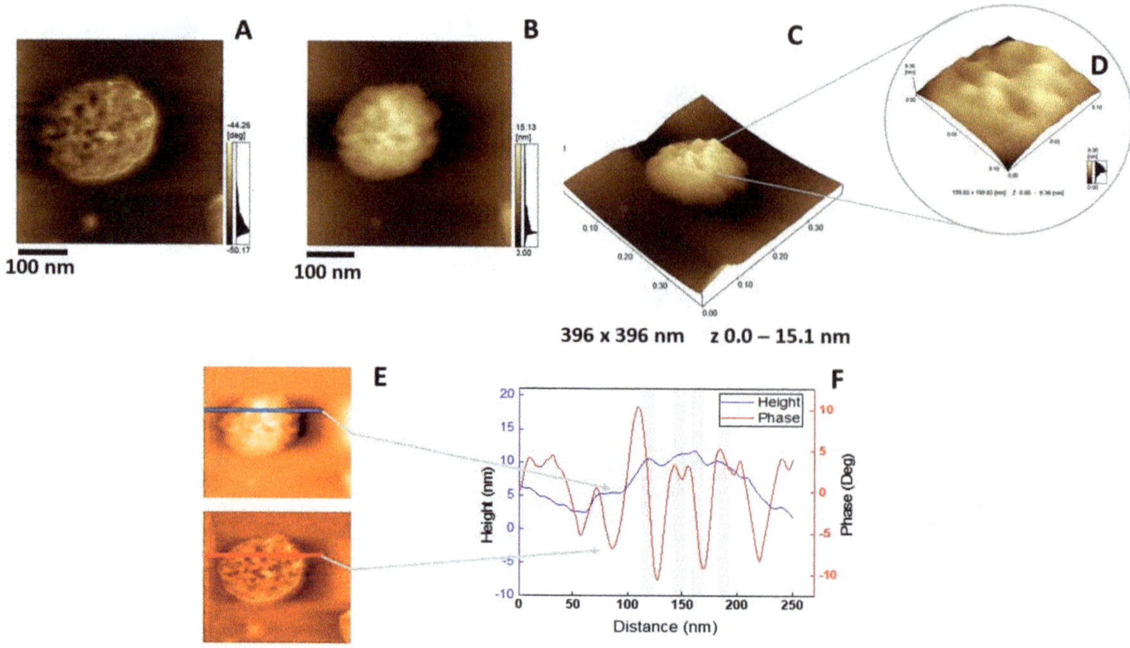

Figure 7. AFM analysis of 6:4:1 DPPC:Chol:DPPS proteoliposomes harboring AnxA6: (**A**) Phase image, (**B**) height image, (**C**) 3D topographic profile, and (**D**) magnification of a 109.83 × 109.83 nm^2 surface area showing protein domains; (**E** and **F**) phase and height line analysis (blue and red lines) of an AnxA6 domain inserted into 6:4:1 DPPC:Chol:DPPS proteoliposomes (the gray shadow highlights the protrusions). (**E**) was prepared using WSxM 4.0 Beta 9.1 software developed by Horcas et al. [33].

3. Materials and Methods

3.1. Materials

All aqueous solutions were prepared using Millipore® DirectQ ultrapure water. Bovine serum albumin (BSA), tris hydroxymethyl-amino-methane (Tris), sodium dodecylsulfate (SDS), and glutaraldehyde (Grade I, purified for use as an electron microscopy fixative) were obtained from Sigma Aldrich (Sigma Aldrich Corp., St. Louis, MO, USA). Bradford reagent was from Bio-Rad, (Bio-Rad Laboratories, Inc., Hercules, CA, USA). DPPC and Chol were purchased from Avanti Polar Lipids (Avanti Polar Lipids, Inc., Alabaster, AL, USA), whereas DPPS was obtained from Sigma Aldrich. All reagents were of analytical grade and used without further purification.

3.2. Expression of AnxA6

Recombinant human AnxA6 protein was expressed and purified as described by Bandorowicz-Pikula et al. [34].

3.3. Isolation of Matrix Vesicles

MVs were isolated from 17–day-old chicken embryos, as described before [3], excluding the trypsin-digestion step. Briefly, leg bones from 20 animals were isolated, cut into 1–3-mm-thick slices and extensively washed in ice cold synthetic cartilage lymph (SCL) containing 1.42 mM Na_2HPO_4, 1.83 mM $NaHCO_2$, 12.7 mM KCl, 0.57 mM $MgCl_2$, 100 mM NaCl, 0.57 mM Na_2SO_4, 5.55 mM glucose, 63.5 mM sucrose, and 16.5 mM 2-[(2-hydroxy-1,1-bis(hydroxymethyl)ethyl)amino]-ethanesulfonic acid (TES), pH 7.4. The bones were digested with collagenase type I from Clostridium histolyticum (200 U of collagenase/g of tissue in SCL medium containing 1 mM calcium) at 37 °C for 3 h in SCL. One μg/mL of protease inhibitor cocktail (Sigma Aldrich) was added and the digested tissue was filtrated and centrifuged at 800× *g* for 30 min at 4 °C. The pellet was discarded, and the supernatant was centrifuged at 30,000× *g* for 30 min at 4 °C. The second supernatant was subjected to a 250,000× *g* centrifugation for 30 min and the supernatant was discarded, whereas the pellet was washed three times in SCL medium without Ca^{2+} to remove calcium and protease inhibitor. The final volume of the MV solution was 400 μL and stored at 4 °C for further analysis.

3.4. Enzyme Activities, Total Protein, and Cholesterol Content

The activity of TNAP in MVs was determined by adding 10 mM *p*-nitrophenyl phosphate in 0.56 M amino 2-methyl 2-propanol and 1mM $MgCl_2$ buffer, pH 10.4, to the MV solution [35]. The appearance of the product (*p*-nitrophenolate) was monitored at 37 °C by measuring the absorbance at 405 nm. For standardization, the total protein content was measured in 2 μL of MV suspension supplemented with 798 μL of pure water and 200 μL of Bradford reagent (Bio-Rad). Total protein concentration was measured at 595 nm. Bovine serum albumin served as protein standard. TNAP (specific) activity was 8–10 U/mg and protein concentration was 1–2 mg/mL. To measure the activity of LDH, aliquots of MV solution were mixed with an equal volume of 0.5% Triton X-100 in order to completely release the enzyme from the vesicles. The reaction mixture contained 1 mM pyruvate and 0.25 mM nicotinamide adenine dinucleotide reduced (NADH). The oxidation of NADH was determined by measuring the absorbance at 340 nm [36]. The content of Chol in MVs after treatment with methyl-β-cyclodextrin (MβCD) was measured by Amplex Red Cholesterol Assay (Thermo Fisher Scientific, Waltham, MA, USA) according to the manufacturer's instruction. The measurement was performed without or with Chol esterase to determine free Chol (FC) to Chol esters (CE) (mol/mol) ratio. The ratio was calculated as follows: $FC/CE = (F_{FC})/(F_T\text{-}F_{FC})$ where F_T and F_{FC} are the fluorescence of resorufin measured at 584 nm in the presence (total Chol) and absence (free Chol) of Chol esterase, respectively.

3.5. Treatment of Matrix Vesicles With Trypsin and Freeze-Thaw Cycles in the Absence and Presence of Calcium

The MV solution was divided in four parts. The samples were mixed with 2 mM $CaCl_2$ without and with 0.1% trypsin and 2 mM EGTA without and with 0.1% trypsin, respectively. Trypsin served to monitor AnxA6 fragments by indicating AnxA6 accessible to protease digestion. Each sample was incubated at room temperature for 1 h, then it was centrifuged at 170,000× *g* for 20 min at room temperature. The supernatants were discarded, whereas the pellets were supplemented with 150 μL of SCL medium and subjected to four freeze-thaw cycles. Each cycle consisted of sample incubation in liquid nitrogen for 5 min and thawing at 37 °C for 5 min followed by vortexing for 1 min. The freeze-thaw cycles were used to disrupt and re-fuse the vesicles, during which time the solute equilibrates between the inside and outside [21]. Each sample was centrifuged to release AnxA6 in the supernatant and the membranous fraction of AnxA6 in the pellet.

3.6. Western Blotting

Proteins were separated by SDS-polyacrylamide gel electrophoresis (SDS-PAGE) on 10% gels [37] and transferred onto nitrocellulose membrane (GE Healthcare Bio-Sciences Corp., Pittsburgh, PA, USA). The membrane was blocked with 5% low-fat milk solution in 20 mM Tris, pH 7.5, and 500 mM NaCl for 60 min at room temperature, washed, and then incubated with mouse anti-AnxA6 monoclonal antibody (1:1000 v/v; BD Biosciences, San Jose, CA; this antibody recognizes two AnxA6 isoforms) or rabbit anti-AnxA6 polyclonal antibody (1:5000 v/v; Abcam, Cambridge, MA, USA; this antibody is nonspecific to AnxA6 isoforms) in 3% low-fat milk, 0.05% Tween-20, 20 mM Tris, pH 7.5, and 500 mM NaCl overnight at 4 °C. The membrane was washed in the same buffer and incubated for 60 min with anti-mouse immunoglobulin (IgG) conjugated with horseradish peroxidase (GE Healthcare Bio-Sciences). The bands were visualized using enhanced chemiluminescence (ECL) Western Blotting Detection Reagents (GE Healthcare Bio-Sciences) according to the manufacturer's instruction and exposing nitrocellulose to Hyperfilm ECL (GE Healthcare Bio-Sciences).

3.7. *Treatment of Matrix Vesicles with Methyl-β-Cyclodextrin*

MVs at a protein concentration of 0.33 mg/mL were incubated with increasing concentrations of methyl-β-cyclodextrin (MβCD) in SCL containing 10 mM EGTA. After 30 min of incubation at room temperature, the samples were centrifuged at 170,000× *g* for 20 min at room temperature. The supernatants were stored at −20 °C for enzyme activity assays, whereas the pellets were washed 3 times with SCL containing 10 mM EGTA and resuspended in 50 μL of the same buffer. One-half of each pellet was utilized to assess the presence of AnxA6, whereas the other half was used to measure TNAP and LDH activities as well as the total protein and Chol concentrations.

3.8. *Preparation of Liposomes*

DPPC, DPPC:DPPS (9:1), DPPC:Chol:DPPS (5:4:1), and DPPC:Chol (6:4) (molar ratios) liposomes (referred to as 9:1 DPPC:DPPS, 5:4:1 DPPC:Chol:DPPS, and 6:4 DPPC:Chol, respectively) were prepared as previously described using a 50 mM Tris-HCl buffer, pH 7.5, containing 2 mM $MgCl_2$ to yield a final solution with 10 mg/mL of lipids [38–41].

3.9. *Preparation of Proteoliposomes*

AnxA6 (0.2 mg/mL) was incorporated into liposomes by direct insertion in 50 mM Tris-HCl buffer, pH 7.5, containing 2 mM $MgCl_2$ with or without 2 mM $CaCl_2$ in a 1:100 protein:lipid ratio. The mixture was incubated for 24 h at 25 °C and ultracentrifuged at 100,000× *g* for 1 h, at 4 °C. The pellet containing proteoliposomes was suspended to the original volume in the same buffer [42]. Protein concentration was measured in presence of 0.2 g/mL SDS [43]. Bovine serum albumin (BSA) was used as a standard.

3.10. *Dynamic Light Scattering (DLS)*

Liposomes and proteoliposomes were filtered through 0.8-μm pore size membranes (Merck Millipore, Burlington, MA), dispersed in deionized water [38–41], and the size distribution and polydispersity index (PI) were determined by DLS using an N5 Submicron Particle Size Analyzer (25mW Helium-Neon Laser-632.8 nm, 90°) (Beckman Coulter, Fullerton, CA). The average value (N = 5) of the vesicles' diameter was obtained at 25 °C by unimodal distribution.

3.11. *Differential Scanning Calorimetry (DSC)*

Transition phase temperature (T_m), enthalpy (ΔH), and phase transition cooperativity ($\Delta t_{1/2}$) of large unilamellar vesicles (LUVs) prepared with different lipid compositions (as described in Section 3.8) were assessed by DSC. LUVs suspensions and reference buffer employed in the experiment were previously degasified under vacuum (140 mbar) for 30 min. The samples and reference (buffer) were scanned from 10 to 90 °C at an average heating rate of 0.5 °C/min, under pressure of 3 atmosphere (atm), and the recorded thermograms were analyzed by means of a Nano-DSC II (Calorimetry Sciences Corporation (CSC), Lindon, UT). The baseline was determined by filling the sample and reference cells with buffer solution. The data were analyzed using the software package CpCalc provided by CSC. The thermograms shown in the figures correspond to the first scan, but at least three heating-cooling cycles were performed for each analysis and ±5% variation [40].

3.12. *Exclusion Pressure (π_{exc})*

The incorporation of AnxA6 in preformed monolayers was monitored using a pendent drop system as described by Andrade et al. [44]. The pendent drop (approximately 11–12 μL) consisted of a microenvironment formed by 50 mM Tris-HCl buffer, pH 7.5, containing 2 mM $MgCl_2$ with or without 2 mM $CaCl_2$. Initially, the monolayers were formed by injection of a few microliters of lipid solution in chloroform into the drop until the monolayer's surface tension (γ_M) of interest was reached. A waiting period of 5 min was applied to stabilize the monolayer. Changes in γ_M were monitored by the axisymmetric drop shape analysis (ADSA) technique following injection of 0.2 μL of Anx6 aqueous solution into the drop (0.25 mg/mL). The interaction of the protein with the monolayer was detected by changes in γ_M. The surface pressure (π) was calculated by subtracting the surface tension of the water drop (γ_0) from γ_M ($\pi = \gamma_0 - \gamma_M$). The changes in surface pressure ($\Delta\pi$) after the injection of AnxA6 were plotted against π_0 and the values of π_{exc} were calculated by the extrapolation of the curve to a nil value of $\Delta\pi$ [44].

3.13. Atomic Force Microscope (AFM) Imaging

Liposomes and proteoliposomes were filtered through 0.8-μm pore size membranes (Merck Millipore) and stabilized by adding 1:1 (v/v) glutaraldehyde (~5% final concentration) to avoid vesicle deformation and disruption. The mixtures were homogenized, and then 8 μL of the sample were dropped onto freshly cleaved mica substrates, left to dry at room temperature, and imaged by an AFM (model SPM-9600 Scanning Probe Microscope, Shimadzu Corporation, Japan) operating in tapping mode as described previously by Bolean et al. [41]. Scanning was performed in air at 25 °C by using silicon probes with a resonance frequency ranging from 324 to 369 kHz (Nanosensors™, Neuchâtel, Switzerland). The scan rate was set at 0.2–0.3 Hz to prevent tip-induced vesicle deformations and/or damages. The used cantilevers had a spring constant of a 38 ± 8 N/m and a resonance frequency of 336 ± 67 kHz. The roughness values and simultaneous analysis of phase and height were carried by using SPM Offline software (Shimadzu) and treated by using WSxM 4.0 Beta 9.1 and Origin 2019 software [33,45]. Three distinct experiments were carried out for each type of vesicles and N = 100 vesicles were analyzed in each experiment.

4. Conclusions

In this study, we investigated the association of AnxA6 with the membrane bilayer of natural MVs and with MV biomimetics. By using biochemical procedures on MVs isolated from 17-day-old chicken embryos, AnxA6 was found to interact with the MV membrane as both an integral and peripheral protein, which suggests distinct lipid–protein interactions and possible functions. Of particular interest is the strong co-localization of AnxA6 with Chol. This result is coherent with the Chol-driven recruitment of AnxA6 from the lumen of the cell toward plasma membrane [46].

Studies based on monolayers and proteoliposomes as MV biomimetics strengthened our findings by shedding the light on the interactions of AnxA6 with different lipid systems in distinct ways. Of particular importance is that AnxA6 induced an increase in the surface pressure and a decrease in the enthalpy of both DPPC (mimicking the outer leaflet of the MV bilayer) and 9:1 DPPC:DPPS (mimicking the inner leaflet of the MV bilayer) lipid monolayers indicating an interaction of AnxA6 with these lipid systems. However, the interaction of AnxA6 with bilayers containing Chol is probably deeper since it led to a strong increase in phase transition cooperativity and increased the packing of the lipid membrane.

The Ca^{2+} dependence of lipid–protein interaction was well evidenced by the increase in surface pressure caused by the addition of AnxA6 to negatively charged lipid monolayers composed of 9:1 DPPC:DPPS in presence of Ca^{2+} with respect to neutral monolayers composed of DPPC and in absence of Ca^{2+}. AnxA6 interacted differently with DPPC and 9:1 DPPC:DPPS monolayers even in the absence of Ca^{2+} as observed by the larger change in phase transition cooperativity in 9:1 DPPC:DPPS vesicles as compared to DPPC vesicles.

AFM images showed protrusions that suggest domains of oligomeric AnxA6 on the surface of DPPC proteoliposomes, thus AnxA6 may adopt several conformations upon interaction with the lipid membrane.

Taken together, these findings substantiate a possible mechanism of AnxA6 translocation within the MV membrane bilayer. First, annexins in the lumen of MVs bind to phosphatidylserine (PS) of the inner leaflet of the MV bilayer in the presence of Ca^{2+} triggering the formation of a nucleational core (NC) with the aid of AnxA5 and other proteins (Figure 8A,B) [22,47]. AnxA6 may also contribute to the accumulation of Ca^{2+} and to stabilize Ca^{2+}-binding to PS, promoting a favorable environment for apatite formation (Figure 8C). Next, a local pH drop during apatite formation may promote the insertion of AnxA6 into the MV bilayer (Figure 8D), which makes AnxA6 resistant to EGTA extraction [48], as confirmed in this work. Whether the transmembrane AnxA6 has the ability to transport Ca^{2+} is not yet settled, although several reports have suggested that AnxA5 [40,49–51] or unspecified annexins [52,53] can induce Ca^{2+} influx. The insertion of AnxA6 induced by acidic pH is reversible upon pH increase [13], which suggests that AnxA6 could be expelled in the extracellular matrix (ECM), where pH is neutral, or it could also bind to phosphatidylcholine (PC) on the MV surface, mimicking the association of AnxA6 with the outer leaflet of the MV membrane bilayer (Figure 8E). This may occur either during translocation of AnxA6 from the lumen toward the extracellular leaflet of MVs or by its presence on the surface of just released MVs to seal membrane holes. Extracellular AnxA6 may contribute to enforce MV interactions with collagen fibers (Figure 8F). The proposed mechanisms of translocation and AnxA6 localizations are consistent with the distinct types of bindings of AnxA6 to DPPS with and without Ca^{2+}, as well as to DPPC with and without Ca^{2+}.

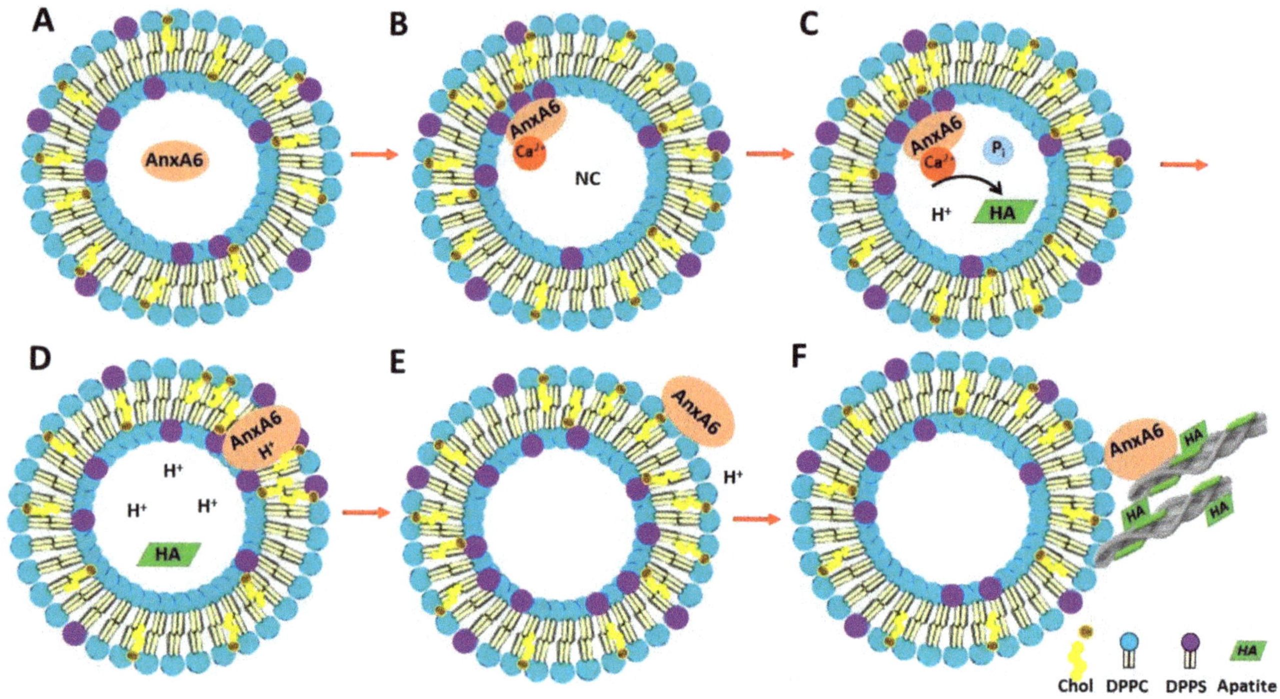

Figure 8. Representation of the mechanism of AnxA6 translocation during MV-mediated mineralization and its putative functions. (**A**) AnxA6 is localized in the lumen of MVs at low calcium concentrations. (**B**) Accumulation of calcium inside MVs favors the binding of AnxA6 to the inner leaflet of the MV membrane bilayer containing phosphatidylserine (PS) in a Ca^{2+}-dependent manner, forming a nucleational core (NC) and initiating apatite formation. (**C**) During apatite formation, protons (H^+) are released, inducing the protonation of AnxA6 and rendering it more hydrophobic due to an ionization potential around 5.5. (**D**) Protonated AnxA6 can translocate within the MV bilayer where it may form an ion pore. (**E**) The pH outside MVs corresponds to that of the extracellular matrix and is neutral. AnxA6 can be deprotonated and released to the external surface of MVs enriched in phosphatidylcholine (PC). (**F**) AnxA6 binding to neutral phospholipids is Ca^{2+}-independent. AnxA6 may also bind to collagen fibers, contributing to the adhesion of MVs to collagen fibers.

Author Contributions: Performed the research, E.A.V., M.B. (Maytê Bolean), and S.M.; conceptualization, M.B. (Maytê Bolean), A.S.-K., S.P., S.M., N.R., R.B., M.B. (Massimo Bottini), and P.C.; funding acquisition, supervision, and validation, A.S.-K., J.B.-P., S.P., D.M., J.L.M., and P.C.; investigation, S.M.; resources, J.B.-P., S.P., and D.M.; writing, original draft, E.A.V., M.B. (Maytê Bolean), A.P.R., R.B., M.B. (Massimo Bottini), and P.C.; writing, review and editing, M.B. (Maytê Bolean), A.S.-K., J.B.-P., S.P., T.G., S.M., D.M., A.P.R., J.L.M., R.B, M.B. (Massimo Bottini), and P.C. All authors have read and agreed to the published version of the manuscript.

Acknowledgments: We thank Marcin Balcerzak (Nencki Institute of Experimental Biology, 3 Pasteur Street, 02-093 Warsaw, Poland) for the assistance during the experimental work and the stimulating discussions. We also thank

Ivana Aparecida Borin (Departamento de Química, Faculdade de Filosofia, Ciências e Letras de Ribeirão Preto da Universidade de São Paulo (FFCLRP-USP), Ribeirão Preto, São Paulo, Brazil) for the acquisition of AFM images.

References

1. Aghajanian, P.; Mohan, S. The art of building bone: Emerging role of chondrocyte-to-osteoblast transdifferentiation in endochondral ossification. *Bone Res.* **2018**, *6*, 19. [CrossRef]
2. Wu, L.N.; Genge, B.R.; Lloyd, G.C.; Wuthier, R.E. Collagen-binding proteins in collagenase-released matrix vesicles from cartilage. Interaction between matrix vesicle proteins and different types of collagen. *J. Biol. Chem.* **1991**, *266*, 1195–1203. [PubMed]
3. Kirsch, T.; Wuthier, R.E. Stimulation of calcification of growth plate cartilage matrix vesicles by binding to type II and X collagens. *J. Biol. Chem.* **1994**, *269*, 11462–11469. [PubMed]
4. Anderson, H.C. Matrix vesicles and calcification. Role of extracellular membrane vesicles in the pathogenesis of various diseases, including cancer, renal diseases, atherosclerosis, and arthritis. *Curr. Rheumatol. Rep.* **2003**, *5*, 222–226. [CrossRef] [PubMed]
5. Anderson, H.C.; Mulhall, D.; Garimella, R. Role of extracellular membrane vesicles in the pathogenesis of various diseases, including cancer, renal diseases, atherosclerosis, and arthritis. *Lab. Invest.* **2010**, *90*, 1549–1557. [CrossRef]
6. Golub, E.E. Role of matrix vesicles in biomineralization. *Biochim. Biophys. Acta* **2009**, *1790*, 1592–1598. [CrossRef]
7. Wuthier, R.E.; Lipscomb, G.F. Matrix vesicles: Structure, composition, formation and function in calcification. *Front. Biosci.* **2011**, *17*, 2812–2902. [CrossRef]
8. Bottini, M.; Mebarek, S.; Anderson, K.L.; Strzelecka-Kiliszek, A.; Bozycki, L.; Simão, A.M.S.; Bolean, M.; Ciancaglini, P.; Pikula, J.B.; Pikula, S.; et al. Matrix vesicles from chondrocytes and osteoblasts. Their biogenesis, properties, functions and biomimetic models. *BBA Gen. Subj.* **2018**, *1862*, 532–546.
9. Balcerzak, M.; Malinowska, A.; Thouverey, C.; Sekrecka, A.; Dadlez, M.; Buchet, R.; Pikula, S. Proteome analysis of matrix vesicles isolated from femurs of chicken embryo. *Proteomics* **2008**, *8*, 192–205. [CrossRef]
10. Thouverey, C.; Malinowska, A.; Balcerzak, M.; Strzelecka-Kiliszek, A.; Buchet, R.; Dadlez, M.; Pikula, S. Proteomic characterization of biogenesis and functions of matrix vesicles released from mineralizing human osteoblast-like cells. *J. Proteom.* **2011**, *74*, 1123–1134. [CrossRef]
11. Ayala–Sanmartin, J. Cholesterol enhances phospholipid binding and aggregation of annexins by their core domain. *Biochem. Biophys. Res. Commun.* **2001**, *283*, 72–79. [CrossRef]
12. Babiychuk, E.B.; Draeger, A. Annexins in cell membrane dynamics. Ca(2+)-regulated association of lipid microdomains. *J. Cell. Biol.* **2000**, *150*, 1113–1124. [CrossRef] [PubMed]
13. Golczak, M.; Kicinska, A.; Bandorowicz–Pikula, J.; Buchet, R.; Szewczyk, A.; Pikula, S. Acidic pH-induced folding of annexin VI is a prerequisite for its insertion into lipid bilayers and formation of ion channels by the protein molecules. *Faseb J.* **2001**, *15*, 1083–1085. [CrossRef]
14. Strzelecka-Kiliszek, A.; Buszewska, M.E.; Podszywalow-Bartnicka, P.; Pikula, S.; Otulak, K.; Buchet, R.; Bandorowicz-Pikula, J. Calcium- and pH-dependent localization of annexin A6 isoforms in Balb/3T3 fibroblasts reflecting their potential participation in vesicular transport. *J. Cell. Bioch.* **2008**, *104*, 418–434. [CrossRef]
15. Genge, B.R.; Cao, X.; Wu, L.N.; Buzzi, W.R.; Showman, R.W.; Arsenault, A.L.; Ishikawa, Y.; Wuthier, R.E. Establishment of the primary structure of the major lipid-dependent Ca^{2+} binding proteins of chicken growth plate cartilage matrix vesicles: Identity with anchorin CII (annexin V) and annexin II. *J. Bone Min.. Res.* **1992**, *7*, 807–819. [CrossRef]
16. Von der Mark, K.; Mollenhauer, J. Annexin V interactions with collagen. *Cell Mol. Life Sci.* **1997**, *53*, 539–545. [CrossRef]
17. Kirsch, T.; Harrison, G.; Golub, E.E.; Nah, H.D. The roles of annexins and types II and X collagen in matrix vesicle-mediated mineralization of growth plate cartilage. *J. Biol. Chem.* **2000**, *275*, 35577–35583. [CrossRef]
18. Chen, N.X.; O'Neill, K.D.; Chen, X.; Moe, S.M. Annexin-mediated matrix vesicle calcification in vascular smooth muscle cells. *J. Bone Min. Res.* **2008**, *23*, 1798–1805. [CrossRef]
19. Huber, R.; Romisch, J.; Pâques, E.P. The crystal and molecular structure of human annexin V, an anticoagulant protein that binds to calcium and membranes. *Embo J.* **1990**, *9*, 3867–3874. [CrossRef] [PubMed]

20. Boye, T.L.; Jeppesen, J.C.; Maeda, K.; Pezesshkian, W.; Solovyeva, V.; Nylandsted, J.; Simonsen, A.C. Annexins induce curvature on free-edge membranes displaying distinct morphologies. *Sci. Rep.* **2018**, *8*, 10309–10322. [CrossRef] [PubMed]
21. R.R.C. New. *Liposomes: A practical approach*; Oxford University Press: Oxford, UK, 1990; p. 58.
22. Wu, L.N.; Yoshimori, T.; Genge, B.R.; Sauer, G.R.; Kirsch, T.; Ishikawa, Y.; Wuthier, R.E. Characterization of the nucleational core complex responsible for mineral induction by growth plate cartilage matrix vesicles. *J. Biol. Chem.* **1997**, *268*, 25084–25094.
23. Mookhtiar, K.A.; Van Wart, H.E. Clostridium histolyticum collagenases: A new look at some old enzymes. *Matrix Suppl.* **1992**, *1*, 116–126.
24. Chen, D.; Golub, E.E. Search for an endogenous collagenase in chicken endochondral bone matrix vesicles. *Penn. Dent. J. Phila.* **2001**, *101*, 12–33. [PubMed]
25. Wuthier, R.E. Lipid composition of isolated epiphyseal cartilage cells, membranes and matrix vesicles. *Biochim. Biophys. Acta* **1975**, *409*, 128–143. [CrossRef]
26. Abdallah, D.; Hamade, E.; Merhi, R.A.; Bassam, B.; Buchet, R.; Mebarek, S. Fatty acid composition in matrix vesicles and in microvilli from femurs of chicken embryos revealed selective recruitment of fatty acids. Fatty acid composition in matrix vesicles and in microvilli from femurs of chicken embryos revealed selective recruitment of fatty acids. *Biochem. Biophys. Res. Commun.* **2014**, *446*, 1161–1164. [PubMed]
27. Kyrikou, I.; Georgopoulos, A.; Hatziantoniou, S.; Mavromoustakos, T.; Demetzos, C. A comparative study of the effects of cholesterol and sclareol, a bioactive labdane type diterpene, on phospholipid bilayers. *Chem. Phys. Lipids* **2005**, *133*, 125–134. [CrossRef] [PubMed]
28. Matsingou, C.; Demetzos, C. Calorimetric study on the induction of interdigitated phase in hydrated DPPC bilayers by bioactive labdanes and correlation to their liposome stability: The role of chemical structure. *Chem. Phys. Lipids* **2007**, *145*, 45–62. [CrossRef]
29. Matsingou, C.; Hatziantoniou, S.; Georgopoulos, A.; Dimas, K.; Terzis, A.; Demetzos, C. Labdane-type diterpenes: Thermal effects on phospholipid bilayers, incorporation into liposomes and biological activity. *Chem. Phys. Lipids* **2006**, *138*, 1–11. [CrossRef]
30. Benz, J.; Bergner, A.; Hofmann, A.; Demange, P.; Göttig, P.; Liemann, S.; Huber, R.; Voges, D. The structure of recombinant human annexin VI in crystals and membrane-bound. *J. Mol. Biol.* **1996**, *260*, 638–643. [CrossRef]
31. Avila-Sakar, A.J.; Kretsinger, R.H.; Creutz, C.E. Membrane-bound 3D structures reveal the intrinsic flexibility of annexin VI. *J. Struct. Biol.* **2000**, *30*, 54–62. [CrossRef]
32. Buzhynskyy, N.; Golczak, M.; Lai-Kee-Him, J.; Lambert, O.; Tessier, B.; Gounou, C.; Bérat, R.; Simon, A.; Granier, T.; Chevalier, J.M.; et al. Annexin-A6 presents two modes of association with phospholipid membranes. A combined QCM-D, AFM and cryo-TEM study. *J. Struct. Biol.* **2009**, *168*, 107–116. [CrossRef]
33. Horcas, I.; Fernández, R.; Gómez-Rodríguez, J.M.; Colchero, J.; Gómez-Herrero, J.; Baró, A.M. WSXM: A software for scanning probe microscopy and a tool for nanotechnology. *Rev. Sci. Instrum.* **2007**, *78*, 013705. [CrossRef] [PubMed]
34. Bandorowicz-Pikula, J.; Kirilenko, A.; Van Deursen, R.; Golczak, M.; Kühnel, M.; Lancelin, J.M.; Pikula, S.; Buchet, R. A Putative Consensus Sequence for the Nucleotide-Binding Site of Annexin A6. *Biochemistry* **2003**, *42*, 9137–9146. [CrossRef] [PubMed]
35. Cyboron, G.W.; Wuthier, R.E. Purification and initial characterization of intrinsic membrane-bound alkaline phosphatase from chicken epiphyseal cartilage. *J. Biol. Chem.* **1981**, *256*, 7262–7268.
36. Hosokawa, R.; Uchida, Y.; Fujiwara, S.; Noguchi, T. Lactate dehydrogenase isoenzymes are present in matrix vesicles. *J. Biol. Chem.* **1988**, *263*, 10045–10047.
37. Laemmli, U.K. Cleavage of structural proteins during the assembly of the head of bacteriophage T4. *Nature* **1970**, *227*, 680–685. [CrossRef]
38. Bolean, M.; Simão, A.M.S.; Favarin, B.Z.; Millán, J.L.; Ciancaglini, P. The effect of cholesterol on the reconstitution of alkaline phosphatase into liposomes. *Biophys. Chem.* **2010**, *152*, 74–79. [CrossRef]
39. Bolean, M.; Simão, A.M.; Favarin, B.Z.; Millán, J.L.; Ciancaglini, P. Thermodynamic properties and characterization of proteoliposomes rich in microdomains carrying alkaline phosphatase. *Biophys. Chem.* **2011**, *158*, 111–118. [CrossRef]
40. Bolean, M.; Simão, A.M.; Kiffer-Moreira, T.; Hoylaerts, M.F.; Millán, J.L.; Itri, R.; Ciancaglini, P. Proteoliposomes with the ability to transport Ca(2+) into the vesicles and hydrolyze phosphosubstrates on their surface. *Arch. Biochem. Biophys.* **2015**, *584*, 79–89. [CrossRef]

41. Bolean, M.; Borin, I.A.; Simão, A.M.S.; Bottini, M.; Bagatolli, L.A.; Hoylaerts, M.F.; Millán, J.L.; Ciancaglini, P. Topographic analysis by atomic force microscopy of proteoliposomes matrix vesicle mimetics harboring TNAP and AnxA5. *Biochim. Biophys. Acta Biomembr.* **2017**, *1859*, 1911–1920. [CrossRef]
42. Bolean, M.; Simão, A.M.S.; Barioni, M.B.; Favarin, B.Z.; Sebinelli, H.B.; Veschi, E.A.; Janku, T.A.B.; Bottini, M.; Hoylaerts, M.F.; Itri, R.; et al. Biophysical aspects of biomineralization. *Biophys. Rev.* **2017**, *9*, 747–760. [CrossRef]
43. Hartree, E.F. Determination of protein: A modification of the Lowry method that gives a linear photometric response. *Anal. Biochem.* **1972**, *48*, 422–427. [CrossRef]
44. Andrade, M.A.R.; Favarin, B.Z.; Derradi, R.; Bolean, M.; Simão, A.M.S.; Millán, J.L.; Ciancaglini, P.; Ramos, A.P. Pendant-drop method coupled to ultraviolet-visible spectroscopy: A useful tool to investigate interfacial phenomena. *Coll. Surf. A.* **2016**, *504*, 305–311. [CrossRef]
45. Sebinelli, H.G.; Borin, I.A.; Ciancaglini, P.; Bolean, M. Topographical and mechanical properties of liposome surfaces harboring Na,K-ATPase by means of atomic force microscopy. *Soft Matter.* **2019**, *15*, 2737–2745. [CrossRef]
46. Sztolsztener, M.E.; Strzelecka-Kiliszek, A.; Pikula, S.; Tylki-Szymanska, A.; Bandorowicz-Pikula, J. Cholesterol as a factor regulating intracellular localization of annexin A6 in Niemann-Pick type C human skin fibroblasts. *Arch. Biochem. Biophys.* **2010**, *493*, 221–233. [CrossRef]
47. Wu, L.N.Y.; Genge, B.R.; Dunkelberger, D.G.; LeGeros, R.Z.; Concannon, B.; Wuthier, R.E. Physicochemical characterization of the nucleational core of matrix vesicles. *J. Biol. Chem.* **1997**, *272*, 4404–4411. [CrossRef]
48. Cmoch, A.; Strzelecka-Kiliszek, A.; Palczewska, M.; Groves, P.; Pikula, S. Matrix vesicles isolated from mineralization-competent Saos-2 cells are selectively enriched with annexins and S100 proteins. *Biochem. Biophys. Res. Commun.* **2011**, *412*, 683–687. [CrossRef]
49. Arispe, N.; Rojas, E.; Genge, B.R.; Wu, L.N.; Wuthier, R.E. Similarity in calcium channel activity of annexin V and matrix vesicles in planar lipid bilayers. *Biophys. J.* **1996**, *71*, 1764–1775. [CrossRef]
50. Kaneko, N.; Ago, H.; Matsuda, R.; Inagaki, E.; Miyano, M. Crystal structure of annexin V with its ligand K-201 as a calcium channel activity inhibitor. *J. Mol. Biol.* **1997**, *274*, 16–20. [CrossRef]
51. Kaneko, N.; Matsuda, R.; Toda, M.; Shimamoto, K. Inhibition of annexin V-dependent Ca^{2+} movement in large unilamellar vesicles by K201, a new 1,4-benzothiazepine derivative. *Biochim. Biophys. Acta* **1997**, *1330*, 1–7. [CrossRef]
52. Hofmann, A.; Escherich, A.; Lewit-Bentley, A.; Benz, J.; Raguenes-Nicol, C.; Russo-Marie, F.; Gerke, V.; Moroder, L.; Huber, R. Interactions of benzodiazepine derivatives with annexins. *J. Biol. Chem.* **1998**, *273*, 2885–2894. [CrossRef]
53. Wang, W.; Xu, J.; Kirsch, T. Annexin-mediated Ca^{2+} influx regulates growth plate chondrocyte maturation and apoptosis. *J. Biol. Chem.* **2003**, *278*, 3762–3769. [CrossRef] [PubMed]

3

(20S)G-Rh2 Inhibits NF-κB Regulated Epithelial-Mesenchymal Transition by Targeting Annexin A2

Yu-Shi Wang, He Li, Yang Li, Shiyin Zhang and Ying-Hua Jin *

Key Laboratory for Molecular Enzymology and Engineering of the Ministry of Education, School of Life Sciences, Jilin University, Changchun 130012, China; wangyushi0317@hotmail.com (Y.-S.W.); lihe0607@163.com (H.L.); liyang915@jlu.edu.cn (Y.L.); zsyshiyin@163.com (S.Z.)

* Correspondence: yhjin@jlu.edu.cn

Abstract: (1) Background: Epithelial-mesenchymal transition (EMT) is an essential step for cancer metastasis; targeting EMT is an important path for cancer treatment and drug development. NF-κB, an important transcription factor, has been shown to be responsible for cancer metastasis by enhancing the EMT process. Our previous studies showed that (20S)Ginsenoside Rh2 (G-Rh2) inhibits NF-κB activity by targeting Anxa2, but it is still not known whether this targeted inhibition of NF-κB can inhibit the EMT process. (2) Methods: In vivo (20S)G-Rh2-Anxa2 interaction was assessed by cellular thermal shift assay. Protein interaction was determined by immuno-precipitation analysis. NF-κB activity was determined by dual luciferase reporter assay. Gene expression was determined by RT-PCR and immuno-blot. EMT was evaluated by wound healing and Transwell assay and EMT regulating gene expression. (3) Results: Anxa2 interacted with the NF-κB p50 subunit, promoted NF-κB activation, then accelerated mesenchymal-like gene expression and enhanced cell motility; all these cellular processes were inhibited by (20S)G-Rh2. In contrast, these (20S)G-Rh2 effect were completely eliminated by overexpression of Anxa2-K301A, an (20S)G-Rh2-binding-deficient mutant of Anxa2. (4) Conclusion: (20S)G-Rh2 inhibited NF-κB activation and related EMT by targeting Anxa2 in MDA-MB-231 cells.

Keywords: Anxa2; epithelial-mesenchymal transition; NF-κB; (20S)G-Rh2

---

1. Introduction

Playing key roles in embryonic development, epithelial-mesenchymal transition (EMT) facilitates body formation and tissue differentiation [1–4]. EMT is triggered by a genetic phenotype shift from epithelial-like to mesenchymal-like in response to pleiotropic signal; cells acquire the migratory and invasive properties by modifying adhesion molecules [3–5]. During this process, EMT specific transcription factors (EMT-TFs), in company with other regulatory factors like histone modifier and non-coding RNA, modify the gene expression through different states along EMT [6–11]. With the transcriptional activation of EMT-TFs, the expression of adherent junction components and tight junction components are negative-regulated, followed by the alteration in cadherin intermediate filament composition and cellular adhesion status [1,3,8,9]. EMT is also implicated in other physiological and pathological processes including would healing, tissue repair, fibrosis and cancer [1–6,11,12]. Tumor metastasis is the leading cause for cancer associated mortality; EMT is the main impetus and the essential access within this duration [11,12]. Despite the enhanced migration capability and invasiveness, EMT protects cancer cells from senescence, apoptosis and immuno-response by enhancing stem-cell properties and triggers the metastasis and dissemination to long distance through circulatory systems [13,14]. Focusing on the diverse regulated molecules, multiple drugs have shown their potential as EMT inhibitors by targeting the tumor microenvironment, corresponding extracellular receptors, intracellular kinases and EMT inducing transcription factors [15]. As its well-described clinical signature of EMT in cancer development, therapeutic strategies and targeted drugs towards EMT have been involved in cancer treatment [16–18].

NF-κB is an important transcription factor involved in multiple biologic processes as immune response, stress response, apoptosis, cell proliferation and cell metastasis [18]. The abnormal activation of NF-κB has been regarded as a hallmark of cancer, which promotes both tumorigenesis and tumor development [19–21]. NF-κB has been shown responsible for the expression of EMT-TFs including SNAIL, TWIST1, SLUG, SIP1 and ZEB1 and promotes EMT progression upon diverse signaling and stimuli, especially in cytokine and chemokine-induced cell growth and migration [22,23]. In the meanwhile, EMT promotes mesenchymal and stem-like phenotype of cancer cells and alters tumor microenvironment via autocrine and paracrine signaling, resulting in a constitutive activation of NF-κB in a ligand-receptor manner rather than genetic alteration [24]. Therefore, inhibition of NF-κB seems to be an important way to inhibit the progression of EMT and related tumors [25–27].

In our previous study, (20S)G-Rh2 has been identified as an NF-κB inhibitor targeting Anxa2, a NF-κB p50 subunit binding protein [28]. Here, we investigated the Anxa2 regulation on NF-κB and EMT. We further studied detailed molecular mechanisms for (20S)G-Rh2-induced NF-κB inhibition and the following EMT inhibition in invasive breast cancer cells. The results provide new evidences for the anticancer activity of ginsenosides.

2. Materials and Methods

2.1. Cell lines and Culture

Human breast cancer cell line MDA-MB-231 (ATCC, HTB-26), MCF-7 (ATCC, HTB-26) and HEK-293-T (ATCC, CRL-11268) were cultured in DMEM high-glucose medium (Gibco) supplemented with 10% fetal bovine serum (Biological Industries, BI), 100 units/mL penicillin and 100 μg/mL streptomycin, in a humidified 5% CO2 atmosphere at 37 °C.

2.2. Chemicals, Antibodies and Plasmids

(20S)G-Rh2, etoposide, phorbol 12-myristate 13-acetate (PMA) were purchased from Sigma-Aldrich.

Antibodies for Twist (25465-1-AP), SIP1 (21672-1-AP), Slug (12129-1-AP), Snail 1 (13099-1-AP), MMP-2 (10373-2-AP), MMP-9 (10375-2-AP), Anxa2 (11256-1-AP), β-actin (66009-1-lg) and myc-tag (16286-1-AP, 60003-2-lg) were purchased from Proteintech (Proteintech Group, Inc.). Antibodies for E-cadherin (sc-8426), N-cadherin (sc-59987), NF-κB p50 subunit (sc-7178), Anxa2 (sc-47696) and IgG (sc-2025) were purchased from Santa Cruz Biotechnology. HRP-conjugated goat anti-rabbit IgG (H + L) secondary antibody (31460) and HRP-conjugated goat anti-mouse IgG (H+L) secondary antibody (31430) were purchased from Invitrogen.

Plasmids for over-expression of wild-type Anxa2 (pcs4-Anxa2-WT-myc) and Anxa2 K031A mutant (pcs4-Anxa2-K301A-myc) were shown as described [28]. A truncation with 1–33 deletion of Anxa2 was amplified by polymerase chain reaction (PCR), followed by a recombination into pcs4-myc vector for over-expression of Anxa2-delta N-terminus (pcs4-Anxa2-dN-myc). For lentivirus package, pLVX-TetOne-Puro (Clontech, 631849), pMD2.G (Addgene plasmid # 12259) and psPAX2 (Addgene plasmid # 12260) were gifts from Professor Zhihua Zou (Collage of Life Sciences, Jilin University). C-terminus-Myc-tagged Anxa2 full-length protein as well as K301A mutant and delta-N truncation were amplified from pcs4 vector and recombined into pLVX-TetOne-Pure vector for myc-tagged Anxa2 over-expression with lentivirus. The primers used within were shown in Table 1. Dual luciferase reporter assay were performed with pNF-κB-TA-luc (Beyotime, Shanghai, China) and pRL-CMV (Promega, WI, USA).

Table 1. Primers for vector construction.

Primer Description	Sequence
pcs4-Anxa2-dN-F	5′- TTGGATCCATGGATGCTGAGCGGGATGCTTTG -3′
pcs4-Anxa2-dN-R	5′- CCGCTCGAGTCATCTCCACCACACAGGTACAG -3′
pLVX-Anxa2-dN-F	5′- GCGTATACATGGATGCTGAGCGGGATGCTTTG -3′
pLVX-Anxa2-F	5′- GCGTATACATGTCTACTGTTCACGAAATCCT -3′
pLVX-Anxa2-R	5′- GCGGATCCTAGCTATCTAGAGGCTCGAGAGG -3′

2.3. *Transfection*

Lipo3000 (Invitrogen) was used for lentivirus package, transient transfection in MCF-7 cells and dual luciferase reporter assay according to the reagent protocol.

For lentivirus transfection, pLVX, pMD2.G and psPAX2 were co-transfected into HEK-293-T cells and supernatant for 24 h and 48 h was collected then added into MDA-MB-231 cells.

2.4. *Immuno-Precipitation*

50 μL of Protein A/G Magnetic Beads (MCE, K0202) was washed with 400 μL of IP lysis buffer (Pierce) for 3 times. 5 μg of antibody for immune-precipitation was diluted with 500 μL of IP lysis buffer, then added to the prepared beads, followed by a rotation for 2 h at 4 °C. MDA-MB-231 cells and MCF-7 cells were collected and lysed with IP lysis buffer supplemented with Protease Inhibitor Cocktail (Roche, 04693124001) and 1-mM phenylmethanesulfonyl fluoride (PMSF). Cell lysis was centrifuged with 12,000× *g* for 20 min at 4 °C and the supernatant was collected. The antibody bonded beads were then collected and combined with cell lysis containing 500 μg of protein with a final volume of 400 μL, followed by another rotation for 2 h at 4 °C. The beads were then washed with IP lysis buffer for 3 times and collected for immuno-blot analysis.

2.5. *Cellular Thermal Shift Assay*

MDA-MB-231 cells and MCF-7 cells were cultured in 100-mm culture plates until the confluence reached 90%. The culture medium was then replaced with new medium supplemented with 15-μM (20S)G-Rh2 (~10 μg/mL) followed by an incubation for 1 h in a humidified 5% CO_2 atmosphere at 37 °C. After digested with trypsin (0.25%, *w/v* in PBS) and counted, cells were collected with centrifugation at 400× *g* for 5 min and re-suspended with PBS containing 1-mM PMSF to a final cell density of 2×10^7 cells/mL. Each 100 μL of cell suspension was added to a 200-μL tube and heated at indicated temperature for 3 min and incubated at 4 °C for another 2 min. After 2-time rapid freeze-thawing from −80 °C to 25 °C, cell suspension was centrifuged with 20,000× *g* for 20 min at 4 °C, the supernatant was collected for immune-blot analysis.

2.6. *Dual Luciferase Reporter Assay*

pNF-κB-TA-luc and pRL-CMV (10:1, w:w) were co-transfected into MDA-MB-231 cells and MCF-7 cells with Lipo3000 and cultured for 24 h before chemical treatment. The activity of luciferase was determined with Dual-Luciferase®Reporter Assay System (Promega, E1910) according to the manufacture's protocol. Luminescence generated by luciferase was collected via Infinite F200 Pro (TECAN).

2.7. *Real-Time Polymerase Chain Reaction*

Whole-cell RNA was isolated with TRIzol (Invitrogen). 2 μg of whole-cell RNA was proceeded with High Capacity cDNA Reverse Transcription Kit (Applied Biosystems, 4368814) for cDNA synthesis followed by real-time PCR analysis via PowerUp SYBR Green Master Mix (Applied Biosystems, A25742) and 7500 Real-Time PCR System (Applied Biosystems). The primers used were shown in Table 2. Gene expression was normalized to that of GAPDH and visualized in histogram format.

Table 2. Primers for RT-PCR.

Gene Name	Sequence
SNAIL-F	5′- TGCCCTCAAGATGCACATCCGA -3′
SNAIL-R	5′- GGGACAGGAGAAGGGCTTCTC -3′
SLUG-F	5′- ATCTGCGGCAAGGCGTTTTCCA -3′
SLUG-R	5′- GAGCCCTCAGATTTGACCTGTC -3′
SIP1-F	5′- GAGTTGATGCCTCGGCTATTGC -3′
SIP1-R	5′- CTGGACATTGAGCTGCTTCGATC -3′
TWIST1-F	5′- GCCAGGTACATCGACTTCCTCT -3′
TWIST1-R	5′- TCCATCCTCCAGACCGAGAAGG -3′
MMP2-F	5′- AGCGAGTGGATGCCGCCTTTAA -3′
MMP2-R	5′- CATTCCAGGCATCTGCGATGAG -3′
MMP9-F	5′- GCCACTACTGTGCCTTTGAGTC -3′
MMP9-R	5′- CCCTCAGAGAATCGCCAGTACT -3′
CDH1-F	5′- GCCTCCTGAAAAGAGAGTGGAAG -3′
CDH1-R	5′- TGGCAGTGTCTCTCCAAATCCG -3′
CDH2-F	5′- CCTCCAGAGTTTACTGCCATGAC -3′
CDH2-R	5′- GTAGGATCTCCGCCACTGATTC -3′
GAPDH-F	5′- GTCTCCTCTGACTTCAACAGCG -3′
GAPDH-R	5′- ACCACCCTGTTGCTGTAGCCAA -3′

2.8. *Cell Migration Analysis*

A culture-insert 2 well (Ibidi, 81176) was utilized for would healing assay to determine the migration capability of MDA-MB-231 cells. MDA-MB-231 cells were digested and re-suspended to a final density of 5×10^5 cells/mL. Cell suspension was added into chambers as well as outer area and cultured for 24 h for cell adhesion before the inserts were removed. Then the culture medium was replaced with new medium supplemented with indicated chemicals. Then the wound healing status was recorded by microscopy at 12 h, 24 h and 48 h after culture medium alteration.

2.9. *Cell Invasion Analysis*

A Matrigel Transwell invasion assay was performed to determine the invasiveness of MDA-MB-231 cells. A total of 10^4 cells in serum-free DMEM medium were seeded into upper chamber (8-μm pore, Corning, 3464) pre-coated with 1 mg/mL Matrigel Matrix (Corning, 354234) and DMEM medium containing 20% FBS was added into 24-well plate. After incubation for 24 h in a humidified 5% CO_2 atmosphere at 37 °C, cells were fixed with pre-cooled methanol for 30 min and stained with 0.1% crystal violet (w/v) in PBS for 10 min. Images were collected by microscopy to determine the cells passing through Matrigel basement.

2.10. *Statistical Analysis*

All data were obtained from independent triple-replicated experiments and presented as the mean ± standard deviation (SD). Significance was determined by a two-tail Student's t-test via GraphPad Prism v.6 (GraphPad Inc., USA) and data with p-value < 0.05 were considered of statistical significance.

3. Results

3.1. *Anxa2 Bound to NF-κB p50 Subunit in MDA-MB-231 Cells and MCF-7 Cells*

It has been well-established that Anxa2 binds to p50 to promote NF-κB activation and cell survival in hepatocellular carcinoma and pancreatic cancer cell lines [28,29]. In order to tell the activity of Anxa2 in NF-κB associated EMT in breast cancer, we first assessed the interaction of Anxa2 with NF-κB p50 subunit in breast cancer MDA-MB-231 and MCF-7 cells by coIP analysis (Figure 1A). As the N-terminus of Anxa2 is responsible for its interaction with p50 [29], we transiently transfected C-terminus myc-tagged full-length Anxa2 and an N-terminus deleted truncated version of Anxa2

(Anxa2-dN). Precipitated by anti-myc antibody, Anxa2-dN failed to interact with p50, demonstrating Anxa2 bound to p50 via its N-terminus in breast cancer cells as does in Hepatoma cells (Figure 1B).

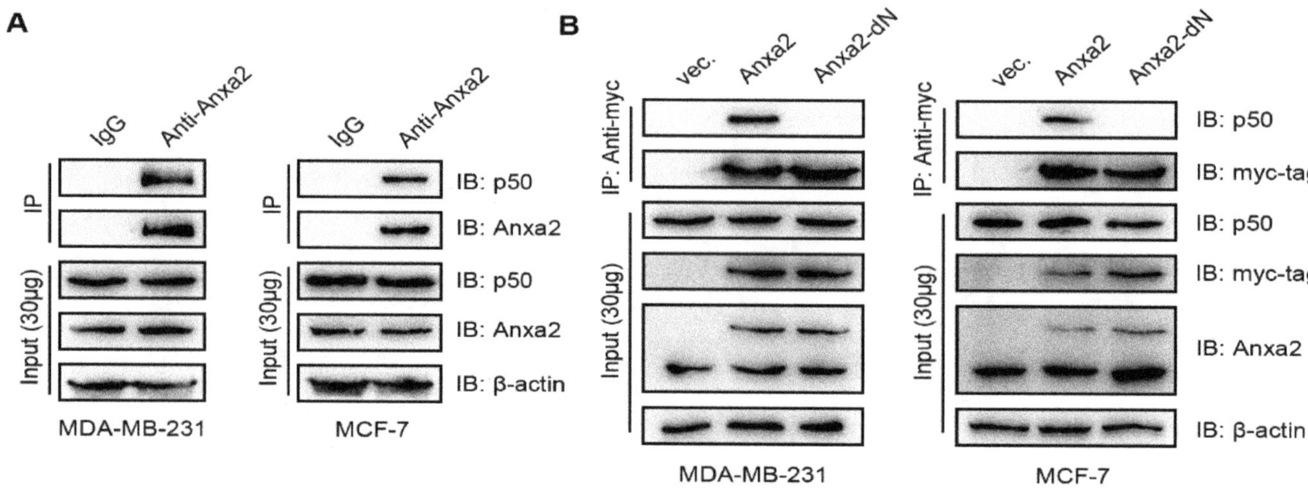

Figure 1. Anxa2 interacts with p50 in breast cancer cell lines. (**A**) An immuno-precipitation was performed with anti-Anxa2 antibody in protein extract from MDA-MB-231 cells and MCF-7 cells and immuno-precipitation with IgG was used as a negative control. (**B**) An immuno-precipitation was performed with anti-myc-tag antibody in protein extract from MDA-MB-231 cells and MCF-7 cells transfected with empty vector (vec.), Anxa2 and Anxa2-dN. A total of 30 µg of whole-cell protein extract was loaded as input. All experiments were repeated for three times.

3.2. Anxa2 Promoted NF-κB Activation and Associated EMT in Invasive Breast Cancer Cells

We co-transfected cells with Anxa2 and dual luciferase reporter system for NF-κB activity analysis, the transfection of full-length Anxa2 enhanced NF-κB activity whereas that of Anxa2-dN failed (Figure 2A). We then analyzed the NF-κB associated EMT-TFs including SNAIL, SLUG, SIP1 and TWIST1, EMT marker (CDH1, gene encoding E-cadherin; CDH2, gene encoding N-cadherin) and pro-invasion matrix metalloproteinases (MMP2, MMP9). The over-expression of full-length Anxa2 increased the expression of three EMT-TFs such as SLUG, SIP1 and TWIST1 and two MMPs (MMP2, MMP9) in invasive breast cancer MDA-MB-231 cells (Figure 2B, left panel). Up-regulated CDH2 and down-regulated CDH1, presenting the accelerated EMT capability were observed in full-length Anxa2 over-expressing MDA-MB-231 cells (Figure 2B, left panel). Correlated with failed activation of NF-κB, the over-expression of Anxa2-dN caused little expression shift in CDH2 and CDH1 (Figure 2B, left panel). In contrast with MDA-MB-231 cells, less-invasive MCF-7 cells presented no obvious alteration in these migration related gene expression under either full-length Anxa2 or Anxa2-dN truncation over-expression (Figure 2B, right panel). Protein levels of these genes in MDA-MB-231 cells were further determined by immune-blot. The up-regulation of protein levels of E-Cadherin and that of down-regulation of N-cadherin, Twist, slug, SIP1, MMP-2 and MMP-9 were observed in accord with the alteration of mRNA levels (Figure 2C).

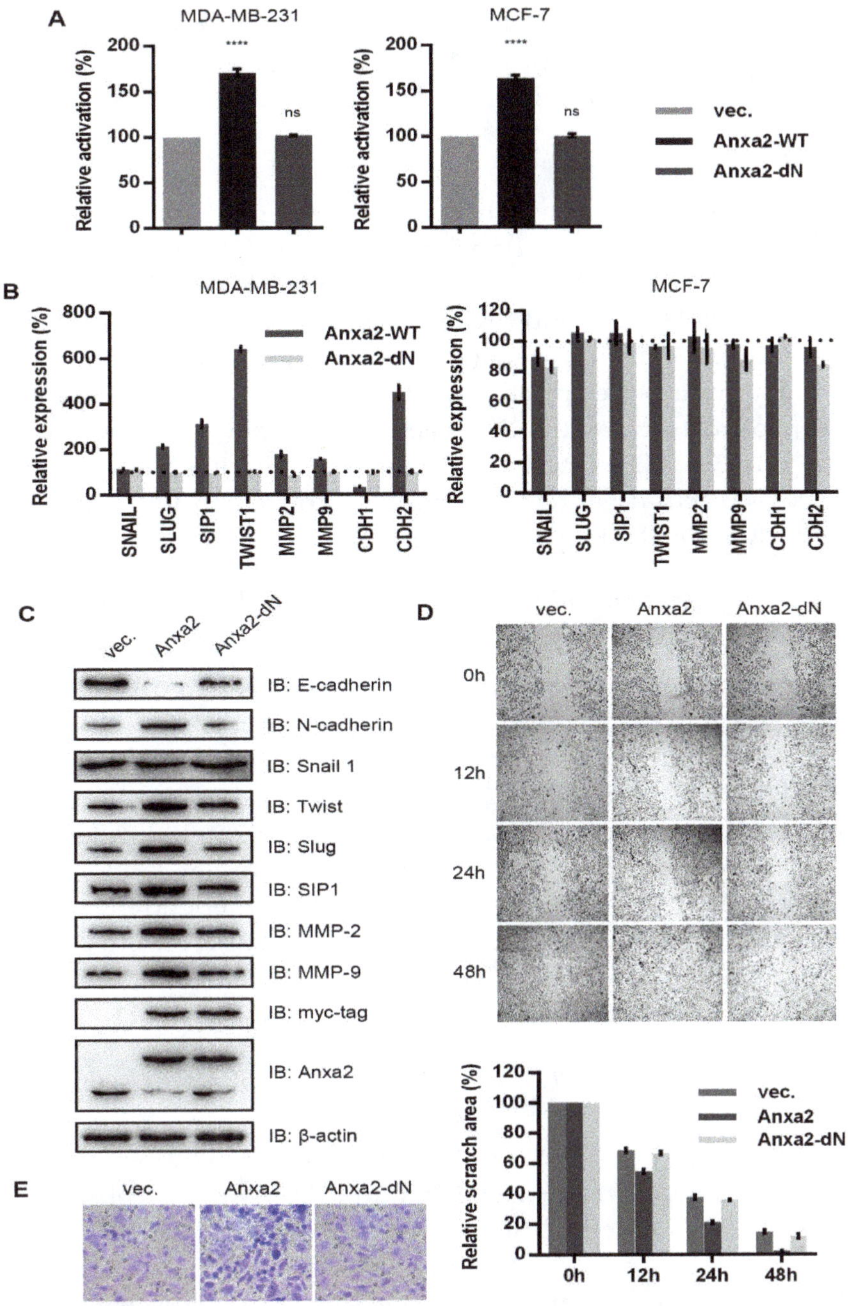

Figure 2. Anxa2 over-expression enhanced NF-κB activation and related cell migration and invasion. (**A**) NF-κB activity was determined via a dual luciferase reporter assay in MDA-MB-231 cells and MCF-7 cells transfected with empty vector, Anxa2 and Anxa2-dN. (**B**) mRNA levels of indicated genes were determined by RT-PCR with RNA extract from MDA-MB-231 cells and MCF-7 cells transfected with Anxa2 and Anxa2-dN t, the dot line at 100% presents mRNA levels in cells transfected with empty vector. (**C**) The protein levels of indicated genes were determined by immuno-blot in MDA-MB-231 cells transfected with empty vector (vec.), Anxa2 and Anxa2-dN. A total of 30 μg of whole-cell protein extract was loaded for immuno-blotting assay. (**D**) A wound healing assay was performed with MDA-MB-231 cells transfected with empty vector (vec.), Anxa2 and Anxa2-dN. Relative scratch area was shown below the microscopy images. (**E**) A Transwell invasion assay was performed with MDA-MB-231 cells transfected with empty vector (vec.), Anxa2 and Anxa2-dN. All experiments were repeated for three times and all data are shown as mean ± SD, with **** presenting $p < 0.0001$, *** presenting $p < 0.001$, ** $p < 0.01$, * presenting $p < 0.05$ and ns presenting $p > 0.05$.

Wound healing assay and Transwell invasion assays were then performed with Anxa2-over-expressing MDA-MB-231 cells. Full-length Anxa2 over-expression enhanced the wound healing efficiency and invasiveness through Matrigel basement whereas Anxa2-dN truncation over-expression showed no facilitation (Figure 2D,E).

3.3. *(20S)G-Rh2 Inhibited NF-κB Activation Targeting Anxa2*

(20S)G-Rh2 was proven as a natural small-molecule ligand for Anxa2 and inhibited NF-κB activation by interfering Anxa2-p50 interaction in HepG2 cells [28]. For the purpose of investigating the inhibitory effect toward NF-κB of (20S)G-Rh2 in breast cancer cells, a cellular thermal shift assay was performed with MDA-MB-231 cells and MCF-7 cells; (20S)G-Rh2 increased the thermal stability of Anxa2 in both cell lines (Figure 3A), indicating (20S)G-Rh2 bound to Anxa2 in MDA-MB-231 cells and MCF-7 cells. A following immune-precipitation showed (20S)G-Rh2 inhibited Anxa2-p50 interaction at resting state or under co-treatment with NF-κB activator etoposide or PMA in either MDA-MB-231 cells or MCF-7 cells (Figure 3B). NF-κB activity was determined via a dual luciferase reporter assay; (20S)G-Rh2 inhibited NF-κB activity at resting state and co-treated with etoposide or PMA in both MDA-MB-231 cells and MCF-7 cells (Figure 3C).

3.4. *(20S)G-Rh2 Inhibited the Migration and Invasion of MDA-MB-231 in a Dose-Dependent Manner*

MDA-MB-231 cells and MCF-7 cells were treated with increasing concentration of (20S)G-Rh2 and the result showed that NF-κB activity was down-regulated in a dose-dependent manner upon (20S)G-Rh2 treatment (Figure 4A). A dose-dependent alteration was also appeared in the gene expression analysis of TWIST1, MMP2, MMP9, CDH1 and CDH2, the expression of SLUG and SIP1 presented stable by (20S)G-Rh2 in MDA-MB-231 cells; only Snail1 presented an unexpected increase (Figure 4B, left panel). No regulated shift was seen in MCF-7 cells under (20S)G-Rh2 treatment (Figure 4B right panel). The protein levels of EMT-TFs, MMPs and N-cadherin were down-regulated and that of E-cadherin was up-regulated in a (20S)G-Rh2-dose-dependent manner in MDA-MB-231 cells (Figure 4C). Wound healing assay and Transwell invasion assay were performed MDA-MB-231 cells upon various concentration of (20S)G-Rh2 treatment. (20S)G-Rh2 inhibited the migration and invasion of MDA-MB-231 cells in a dose-dependent manner (Figure 4D,E).

3.5. *(20S)G-Rh2 Binding-Deficient Mutant of Anxa2 Protected MDA-MB-231 Cells from (20S)G-Rh2 Induced NF-κB Inhibition*

To provide further evidence for the (20S)G-Rh2 effect on Anxa2 function, we over-expressed the (20S)G-Rh2-binding-deficient mutant, Anxa2-K301A, in MDA-MB-231 cells. A cellular thermal shift assay showed that Anxa2-K301A failed to interact with (20S)G-Rh2 (Figure 5A). (20S)G-Rh2 failed to interfere the interaction between Anxa2-K301A and p50; it even enhanced it (Figure 5B). (20S)G-Rh2 induced NF-κB inhibition and downstream gene expression shift were also unchanged or reversed in Anxa2-K301A over-expressing cells (Figure 5C,D). Cell migration capability and invasiveness were unaltered with Anxa2-K301A overexpression under (20S)G-Rh2 treatment (Figure 5E,F).

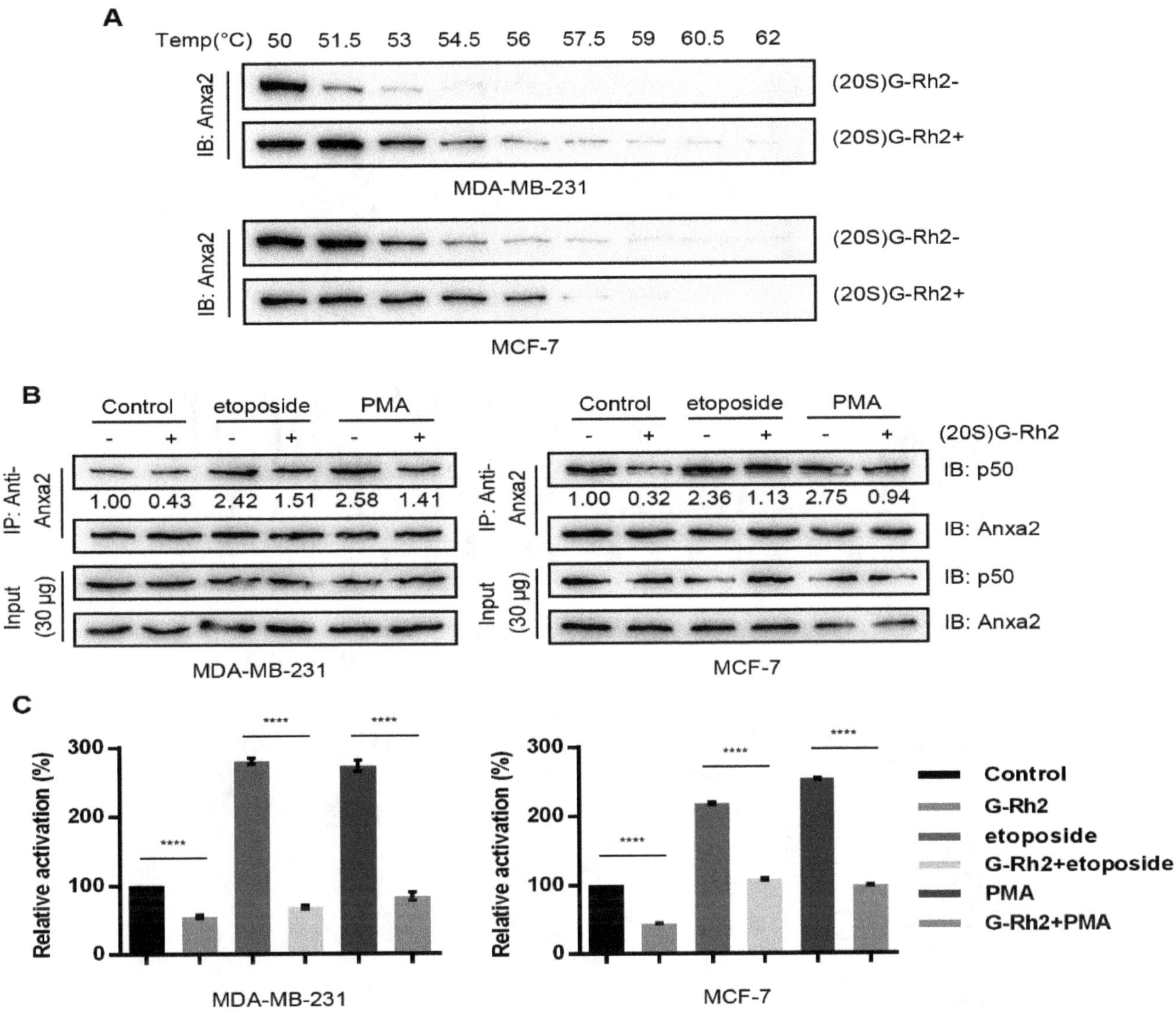

Figure 3. (20S)G-Rh2 inhibits NF-κB activation by binding to Anxa2. (**A**) A cellular thermal shift was performed in MDA-MB-231 cells and MCF-7 cells under 10-μM-(20S)G-Rh2 treatment or not. (**B**) An immuno-precipitation was performed with anti-Anxa2 antibody in protein extract from MDA-MB-231 cells and MCF-7 cells under treatment or 6-μM (20S)G-Rh2, 25 μg/mL etoposide,100 ng/mL PMA and combined chemicals for 12 h. (**C**) NF-κB activity was determined via a dual luciferase reporter assay in MDA-MB-231 cells and MCF-7 cells under treatment or 6-μM (20S)G-Rh2, 25 μg/mL etoposide,100 ng/mL PMA and combined chemicals for 12 h. A total of 30 μg of whole-cell protein extract was loaded as input. All experiments were repeated for three times and all data are shown as mean ± SD, with **** presenting $p < 0.0001$, *** presenting $p < 0.001$, ** $p < 0.01$, * presenting $p < 0.05$ and ns presenting $p > 0.05$.

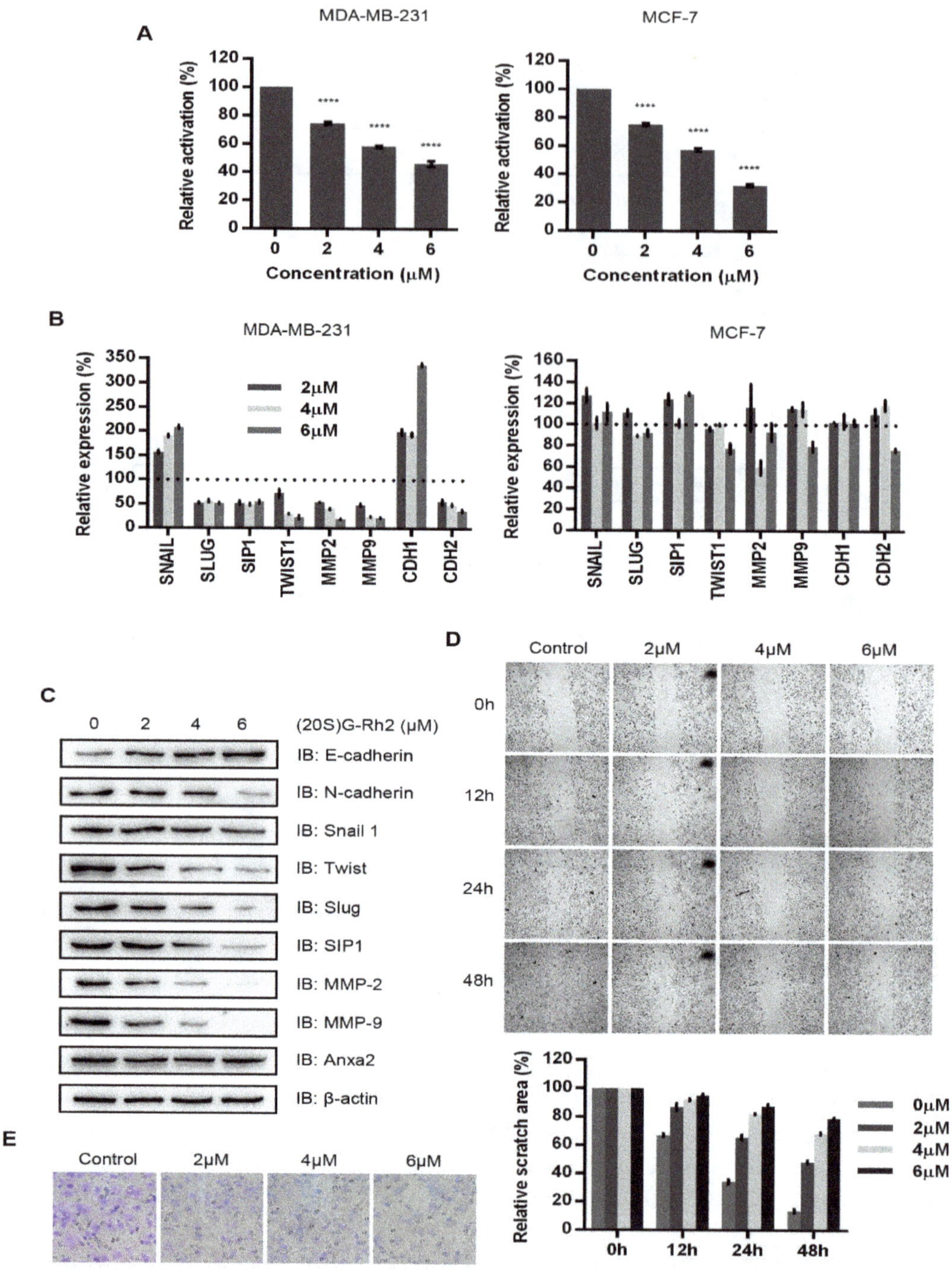

Figure 4. (20S)G-Rh2 inhibits NF-κB related cell migration and invasion. (**A**) NF-κB activity was determined via a dual luciferase reporter assay in MDA-MB-231 cells and MCF-7 cells under (20S)G-Rh2 treatment for 12 h. (**B**) mRNA levels of indicated genes were determined by RT-PCR with RNA extract from MDA-MB-231 cells under (20S)G-Rh2 treatment for 12 h. (**C**) The protein levels of indicated genes were determined by immuno-blot in MDA-MB-231 cells under (20S)G-Rh2 treatment for 12 h. A total of 30 μg of whole-cell protein extract was loaded for immuno-blotting assay. (**D**) A wound healing assay was performed with MDA-MB-231 cells under (20S)G-Rh2 treatment for 12 h. Relative scratch area was shown below the microscopy images. (**E**) A Transwell invasion assay was performed with MDA-MB-231 cells t under (20S)G-Rh2 treatment for 12 h. All experiments were repeated for three times and all data are shown as mean ± SD, with **** presenting $p < 0.0001$, *** presenting $p < 0.001$, ** $p < 0.01$, * presenting $p < 0.05$ and ns presenting $p > 0.05$.

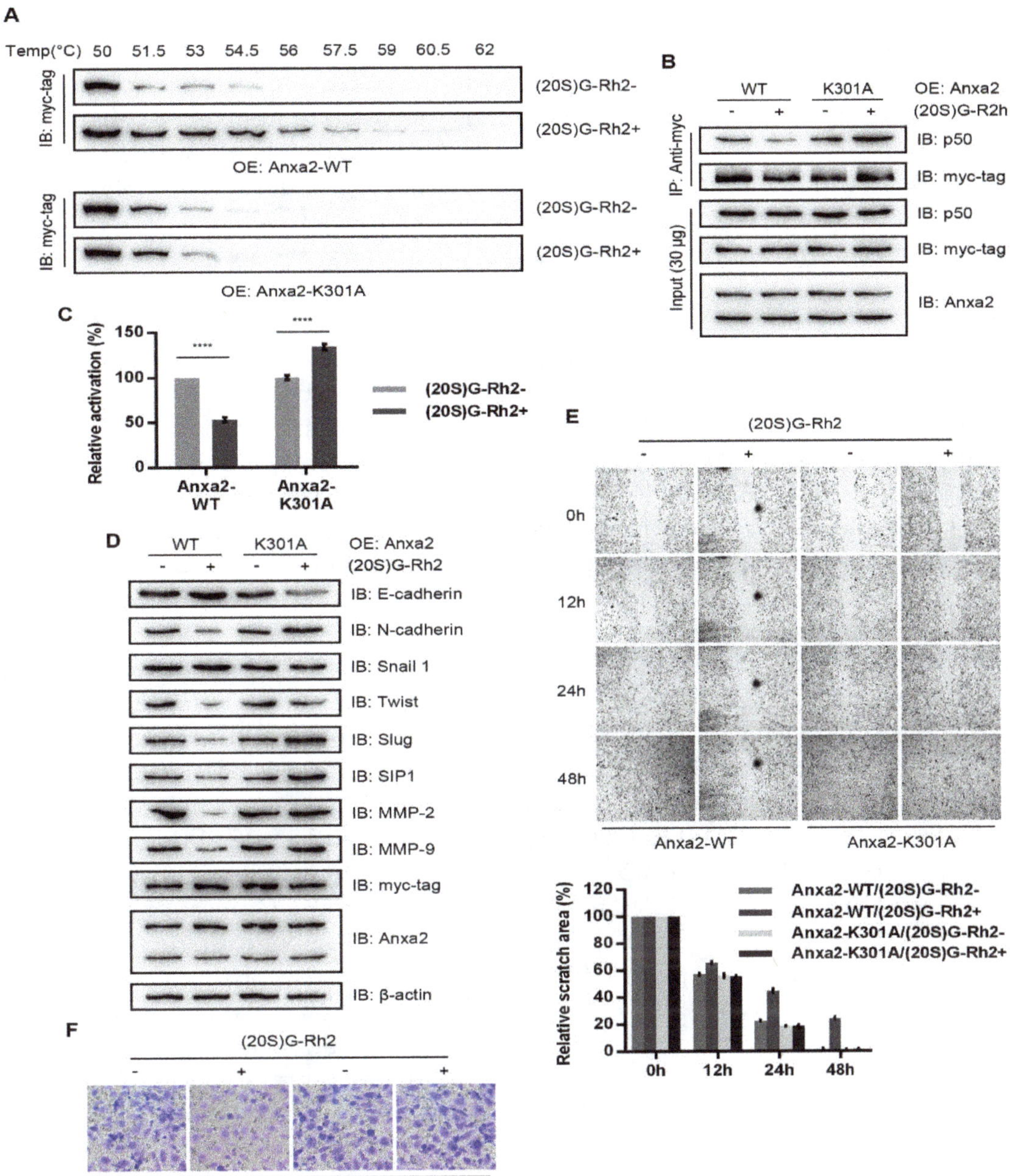

Figure 5. Anxa2-K301A protects MDA-MB-231 cells from (20S)G-Rh2-induced NF-κB inhibition. (**A**) A cellular thermal shift was performed in MDA-MB-231 cells transfected with Anxa2-WT and Anxa2-K301A under 10-μM-(20S)G-Rh2 treatment or not. (**B**) An immuno-precipitation was performed with anti-myc-tag antibody in protein extract from MDA-MB-231 cells transfected with Anxa2-WT and Anxa2-K301A under 6-μM-(20S)G-Rh2 treatment or not. A total of 30 μg of whole-cell protein extract was loaded as input. (**C**) NF-κB activity was determined via a dual luciferase reporter assay in MDA-MB-231 cells transfected with Anxa2-WT and Anxa2-K301A under 6-μM-(20S)G-Rh2 treatment for 12 h or not. (**D**) The protein levels of indicated genes were determined by immuno-blot in MDA-MB-231 cells transfected with Anxa2-WT and Anxa2-K301A under 6-μM-(20S)G-Rh2 treatment or not. A total of 30 μg of whole-cell protein extract was loaded for immuno-blotting assay. (**E**) A wound healing assay was performed with MDA-MB-231 cells transfected with Anxa2-WT and Anxa2-K301A under 6-μM-(20S)G-Rh2 treatment or not. Relative scratch area was shown below the microscopy images. (**F**) A Transwell invasion assay was performed with MDA-MB-231 cells transfected with Anxa2-WT and Anxa2-K301A under 6-μM-(20S)G-Rh2 treatment or not. All experiments were repeated for three times and all data are shown as mean ± SD, with **** presenting $p < 0.0001$, *** presenting $p < 0.001$, ** $p < 0.01$, * presenting $p < 0.05$ and ns presenting $p > 0.05$.

4. Discussion

Cancer metastasis present responsibility for over 90% cancer mortality. EMT—one of the main generators for cancer metastasis—has become an assignable factor in both cancer research and treatment [1,11]. Derived by EMT promoting genes, epithelial-like cancer cells re-program gene expression phenotype, lose epithelial feature, dissociate from primary tumor tissue and spread with circulatory system [2,4–6]. A variety of gene transcriptional regulation was involved in this program, among which EMT-TFs and up-stream transcriptional factors like NF-κB and STAT3, directly alter the expression of key effector molecules [22,23,30,31].

Among them, multi-functional protein Anxa2 behaves as an implicated promoter for EMT and related cellular event. Anxa2 was first associated with cell migration for its heterotetramer with S100A10 [32,33], which promotes extracellular matrix degradation by activating plasmin and following researches indicate Anxa2 is responsible for membrane dynamics in invasive cells by binding to other S100 protein like S100A4 [34], S100A6 [35] and S100A11 [36,37]. Implicated in signaling transduction, phosphorylated Anxa2 at Tyr23 binds to STAT3 and promotes its phosphorylation at Tyr705 as well as following dimerization and nuclear translocation, resulting in up-regulated MMPs and enhanced EMT [38,39]. In addition, couple of researches have demonstrated Anxa2 content is correlated to cell migration capability [40–43].

It has been well-established that NF-κB bound to the promoter region of ZEB-1/2, SNAIL, SLUG, SIP1 and TWIST1 and the transcription of ZEB-1/2, SLUG, SIP1 and TWIST1 accords with NF-κB activation [22,23]. Anxa2, acting as an NF-κB co-activator in hepatocellular carcinoma and pancreatic cancer [28,29], was first identified bound to p50 subunit of NF-κB in breast cancer cells (Figure 1A,B). Over-expression of Anxa2 in invasive breast cancer MDA-MB-231 cells accelerated NF-κB activation (Figure 2A) and the expression EMT-TFs including SLUG, SIP1 and TWIST1 (Figure 2B,C), followed by the down-regulation of E-cadherin and up-regulation of N-cadherin (Figure 2B,C). Though containing NF-κB binding site, SNAIL somehow failed to follow NF-κB regulation, presenting differential variation trend as other EMT-TFs (Figures 2B and 4B) [23]. MMPs responsible for the degradation of extracellular matrix and promote cell motility appeared to be up-regulated in according with EMT phenotype in Anxa2-over-expressing cells (Figure 2B,C). These gene expression alteration resulted in the enhanced migratory properties and invasiveness (Figure 2D,E). In contrast, no altered expression of EMT-TFs was observed in company with the activation of NF-κB in less aggressive MCF-7 cells (Figure 2A,B), indicating EMT-TFs expression requires additional regulatory mechanism, which is necessary to perform further investigation in the future study.

(20S)G-Rh2 is one of the most famous ginsenosides from ginseng extract for its remarkable anti-cancer activity. (20S)G-Rh2 has been proven efficiently inducing growth arrest and apoptosis [44–47]. Moreover, in-vivo study indicates that (20S)G-Rh2 inhibits tumor growth by inhibiting tumor associated angiogenesis and enhancing anti-tumor immunological response [48–50]. In addition, couples of studies have indicated (20S)G-Rh2 presents inhibitory effect towards EMT. Based on our previous research on (20S)G-Rh2-Anxa2 interaction, we supposed that (20S)G-Rh2 inhibited EMT by targeting Anxa2. A cellular thermal shift assay verified the intracellular interaction between Anxa2 and (20S)G-Rh2 (Figure 3A): an immuno-precipitation as well as NF-κB luciferase reporter assay confirmed the inhibitory impact on NF-κB by (20S)G-Rh2 (Figure 3B,C and Figure 4A) in breast cancer cell line MDA-MB-231 and MCF-7. As a result, down-stream pro-EMT genes were down-regulated with epithelial phenotype marker E-cadherin accelerated in MDA-MB-231 cells under (20S)G-Rh2 treatment (Figure 4B,C); cell migration and invasion were also apparently inhibited (Figure 4D,E).

In our previous work, we have identified the Lys301 residue responsible for (20S)G-Rh2 binding; Anxa2-K301A has been shown to be a (20S)G-Rh2 binding deficiency mutant [28]. In addition,

Anxa2-K301A mutants maintained p50 binding capacity and promoted NF-κB under (20S)G-Rh2 treatment [28]. Anxa2a-WT and K301A mutants were transfected with MDA-MB-231 cells to examine the targeted inhibition of NF-κB and related EMTs. (20S)G-Rh2 failed to interact with Anxa2-K301A (Figure 5A); p50-Anxa2-K301A interaction and NF-κB activation were surprisingly enhanced under (20S)G-Rh2 treatment (Figure 5B,C). Functioning as a multi-targeted chemical, (20S)G-Rh2 induced various cellular stress and NF-κB stayed inactive under the targeted inhibition [28,44–50]. This status was reversed when Anxa2 mutant maintained the activation property regardless of (20S)G-Rh2 and stress related NF-κB activation was generated. Under the constitutive NF-κB activation by Anxa2-K301A over-expression presented higher mesenchymal-like phenotype (Figure 5D) and equivalent migratory and invasive properties (Figure 5E,F).

5. Conclusions

Our data suggest that Anxa2 is an important co-activator of NF-κB in breast cancer cells. Over-expression of Anxa2 promoted NF-κB activation and pro-EMT gene expression. With the up-regulation of EMT-TFs and MMPs, MDA-MB-231 cells presented a mesenchymal-like phenotype and achieved enhanced migratory properties as well as invasiveness. (20S)G-Rh2 induced NF-κB inhibition by targeting Anxa2 and led to the suppression in pro-EMT gene expression, resulting in the EMT arrest at the epithelial state. Over expression of Anxa2-K301A maintained NF-κB activation and the mesenchymal-like phenotype under (20S)G-Rh2 treatment in MDA-MB-231 cells. Taken together, (20S)G-Rh2 targets Anxa2 in invasive breast cancer cells, inhibits NF-κB activity and EMT and may be a potent candidate for breast cancer treatment and related research.

Author Contributions: Conceptualization, Y.-S.W. and Y.-H.J.; methodology, Y.-S.W., H.L. and S.Z.; software, Y.-S.W.; formal analysis, Y.-S.W. and H.L.; investigation, Y.-S.W. and Y.L.; resources, Y.S.W. and Y.-H.J.; data curation, Y.-S.W.; writing—original draft preparation, Y.-S.W.; writing—review and editing, Y.-H.J.; visualization, Y.-S.W.; supervision, Y.-H.J.; project administration, Y.-H.J.; funding acquisition, Y.-H.J. All authors have read and agreed to the published version of the manuscript.

References

1. Thiery, J.P.; Acloque, H.; Huang, R.Y.J.; Nieto, M.A. Epithelial-mesenchymal transitions in development and disease. *Cell* **2009**, *139*, 871–890. [CrossRef] [PubMed]
2. Nieto, M.A.; Huang, R.Y.J.; Jackson, R.A.; Thiery, J.P. EMT: 2016. *Cell* **2016**, *166*, 21–45. [CrossRef] [PubMed]
3. Lamouille, S.; Xu, J.; Derynck, R. Molecular mechanisms of epithelial-mesenchymal transition. *Nat. Rev. Mol. Cell Biol.* **2014**, *15*, 178–196. [CrossRef]
4. Nieto, M.A. Epithelial plasticity: A common theme in embryonic and cancer cells. *Science* **2013**, *342*, 1234850. [CrossRef] [PubMed]
5. Pastushenko, I.; Brisebarre, A.; Sifrim, A.; Fioramonti, M.; Revenco, T.; Boumahdi, S.; Van Keymeulen, A.; Brown, D.; Moers, V.; Lemaire, S.; et al. Identification of the tumour transition states occurring during EMT. *Nature* **2018**, *556*, 463–468. [CrossRef] [PubMed]
6. Beck, B.; Lapouge, G.; Rorive, S.; Drogat, B.; Desaedelaere, K.; Delafaille, S.; Dubois, C.; Salmon, I.; Willekens, K.; Marine, J.C.; et al. Different levels of Twist1 regulate skin tumor initiation, stemness, and progression. *Cell Stem Cell* **2015**, *16*, 67–79. [CrossRef]
7. Derynck, R.; Weinberg, R.A. EMT and cancer: More than meets the eye. *Dev. Cell* **2019**, *49*, 313–316. [CrossRef]
8. Aiello, N.M.; Maddipati, R.; Norgard, R.J.; Balli, D.; Li, J.; Yuan, S.; Yamazoe, T.; Black, T.; Sahmoud, A.; Furth, E.E.; et al. EMT subtype influences epithelial plasticity and mode of cell migration. *Dev. Cell* **2018**, *45*, 681–695. [CrossRef]
9. Meyer-Schaller, N.; Cardner, M.; Diepenbruck, M.; Saxena, M.; Tiede, S.; Lüönd, F.; Ivanek, R.; Beerenwinkel, N.; Christofori, G. A hierarchical regulatory landscape during the multiple stages of EMT. *Dev. Cell* **2019**, *48*, 539–553. [CrossRef]
10. Yousefi, H.; Maheronnaghsh, M.; Molaer, F.; Mashouri, L.; Reza Aref, A.; Momeny, M.; Alashari, S.K. Long noncoding RNAs and exosomal lncRNAs: Classification, and mechanisms in breast cancer metastasis and drug resistance. *Oncogene* **2020**, *39*, 953–974. [CrossRef]

11. Brabletz, T. EMT and MET in metastasis: Where are the cancer stem cells? *Cancer Cell* **2012**, *22*, 699–701. [CrossRef] [PubMed]
12. Pastushenko, I.; Blanpain, C. EMT transition states during tumor progression and metastasis. *Trends Cell Biol.* **2019**, *29*, 212–226. [CrossRef] [PubMed]
13. Puram, S.V.; Tirosh, I.; Parikh, A.S.; Patel, A.P.; Yizhak, K.; Gillespie, S.; Rodman, C.; Luo, C.L.; Mroz, E.A.; Emerick, K.S.; et al. Single-cell transcriptomic analysis of primary and metastatic tumor ecosystems in head and neck cancer. *Cell* **2017**, *171*, 1611–1624. [CrossRef] [PubMed]
14. Schliekelman, M.J.; Taguchi, A.; Zhu, J.; Dai, X.; Rodriguez, J.; Celiktas, M.; Zhang, Q.; Chin, A.; Wong, C.H.; Wang, H.; et al. Molecular portraits of epithelial, mesenchymal, and hybrid states in lung adenocarcinoma and their relevance to survival. *Cancer Res.* **2015**, *75*, 1789–1800. [CrossRef]
15. Marcucci, F.; Stassi, G.; De Maria, R. Epithelial-mesenchymal transition: A new target in anticancer drug discovery. *Nat. Rev. Drug Discov.* **2016**, *15*, 311–325. [CrossRef]
16. Lo, H.C.; Zhang, X.H. EMT in metastasis: Finding the right balance. *Dev. Cell* **2018**, *45*, 663–665. [CrossRef]
17. Acharyya, S.; Oskarsson, T.; Yanharanta, S.; Madlladi, S.; Kim, J.; Morris, P.G.; Manova-Todorova, K.; Leversha, M.; Hogg, N.; Seshan, V.E.L.; et al. A CXCL1 paracrine network links cancer chemoresistance and metastasis. *Cell* **2012**, *150*, 165–178. [CrossRef]
18. Yang, Y.R.; Kim, D.H.; Seo, Y.K.; Park, D.; Jang, H.J.; Choi, S.Y.; Lee, Y.H.; Lee, G.H.; Nakajima, K.; Taniguchi, N.; et al. Elevated O-GlcNAcylation promotes colonic inflammation and tumorigenesis by modulating NF-κB signaling. *Oncotarget* **2015**, *6*, 12529–12542. [CrossRef]
19. Hanahan, D.; Weinberg, R.A. The hallmarks of cancer. *Cell* **2000**, *100*, 57–70. [CrossRef]
20. AlQathama, A.; Prieto, J.M. Natural products with therapeutic potential in melanoma metastasis. *Nat. Prod. Rep.* **2015**, *32*, 1170–1182. [CrossRef]
21. Karin, M. Nuclear factor-kappaB in cancer development and progression. *Nature* **2006**, *441*, 431–436. [CrossRef] [PubMed]
22. Chua, H.L.; Bhar-Nakshatri, P.; Clare, S.E.; Morumiya, A.; Badve, S.; Nakshatri, H. NF-κB represses E-cadherin expression and enhances epithelial to mesenchymal transition of mammary epithelial cells: Potential involvement of ZEB-1 and ZEB-2. *Oncogene* **2007**, *26*, 711–724. [CrossRef] [PubMed]
23. Pires, B.R.; Mencalha, A.L.; Ferreira, G.M.; de Souza, W.F.; Morgado-Díaz, J.A.; Maia, A.M.; Corrêa, S.; Abdelhay, E.S. NF-KappaB is involved in the regulation of EMT genes in breast cancer cells. *PLoS ONE* **2017**, *12*, e0169622. [CrossRef] [PubMed]
24. Liu, J.; Wu, Z.; Han, D.; Wei, C.; Liang, Y.; Jiang, T.; Chen, L.; Sha, M.; Cao, Y.; Huang, F.; et al. Mesencephalic astrocyte-derived neurotrophic factor inhibits liver cancer through small ubiquitin-related modifier (SUMO)ylation-related suppression of NF-κB/Snail signaling pathway and epithelial-mesenchymal transition. *Hepatology* **2020**. [CrossRef]
25. Shin, S.R.; Sanchez-Velar, N.; Sherr, D.H.; Sonenshein, G.E. 7,12-dimethylbenz(a)anthracene treatment of a c-rel mouse mammary tumor cell line induces epithelial to mesenchymal transition via activation of nuclear factor-κB. *Cancer Res.* **2006**, *66*, 2570–2575. [CrossRef]
26. Liu, P.; Yang, P.; Zhang, Z.; Liu, M.; Hu, S. Ezrin/NF-κB pathway regulates EGF-induced epithelial-mesenchymal transition (EMT), metastasis, and progression of osteosarcoma. *Med. Sci. Monit.* **2018**, *24*, 2098–2108. [CrossRef]
27. Fang, H.; Wu, Y.; Huang, X.; Wang, W.; Ang, B.; Cao, X.; Wan, T. Toll-like receptor 4 (TLR4) is essential for Hsp70-like protein 1 (HSP70L1) to activate dendritic cells and induce Th1 response. *J. Biol. Chem.* **2011**, *286*, 30393–30400. [CrossRef]
28. Wang, Y.S.; Lin, Y.; Li, Y.; Song, Z.; Jin, Y.H. The identification of molecular target of (20S)Ginsenoside Rh2 for its anti-cancer activity. *Sci. Rep.* **2017**, *7*, 12408. [CrossRef]
29. Jung, H.; Kim, J.S.; Kim, W.K.; Oh, K.J.; Lee, H.J.; Han, B.S.; Kim, D.S.; Seo, Y.S.; Lee, S.C.; Park, S.G.; et al. Intracellular annexin A2 regulates NF-κB signaling by binding to the p50 subunit implications for gemcitabbine resistance in pancreatic. *Cell Death Dis.* **2015**, *6*, e1606. [CrossRef]
30. Chung, S.S.; Giehl, N.; Wu, Y.; Vadgama, J.V. STAT3 activation in HER2-overexpressing breast cancer promotes epithelial–mesenchymal transition and cancer stem cell traits. *Int. J. Oncol.* **2014**, *44*, 403–411. [CrossRef]

31. Xiong, H.; Hong, J.; Du, W.; Lin, Y.W.; Ren, L.L.; Wang, Y.C.; Su, W.Y.; Wang, J.L.; Cui, Y.; Wang, Z.H.; et al. Roles of STAT3 and ZEB1 proteins in E-cadherin down-regulation and human colorectal cancer epithelial–mesenchymal transition. *J. Biol. Chem.* **2012**, *287*, 5819–5832. [CrossRef]
32. Bydoun, M.; Waisman, D.M. On the contribution of S100A10 and annexin A2 to plasminogen activation and ongogenesis: An enduring ambiguity. *Future Oncol.* **2014**, *10*, 2469–2479. [CrossRef]
33. Myrvang, H.K.; Guo, X.; Li, C.; Dekker, L.V. Protein interactions between surface annexin A2 and S100A10 mediate adhesion of breast cancer cells to microvascular endothelial cells. *Febs Lett.* **2013**, *587*, 3210–3215. [CrossRef]
34. Semo, A.; Moreno, M.J.; Onichtchenko, A.; Abulrob, A.; Ball, M.; Ekiel, I.; Pietrzynski, G.; Stanimirovic, D.; Alakhov, V. Metastasis-associated protein S100A4 induces angiogenesis through interaction with Annexin II and accelerated plasmin formation. *J. Biol. Chem.* **2005**, *280*, 20833–20841. [CrossRef] [PubMed]
35. Nedjadi, T.; Kitteringham, N.; Campbell, F.; Jenkins, R.E.; Park, B.K.; Navarro, P.; Ashcroft, F.; Tepikin, A.; Neoptolemos, J.P.; Costello, E. S100A6 binds to annexin 2 in pancreatic cancer cells and promotes pancreatic cancer cell motility. *Br. J. Cancer* **2009**, *101*, 1145–1154. [CrossRef] [PubMed]
36. Jaiswal, J.K.; Nylandsted, J. S100 and annexin proteins identify cell membrane damage as the Achilles heel of metastatic cancer cells. *Cell Cycle* **2015**, *14*, 502–509. [CrossRef] [PubMed]
37. Jaiswal, J.K.; Lauritzen, S.P.; Scheffer, L.; Sakaguchi, M.; Bunkenborg, J.; Simon, S.M.; Kallunki, T.; Jäättelä, M.; Nylandsted, J. S100A11 is required for efficient plasma membrane repair and survival of invasive cancer cells. *Nat. Commun.* **2014**, *5*, 3795. [CrossRef] [PubMed]
38. Yuan, J.; Yang, Y.; Gao, Z.; Wang, Z.; Ji, W.; Song, W.; Zhang, F.; Niu, R. Tyr23 phosphorylation of Anxa2 enhances STAT3 activation and promotes proliferation and invasion of breast cancer cells. *Breast Cancer Res. Treat.* **2017**, *164*, 327–340. [CrossRef]
39. Wang, T.; Yuan, J.; Zhang, J.; Tian, R.; Ji, W.; Zhou, Y.; Yang, Y.; Song, W.; Zhang, F.; Niu, R. Anxa2 binds to STAT3 and promotes epithelial to mesenchymal transition in breast cancer cells. *Oncotarget* **2015**, *6*, 30975–30992. [CrossRef]
40. Yan, X.; Zhang, D.; Wu, W.; Wu, S.; Qian, J.; Hao, Y.; Yan, F.; Zhu, P.; Wu, J.; Huang, G.; et al. Mesenchymal Stem Cells Promote Hepatocarcinogenesis via lncRNA-MUF Interaction with ANXA2 and miR-34a. *Cancer Res.* **2017**, *77*, 6704–6716. [CrossRef]
41. Wu, M.; Sun, Y.; Xu, F.; Liang, Y.; Liu, H.; Yi, Y. Annexin A2 Silencing Inhibits Proliferation and Epithelial-to-mesenchymal Transition through p53-Dependent Pathway in NSCLCs. *J. Cancer* **2019**, *10*, 1077–1085. [CrossRef] [PubMed]
42. Liu, Y.; Li, H.; Ban, Z.; Nai, M.; Yang, L.; Chen, Y.; Xu, Y. Annexin A2 inhibition suppresses ovarian cancer progression via regulating β-catenin/EMT. *Oncol. Rep.* **2017**, *37*, 3643–3650. [CrossRef] [PubMed]
43. Cui, L.; Song, J.; Wu, L.; Cheng, L.; Chen, A.; Wang, Y.; Huang, Y.; Huang, L. Role of Annexin A2 in the EGF-induced epithelial-mesenchymal transition in human CaSki cells. *Oncol. Lett.* **2017**, *13*, 377–383. [CrossRef] [PubMed]
44. Guo, X.X.; Guo, Q.; Li, Y.; Lee, S.K.; Wei, X.N.; Jin, Y.H. Ginsenoside Rh2 induces human hepatoma cell apoptosis via bax/bak triggered cytochrome C release and caspase-9/ caspase-8 activation. *Int. J. Mol. Sci.* **2012**, *13*, 15523–15535. [CrossRef] [PubMed]
45. Lee, H.; Lee, S.; Jeong, D.; Kim, S.J. Ginsenoside Rh2 epigenetically regulates cell-mediated immune pathway to inhibit proliferation of MCF-7 breast cancer cells. *J. Ginseng Res.* **2018**, *42*. [CrossRef] [PubMed]
46. Mathiyalagan, R.; Wang, C.; Kim, Y.J.; Castro-Aceituno, V.; Ahn, S.; Subramaniyam, S.; Simu, S.Y.; Jiménez-Pérez, Z.E.; Yang, D.C.; Jung, S.K. Preparation of polyethylene glycol-ginsenoside Rh1 and Rh2 conjugates and their efficacy against lung cancer and inflammation. *Molecules* **2019**, *24*, 4367. [CrossRef]
47. Hou, J.G.; Jeon, B.M.; Yun, Y.J.; Cui, C.H.; Kim, S.C. Ginsenoside Rh2 ameliorates doxorubicin-induced senescence bystander effect in breast carcinoma cell MDA-MB-231 and nomal epithelial cell MCF-10A. *Int. J. Mol. Sci.* **2019**, *20*, 1244. [CrossRef]
48. Qi, Z.; Chen, L.; Li, Z.; Shao, Z.; Qi, Y.; Gao, K.; Liu, S.; Sun, Y.; Li, P.; Liu, J. Immunomodulatory Effects of (24R)-Pseudo-Ginsenoside HQ and (24S)-Pseudo-Ginsenoside HQ on Cyclophosphamide-Induced Immunosuppression and Their Anti-Tumor Effects Study. *Int. J. Mol. Sci.* **2019**, *20*, 836. [CrossRef]

49. Wang, M.; Yan, S.J.; Zhang, H.T.; Li, N.; Liu, T.; Zhang, Y.L.; Li, X.X.; Ma, Q.; Qiu, X.C.; Fan, Q.Y.; et al. Ginsenoside Rh2 enhances the antitumor immunological response of a melanoma mice model. *Onco. Lett.* **2017**, *13*, 681–685. [CrossRef]
50. Huang, Y.; Huang, H.; Han, Z.; Li, W.; Mai, Z.; Yuan, R. Ginsenoside Rh2 Inhibits Angiogenesis in Prostate Cancer by Targeting CNNM1. *J. Nanosci. Nanotechnol.* **2019**, *19*, 1942–1950. [CrossRef]

4

Exploring Biased Agonism at FPR1 as a Means to Encode Danger Sensing

Jieny Gröper [1,2], Gabriele M. König [3], Evi Kostenis [3], Volker Gerke [1,2], Carsten A. Raabe [1,4,*] and Ursula Rescher [1,2,*]

1 Institute of Medical Biochemistry, Center for Molecular Biology of Inflammation, University of Muenster, Von-Esmarch-Str. 56, D-48149 Muenster, Germany; Jieny.Groeper@ukmuenster.de (J.G.); gerke@uni-muenster.de (V.G.)
2 Cells in Motion" Interfaculty Centre, University of Muenster, 48149 Muenster, Germany
3 Institute for Pharmaceutical Biology, University of Bonn, Nussallee 6, 53115 Bonn, Germany; G.Koenig@uni-bonn.de (G.M.K.); kostenis@uni-bonn.de (E.K.)
4 Institute of Experimental Pathology, Center for Molecular Biology of Inflammation, University of Muenster, Von-Esmarch-Str. 56, D-48149 Muenster, Germany
* Correspondence: raabec@uni-muenster.de (C.A.R.); rescher@uni-muenster.de (U.R.)

† Running title: Evaluation of biased agonism at FPR1.

Abstract: Ligand-based selectivity in signal transduction (biased signaling) is an emerging field of G protein-coupled receptor (GPCR) research and might allow the development of drugs with targeted activation profiles. Human formyl peptide receptor 1 (FPR1) is a GPCR that detects potentially hazardous states characterized by the appearance of N-formylated peptides that originate from either bacteria or mitochondria during tissue destruction; however, the receptor also responds to several non-formylated agonists from various sources. We hypothesized that an additional layer of FPR signaling is encoded by biased agonism, thus allowing the discrimination of the source of threat. We resorted to the comparative analysis of FPR1 agonist-evoked responses across three prototypical GPCR signaling pathways, i.e., the inhibition of cAMP formation, receptor internalization, and ERK activation, and analyzed cellular responses elicited by several bacteria- and mitochondria-derived ligands. We also included the anti-inflammatory annexinA1 peptide Ac2-26 and two synthetic ligands, the W-peptide and the small molecule FPRA14. Compared to the endogenous agonists, the bacterial agonists displayed significantly higher potencies and efficacies. Selective pathway activation was not observed, as both groups were similarly biased towards the inhibition of cAMP formation. The general agonist bias in FPR1 signaling suggests a source-independent pathway selectivity for transmission of pro-inflammatory danger signaling.

Keywords: bias analysis; G protein-coupled receptor (GPCR); formyl peptide receptor 1; danger-associated molecular pattern (DAMP); pathogen-associated molecular pattern (PAMP); annexin A1 peptide Ac2-26

---

1. Introduction

A key feature of the eukaryotic immune defense is its capability to sense danger signals. Microbes are a source of potentially deleterious infections, and consequently, 'pathogen-associated molecular patterns' (PAMPs) and 'danger-associated molecular patterns' (DAMPs) represent chemical signatures that are sensed and transduced via the corresponding 'pattern recognition receptors' (PRRs) [1,2].

One such molecular pattern is the characteristic N-formylated methionine at the N-terminus of bacterial proteins. This modified amino acid is not utilized for the initiation of eukaryotic protein translation. However, mitochondria, which are considered to represent endosymbionts of bacterial origin [3], initiate protein biosynthesis with N-formylated methionine and might release detectable levels of formylated peptides in situations of enhanced cell death or trauma [4,5]. Therefore, the appearance of N-formylated peptides signals potentially hazardous states caused by either bacterial threats (PAMPs) or tissue destruction (DAMPs).

In higher eukaryotes, this unique pattern is sensed by the family of formyl peptide receptors (FPRs), which belong to the superfamily of G protein-coupled receptors (GPCRs) [5,6]. Thus, a shared sensor system detects bacteria-derived ligands and those that are liberated by non-infectious tissue destruction from mitochondria. The corresponding formylated peptides elicit strong signals via the activation of formyl peptide receptor 1 (FPR1) [7,8]. Here, we addressed whether human FPR1 is capable to decode the actual source of threat via different, i.e., ligand-specific signaling, and consequently modify the elicited cellular responses. We hypothesized that these layers of signal information are encoded by distinguishing ligand perception and potentially are governed by biased agonism. This emerging concept in GPCR-mediated signal transduction [9] emphasizes that ligands selectively stabilize specific receptor conformations that ultimately favor one (or more) signaling pathways out of the many the receptor is linked to. The final cellular response elicited by the receptor/ligand interaction is therefore biased towards specific signaling pathways [9–11]. We also considered that the different classes of agonists might group due to their efficacies and potencies for the same cellular pathways, i.e., apart from the selective activation of entirely different signaling pathways. In this scenario, different classes of agonists would not be associated with their own unique profile of qualitatively distinct cellular responses but instead would cause the same effect, as revealed by similar bias factors, yet on a different scale.

In general, agonist-activated GPCRs stimulate a broad set of heterotrimeric $\alpha\beta\gamma$ guanine nucleotide-binding proteins (G proteins), which in turn regulate adenylyl cyclase or phospholipase C (PLC) activities, ultimately leading to effective changes in second messengers cyclic AMP (cAMP) and inositol triphosphate (IP3) levels [12]. Apart from regulating enzymes to control the generation of second messengers, key intracellular signaling pathways are also activated via the GPCR signaling axis. For instance, GPCR-linked stimulation of the mitogen-activated protein kinase (MAPK) pathway, which plays important functional roles in, e.g., the regulation of (chronic) inflammation or even the development of cancer [13,14], is well established. Receptor internalization, which is linked with desensitization and signal termination, is regulated via GPCR/agonist interactions. Notably, emerging novel findings suggest that this mechanism might also be linked to prolonged receptor signaling [15,16].

Here, we compared the signaling profiles of representative microbial, endogenous, and synthetic FPR1 agonists. Our results reveal that the FPR1 activators cluster depending on their origin: the bacterial formylated peptides are strong, potent, and efficacious superagonists, whereas the mitochondrial peptides are less effective. Bias calculation uncovered that although the agonists operate on different levels as defined by their logistic parameters, FPR1 signal transmission generally is biased toward inhibition of cAMP formation.

2. Materials and Methods

2.1. FPR1 Ligands and Reagents

Formylated peptides corresponding to the N-termini of the human mitochondrially encoded proteins NADH:ubiquinone oxidoreductase core subunit 2 (MT-ND2; fMNPLAQ), NADH:ubiquinone oxidoreductase core subunit 6 (MT-ND6; fMMYALF), and cytochrome b (CYTB; fMTPMRKTNPLMKLIN), the formylated pentapeptide fMIVIL from *Listeria monocytogenes*, and the gG-2p20 peptide GLLWVEVGGEGPGPT derived from the secreted glycoprotein sgG-2 of herpes simplex virus type 2 (HSV-2) were custom-synthesized (Biomatik, Cambridge, ON, Canada). The acetylated peptide Ac2-26 corresponding to the N-terminus of human annexin A1 (AcAMVSEFLKQAWFIENEEQEYVQTVK) and the synthetic agonist W-peptide (WKYMVm) were purchased from Tocris (Wiesbaden-Nordenstadt, Germany). The prototype FPR1 agonist fMLF derived from *E. coli* was purchased from Sigma. Stock solutions were prepared as indicated in Table A1. The mouse monoclonal anti-FLAG antibody M1 (Sigma-Aldrich, Darmstadt, Germany), which recognizes the FLAG epitope only when present at the very N-terminus of the FPR1 receptor, i.e., after successful cleavage of the hemagglutinin signal sequence (see below) in the endoplasmic reticulum (ER), was labeled with DyLight488 antibody labeling kit (Thermo Fisher Scientific, Waltham, MA, USA) according to the manufacturer's protocol. Pertussis toxin (PTX) from *Bordetella pertussis* was purchased from Tocris, the $G\alpha_q$ inhibitor FR900359 (FR, formerly known as UBO-QIC), a cyclic depsipeptide from the plant *Ardisia crenata sims sims* was purified following a previously published protocol [17]. Reversed-phase high-performance liquid chromatography separation of the FR-containing fraction (column: YMC C_{18} Hydrosphere, 250 × 4.6 mm, 3 μm; MeOH:H_2O (8:2), 0.7 mL min^{-1}) afforded FR with a purity of 95%. For G protein inhibition experiments, cells were pretreated for 16 h with 100 ng/mL PTX or for 1 h with 1 μM FR preincubation in cell culture medium at 37 °C.

2.2. FPR1-Encoding Plasmid, HeLa-FPR1- Cell Line, and Cell Culture Conditions

The FPR1 expression vector, containing the N-terminally FLAG-tagged human FPR1, was generated as previously described [18]. The FPR1 coding sequence was PCR-amplified from a cDNA library representing human total leukocyte RNA (Takara Bio, Saint-Germain-en-Laye, France). The FLAG-epitope was introduced immediately upstream to the original FPR1 start codon and is preceded by a cleavable influenza hemagglutinin signal sequence to facilitate cell surface presentation. This tagged FPR1 CDS was transferred into the mammalian expression vector pcDNA3.1 (-) (Thermo Fisher Scientific) via XhoI and EcoRI restriction sites. HeLa cells cultured in Dulbecco's modified Eagle's medium (DMEM, Sigma), supplemented with 10% standardized fetal bovine serum (FBS Superior, Biochrom, Cambridge, UK), 100 U/mL penicillin, and 0.1 mg/mL streptomycin) at 37 °C in a 7% CO_2 atmosphere were transfected using Lipofectamine 2000 (Thermo Fisher Scientific). Clonal lines were selected with 800 ng/mL geneticin (G418, AppliChem, Darmstadt, Germany).

2.3. FPR1 Expression Analysis in Parental and Recombinant HeLa Cells by qPCR and Immunofluorescence Microscopy

qRT-PCR was employed to confirm that parental, i.e., non-transfected HeLa cells do not express members of the FPR family at detectable levels and to confirm FPR1 expression in the stably expressing HeLa-FPR1 cell lines. Total RNA from HeLa cells was isolated with the RNeasy mini kit (Qiagen, Hilden, Germany) according to the manufacturers' instructions; 1 μg of RNA starting material was converted into cDNA with the high-capacity cDNA reverse transcription kit and random hexamer primers (Thermo Fisher Scientific). Subsequent qPCR analysis was performed with QuantiTect primer assays (Qiagen) for FPR1 (Hs_FPR_1_SG, QT00199745) and custom-designed sets of primers (Microsynth, Lindau, Germany) for amplification of FPR2 (for: 5′-TTGGTTTCCCTTTCAACTGG-3′ rev: 5′-AGACGTAAAGCATGGGGTTG-3′) and FPR3 (for: 5′-GGTTGAACGTGTTCATTACC -3′ rev: 5′-TGGTTTCTGTGAATTTTGGC-3′). Housekeeping genes actin (Hs_ACTB_1_SG, QT00095431) and glyceraldehyde 3-phosphate dehydrogenase (Hs_GAPDH_2_SG, QT01192646) served as references. All qPCR reactions were conducted with the Brilliant III Ultra-Fast SYBR Green qPCR Master Mix (Agilent Technologies, Santa Clara, CA, USA). Four independent cell samples were analyzed in technical replicates and amplified for 45 cycles on a CFX 384 real-time PCR cycler. The PCR amplification was analyzed with the CFX Manager Software v.2.1 (Bio-Rad, Hercules, CA, USA).

Expression and correct localization of tagged FPR1 were confirmed by immunofluorescence imaging. HeLa-FPR1 cells were cultured on glass coverslips and fixed with 4% paraformaldehyde for 10 min at room temperature. After incubation with anti-FLAG M1 antibody (diluted 1:100 in 2% BSA in PBS containing Ca^{2+} and Mg^{2+} (PBS^{++}, Sigma) for 60 min, cells were treated with anti-mouse Alexa 594-coupled secondary antibody (Invitrogen, Carlsbad, CA, USA) for an additional 45 min at room temperature. To visualize cell nuclei, Hoechst 33,342 stain (Thermo Fisher Scientific, diluted 1:100) was added during the incubation with the secondary antibody. Oregon Green 488 conjugated wheat germ agglutinin (Invitrogen, WGA, 5 μg/mL in Hank's balanced salt solution containing Ca^{2+} and Mg^{2+} for 10 min at room temperature) was used to label the plasma membrane. Samples were imaged with a LSM800 (Zeiss, Oberkochen, Germany) confocal microscope using a 63× objective.

2.4. *Flow Cytometric Analysis of Agonist-Induced Receptor Internalization*

HeLa-FPR1 cells were treated with vehicle or agonists diluted in internalization medium (IM; DMEM, 20 mM HEPES, 1 mg/mL BSA, pH 7.2) for 15 min at 37 °C. Cells were washed in PBS (Sigma), and detached in PBS/5 mM EDTA for 3 min at 37 °C. For the detection of the cell surface receptor pool, cells were washed with ice-cold PBS containing 5% BSA and 1 mM $CaCl_2$ and subsequently incubated with 5 μg/mL DyLight488-conjugated anti-FLAG M1 antibody for 45 min. 7AAD (eBioscience, San Diego, CA, USA) allowed the exclusion of compromised cells. Median fluorescence intensities (MFI) of 10,000 viable cells per condition were measured with a Guava easyCyte flow cytometer and the InCyte^{TM} Software (Merck-Millipore, Darmstadt, Germany). Agonist-induced internalization was defined as the difference between the MFI of vehicle-treated controls and the MFI detected in agonist-treated cells. For each measurement, agonist-induced internalization was normalized to the mean internalization induced by 10^{-4} M W-peptide, which consistently represented the maximum system response for internalization.

2.5. *HTRF-Based Quantification of cAMP Levels*

A competitive immunoassay based on time-resolved measurement of fluorescence resonance energy transfer (HTRF, homogeneous time-resolved fluorescence) between a cryptate-labeled specific antibody (donor) and a d2-coupled cAMP acceptor molecule (cAMP-G_i Kit, Cisbio, Codolet, France) was used to measure cyclic AMP (cAMP) formation in cells, as described elsewhere earlier [19]. In brief, cells cultured as described above on 96 well plates (50k cells/well) were incubated with 5 μM of the adenylyl cyclase activator forskolin (Sigma) in complete medium supplemented with 500 μM of the phosphodiesterase inhibitor 3-isobutyl-1-methylxanthine (IBMX, Sigma), together with the agonists at the indicated concentrations, for 30 min at 37 °C. Samples were transferred onto a 384 well low volume plate, conjugates were added, and samples were analyzed with the CLARIOstar reader (BMG Labtech, Ortenberg, Germany) (200 flashes/well, integration start 60 μsec, integration time 400 μsec, settling time 100 μsec). Luminescence signals were expressed as the ratio of 10,000× (acceptor signal/donor signal). Values were normalized to the maximum system output, which was obtained as forskolin-induced cAMP formation.

2.6. *HTRF-Based Quantification of MAPK/ERK Phosphorylation Levels*

The advanced Phospho-ERK1/2 (Thr202/Tyr204) plate-based assay (Cisbio) was used to measure ERK activation through HTRF using a sandwich assay format of a donor-coupled antibody and an acceptor antibody as described before in [20]. Briefly, cells grown on 96 well plates (45k cells/well) were serum-starved for 2 h and stimulated with the agonists at the indicated concentrations for 5 min at 37 °C. Lysates were transferred to a 384 well low volume plate, incubated with the antibodies for 4 h at room temperature, and luminescence was recorded with a CLARIOstar reader (BMG Labtech) (200 flashes/well, integration start 60 μsec, integration time 400 μsec, settling time 100 μsec). HTRF ratios were normalized to maximum system output obtained through stimulation with 100 nM phorbol 12-myristate 13-acetate (PMA).

2.7. *HTRF-Based Quantification of IP1 Levels*

The IP-One-Gq assay (Cisbio), that detects inositol monophosphate, a stable downstream metabolite of IP3 induced by phospholipase C activation, was utilized to establish FR pretreatment conditions, as described elsewhere earlier [21]. In brief, cells grown overnight on 384 well plates (5k cells/well) were pretreated with 1 μM FR for 1 h at 37 °C prior to stimulation with 100 μM phospholipase C activator 2,4,6-trimethyl-N-[3-(trifluoromethyl)phenyl]benzenesulfonamide (m-3M3FBS, Tocris) for 2 h at 37 °C. LiCl present in the stimulation buffer provided in the kit prevented the degradation of IP1. After addition of the conjugates, samples were incubated for 1 h at room temperature and read on a CLARIOstar plate reader (BMG Labtech) (200 flashes/well, integration start 60 μsec, integration time 400 μsec, settling time 100 μsec).

2.8. Curve Fitting

For each individual curve, ligand concentrations were log-transformed, normalized, and expressed as fractions of the maximum system response per pathway. Concentration-response curves were analyzed with Graphpad Prism 6 and the in-built four parameters sigmoidal model with a Hill coefficient of 1. EC_{50} values were derived from individual sigmoidal curve fits. If the data fitting was ambiguous without further constraints (very weak partial agonists, no inflection point), the minimum was set to zero and the highest experimental response was considered to represent E_{max}. The ROUT method [22] was used to detect and eliminate outliers. An agonist-induced response was defined by a curve span >3 SEM. If no fitting was possible (i.e., no detection of agonist-elicited responses), the 'no response' (NR) label was assigned. To analyze the relationship between data, linear regression and Spearman's correlation coefficients were calculated (Graphpad Prism 8).

Bias was calculated with the operational model [23]. The logarithm of the activity ratio E_{max}/EC_{50} [24–26] served as a surrogate for the actual transduction coefficient log(t/KA) [27] and was calculated individually for each agonist and pathway. To visualize the ligand "texture", agonist-specific ΔlogR values were plotted for the analyzed pathways. To compare the relative effectiveness of an agonist at a given pathway with the reference agonist, ΔlogR values were calculated as differences of the respective pathway-specific logR of the agonist of interest and the endogenous MT-ND6 peptide as the reference agonist. To rank the relative signaling preferences of a ligand for one pathway over another, ΔΔlogR values were calculated.

3. Results

To address in a systematic fashion whether FPR1 agonists differ in their corresponding signaling fingerprints, we established a heterologous FPR1 expression system [18]. The resulting transgenic HeLa-FPR1 cell line was analyzed via qRT-PCR to confirm the stable expression of FPR1 (Figure 1a). As parental HeLa cells did not express any detectable levels of FPR1-3, this experimental design enabled monitoring specific FPR1-mediated responses [8,28]. The correct localization and incorporation of the FLAG-tagged receptor in the plasma membrane were validated by immunofluorescence imaging (Figure 1b). In order to avoid potentially distorting influences resulting from agonist effects other than those elicited via the FPR1 signaling axis, we used parental HeLa cells as a negative control. No signal could be observed when these cells were stained with the M1 antibody, and none of the tested agonists caused any significant response for the analyzed pathways, thereby confirming the specificity of our results (Figure A1).

3.1. FRET-Based Analysis of Agonist-Induced Changes in cAMP Levels

Agonist-induced cellular responses mediated via FPR1 previously have been reported to depend on $G\alpha_i$ for signal transduction [29]. We, therefore, resorted to a FRET-based system to analyze agonist-mediated changes in cAMP de novo generation. Notably, none of the analyzed FPR1 agonists caused any detectable increase of cellular cAMP levels, thereby confirming that the tested ligands did not trigger FPR1 responses via $G\alpha_s$ (Figure A2). To stimulate the cellular cAMP formation to its maximum, we utilized forskolin, a known activator of adenylyl cyclase activity [30] and determined the inhibitory potential of the different FPR1 agonists (exemplary raw data are shown in Figure A3a). Relative potencies and efficacies of the analyzed agonists differed significantly in our assay (Figure 2). Bacterial peptides [31,32] and the W-peptide [15,33,34], formed a distinct group of agonists with high potencies and efficacies, mitochondrial peptides—on the other hand—were characterized by significantly lower potencies. However, differences in the respective efficacies for bacterial and mitochondrial peptides were not as pronounced. Notably, the annexin A1 peptide Ac2-26 [35], which displayed the lowest potency in our assay, was still able to suppress de novo cAMP formation to a level similar to the mitochondrial agonists. Interestingly, gG2p20, an exogenous ligand-derived from herpes virus [36], and the synthetic small molecule FPRA14 [37,38] mimicked the mitochondrial peptides to some extent. To analyze agonist profiles in more mechanistic detail, we investigated whether the magnitude of the agonist-elicited responses was linked with the respective potencies. Generally, EC_{50} values of the agonists displayed a very strong negative monotonic relationship with their respective ability to elicit E_{max}, as demonstrated by the Spearman's rank correlation coefficient ($r = -0.917$, $p = 0.001$, n 9).

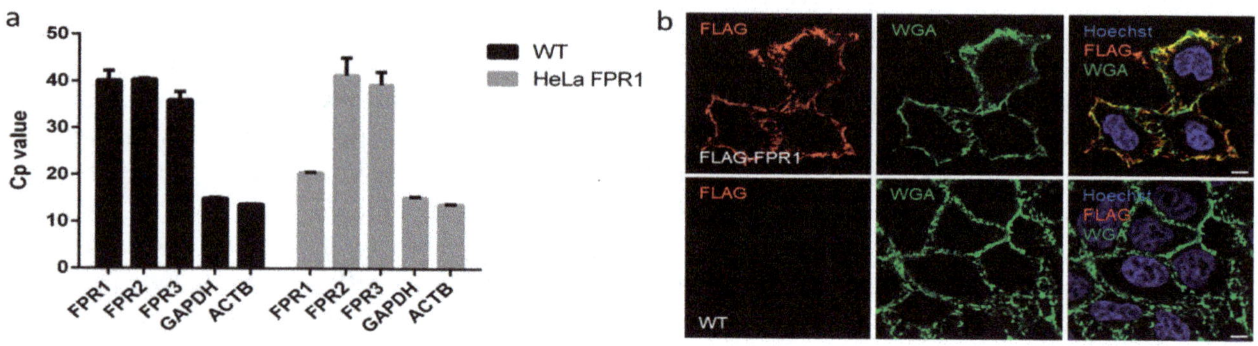

Figure 1. Evaluation of FPR1 (formyl peptide receptor 1) expression and localization in the heterologous HeLa expression system. (**a**) qRT-PCR revealed that none of the three FPR paralogs is expressed in parental HeLa cells (WT), whereas FPR1 is readily detectable in the HeLa-FPR1 transgenic cell line. Housekeeping genes ACTB (beta actin) and GAPDH (glyceraldehyde 3-phosphate dehydrogenase) served as internal controls. (**b**) Confocal immunofluorescence microscopy with the M1 antibody (red channel) confirmed the plasma membrane localization of FLAG-tagged FPR1. The plasma membrane was stained with WGA (green channel), and nuclei were visualized using Hoechst stain (blue channel). Upper panel: HeLa-FPR1 cell line; lower panel: WT HeLa cells, scale bar, 5 μm.

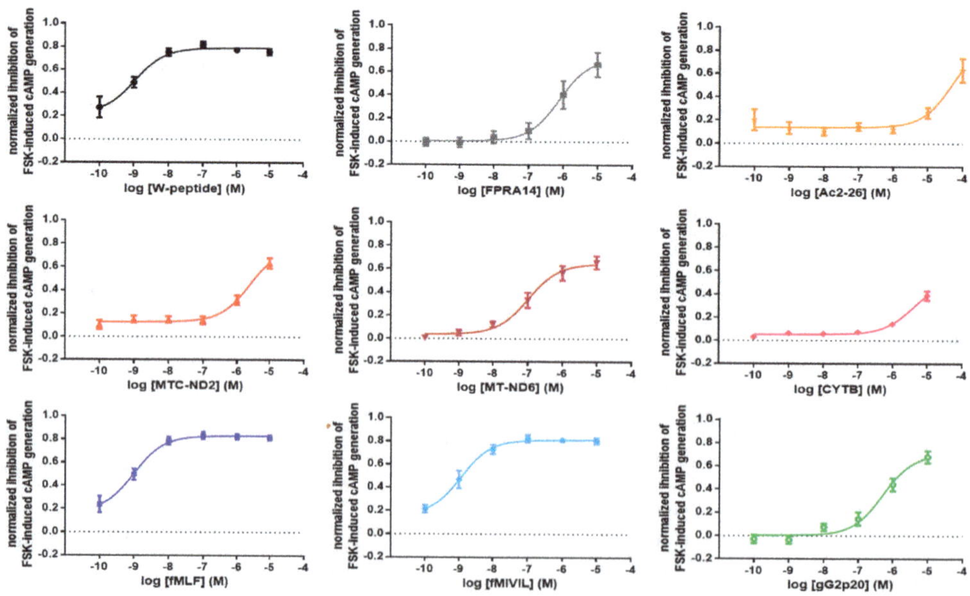

Figure 2. Concentration–response curves for FPR1-mediated inhibition of cAMP production. Cells were stimulated with forskolin (FSK) and the respective FPR1 agonists. Responses recorded 30 min after agonist addition were normalized to the maximum system response obtained with forskolin. Data points represent the mean ± SEM of at least 7 independent measurements.

3.2. FRET-Based Analysis of Agonist-Induced Changes in MAPKinase Activation

The MAPKinase cascade is a common effector of GPCR activation [39–41]. As shown in Figure 3, all agonists were able to increase ERK1/2 phosphorylation (exemplary raw data are shown in Figure A3b). Interestingly, the mitochondrial agonist MT-ND6 performed comparably to the bacterial agonists and the W-peptide, whereas the other mitochondrial peptides, as well as Ac2-26, only weakly activated ERK1/2 phosphorylation. As in the case of the adenylyl cyclase inhibition, gG2p20 and the synthetic small molecule FPRA14 displayed almost identical profiles. Spearman's rank correlation uncovered a strong negative monotonic correlation between EC_{50} and E_{max} values ($r = -0.8$, $p = 0.014$, n 9).

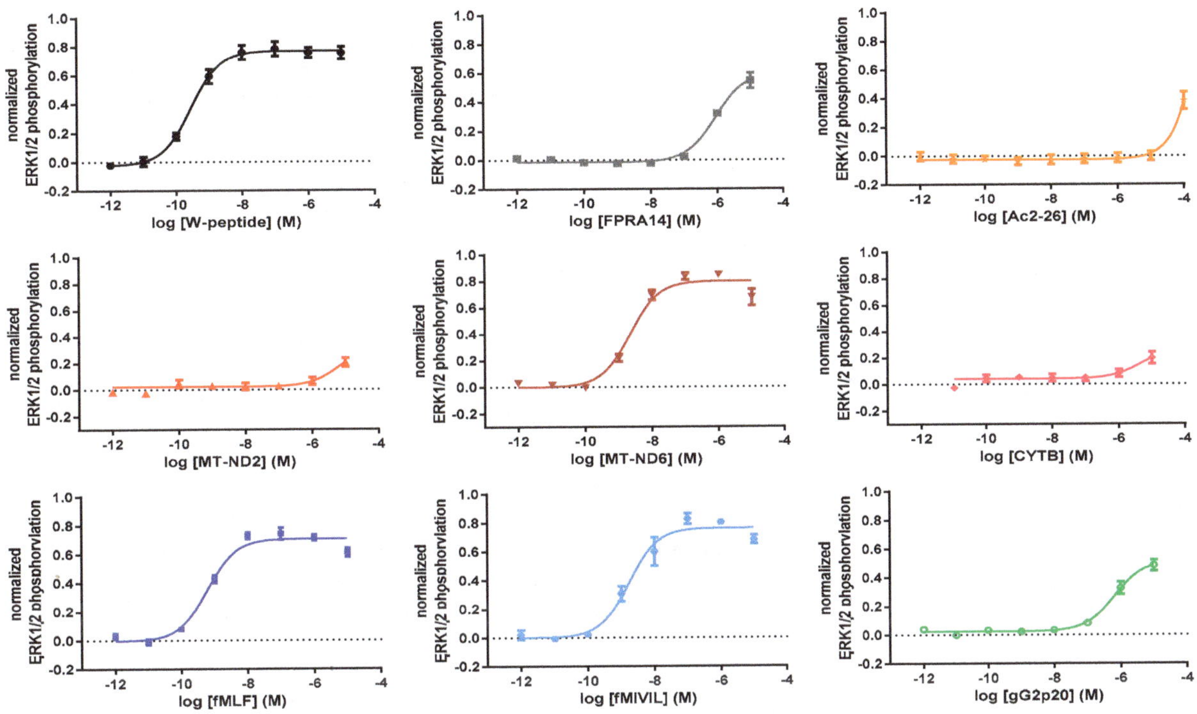

Figure 3. Concentration–response curves for FPR1-mediated phosphorylation of ERK1/2 on Thr202/Tyr204. Responses monitored 5 min after the addition of agonist were normalized to the maximum system response, which was obtained with PMA (phorbol 12-myristate 13-acetate). Data points represent the mean ± SEM of at least 5 independent measurements.

3.3. Agonist-Mediated Internalization of FPR1

Activated GPCRs usually are removed from the cell surface and are internalized into endosomes [15,16]. To test the potential of our agonists to induce FPR1 internalization, we determined the agonist-dependent decrease in FPR1 cell surface presentation, based on the detection of the N-terminal FLAG-epitope by flow cytometry [42,43]. The actual amount of internalized receptor was calculated via the difference of cell surface signals measured in untreated and agonist-treated samples after 15 min of agonist addition. Administration of W-peptide, bacterial peptides fMLF and fMIVIL, or the endogenous mitochondrial peptide MT-ND6 decreased the cell surface-associated FPR1 pool in a concentration-dependent manner. In contrast to that, the mitochondria-derived peptides MT-ND2, CYTB, the synthetic agonist FPRA14, and the annexin A1 peptide Ac2-26 did not elicit detectable internalization and were therefore deemed non-responders. Importantly, no significant difference was observed between the two formylated ligands fMLF and MT-ND6, indicating that the source of signal (PAMP vs. DAMP) was not encoded by these ligands (Figure 4).

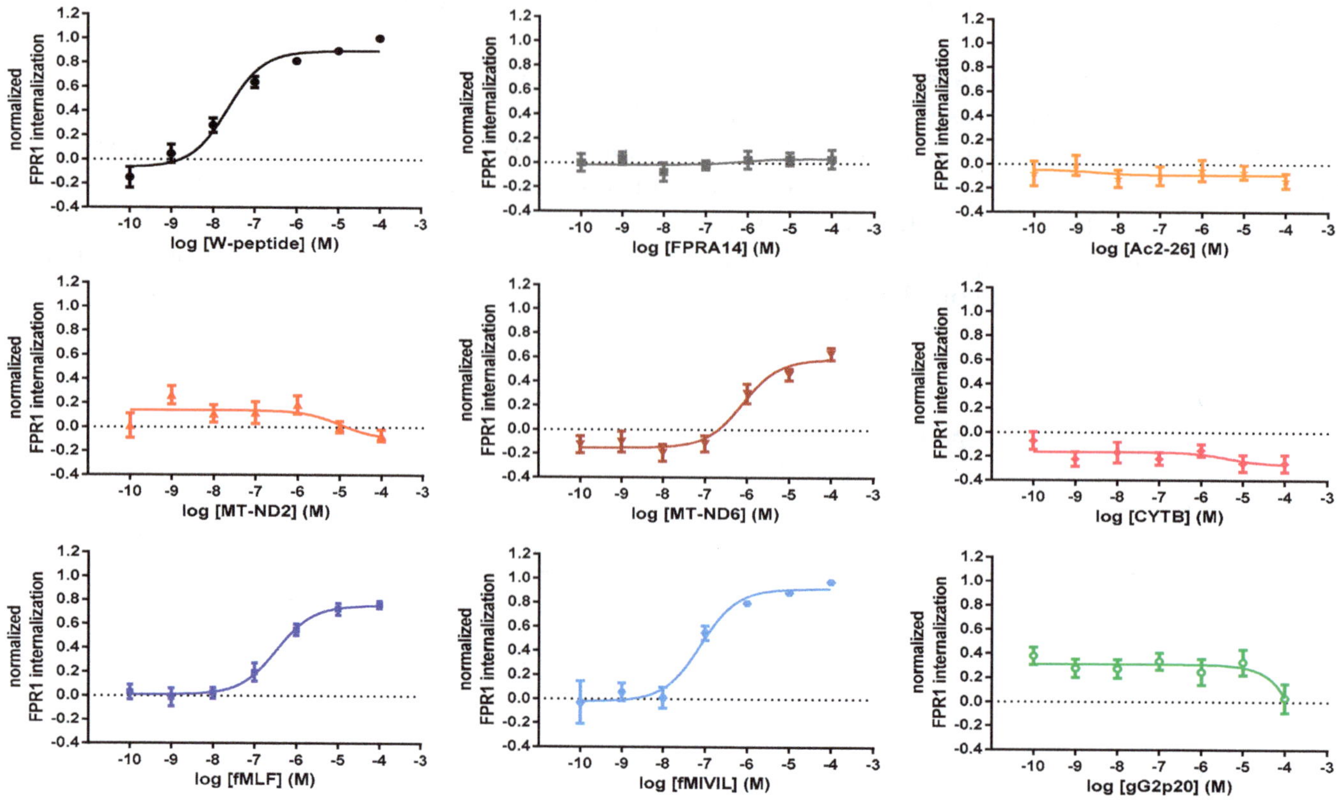

Figure 4. Concentration–response curves for agonist-induced FPR1 internalization. Receptor internalization was calculated as the difference between FLAG signals of unstimulated cells and signals recorded 15 min after agonist addition. Results were normalized to the maximum system response obtained with 10^{-4} M W-peptide. Data points represent the mean ± SEM of 6 independent measurements.

3.4. G protein Dependency of Agonist-Mediated Responses

We considered that the correlation for E_{max} and EC_{50} values for individual agonists and pathways, as revealed by the Spearman's rank coefficients, might—albeit indirectly—provide insights into common effector proteins interacting with the receptor to transduce agonist-elicited responses. We detected a very strong positive monotonic correlation between EC_{50} values for cAMP inhibition and ERK phosphorylation ($r = 0.933$, $p = 0.001$, n 9). A similar trend could also be established for the corresponding E_{max} values, which again revealed a strong positive correlation (0.733, $p = 0.031$, n 9; for an overview of $logEC_{50}$ and E_{maxA} values, see Figure A4 and Table A2.). To directly assess whether the different molecular pathways are driven by $G\alpha_i$-protein-dependent signal transduction, we repeated our experiments in the presence of the $G\alpha_i$-specific inhibitor pertussis toxin (PTX) at a concentration that completely abolished the inhibition of cAMP formation induced by W-peptide (Figure A4a). PTX treatment of HeLa-FPR1 cells did not interfere with ERK activation per se, however, agonist-mediated ERK1/2 activation was suppressed, thus strongly implying that both pathways, (at least) in our cellular system, relied on the $G\alpha_i$-protein (Figure 5). In stark contrast, receptor internalization was not dependent on $G\alpha_i$ activation, as agonist-evoked FPR1 internalization was not affected in PTX-pretreated cells (Figure 6a) and was equally undisturbed in cells pretreated with the $G\alpha_q$ inhibitor FR (Figure 6b).

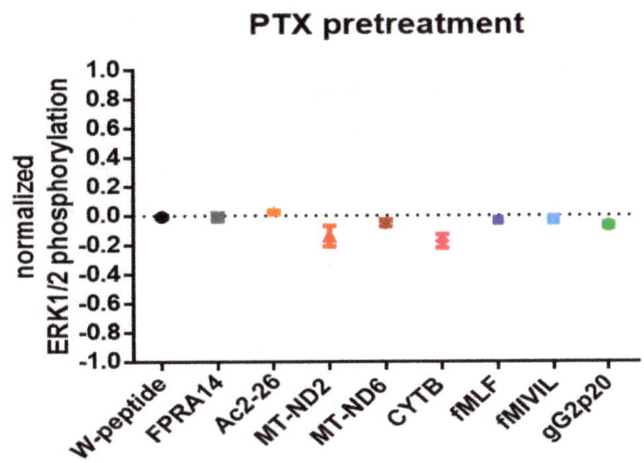

Figure 5. FPR1-mediated ERK1/2 phosphorylation is abolished in PTX (pertussis toxin)-pretreated cells. Cells were pretreated for 16 h with 100 ng/mL PTX and subsequently stimulated with agonist concentrations eliciting the respective E_{max}. Results were normalized to the maximum system response obtained with PMA. Data points represent the mean ± SEM of 6 independent measurements.

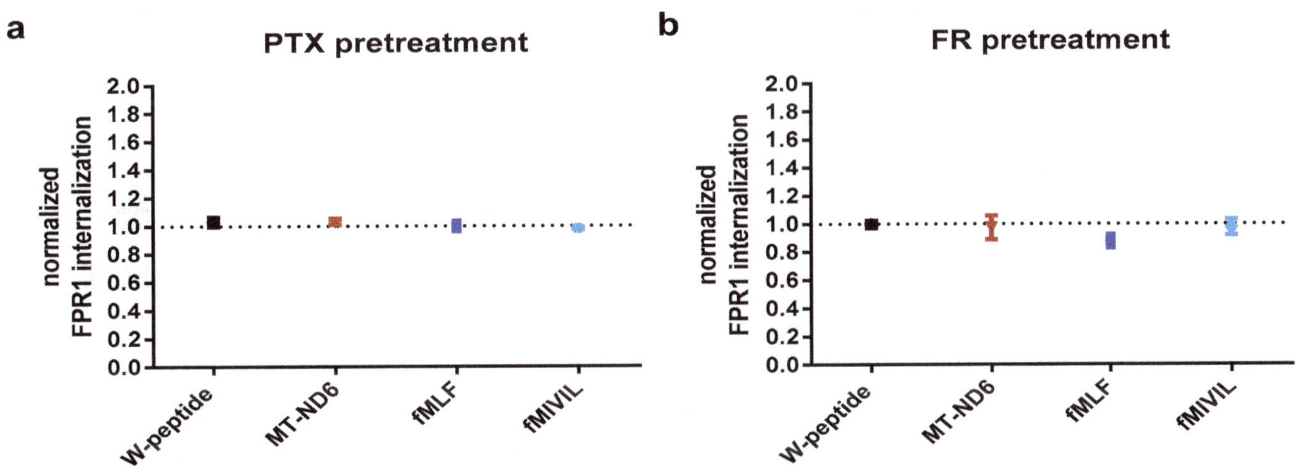

Figure 6. Agonist-induced FPR1 internalization is not dependent on Gα-mediated signal transmission. (**a**) Cells were pretreated either for 16 h with PTX or (**b**) for 1 h with FR and subsequently stimulated with agonist concentrations eliciting the respective E_{max}. Results were normalized to the 10^{-4} M W-peptide-induced maximum response. Data points represent the mean ± SEM of 6 independent measurements.

3.5. Analysis of Agonist Bias on FPR1

For the identification and analysis of ligand bias [23] associated with the FPR1 signaling axis, we selected the mitochondria-derived MT-ND6 as endogenous reference agonist for all tested pathways. For each agonist and pathway, we determined the corresponding logR (the logarithm of the activity ratio) values [24] and normalized the activities to MT-ND6 ($\Delta logR = logR_{agonist} - logR_{MT\text{-}ND6}$, see Table I) [23]. Figure 7 summarizes the resulting agonist ranking for each pathway. Overall, two agonist clusters were established: bacterial ligands, along with the W-peptide, which resembles a conserved spatial structure of bacterial agonists [44], constituted the group of "high-performers". Endogenous mitochondria-derived agonists, the herpes virus-derived peptide, the small synthetic agonist FPRA14, and annexin A1 peptide Ac2-26 represented the second group of considerably lower activity (Figure 7a). Interestingly, the endogenous agonist MT-ND6 featured balanced characteristics, with a profile "in-between" both groups. Graphical representations of the intrinsic activity profiles, i.e., in terms of ΔlogR, further highlight the tendency of our tested ligands to segregate into two clusters, which was observed independently of pathway but in relation to the agonist origin (Figure 7b).

However, the segregation was less pronounced when the magnitude of the respective response was taken into consideration (Figure 8, Figure A5, and Table A2). Analysis of pathway preferences, i.e., in terms of ΔlogR values, revealed that FPR1 activation was generally biased towards the inhibition of cAMP formation compared to the activation of the MAPKinase pathway or receptor internalization (Figure 9). However, the direct comparison of the weak agonist Ac2-26 with the strong agonist fMIVIL impressively revealed that bias was realized at different logistic levels (Figure 9).

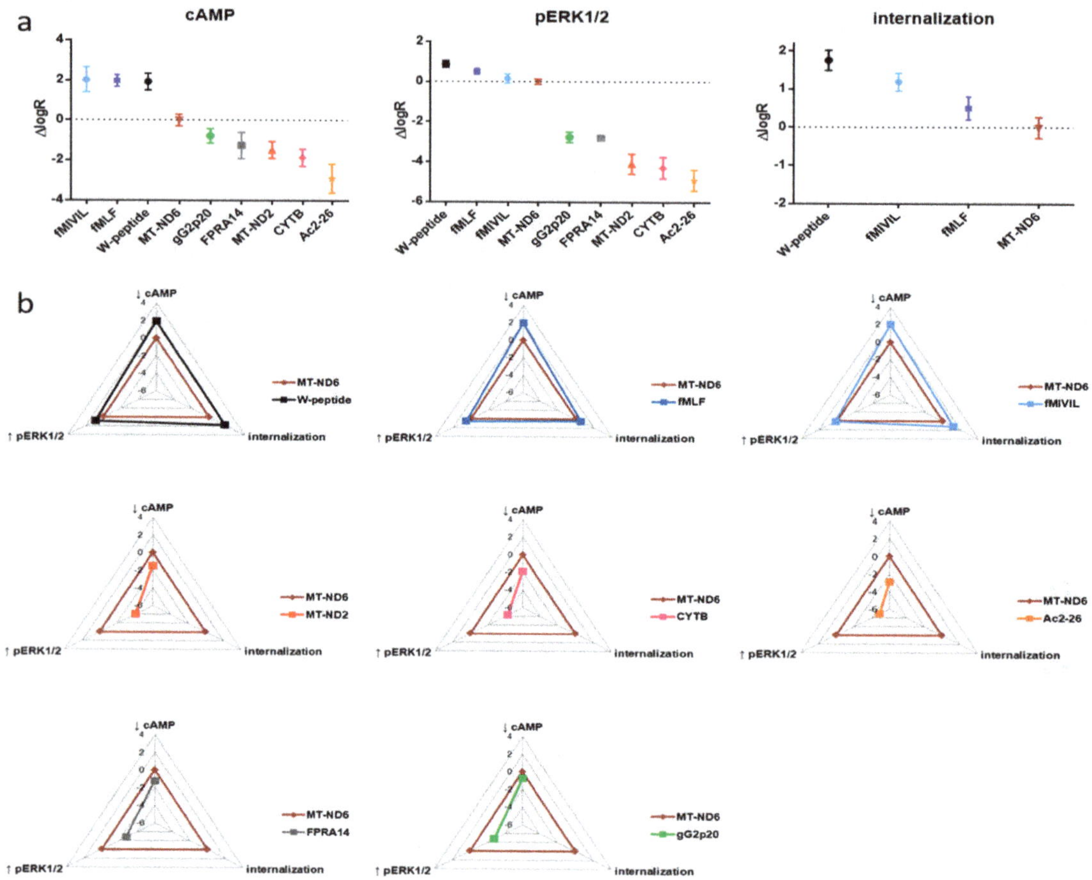

Figure 7. Relative agonist profiles compared to the endogenous reference agonist. (**a**) ΔlogR values were calculated from logR values with the mitochondrial peptide MT-ND6 as reference agonist. Data points represent the mean and 95% confidence intervals. (**b**) Relative agonist activities corresponding to the ΔlogR values for the respective pathways are depicted in radial graphs. Each radius is displayed in the logarithmic scale.

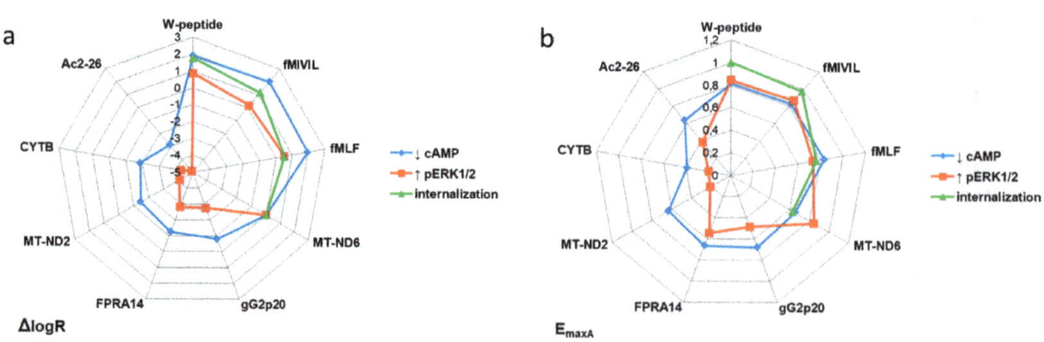

Figure 8. Graphical representation of the agonist profiles. Relative agonist activities corresponding to the ΔlogR values and the logistic parameters (**a**) EC_{50} and (**b**) E_{maxA} for the respective pathways are depicted in radial graphs. Each radius for ΔlogR is displayed in the logarithmic scale, radiuses for E_{max} are scaled linearly.

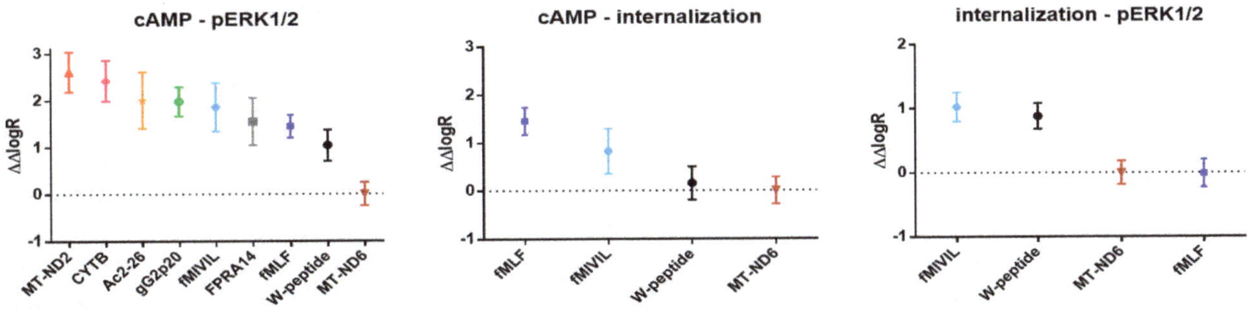

Figure 9. Comparison of agonist pathway preferences. ΔΔlogR values were calculated from ΔlogR values. Data points represent the mean and 95% confidence intervals.

4. Discussion

FPR1 and the corresponding large repertoire of FPR1 agonists, derived from various cellular and pathogenic sources, constitute a powerful system for the detection of insults that are linked with tissue destruction under infectious and sterile conditions, such as pathogenic challenges or trauma caused by burns or injury [9,45]. Because overactivation of the innate immune response is often correlated with excessive and deleterious tissue damage, e.g., in influenza A virus (IAV) infection, targeting the FPR family might represent a novel approach to balance innate immunity. Indeed, our previous studies revealed the advantageous use of the host-derived anti-inflammatory N-terminal annexin A1 peptide Ac-2-26, a pan-FPR agonist [46] to counteract viral load and mortality in a preclinical murine IAV infection model [47], thus encouraging the development of novel FPR-based therapies.

Generally, the often observed signaling diversity elicited by GPCR agonists that are acting on the same receptor is based on the ligand preference for certain receptor conformational states linked to a subset of all possible signaling responses. This diversity has been termed "functional selectivity" and has led to the concept of "biased agonism" and helps to explain different regulatory outcomes via the activation of the same receptor [9,11]. Therapeutically, bias analysis might help to identify compounds that direct receptor signaling toward desired responses, thus aiding in the development of novel therapeutics with an effective pharmacological profile that avoids activation of unwanted signaling pathways and hence side effects [48,49]. Indeed, a few biased agonists have been developed and are currently in various stages of clinical trials; this is true even for FPR-targeting compounds [50].

To further explore the potential of biased agonism acting on FPRs, we selected human FPR1, the founding member of the FPR family [8], for broader analysis. FPR1 represents the sensor platform for short formylated peptides, a pattern commonly associated with bacterial PAMPs and mitochondria-derived DAMPs [51,52]. However, the preference for such modified peptides is not exclusive, and even non-formylated derivatives (such as Ac2-26) are known to activate FPR1 effectively. The structural diversity of FPR1 agonists led us to hypothesize that potentially different classes of agonists might group, based on their origin, i.e., as PAMPs, otherwise endogenous ligands or DAMPs, thus enabling the receptor to decode the actual source of danger and in turn to channel the cellular responses.

Commonly, agonists are classified according to their ability to invoke the maximum receptor-mediated response of a given pathway. These agonists represent "full" agonists, whereas "partial" agonists only elicit a fraction of the cellular responses caused by a full agonist. This classification is intuitive and seems to be well suited to describe the properties of most ligands; it inherently suffers, however, from the disadvantage that potentially better—yet unidentified—agonists (evoking a higher response) cannot be accounted for. An alternative approach classifies agonist efficacies in relation to an endogenous reference agonist of high efficacy. Therefore, some agonists might be identified, which are even capable to elicit stronger cellular responses than those associated with the reference agonist; consequently, these ligands are referred to as "superagonists", although this term still has to be defined in broader detail [53]. The molecular explanation for this phenomenon might lie in the observation that simultaneous binding of an efficacious agonist and a G protein is required to induce the full receptor response [54,55]. Our results imply that bacterial agonists function as bona fide superagonists at FPR1. Of note, FPR1 agonist clusters clearly grouped based on ligand efficacies and potencies;

these logistic properties were strongly correlated as revealed by the analysis "within" as well as "across" $G\alpha_i$-dependent pathways. Moreover, our results argued in favor of a shared signal transmission selectivity for a given pathway across these structurally unrelated agonist classes. This was most obvious in the case of agonist-mediated ERK activation [56,57], which was entirely G protein-dependent, because in no instance did we observe Gαi-independent activation of the ERK signaling cascade. Yet, lack of evidence for G protein-independent signals, as far as ERK phosphorylation is concerned, does by no means exclude that FPR1 agonists display preferences for supposedly distinct subsets of FPR1-elicited signaling pathways over others. Internalization might—at first glance—appear G protein-independent. However, the most likely scenario probably is that endocytosis requires an active receptor conformation rather than an active signaling pathway and therefore occurs even when G proteins are precluded from active signal transmission. It is therefore perhaps not surprising that particularly high-efficacy ligands are the most effective at causing receptor internalization. It also cannot be ruled out that a given cellular environment is hard-wired for a set of pathways [58]. Hence, the decoding capacity of our heterologous expression system might cause the cells' inability to distinguish the source of these ligands, leading to so-called "system bias". Our data suggests that peptide-based FPR1 pharmacotherapy might be worthwhile exploring, however, peptide aggregation is a huge challenge [59].

Based on the bias calculations, all of the agonists preferentially inhibited cAMP over ERK phosphorylation. cAMP is a prominent second messenger in the PKA signaling pathway and a potent regulator of immunity. Because increasing and decreasing cAMP levels are correlated to dampening or stimulating immune responses, respectively, the cellular cAMP balance is considered a bona fide druggable target [60]. Surprisingly, the Ac2-26 peptide was also biased toward adenylyl cyclase inhibition, similar to the endogenous mitochondrial peptides, and the agonistic profile did not resonate with the established anti-inflammatory properties.

The bias in FPR1 activation toward inhibition of cAMP production may signal imminent danger, regardless of the source. However, similar bias factors can be associated with entirely different logistic parameters, and therefore, might cause different levels of physiological response. Our data do not support a selective activation of entirely different signaling pathways, as typically associated with biased agonists. Rather, our findings identified that the actual differences of the danger signals are encoded by the different levels of response (described by the logistic parameters EC_{50} and the maximum agonist-elicited response E_{max}).

Another means by which GPCR signaling can be diversified is the organization of GPCR-based signaling platforms via homo- and hetero-oligomerization [61]. Indeed, there is emerging functional evidence for FPR higher-order structures [62,63]. Whether such supramolecular sensor complexes are able to decode additional information is certainly an important line of future research.

Author Contributions: Conceptualization, C.A.R. and U.R.; methodology, J.G., C.R. and U.R.; validation, C.A.R., J.G. and U.R.; formal analysis, J.G.; investigation, J.G.; resources, U.R., G.M.K., E.K.; writing—original draft preparation and methodology, J.G., V.G., C.A.R. and U.R.; writing—review and editing, J.G., C.A.R., U.R., G.M.K., E.K.; visualization, J.G.; supervision, C.A.R.; project administration, U.R.; funding acquisition, V.G., U.R. All authors have read and agreed to the published version of the manuscript.

Acknowledgments: We thank Rod Flower, Terry Kenakin, and Henry Showell for thought-provoking impulses, insightful discussions and comments.

Appendix A

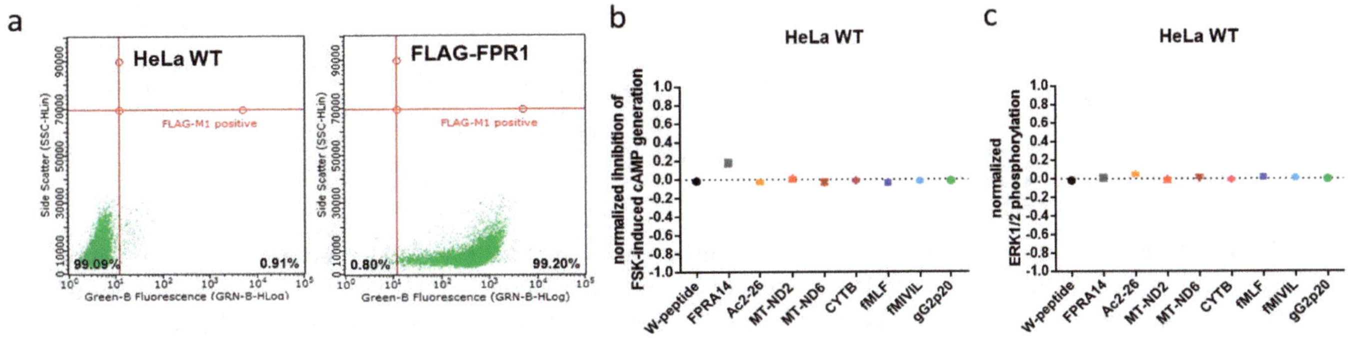

Figure A1. Agonist effects are specific for the FPR1 signaling axis. Parental HeLa cells were subjected to the assays analyzing (**a**) internalization, (**b**) cAMP formation, and (**c**) MAPKinase activation, at agonist concentrations eliciting the respective E_{max} response. (**a**) A representative result, (**b**), (**c**) mean response ± SEM, n 6.

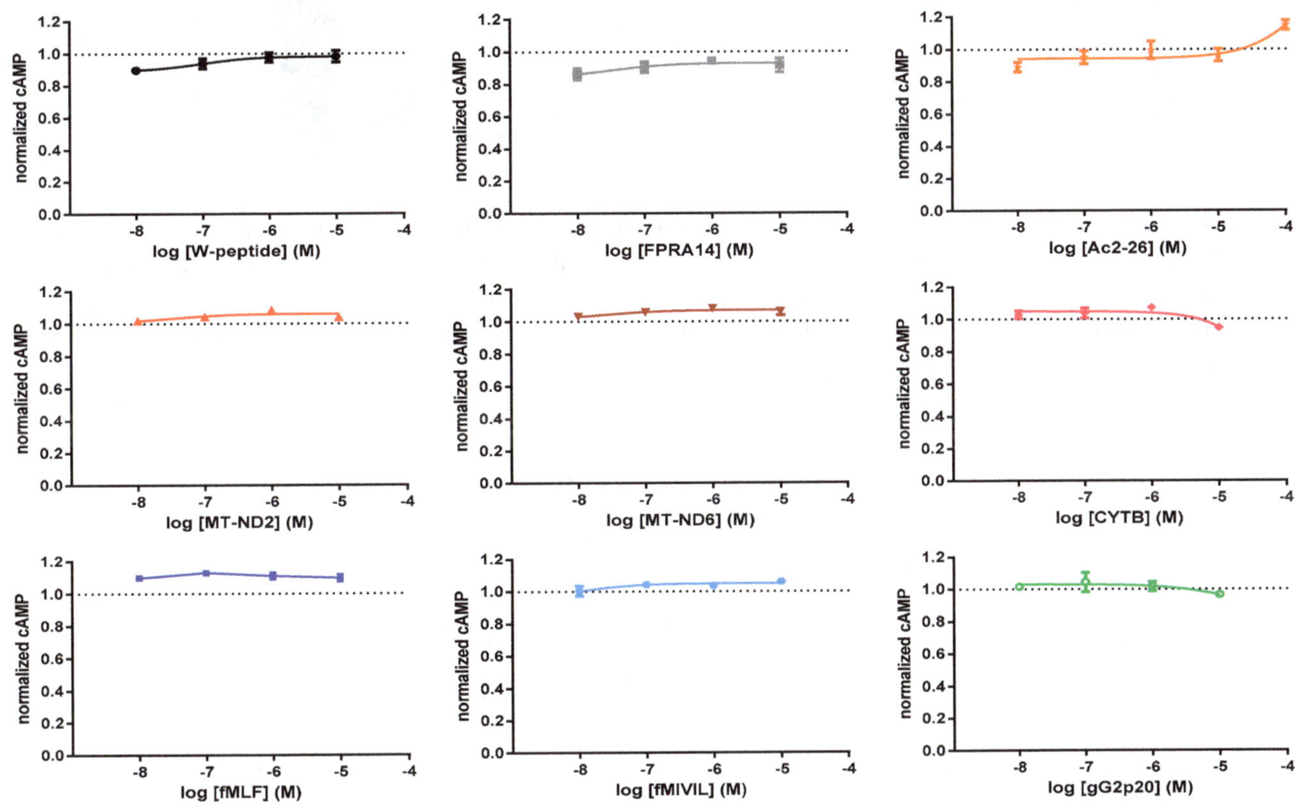

Figure A2. Concentration-response curves for FPR1-mediated stimulation of cAMP production. None of the agonists increased intracellular cAMP levels, in line with the reported FPR1-mediated activation of $G\alpha_i$. Mean response ± SEM, n 3.

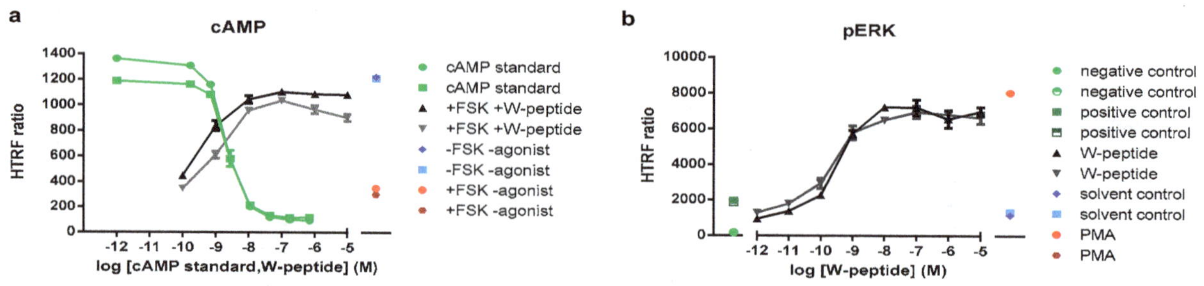

Figure A3. Exemplary raw data of concentration-response curves for FPR1-mediated stimulation of (**a**) cAMP production and (**b**) MAPKinase activation. In addition to the W-peptide measurements, data points for the cAMP standards and the ERK positive and negative controls included in the kit are shown. Data were normalized to the maximum output as obtained with forskolin and PMA.

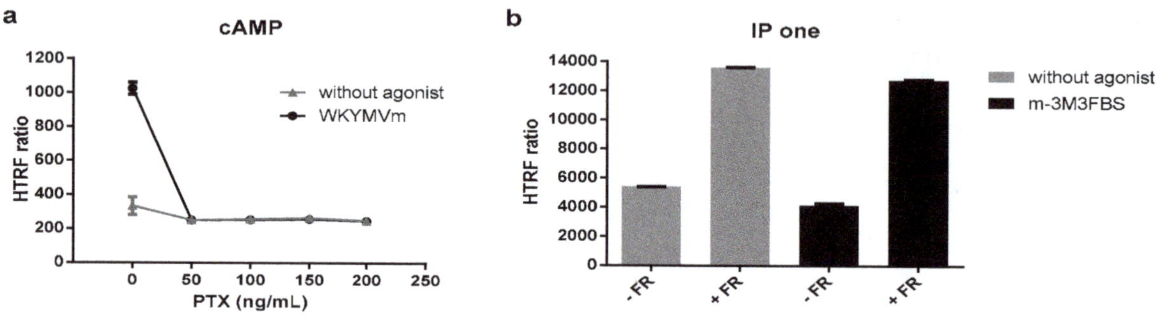

Figure A4. Analysis of G protein dependency. (**a**) PTX-mediated inhibition of FPR1-mediated changes in cAMP generation. (**a**) Maximum FPR1-mediated inhibition of forskolin-induced cAMP formation was completely prevented in cells pretreated for 16 h with 100 ng/mL PTX. (**b**) Effect of 1 μM FR preincubation for 1 h on IP-1 production induced by the phospholipase C activator m-3M3FBS. Data points represent the mean ± SEM.

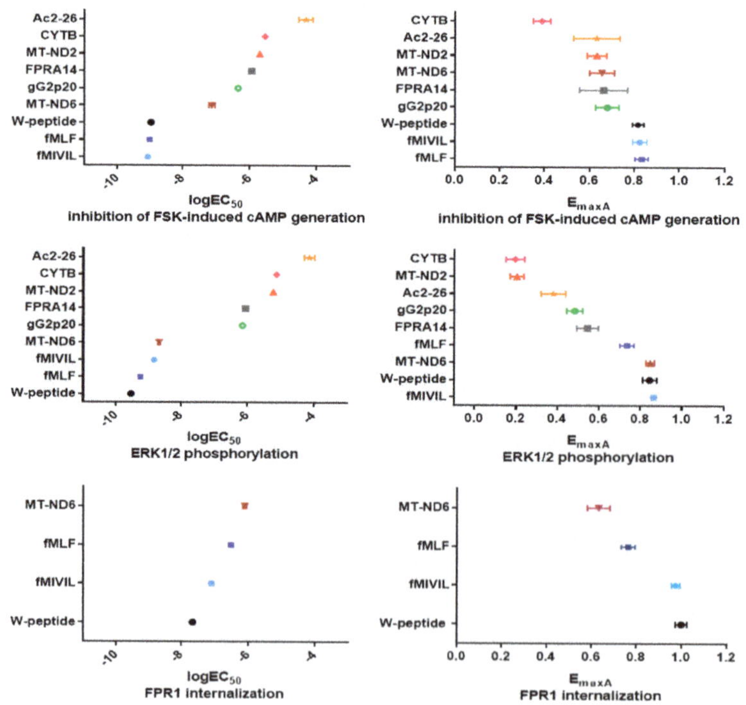

Figure A5. Depiction of $logEC_{50}$ and E_{maxA} values of each pathway. Ligands are arranged according to the intensity of the response they elicited in the respective pathway. Data are expressed as mean ± SEM after outlier removal.

Table A1. Ligand stock solutions. Concentrations of stock solutions were calculated based on the solubility of the compound as indicated by the manufacturer and its molecular weight (M). Stock solutions were prepared using the solvents recommended by the manufacturer.

Ligand	Stock Concentration	Solvent	Solubility	M
Ac2-26	324 µM	PBS	1 mg/mL	3089.46 g/mol
CYTB	1.0 mM	DMSO	2 mg/mL	1816.26 g/mol
fMIVIL	3.25 mM	DMSO	2 mg/mL	615.83 g/mol
fMLF	10 mM	DMSO	4 mg/mL	437.60 g/mol
FPRA14	100 mM	DMSO	100 mM	408.92 g/mol
gG2p20	1.36 mM	DMSO	2 mg/mL	1467.62 g/mol
MT-ND2	2.85 mM	DMSO	2 mg/mL	700.79 g/mol
MT-ND6	2.49 mM	DMSO	2 mg/mL	803.00 g/mol
WKYMVm	2.3 mM	H_2O	2 mg/mL	856.11 g/mol

Table A2. Overview of logistic parameters for FPR1 agonists. $LogEC_{50}$ and E_{max} were determined for FPR1 ligands for the analyzed pathways. Data are expressed as mean ± SEM values. NR, no response.

	Ligand	cAMP Formation		ERK1/2 Activation		Internalization	
		$logEC_{50}$	E_{maxA}	$logEC_{50}$	E_{maxA}	$logEC_{50}$	E_{maxA}
synthetic	FPR14	−5.94 0.19	0.66 0.11	−6.06 0.05	0.55 0.05	NR	NR
	W-peptide	−8.96 0.15	0.82 0.03	−9.54 0.06	0.85 0.04	−7.66 0.08	1.00 0.03
endogenous	Ac2-26	−4.31 0.20	0.63 0.10	−4.13 0.16	0.38 0.06	NR	NR
mitochondrial	CYTB	−5.52 0.11	0.39 0.04	−5.11 0.24	0.20 0.05	NR	NR
	MT-ND2	−5.69 0.13	0.63 0.04	−5.21 0.16	0.21 0.03	NR	NR
	MT-ND6	−7.13 0.09	0.66 0.06	−8.66 0.04	0.85 0.02	−6.11 0.06	0.63 0.05
bacterial	fMIVIL	−9.06 0.24	0.82 0.03	−8.81 0.11	0.86 0.01	−7.10 0.06	0.97 0.02
	fMLF	−9.00 0.10	0.83 0.03	−9.24 0.04	0.74 0.04	−6.53 0.10	0.76 0.03
viral	gG2p20	−6.36 0.12	0.68 0.05	−6.15 0.08	0.48 0.04	NR	NR

References

1. Takeuchi, O.; Akira, S. Pattern Recognition Receptors and Inflammation. *Cell* **2010**, *140*, 805–820. [CrossRef] [PubMed]
2. Mogensen, T.H. Pathogen recognition and inflammatory signaling in innate immune defenses. *Clin. Microbiol. Rev.* **2009**, *22*, 240–273, Table of Contents. [CrossRef] [PubMed]
3. Calfee, C.S.; Matthay, M.A. Clinical immunology: Culprits with evolutionary ties. *Nature* **2010**, *464*, 41–42. [CrossRef] [PubMed]
4. Dyall, S.D.; Brown, M.T.; Johnson, P.J. Ancient invasions: From endosymbionts to organelles. *Science (New York N.Y.)* **2004**, *304*, 253–257. [CrossRef]
5. Zhang, Q.; Raoof, M.; Chen, Y.; Sumi, Y.; Sursal, T.; Junger, W.; Brohi, K.; Itagaki, K.; Hauser, C.J. Circulating mitochondrial DAMPs cause inflammatory responses to injury. *Nature* **2010**, *464*, 104–107. [CrossRef]

6. Chen, G.Y.; Nunez, G. Sterile inflammation: Sensing and reacting to damage. *Nat. Rev. Immunol.* **2010**, *10*, 826–837. [CrossRef]
7. Le, Y.; Murphy, P.M.; Wang, J.M. Formyl-peptide receptors revisited. *Trends Immunol.* **2002**, *23*, 541–548. [CrossRef]
8. Boulay, F.; Tardif, M.; Brouchon, L.; Vignais, P. The human N-formylpeptide receptor. Characterization of two cDNA isolates and evidence for a new subfamily of G-protein-coupled receptors. *Biochemistry* **1990**, *29*, 11123–11133. [CrossRef]
9. Raabe, C.A.; Groper, J.; Rescher, U. Biased perspectives on formyl peptide receptors. *Biochim. Et Biophys. Acta. Mol. Cell Res.* **2019**, *1866*, 305–316. [CrossRef]
10. Rankovic, Z.; Brust, T.F.; Bohn, L.M. Biased agonism: An emerging paradigm in GPCR drug discovery. *Bioorganic Med. Chem. Lett.* **2016**, *26*, 241–250. [CrossRef]
11. Wootten, D.; Christopoulos, A.; Marti-Solano, M.; Babu, M.M.; Sexton, P.M. Mechanisms of signalling and biased agonism in G protein-coupled receptors. *Nat. Rev. Mol. Cell Biol.* **2018**, *19*, 638–653. [CrossRef] [PubMed]
12. Flock, T.; Hauser, A.S.; Lund, N.; Gloriam, D.E.; Balaji, S.; Babu, M.M. Selectivity determinants of GPCR-G-protein binding. *Nature* **2017**, *545*, 317–322. [CrossRef] [PubMed]
13. Jain, R.; Watson, U.; Vasudevan, L.; Saini, D.K. ERK Activation Pathways Downstream of GPCRs. *Int. Rev. Cell Mol. Biol.* **2018**, *338*, 79–109. [CrossRef] [PubMed]
14. Huang, P.; Han, J.; Hui, L. MAPK signaling in inflammation-associated cancer development. *Protein Cell* **2010**, *1*, 218–226. [CrossRef]
15. Calebiro, D.; Godbole, A. Internalization of G-protein-coupled receptors: Implication in receptor function, physiology and diseases. *Best Pract. Res. Clin. Endocrinol. Metab.* **2018**, *32*, 83–91. [CrossRef]
16. Pavlos, N.J.; Friedman, P.A. GPCR Signaling and Trafficking: The Long and Short of It. *Trends Endocrinol. Metab.* **2017**, *28*, 213–226. [CrossRef]
17. Schrage, R.; Schmitz, A.L.; Gaffal, E.; Annala, S.; Kehraus, S.; Wenzel, D.; Bullesbach, K.M.; Bald, T.; Inoue, A.; Shinjo, Y.; et al. The experimental power of FR900359 to study Gq-regulated biological processes. *Nat. Commun.* **2015**, *6*, 10156. [CrossRef]
18. Ernst, S.; Zobiack, N.; Boecker, K.; Gerke, V.; Rescher, U. Agonist-induced trafficking of the low-affinity formyl peptide receptor FPRL1. *Cell. Mol. Life Sci.* **2004**, *61*, 1684–1692. [CrossRef]
19. Conche, C.; Boulla, G.; Trautmann, A.; Randriamampita, C. T cell adhesion primes antigen receptor-induced calcium responses through a transient rise in adenosine 3′,5′-cyclic monophosphate. *Immunity* **2009**, *30*, 33–43. [CrossRef]
20. Grundmann, M.; Merten, N.; Malfacini, D.; Inoue, A.; Preis, P.; Simon, K.; Ruttiger, N.; Ziegler, N.; Benkel, T.; Schmitt, N.K.; et al. Lack of beta-arrestin signaling in the absence of active G proteins. *Nat. Commun.* **2018**, *9*, 341. [CrossRef]
21. Gao, Z.G.; Jacobson, K.A. On the selectivity of the Galphaq inhibitor UBO-QIC: A comparison with the Galphai inhibitor pertussis toxin. *Biochem. Pharmacol.* **2016**, *107*, 59–66. [CrossRef] [PubMed]
22. Motulsky, H.J.; Brown, R.E. Detecting outliers when fitting data with nonlinear regression-a new method based on robust nonlinear regression and the false discovery rate. *BMC Bioinform.* **2006**, *7*, 123. [CrossRef] [PubMed]
23. Kenakin, T.; Watson, C.; Muniz-Medina, V.; Christopoulos, A.; Novick, S. A simple method for quantifying functional selectivity and agonist bias. *ACS Chem. Neurosci.* **2012**, *3*, 193–203. [CrossRef] [PubMed]
24. Ehlert, F.J. Analysis of allosterism in functional assays. *J. Pharmacol. Exp. Ther.* **2005**, *315*, 740–754. [CrossRef]
25. Tran, J.A.; Chang, A.; Matsui, M.; Ehlert, F.J. Estimation of relative microscopic affinity constants of agonists for the active state of the receptor in functional studies on M2 and M3 muscarinic receptors. *Mol. Pharmacol.* **2009**, *75*, 381–396. [CrossRef]
26. Figueroa, K.W.; Griffin, M.T.; Ehlert, F.J. Selectivity of agonists for the active state of M1 to M4 muscarinic receptor subtypes. *J. Pharmacol. Exp. Ther.* **2009**, *328*, 331–342. [CrossRef]
27. Griffin, M.T.; Figueroa, K.W.; Liller, S.; Ehlert, F.J. Estimation of agonist activity at G protein-coupled receptors: Analysis of M2 muscarinic receptor signaling through Gi/o,Gs, and G15. *J. Pharmacol. Exp. Ther.* **2007**, *321*, 1193–1207. [CrossRef]
28. Muto, Y.; Guindon, S.; Umemura, T.; Kohidai, L.; Ueda, H. Adaptive evolution of formyl peptide receptors in mammals. *J. Mol. Evol.* **2015**, *80*, 130–141. [CrossRef]

29. Tsu, R.C.; Lai, H.W.; Allen, R.A.; Wong, Y.H. Differential coupling of the formyl peptide receptor to adenylate cyclase and phospholipase C by the pertussis toxin-insensitive Gz protein. *Biochem. J.* **1995**, *309 Pt 1*, 331–339. [CrossRef]

30. Seamon, K.B.; Padgett, W.; Daly, J.W. Forskolin: Unique diterpene activator of adenylate cyclase in membranes and in intact cells. *Proc. Natl. Acad. Sci. USA* **1981**, *78*, 3363–3367. [CrossRef]

31. Marasco, W.A.; Phan, S.H.; Krutzsch, H.; Showell, H.J.; Feltner, D.E.; Nairn, R.; Becker, E.L.; Ward, P.A. Purification and identification of formyl-methionyl-leucyl-phenylalanine as the major peptide neutrophil chemotactic factor produced by Escherichia coli. *J. Biol. Chem.* **1984**, *259*, 5430–5439. [PubMed]

32. Southgate, E.L.; He, R.L.; Gao, J.L.; Murphy, P.M.; Nanamori, M.; Ye, R.D. Identification of formyl peptides from Listeria monocytogenes and Staphylococcus aureus as potent chemoattractants for mouse neutrophils. *J. Immunol. (Baltim. Md. 1950)* **2008**, *181*, 1429–1437. [CrossRef] [PubMed]

33. Bae, Y.S.; Kim, Y.; Park, J.C.; Suh, P.G.; Ryu, S.H. The synthetic chemoattractant peptide, Trp-Lys-Tyr-Met-Val-D-Met, enhances monocyte survival via PKC-dependent Akt activation. *J. Leukoc. Biol.* **2002**, *71*, 329–338. [PubMed]

34. Katada, T.; Ui, M. Accelerated turnover of blood glucose in pertussis-sensitized rats due to combined actions of endogenous insulin and adrenergic beta-stimulation. *Biochim. Et Biophys. Acta* **1976**, *421*, 57–69. [CrossRef]

35. Perretti, M.; Ahluwalia, A.; Harris, J.G.; Goulding, N.J.; Flower, R.J. Lipocortin-1 fragments inhibit neutrophil accumulation and neutrophil-dependent edema in the mouse. A qualitative comparison with an anti-CD11b monoclonal antibody. *J. Immunol. (Baltim. Md. 1950)* **1993**, *151*, 4306–4314.

36. Bellner, L.; Thoren, F.; Nygren, E.; Liljeqvist, J.A.; Karlsson, A.; Eriksson, K. A proinflammatory peptide from herpes simplex virus type 2 glycoprotein G affects neutrophil, monocyte, and NK cell functions. *J. Immunol. (Baltim. Md. 1950)* **2005**, *174*, 2235–2241. [CrossRef] [PubMed]

37. Cussell, P.J.G.; Howe, M.S.; Illingworth, T.A.; Gomez Escalada, M.; Milton, N.G.N.; Paterson, A.W.J. The formyl peptide receptor agonist FPRa14 induces differentiation of Neuro2a mouse neuroblastoma cells into multiple distinct morphologies which can be specifically inhibited with FPR antagonists and FPR knockdown using siRNA. *PLoS ONE* **2019**, *14*, e0217815. [CrossRef]

38. Schepetkin, I.A.; Kirpotina, L.N.; Khlebnikov, A.I.; Quinn, M.T. High-throughput screening for small-molecule activators of neutrophils: Identification of novel N-formyl peptide receptor agonists. *Mol. Pharmacol.* **2007**, *71*, 1061–1074. [CrossRef]

39. Lopez-Ilasaca, M. Signaling from G-protein-coupled receptors to mitogen-activated protein (MAP)-kinase cascades. *Biochem. Pharmacol.* **1998**, *56*, 269–277. [CrossRef]

40. Goldsmith, Z.G.; Dhanasekaran, D.N. G protein regulation of MAPK networks. *Oncogene* **2007**, *26*, 3122–3142. [CrossRef]

41. Gutkind, J.S. Regulation of mitogen-activated protein kinase signaling networks by G protein-coupled receptors. *Sci. Stke Signal Transduct. Knowl. Environ.* **2000**, *2000*, re1. [CrossRef] [PubMed]

42. Vines, C.M.; Revankar, C.M.; Maestas, D.C.; LaRusch, L.L.; Cimino, D.F.; Kohout, T.A.; Lefkowitz, R.J.; Prossnitz, E.R. N-formyl peptide receptors internalize but do not recycle in the absence of arrestins. *J. Biol. Chem.* **2003**, *278*, 41581–41584. [CrossRef] [PubMed]

43. Wagener, B.M.; Marjon, N.A.; Prossnitz, E.R. Regulation of N-Formyl Peptide Receptor Signaling and Trafficking by Arrestin-Src Kinase Interaction. *PLoS ONE* **2016**, *11*, e0147442. [CrossRef] [PubMed]

44. Bufe, B.; Zufall, F. The sensing of bacteria: Emerging principles for the detection of signal sequences by formyl peptide receptors. *Biomol. Concepts* **2016**, *7*, 205–214. [CrossRef] [PubMed]

45. He, H.Q.; Ye, R.D. The Formyl Peptide Receptors: Diversity of Ligands and Mechanism for Recognition. *Mol. (BaselSwitz.)* **2017**, *22*. [CrossRef] [PubMed]

46. Ernst, S.; Lange, C.; Wilbers, A.; Goebeler, V.; Gerke, V.; Rescher, U. An annexin 1 N-terminal peptide activates leukocytes by triggering different members of the formyl peptide receptor family. *J. Immunol. (Baltim. Md. 1950)* **2004**, *172*, 7669–7676. [CrossRef]

47. Schloer, S.; Hubel, N.; Masemann, D.; Pajonczyk, D.; Brunotte, L.; Ehrhardt, C.; Brandenburg, L.O.; Ludwig, S.; Gerke, V.; Rescher, U. The annexin A1/FPR2 signaling axis expands alveolar macrophages, limits viral replication, and attenuates pathogenesis in the murine influenza A virus infection model. *Off. Publ. Fed. Am. Soc. Exp. Biol.* **2019**, *33*, 12188–12199. [CrossRef]

48. Kenakin, T. Biased Receptor Signaling in Drug Discovery. *Pharmacol. Rev.* **2019**, *71*, 267–315. [CrossRef]

49. Gesty-Palmer, D.; Luttrell, L.M. Refining efficacy: Exploiting functional selectivity for drug discovery. *Adv. Pharmacol. (San DiegoCalif.)* **2011**, *62*, 79–107. [CrossRef]
50. Qin, C.X.; May, L.T.; Li, R.; Cao, N.; Rosli, S.; Deo, M.; Alexander, A.E.; Horlock, D.; Bourke, J.E.; Yang, Y.H.; et al. Small-molecule-biased formyl peptide receptor agonist compound 17b protects against myocardial ischaemia-reperfusion injury in mice. *Nat. Commun.* **2017**, *8*, 14232. [CrossRef]
51. Meyer, A.; Laverny, G.; Bernardi, L.; Charles, A.L.; Alsaleh, G.; Pottecher, J.; Sibilia, J.; Geny, B. Mitochondria: An Organelle of Bacterial Origin Controlling Inflammation. *Front. Immunol.* **2018**, *9*, 536. [CrossRef] [PubMed]
52. Rabiet, M.J.; Huet, E.; Boulay, F. Human mitochondria-derived N-formylated peptides are novel agonists equally active on FPR and FPRL1, while Listeria monocytogenes-derived peptides preferentially activate FPR. *Eur. J. Immunol.* **2005**, *35*, 2486–2495. [CrossRef] [PubMed]
53. Schrage, R.; De Min, A.; Hochheiser, K.; Kostenis, E.; Mohr, K. Superagonism at G protein-coupled receptors and beyond. *Br. J. Pharmacol.* **2016**, *173*, 3018–3027. [CrossRef] [PubMed]
54. Rasmussen, S.G.; DeVree, B.T.; Zou, Y.; Kruse, A.C.; Chung, K.Y.; Kobilka, T.S.; Thian, F.S.; Chae, P.S.; Pardon, E.; Calinski, D.; et al. Crystal structure of the beta2 adrenergic receptor-Gs protein complex. *Nature* **2011**, *477*, 549–555. [CrossRef]
55. Rasmussen, S.G.; Choi, H.J.; Fung, J.J.; Pardon, E.; Casarosa, P.; Chae, P.S.; Devree, B.T.; Rosenbaum, D.M.; Thian, F.S.; Kobilka, T.S.; et al. Structure of a nanobody-stabilized active state of the beta(2) adrenoceptor. *Nature* **2011**, *469*, 175–180. [CrossRef] [PubMed]
56. Pyne, N.J.; Pyne, S. Receptor tyrosine kinase-G-protein-coupled receptor signalling platforms: Out of the shadow? *Trends Pharmacol. Sci.* **2011**, *32*, 443–450. [CrossRef] [PubMed]
57. Eishingdrelo, H.; Kongsamut, S. Minireview: Targeting GPCR Activated ERK Pathways for Drug Discovery. *Curr. Chem. Genom. Transl. Med.* **2013**, *7*, 9–15. [CrossRef] [PubMed]
58. Malhotra, D.; Shin, J.; Solnica-Krezel, L.; Raz, E. Spatio-temporal regulation of concurrent developmental processes by generic signaling downstream of chemokine receptors. *eLife* **2018**, *7*. [CrossRef]
59. Zapadka, K.L.; Becher, F.J.; Gomes Dos Santos, A.L.; Jackson, S.E. Factors affecting the physical stability (aggregation) of peptide therapeutics. *Interface Focus* **2017**, *7*, 20170030. [CrossRef]
60. Raker, V.K.; Becker, C.; Steinbrink, K. The cAMP Pathway as Therapeutic Target in Autoimmune and Inflammatory Diseases. *Front. Immunol.* **2016**, *7*, 123. [CrossRef]
61. Smith, N.J.; Milligan, G. Allostery at G protein-coupled receptor homo- and heteromers: Uncharted pharmacological landscapes. *Pharmacol. Rev.* **2010**, *62*, 701–725. [CrossRef] [PubMed]
62. Kasai, R.S.; Suzuki, K.G.; Prossnitz, E.R.; Koyama-Honda, I.; Nakada, C.; Fujiwara, T.K.; Kusumi, A. Full characterization of GPCR monomer-dimer dynamic equilibrium by single molecule imaging. *J. Cell Biol.* **2011**, *192*, 463–480. [CrossRef] [PubMed]
63. Cooray, S.N.; Gobbetti, T.; Montero-Melendez, T.; McArthur, S.; Thompson, D.; Clark, A.J.; Flower, R.J.; Perretti, M. Ligand-specific conformational change of the G-protein-coupled receptor ALX/FPR2 determines proresolving functional responses. *Proc. Natl. Acad. Sci. USA* **2013**, *110*, 18232–18237. [CrossRef] [PubMed]

5

Annexin A1/Formyl Peptide Receptor Pathway Controls Uterine Receptivity to the Blastocyst

Cristina B. Hebeda [1], Silvana Sandri [1], Cláudia M. Benis [1], Marina de Paula-Silva [1], Rodrigo A. Loiola [1], Chris Reutelingsperger [2], Mauro Perretti [3] and Sandra H. P. Farsky [1,*]

1 Department of Clinical and Toxicological Analyses, School of Pharmaceutical Sciences, University of Sao Paulo, São Paulo CEP 05508-000, Brazil; crisbh@gmail.com (C.B.H.); ssandris@gmail.com (S.S.); claudiabenis@usp.br (C.M.B.); mpsilva.bio@gmail.com (M.d.P.-S.); rodrigoazl@gmail.com (R.A.L.)
2 Faculty of Health, Medicine and Life Sciences, Part of Maastricht University Medical Center, Part of Maastricht University, 6211 LK Maastricht, The Netherlands; c.reutelingsperger@maastrichtuniversity.nl
3 The William Harvey Research Institute, Queen Mary University of London, London EC1M 6BQ, UK; m.perretti@qmul.ac.uk
* Correspondence: sfarsky@usp.br

Abstract: Embryo implantation into the uterine wall is a highly modulated, complex process. We previously demonstrated that Annexin A1 (AnxA1), which is a protein secreted by epithelial and inflammatory cells in the uterine microenvironment, controls embryo implantation in vivo. Here, we decipher the effects of recombinant AnxA1 in this phenomenon by using human trophoblast cell (BeWo) spheroids and uterine epithelial cells (Ishikawa; IK). AnxA1-treated IK cells demonstrated greater levels of spheroid adherence and upregulation of the tight junction molecules claudin-1 and *zona occludens-1*, as well as the glycoprotein mucin-1 (Muc-1). The latter effect of AnxA1 was not mediated through IL-6 secreted from IK cells, a known inducer of Muc-1 expression. Rather, these effects of AnxA1 involved activation of the formyl peptide receptors FPR1 and FPR2, as pharmacological blockade of FPR1 or FPR1/FPR2 abrogated such responses. The downstream actions of AnxA1 were mediated through the ERK1/2 phosphorylation pathway and F-actin polymerization in IK cells, as blockade of ERK1/2 phosphorylation reversed AnxA1-induced Muc-1 and claudin-1 expression. Moreover, FPR2 activation by AnxA1 induced vascular endothelial growth factor (VEGF) secretion by IK cells, and the supernatant of AnxA1-treated IK cells evoked angiogenesis in vitro. In conclusion, these data highlight the role of the AnxA1/FPR1/FPR2 pathway in uterine epithelial control of blastocyst implantation.

Keywords: mucin-1; claudin-1; *zona occludens*; ERK1/2 pathway; angiogenesis; F-actin polymerization; BeWo spheroids; Ishikawa cells

---

1. Introduction

The endometrium is a critical tissue for the establishment and maintenance of pregnancy, during which it undergoes extensive physiological changes and demonstrates extraordinary plasticity. Cyclic changes in its tissues enable the endometrium to convert to a receptive state, allowing implantation, attachment, and invasion by the embryo through the epithelium into the underlying stromal compartment [1,2]. Embryo implantation is a highly-organized process that involves a receptive epithelium as well as a competent embryo for attachment, which is finely controlled by soluble and membrane-bound factors such as cytokines, prostaglandins, growth factors, and matrix-degrading enzymes as well as adhesion molecules, and transcription factors [2,3]. In order to achieve successful implantation, crosstalk between a receptive uterus and a competent blastocyst can only occur during a limited time span, known as the "window of implantation" [4,5]. During this short period of time, which occurs approximately 6 to 10 days after ovulation in humans [1,6], the uterus undergoes structural and functional remodeling, mainly via the modulation of estradiol and progesterone. Basically, uterine receptivity is improved when estradiol levels decrease and high levels of progesterone are present [3]. Under these specific conditions, glycoproteins are expressed and highly secreted to prepare the endothelium for embryo attachment, tight junctions are reinforced, and angiogenesis occurs [2,7].

Annexin A1 (AnxA1) is a 37 KDa protein that belongs to the calcium and phospholipid-binding protein family within the annexin superfamily. A wide range of cells secretes AnxA1, including those of the innate immune system, as well as epithelial and cancer cells. AnxA1 mediates physiological processes in the body [8,9] although its secretion is highly augmented during challenging processes, such as inflammation and cancer [10,11]. The most well-recognized functions of AnxA1 are its potent anti-inflammatory and pro-resolution activities in the context of the innate immune response, during which glucocorticoids and cytokines induce synthesis and secretion of AnxA1 to halt inflammation and induce its resolution [12,13]. The binding of extracellular AnxA1 to G-protein coupled seven-domain transmembrane formylated peptide receptors (FPRs), especially type 2 (FPR2), is the most described and proven anti-inflammatory mechanism of AnxA1 [14–17]. Binding of AnxA1 to FPR2 induces rapid heterotrimeric G protein dissociation into the α and $\beta\gamma$ subunits. $\beta\gamma$ subunit downstream transduction signals, such as those mediated via phospholipase Cγ (PLCγ), result in activation of Ras family proteins and, in turn, activation of the mitogen-activated protein kinase (MAPK) pathway, particularly that of the extracellular signal-regulated kinases (ERK)-1/2. Activation of these later pathways leads to Ca^{2+} mobilization and activation of protein kinase C (PKC) [18,19].

Previous data have correlated high levels of AnxA1 in human uterine tissue during gestation and in the seminal fluid [20,21], while lower amounts of AnxA1 have been found in the amnion and placenta at delivery [22–26]. More recently, our group has linked AnxA1 to pregnancy. Specifically, AnxA1-genetically deficient mice (AnxA1$^{-/-}$) presented an increased number of blastocysts and implantation sites, resulting in an increased number of pups delivered [27]. Additionally, the uterine microenvironment of AnxA1$^{-/-}$ mice displayed an inflammatory profile, including a higher content of neutrophils and M1 macrophages as well as enhanced levels of pro-inflammatory cytokines, especially IL-6 [28]. These findings suggest that AnxA1 may play a crucial role in the maintenance of the uterine microenvironment, particularly in relation to maintenance of a receptive environment during implantation [27]. To further understand this process, in the current study we have elucidated the direct actions of AnxA1 on the initial events of blastocyst implantation in cultured human uterine epithelial cells.

2. Materials and Methods

2.1. Cell Lines

The human uterine epithelial cell line Ishikawa (IK) was purchased from Banco de Células do Rio de Janeiro. IK cells were maintained in Dulbecco's Modified Eagle Medium (DMEM; #12100046, Gibco, Carlsbad, CA, USA) supplemented with 10% heat-inactivated fetal bovine serum (FBS; #2024-06, Gibco), 2 mM L-glutamine (#25030081, Gibco) 1 mM pyruvate (#11360-070, Gibco) and 1% antibiotic solution containing streptomycin and penicillin (#15140-122, Gibco). Human umbilical vein endothelial cells (HUVECs) were donated by Dr. Ricardo José Giordano from the Chemical Institute of the University of Sao Paulo. HUVEC cells were maintained in Roswell Park Memorial Institute (RPMI) 1640 medium (#31800089, Gibco) containing 10% FBS and 1% antibiotic solution containing streptomycin and penicillin. BeWo cells were kindly donated by Professor Ana Campa from the Faculty of Pharmaceutic Sciences, University of Sao Paulo, and maintained in DMEM/F12 medium (#12500-062, Gibco) supplemented with 10% FBS and 1% antibiotic solution containing streptomycin and penicillin. All cells were maintained in an atmosphere of 5% CO_2 at 37 °C and sub-cultured every 3 days by trypsinization, if necessary.

2.2. Cell Treatments

Uterine epithelial cells were seeded in 24-well plates (Corning, New York, NY, USA) and cultured for adhesion over 18 h. Once cells had adhered, the medium was replaced and the cells were either pre-incubated with the culture medium (non-treated [NT], i.e., control) or medium supplemented with Boc-2 (1 μM; #SKU 0215276005, MP Biomedicals, Santa Ana, CA, USA), cyclosporine H (1 μM; #AG CN2 0447-M005, Adipo Gen Life Sciences, San Diego, CA, USA) or WRW_4 (1 μM; #2262, Tocris Bioscience, Bristol, UK) for 30 min. Following the pre-incubation, AnxA1 (1.35 nM; donated by Professor Chris Reutelingsperger from Faculty of Health, Medicine and Life Sciences, Maastricht University) was added to the cell culture, either in the absence or presence of inhibitors, and incubated for a time period according to the specific assay performed.

For tube formation assay, uterine epithelial cells were washed three times with warm PBS and pre-incubated with FPR inhibitors (30 min) followed by addition of AnxA1, and cultured for 18 h in RPMI 1640 medium supplemented with 1% bovine serum albumin (BSA; #A9418-10G, Sigma-Aldrich). Afterwards, the supernatant was collected in sterile conditions and used to perform the assay.

2.3. Cell Viability Assay

Uterine epithelial cells were seeded at 2.5×10^4 cells/well in 24-well plates and incubated in the absence or presence of different concentrations of AnxA1, Boc-2, cyclosporine H, or WRW_4 over either 24 or 48 h. Following the incubation period, the medium was carefully removed and 300 μL of 3-(4,5-dimethylthiazol-2-yl)-2,5-diphenyltetrazolium bromide (MTT, 0.5 mg/mL; #M5655, Sigma-Aldrich) was added in each well. Cells were maintained at 37 °C for 3 h, after which the supernatant was removed and 200 μL of dimethyl sulfoxide (DMSO; #276855, Sigma-Aldrich) was added into each well and homogenized for 15 min. Absorbance was determined using a spectrophotometer at 575 nm (SpectraMax M Series, Molecular Devices, San Jose, CA, USA). Results were expressed as the percentage of viable cells relative to NT cells (control).

2.4. Flow Cytometry

Flow cytometry experiments were performed to characterize the expression levels of AnxA1, FPR1, and FPR2 in uterine epithelial cells, as well as to investigate the role of AnxA1 on CD61, signal transducer and activator of transcription (STAT)1α, nuclear factor (NF)-κB, ERK1/2, homeobox A-10 (HOXA10), progesterone, and estrogen receptor expressions. Briefly, uterine epithelial cells were seeded at 5×10^4 cells/well and treated as mentioned above. Cells were trypsinized (#T1757, Vitrocell, Campinas, SP, BRA), washed twice in PBS containing 1% BSA (collectively referred to as PBS/BSA). To investigate the expression of AnxA1, HOXA10, progesterone, and estrogen receptors, the cells were fixed overnight at 4 °C using FACS lysing solution (#349202, BD Biosciences, San Jose, CA, USA), then washed with PBS containing 1% glycine (#01A1021.01.AG, Synth, Diadema, SP, BRA), permeabilized with Triton-X (0.001%; #T8787, Sigma-Aldrich), washed with PBS/BSA, and incubated with primary anti-human rabbit antibodies to AnxA1 (1:100; #713400, Thermo Fisher, Waltham, MA, USA), HOXA10 (1:100; #720220, Thermo Fisher), the progesterone receptor (1:500; #IM-0558, Rhea Biotech, Campinas, SP, BRA), or the estrogen receptor (1:250; #IM-0557, Rhea Biotech) overnight at 4 °C. Next, cells were washed with PBS/BSA and incubated with secondary goat anti-rabbit antibodies conjugated to Alexa Fluor 488 (1:500; #A11008 Invitrogen) for 40 min in the dark at room temperature (RT). To investigate STAT1α and ERK1/2 expressions, the cells were fixed in cold methanol for 30 min at RT, permeabilized in 0.1% Triton-X for 20 min at RT, and then incubated with primary anti-total STAT1α (1:50; #9172, Cell Signaling, Boston, MA, USA), anti-phospho-STAT1α (1:50; #9167, Cell Signaling), anti-total ERK1/2 (1:200; #ab54230, Abcam, Burlingame, CA, USA) and/or anti-phospho ERK1/2 (1:200; #ab214036, Abcam) antibodies overnight at 4 °C. After this incubation, the cells were blocked with PBS containing 2% FBS, and incubated with secondary anti-rabbit-phycoerythrin (1:200; #ab97070, Abcam) or anti-mouse-fluorescein isothiocyanate (FITC, 1:200; #ab6785, Abcam) goat antibodies for 1 h in the dark at RT. In order to analyze FPR1, FPR2, CD61 and NF-kB expression, the cells were washed twice in PBS/BSA and incubated with specific antibodies as follows: FPR1-PE (1:100; #FAB3744P BD Biosciences, Minneapolis, MN, USA), FPR2-FITC (1:100; #bs3654R FITC; Bioss, Boston, MA, USA), CD61-FITC (1:50; #555753; BD Biosciences) or NF-kB (1:100; #0465R, Biolegend, San Diego, CA, USA) for 40 min in the dark at RT. The cells were then washed and resuspended in PBS at the end of each protocol. Samples were subjected to flow cytometric analysis in a BD Accuri C6 flow cytometer taking 10,000 events into consideration and using CSampler software (BD Pharmingen, CA, USA).

2.5. *Proliferation Assay*

The proliferation assay was performed using a Live/Dead Viability/Cytotoxicity Kit for mammalian cells (#L3224, Thermo Fisher). Briefly, uterine epithelial cells were seeded in 24-well plates at a concentration of 1×10^4 cells/well and incubated in culture medium containing 0.5% BSA for 24 h prior to treatment. Following the incubation, the cells were washed and treated with medium containing either 0.5%, 2%, or 10% BSA in the absence or presence of AnxA1 (1.35 nM) for 24 or 48 h. Next, the cells were washed with PBS, trypsinized, and then incubated using the Live/Dead Viability/Cytotoxicity Kit according to the manufacturer's instructions. Samples were analyzed using a BD Accuri C6 flow cytometer and 10,000 events were considered in the analysis using CSampler software (BD Pharmingen).

2.6. *Immunofluorescence*

Uterine epithelial cells were seeded at a concentration of 5×10^4 cells/well on glass coverslips inside the wells of 24-well plates and treated as mentioned above. Cells were fixed in cold methanol for 30 min at −20 °C, after which the methanol was removed, and the cells were maintained at −20 °C until the assay was performed. Briefly, cells were washed in PBS and incubated overnight at 4 °C in the presence of anti-Muc-1 (rabbit anti-human; #bs-4763R, Bioss), anti-claudin-1 (rabbit anti-human; #ab15098, Abcam), anti-*zona occludens*-1 (ZO-1; goat anti-human; #PA5-19090, Thermo Fischer), or anti-AnxA1 primary antibodies. Following this incubation, the cells were washed with PBS/BSA and incubated with donkey anti-goat and goat anti-rabbit secondary antibodies conjugated to either Alexa Fluor 568 (#A11011, Thermo Fisher) or FITC (#A11008, Thermo Fisher), respectively. The coverslips containing the cells were removed from the 24-well plates, inverted, and mounted on glass slides in 5 μL of Vectashield (#H-1200, Vectorlabs, Burlingame, CA, USA). The slides were maintained at 4 °C and images were acquired using an Axio Zeiss microscope and analyzed with ImageJ software (NIH, Bethesda, MD, USA).

2.7. *ELISA*

Uterine epithelial cells were seeded at 3–5 $\times 10^4$ cells/well and treated as mentioned above. The supernatant from these preparations was used to quantify the expression of IL-6 (#555220 BD OptEIA, BD Biosciences Pharmingen), AnxA1 (#MBS704042, MyBiosource, San Diego, CA, USA) and vascular endothelial growth factor (VEGF; #KHG0111, Thermo Fisher) through ELISA, according to the manufacturer's instructions. Expression level results from the ELISA were expressed in terms of pg/mL.

2.8. *Trophoblast Spheroid*

The method of trophoblast spheroid growth was adapted from a previous study [29]. Briefly, in order to obtain spheroids, 50 μL agarose solution (1.5%; #A9539, Sigma-Aldrich) was added to 96-well plates. After solidifying, BeWo cells were seeded at a concentration of 1 x 10^4 cells/well and maintained in an atmosphere of 5% CO_2 at 37 °C for 72 h. Following the incubation, individual spheroids were visualized through optic microscopy. The viability of spheroids was determined using the Live/Dead Viability/Cytotoxicity Kit (Thermo Fisher).

2.9. Implantation Assay

The BeWo spheroid implantation model used here was adapted from previous studies [29–31]. Briefly, uterine epithelial cells were plated in 96-well plates at a concentration of 2.5×10^4 cells/well and, after adherence, the cells were incubated for 1 h in either the control medium or media containing Boc-2 (1 μM), cyclosporine H (1 μM), or WRW4 (1 μM), and then incubated with AnxA1 (1.35 nM) throughout the implantation time. The spheroids (one spheroid/well) were gently transferred onto adhered uterine epithelial cells and this co-culture was maintained in a humid atmosphere at 5% CO_2 and 37 °C for 2 h. Following this incubation period, the wells were filled up to the brim with culture medium and the plates were sealed with an adhesive film for microplates, inverted, and then centrifuged at 30× *g* at RT for 5 min. After centrifugation, the plates were kept inverted while they were taken from the centrifuge and examined under a Leica DMi1 inverted microscope (Leica, Shinagawa, Tokyo, Japan) for the presence of the spheroids. The spheroids that disappeared during the centrifugation process were considered to be unattached, and the results were expressed as the percentage of attached spheroids.

2.10. Confocal Microscopy

2.10.1. F-Actin Expression

Uterine epithelial cells were seeded at a concentration of 5×10^4 cells/well on a 24-well plate and then treated as mentioned above. Next, the cells were washed in PBS and fixed in 2% paraformaldehyde for 20 min at RT. The cells were then washed in PBS, incubated with rhodamine phalloidin (#R415, Thermo Fisher) for 20 min in the dark at RT, and then washed in PBS. The intensity of fluorescence was detected using high-content imaging with a GE IN Cell Analyzer 2200 (GE Healthcare Life Sciences, Chicago, IL, USA) and quantified with IN Carta^{TM} image analysis software (GE Healthcare Life Sciences).

2.10.2. AnxA1 Expression

AnxA1 expression at the implantation site was evaluated in C57bl/6 mice of 5 to 6 weeks of age. For this purpose, female mice were caged overnight with male mice (3:1) and successful mating was verified the following morning. The presence of a vaginal plug was designated as day 0.5 of gestation. The animals were maintained and bred at the Animal House at the School of Pharmaceutical Sciences, University of Sao Paulo (Brazil). Chow (Quimtia, Colombo, PR, Brazil) and water were made available to the mice *ad libitum*. All animals were housed in a temperature-controlled room (22–25 °C and 70% humidity) with a 12-h light–dark cycle. All procedures were performed according to the Brazilian Society for the Science of Laboratory Animals (SBCAL) and approved by the Institutional Animal Care and Use Committee from the Faculty of Pharmaceutical Sciences of the University of Sao Paulo (Protocol number 557).

For confocal microscopy, the uterus was removed at day 5.5 of gestation following euthanasia of the mice via isoflurane overdose. Samples of the implantation sites were fixed in 4% buffered paraformaldehyde for 72 h at 4 °C, washed in Tris-buffered saline (TBS) and incubated with AnxA1 polyclonal antibody (#713400, Thermo Fisher) or only TBS (negative control) for 24 h. Next, these tissues were incubated with goat anti-rabbit antibodies conjugated with Alexa Fluor 488 (#A11008, Thermo Fisher) and DAPI (10 μg/mL; #D9542, Sigma-Aldrich) for 4 h, at RT in the dark, and then analyzed using a Confocal Zeiss LSM-780-NLO microscope (Carl Zeiss, Jena, Germany).

2.11. Tube Formation

The tube formation assay was performed as detailed in previous studies [32,33]. Briefly, HUVEC cells were serum-starved for 24 h in RPMI 1640 medium with 1% BSA. Next, the cells were trypsinized, harvested, and plated on a 96-well plate at a density of 2.5×10^4 cells/well on 100 μL Matrigel coating (#356237, Corning). The cells were incubated for 4 h with uterine epithelial cell-free supernatant that was previously obtained according to item 2.2. HUVEC cells, maintained in RPMI 1640 containing

10% FBS, were used as the positive control (data not shown). Photomicrographs (5X magnification) were obtained using a Leica DMI1 optical microscope (Shinagawa, Tokyo, Japan), and closed units (polygons) were considered in the count. Twelve fields were count per well.

2.12. *Statistical Analyses*

The data were expressed as mean ± standard error of the mean (SEM) and comparisons were made between the experimental groups using a one-way ANOVA followed by either the Tukey test or Bonferroni test for multiple comparisons using GraphPad software version 5. A p value < 0.05 was used to denote statistically significant differences.

3. Results

3.1. *Uterine Epithelial Cells Express FPRs 1 and 2 and Secrete AnxA1*

To validate our study, we first confirmed that uterine epithelial cells express and secrete AnxA1, and express its receptors, FPR1 and FPR2 (Figure S1). The secretion of AnxA1 was not detected from other epithelial cell lineages, such as Caski and Siha cells, and low levels were detected for HeLa (Figure S1B). Additionally, the concentration-response curves demonstrated that AnxA1, Boc-2, cyclosporine H, and WRW_4 did not affect the cellular viability under any of the concentrations employed in our studies following either 24 or 48 h of incubation (Figure S2A,C–E). Moreover, AnxA1 did not alter the cellular proliferation (Figure S2B). Using these data, effective concentrations of FPR agonists and antagonists were chosen to proceed with the further investigations, specifically 1 μM of Boc-2, cyclosporine H, and WRW_4, and 1.35 nM of AnxA1.

3.2. *AnxA1 Increased the Number of Implanted Trophoblast Spheroids*

BeWo spheroids were cultured on uterine epithelial cells in order to mimic embryo implantation in vitro (Figure S3A). Of note, BeWo spheroid viability was confirmed by observation of both a higher number of viable (green; Figure S3B,D) and lower number of dead cells (red; Figure S3C,D).

The in vitro implantation assay showed that NT (i.e., control) uterine epithelial cells demonstrated 36.4% spheroid adherence after 2 h of incubation. Similar adherence is observed when cells were treated with Boc-2, cyclosporine H or WRW_4. AnxA1 treatment evoked a large increase in spheroid adherence, as 85.4% of the spheroids attached to the uterine epithelial cells following the treatment. This effect was reversed when cells were co-incubated with either cyclosporine H or Boc-2 with AnxA1. WRW_4 did not affect the improved adherence evoked by AnxA1 (Figure 1A). A representative image of the in vitro spheroid adhesion assay is shown in Figure 1B.

3.3. *AnxA1 Induced Muc-1 Expression in Uterine Epithelial Cells via FPR1 and FPR2*

Mucins are glycoproteins that line the surfaces of organs exposed to the external environment, including the lung, gut, eyes, and uterus [34]. It has been shown that, in humans, mucin-1 (Muc-1) acts as a scaffold and ligand for selectins present on the blastocyst in order to facilitate attachment [35,36]. The data obtained here show that, in uterine epithelial cells, AnxA1-induced expression of Muc-1 was abrogated by simultaneous incubation with Boc-2 or WRW4 (Figure 2A). Representative images of the immunofluorescence studies are shown in Figure 2B.

Since IL-6 is a key cytokine involved in blastocysts implantation and Muc-1 expression [37,38], we hypothesized that AnxA1 may control Muc-1 expression via IL-6. Indeed, we confirmed IL-6 increased Muc-1 expression by uterine epithelial cells (Figure 2C,D), although IL-6 secretion was impaired in cells treated with AnxA1 (Figure 2E). Taken together, these data show an IL-6-independent mechanism by which AnxA1 impacts on Muc-1 expression, and that such actions are mediated via FPRs.

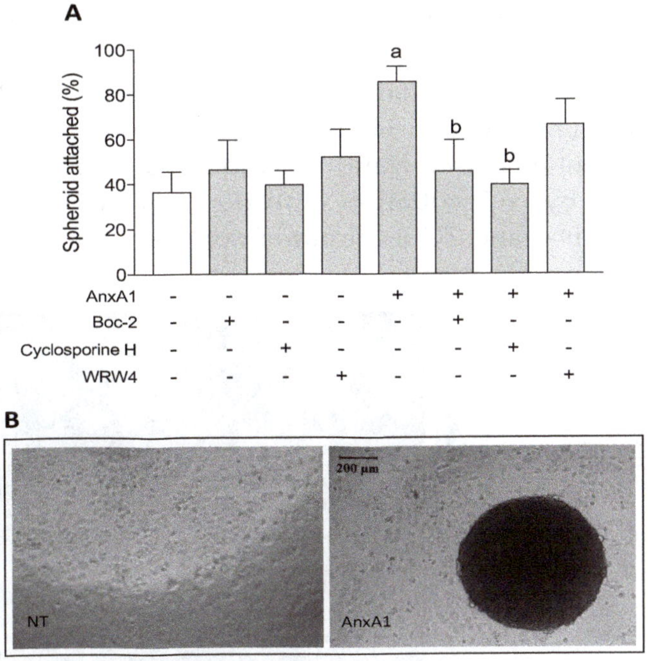

Figure 1. AnxA1 increased BeWo spheroid attachment via FPR1 on uterine epithelial cells. Uterine epithelial cells were treated with FPRs antagonists during 1 h and AnxA1 was added with spheroids. Uterine epithelial and spheroids were co-cultured during 2 h, and the percentage of adhered spheroids were calculated and considered as attached. (−) means absence and (+) means presence of treatments (**A**). Representative image of non-treated (NT) and AnxA1-treated uterine epithelial cells containing or not a spheroid is shown in (**B**). The data are expressed as mean ± standard error of 10 experiments. ^{a} $p < 0.05$ vs. NT; ^{b} $p < 0.05$ vs. AnxA1.

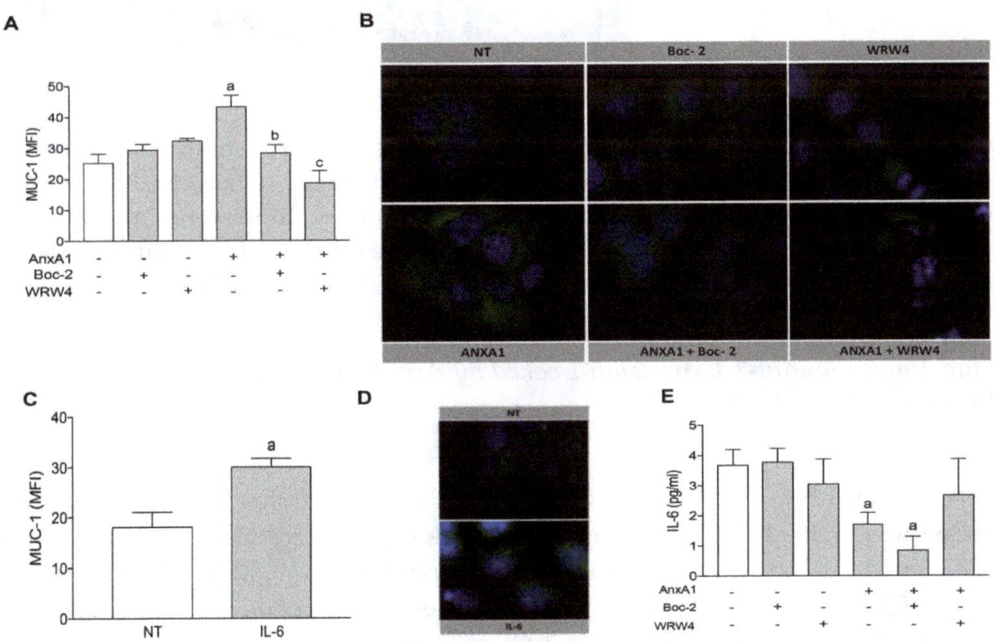

Figure 2. AnxA1 controlled Muc-1 expression on uterine epithelial cells via FPR1/FPR2, independent of IL-6 secretion. Muc-1 expression on uterine epithelial cells was determined 24 h after incubations (**A**,**B**). Muc-1 expression was determined 24 h after IL-6 treatment (**C**). IL-6 secretion was determined in the supernatant of uterine epithelial cells 24 h after treatments (**D**). (−) means absence and (+) means presence of treatments. The data are expressed as mean ± standard error of mean of three to five independent experiments. ^{a} $p < 0.05$ vs. NT; ^{b,c} $p < 0.05$ vs. AnxA1.

3.4. AnxA1 Induced Claudin-1 and Zona Occludens-1 Expression in Uterine Epithelial Cells via FPR1 and FPR2

Claudin-1 is a member of the junctional complex and is associated with cytoplasmic plaque proteins in the *zona occludens* (ZO), which is crucial for maintaining the integrity of the uterine epithelium [7]. AnxA1 treatment increased claudin-1 and ZO-1 expression in uterine epithelial cells, and this effect was abrogated by co-incubation with Boc-2 or WRW_4 (Figure 3A,B, respectively). Representative images of claudin-1 and ZO-1 immunofluorescence are shown in Figure 3.

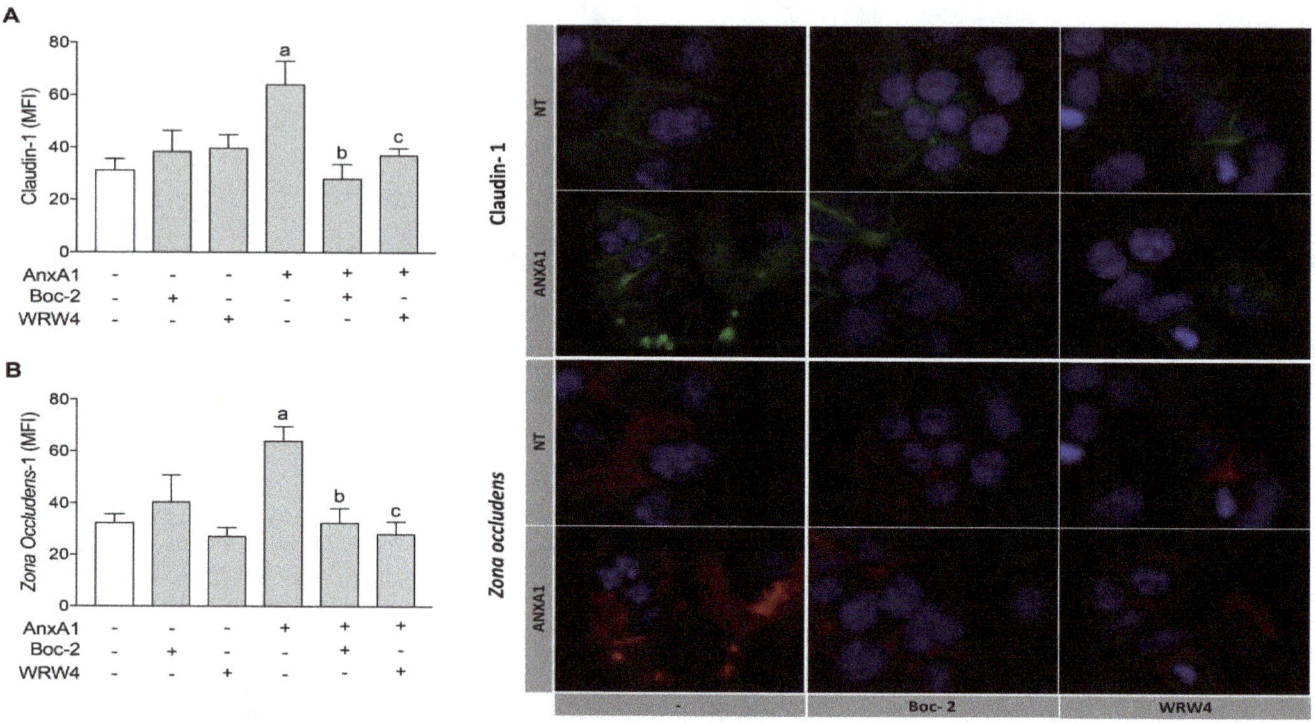

Figure 3. AnxA1 controlled claudin-1 and ZO-1 expressions on uterine epithelial cells via FPR1 and FPR2. Claudin-1 (**A**) and ZO-1 (**B**) expressions on uterine epithelial cells were determined 24 h after incubations. Representative images of claudin-1 and ZO-1 immunofluorescence are shown. (−) means absence and (+) means presence of treatments. The data are expressed as mean ± standard error of mean of three to five independent experiments. ^{a} $p < 0.05$ vs. NT; ^{b} $p < 0.01$ and ^{c} $p < 0.05$ vs. AnxA1.

3.5. Increased Muc-1 and Claudin-1 Expression Evoked by AnxA1 Was Supported by the MAPK Pathway Activation in Uterine Epithelial Cells

ERK1/2, STAT1α, and NF-κB are some of the signaling molecules connected to Muc-1, claudin-1, and ZO-1 expression [31,39–41]. Furthermore, AnxA1 produces its actions mainly through the MAPK, JAK/STAT, and NF-κB signaling transduction pathways [9,42,43]. We observed that AnxA1 treatment increased ERK1/2 phosphorylation in uterine epithelial cells (Figure 4A) but did not modify STAT1α phosphorylation (Figure 4B) or the p65 subunit of NF-κB (Figure 4C) compared to the control cells (NT). Moreover, our findings showed that the pharmacological blockade of ERK1/2 phosphorylation, by pre-incubation with PD98059, abrogated the increment of Muc-1 and claudin-1 expressions induced by AnxA1 in uterine epithelial cells (Figure 4D,F, respectively). In contrast, pre-incubation of uterine epithelial cells with PD98059 did not block the ZO-1 expression evoked by AnxA1 (Figure 4H). Representative images of Muc-1, claudin-1 and ZO-1 immunofluorescence are shown in Figure 4E,G,I, respectively.

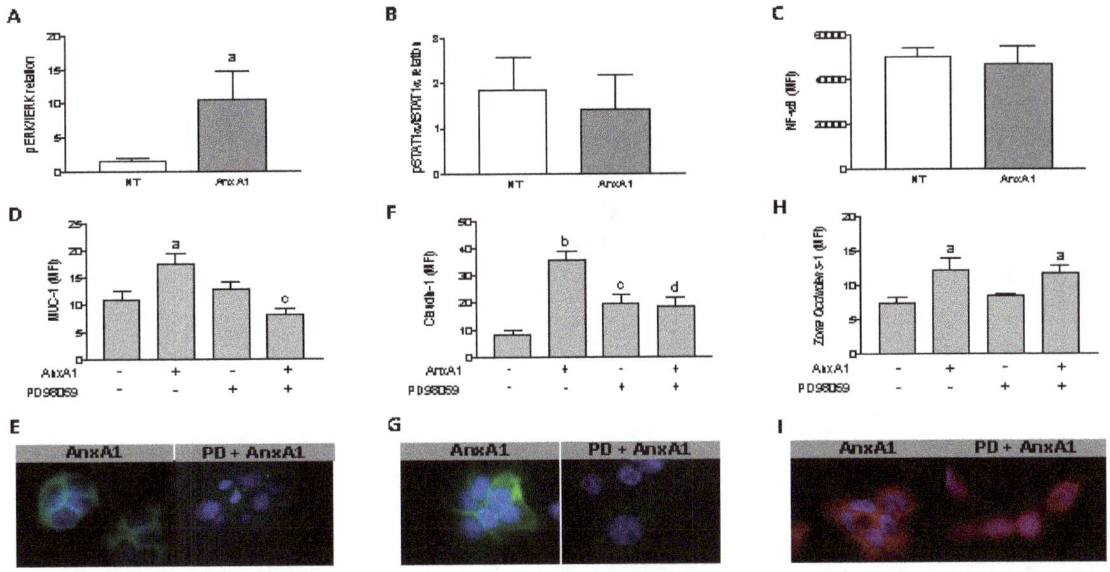

Figure 4. AnxA1 controlled Muc-1 and Claudin-1 expressions on uterine epithelial cells via ERK1/2 phosphorylation. The effect of AnxA1 on ERK (**A**) and STAT1α (**B**) phosphorylation and NF-κB expression (**C**) were investigated by flow cytometry. The inhibition on ERK1/2, evoked by PD98059 incubation, was investigated on Muc-1 (**D**), Claudin-1 (**F**) and ZO-1 (**H**) expressions using immunofluorescence. Representative images of Muc-1, Claudin-1 and ZO-1 are shown in (**E**), (**G**) and (**I**), respectively. (−) means absence and (+) means presence of treatments. The data are expressed as mean ± standard error of mean of three to five independent experiments. ^{a} $p < 0.05$ and ^{b} $p < 0.01$ vs. NT; ^{c} $p < 0.05$ and ^{d} $p < 0.01$ vs. AnxA1.

3.6. *AnxA1 Increased F-Actin Polymerization in Uterine Epithelial Cells via FPR1*

F-actin is connected to the ZO-1 and is important in stabilizing the tight junctions [44,45]. AnxA1 treatment increased F-actin polymerization in comparison to the control treatments (NT; dotted line), and this effect was inhibited by co-treatment with Boc-2 or cyclosporine H. Co-incubation of cells with WRW_4 and AnxA1 did not modify F-actin polymerization (Figure 5A). Representative images are shown in Figure 5B.

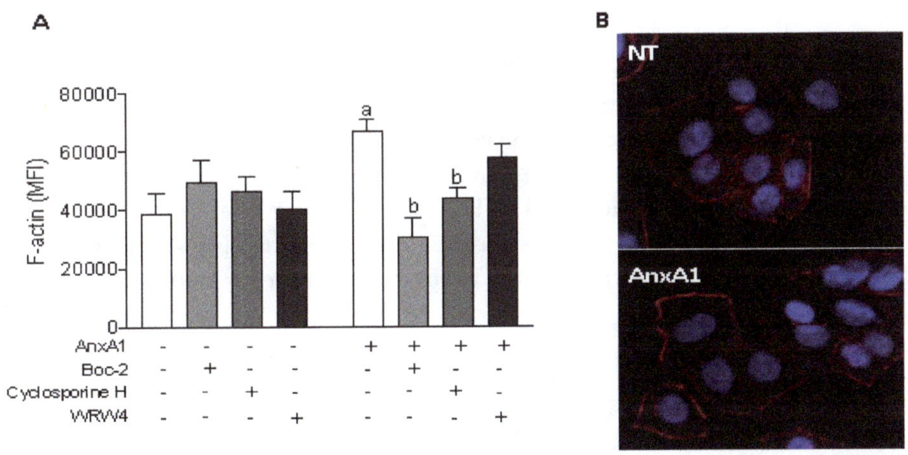

Figure 5. AnxA1 induced F-actin polymerization via FPR1. F-actin was quantified on uterine epithelial cells 2 h after incubations, using phalloidin-rhodamine by confocal microscopy. (−) means absence and (+) means presence of treatments (**A**). Representative images of F-actin polymerization are shown (**B**). The data are expressed as mean ± standard error of mean of three to five independent experiments. ^{a} $p < 0.05$ vs. NT; ^{b} $p < 0.05$ vs. AnxA1.

3.7. *AnxA1 Controls Endothelial Tube Formation and VEGF Secretion via FPR2*

Angiogenesis is a fundamental step in the implantation process, and thus, required for pregnancy to progress [46]. Therefore, we investigated the role of substances secreted by uterine epithelial cells after treatment with AnxA1 and/or FPR inhibitors on tube formation of HUVECs, referred to as in vitro angiogenesis. Cyclosporine H or WRW_4 did not alter the number of HUVEC tubes *per se*. The supernatant of uterine epithelial cells, which had previously been incubated with AnxA1 or AnxA1 plus the FPR1 antagonist cyclosporine H, did not modify the formation of tubes by HUVECs. In contrast, the supernatant of uterine epithelial cells treated with AnxA1 and WRW_4 did markedly reduce the number of tubes formed (Figure 6A). Representative images demonstrating this are depicted in Figure 6B.

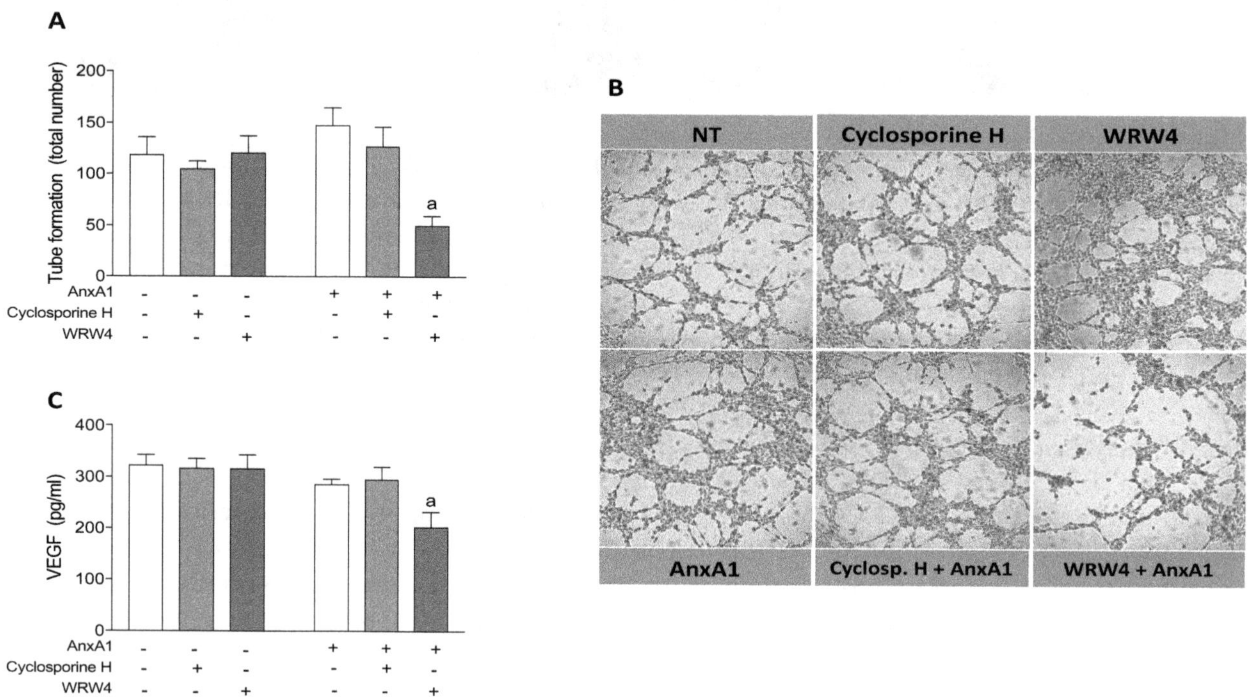

Figure 6. Supernatant of uterine epithelial cells treated with AnxA1 plus WRW_4 presented reduced levels of VEGF and reduced HUVEC tube formation. Supernatant was obtained from uterine epithelial cells previously treated with cyclosporine H or WRW_4 in the absence or presence of AnxA1. HUVEC was incubated with supernatants and tube formation was evaluated 4 h later (**A**). Representative images of tube formation are shown in (**B**). The levels of VEGF in the uterine cells supernatant were quantified 4 h after incubations (**C**). (−) means absence and (+) means presence of treatments. The data are expressed as mean ± standard error of mean of at least three to five independent experiments. [a] $p < 0.05$ vs. AnxA1.

VEGF is a fundamental growth factor in angiogenesis. Therefore, to understand the mechanisms linked to the altered tube formation observed, we quantified the levels of VEGF in the supernatant previously obtained from uterine epithelial cells incubated with AnxA1 and FPR blockers. In accordance with the results obtained from the HUVEC tube formation experiments, VEGF levels were not altered by cyclosporine H, WRW_4, AnxA1, or co-incubation with cyclosporine H and AnxA1. In contrast, significantly lower levels of VEGF were quantified in the supernatants from uterine epithelial cells treated with both WRW_4 and AnxA1 (Figure 6C).

3.8. AnxA1 Is Physiologically Expressed on Uterine Epithelial Cells and the Blastocyst during In vivo Implantation

To corroborate our results on the functional involvement of AnxA1 in uterine receptivity and embryo implantation, the expression of AnxA1 was monitored by confocal microscopy of implantation sites obtained from C57bl/6 mice on gestational day 5.5 (Figure 7A). The data obtained showed that AnxA1 was broadly expressed by both the luminal epithelium and the embryo (Figure 7B). A representative image of the negative control is shown in Figure S5.

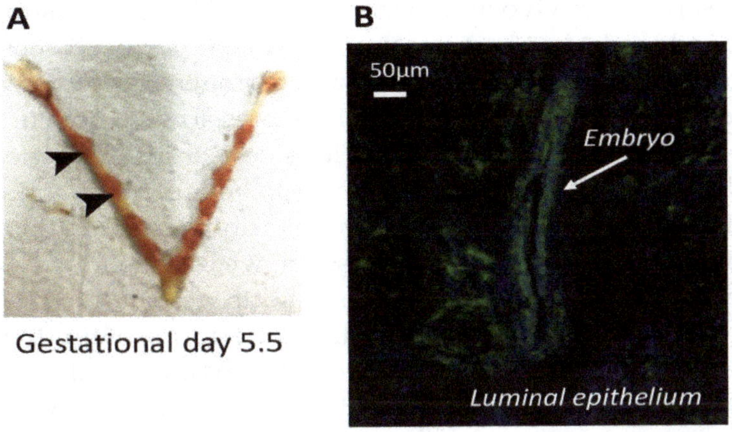

Figure 7. AnxA1 is expressed at embryo-implantation site in vivo. Implantations sites were obtained from C57bl/6 at gestational day 5.5 (**A**). Representative image of AnxA1 expression at implantation site (arrow) and luminal tissue by confocal microscopy is shown (**B**).

4. Discussion

Failure of the blastocyst to implant in the uterine wall is a putative cause of unsuccessful pregnancy. Although large volumes of data have been published regarding this process, research thus far has failed to produce an effective treatment that supports the uterus' receptivity to implantation. Blastocyst apposition and attachment to the uterus involve the actions of a diverse range of molecules and intracellular signaling processes that are not unique in themselves, but play unique roles in each step of the process [2]. Moreover, ethical concerns limit the ability for human studies, and thus, our current understanding of human embryo–endometrium interactions is limited, although experimental models and in vitro cell systems have been designed and are currently being used to help understand the complexity of this phenomenon [29,30]. About 30% of human pregnancies end in miscarriages, the vast majority of which occur in the early phase of gestation [47], thus emphasizing the crucial importance of further studies in this field. Here, we highlight the role that the AnxA1 protein plays, via activation of G-protein coupled receptors (specifically FPRs), in preparing the uterine epithelium for blastocyst implantation using an in vitro model. We found that, indeed, AnxA1 application favors blastocyst attachment by inducing the expression of proteins by the uterine epithelium that control the paracellular flux, structure, and adhesiveness of the uterine wall, as well as promoting angiogenesis.

Progesterone binds to nuclear receptors in the cells of the uterine epithelium and stroma, and activates the transcription factor HOXA10 to induce cell proliferation in the uterine wall, a process which is required for implantation of the blastocyst [48,49]. It has been shown the HOXA10 gene promoter contains a progesterone-responsive element, implying that HOXA10 is a direct target of the progesterone receptor [50,51]. Moreover, activation of the transcription factor HOXA10 induces the expression of the adhesion molecule CD61, a pivotal beta-3 integrin for blastocyst attachment in the epithelium [52]. In humans, high plasma progesterone levels have been associated with increased levels of CD61 in uterine biopsy samples, suggesting a role for CD61 as a biomarker of uterine receptivity [53]. A role for AnxA1 in the fundamental progesterone/HOXA10/CD61 pathway of blastocyst implantation was initially disregarded after studies demonstrated that AnxA1 treatment failed to induce intracellular expression of the progesterone receptor or HOXA10 (Figure S4), epithelial proliferation (Figure S2) or expression of CD61 (Figure S4).

Conversely, our data show that AnxA1, via activation of FPRs, induces expression of Muc-1. Muc-1 is a glycoprotein that is highly expressed in the receptive human endometrium and, subsequently, is removed from the apical epithelium at the location of implantation, while continuing to be expressed by cells neighboring the implantation site [2,54–56]. Moreover, abnormal Muc-1 expression has been detected in women suffering from recurrent miscarriages or fertility problems [57]. Muc-1 expression is controlled by various molecules, such as IL-6 [37,38] and high levels of progesterone, either alone or in combination with estradiol in vivo and in vitro [58,59]. Here, we assumed that other pathways are involved in AnxA1-induced Muc-1 expression, as AnxA1 reduced secretion of IL-6 by uterine epithelial cells while no alterations of progesterone and estrogen receptors were observed in this cell line. Data regarding the action(s) of AnxA1 on mucin expression are not currently available, and the results presented here may help further investigations concerning the effects of AnxA1 on Muc-1 expression in other epithelial cells, since Muc-1 is known to be present on the surface of most epithelial cells and to be involved in processes such as microbe invasion, inflammation, and fibrosis [60,61].

Similar to the results observed for Muc-1 expression, we detected increase in claudin-1 and ZO-1 expression following treatment with AnxA1, and these effects were blocked by co-treatment with FPR1 and FPR2 antagonists. Claudin-1 is a transmembrane adhesion protein that binds to cytoplasmatic adapter proteins of the ZO-1 and forms tight junctions [62–64]. Crucial roles for both proteins have been described in regulating transepithelial permeability to small molecules and ions [63,65], as well as cell growth and differentiation under various conditions [42,66–69]. Recently, claudin-1 and ZO-1 have been revealed to be key components involved in uterine receptivity, as tight junctions become reinforced during blastocyst implantation [7,70]. Here, we show that the AnxA1/FPR1/2 pathway contributes to strengthening claudin-1 and ZO-1 between uterine epithelial cells, suggesting a role by AnxA1 in controlling paracellular flow. Accordingly, the reinforcement of tight junctions contributes to the retention of uterine secretions within the lumen of the uterus, which is important for nourishing the embryo and favoring implantation [7,70,71].

The downstream effects of AnxA1 on intracellular pathways in uterine epithelial cells involve MAPK signaling, but not engagement of the transcription factors STAT1α or NF-κB. Increased ERK1/2 phosphorylation was detected following AnxA1 treatment, while inhibition of ERK1/2 phosphorylation reverted the changes in expression of Muc-1 and claudin-1 induced by this AnxA1 treatment. In fact, it has been shown that both Muc-1 and claudin-1 are pre-transcriptionally induced via the MAPK/ERK pathway [72–74]. Furthermore, MAPK signaling is a common pathway implicated in various actions induced by AnxA1 [9,13,75,76]. However, AnxA1 controls ZO-1 expression independently of ERK1/2 phosphorylation, potentially via improving F-actin polymerization as interactions between ZO-1 and the actin cytoskeleton have been reported in epithelial cells [44,77]. The pivotal role of AnxA1 on cytoskeletal reorganization of epithelial cells has been demonstrated under physiological conditions, and a deficiency of AnxA1 expression with consequent destabilization of focal adhesions and tight junctions has been implicated as a possible mechanism of various diseases [78–83].

Angiogenesis is crucial for successful implantation and placentation, and is critical during decidualization [46]. Angiogenesis begins early in the course of implantation and is supported by pro-angiogenic molecules [84]. The role of AnxA1 in angiogenesis has been shown in different models of cancer [85,86]. Moreover, the N-terminal AnxA1 peptide $Ac_{2\text{-}26}$ has been demonstrated to increase endothelial tube formation by increasing proliferation, migration, and actin polymerization in a manner similar to that induced by VEGF-A [83]. Our data show AnxA1, via FPR2, induces the secretion of angiogenesis factors by uterine epithelial cells, resulting in HUVEC tube formation. In our model, VEGF was identified as one of these angiogenic factors as its concentrations were reduced if uterine epithelial cells were treated with FPR2 antagonist in association with AnxA1. This is in line with recent work by Ferraro and co-workers [87], who demonstrated a functional link between AnxA1 and reparative macrophage phenotype in settings of heart failure, consequent to VEGF release from the immune cell. Therefore, we conclude that AnxA1, via FPR2, induced epithelial secretion of VEGF which, subsequently, modulated angiogenesis in the endothelium, demonstrating a complementary mechanism by which AnxA1 is able to support embryo implantation.

The direct action of AnxA1 on uterus epithelium demonstrated here seems to sound controversial to the high number of pups delivered by AnxA1 knockout mice [27]. In addition to the differences found in epithelium of mice and humans, we infer the exacerbated inflammation detected in the uterus microenvironment in AnxA1-deficient mice may favor the implantation process. It is known that elevated levels of pro-inflammatory cytokines, such as IL-6, favor the implantation of blastocysts [37]. Therefore, we infer AnxA1 is pivotal in the implantation phase, by controlling the inflammation that maintains the microenvironment to support a compatible blastocyst implantation, and also, by inducing the required signaling in the epithelium to trigger the adhesive properties.

In conclusion, this study unveils the existence of an intricate mechanism by which AnxA1 controls embryo implantation through regulating expressions and functions of key molecules linked to uterine receptivity, integrity, and angiogenesis. AnxA1 interacts with FPRs to activate members of MAP kinases and modulate the epithelial cytoskeleton, resulting in a uterine environment conducive for embryo implantation in the epithelium. Moreover, AnxA1 is also connected to the dynamic interplay between the uterine epithelium and endothelium, crucial for embryo implantation, posterior decidualization, and consequently, successful pregnancy.

Author Contributions: C.B.H., S.S., R.A.L., M.d.P.-S., C.R., M.P., and S.H.P.F. were involved in the conceptualization of the study; C.B.H., S.S., C.M.B., R.A.L., and S.H.P.F. collected, analyzed and interpreted the data; C.B.H., S.S., and S.H.P.F. wrote the manuscript; S.H.P.F. was the project administrator; S.H.P.F. acquired the funding acquisition. All authors critically reviewed the paper and approved its final version. All authors have read and agreed to the published version of the manuscript.

Acknowledgments: The authors acknowledge Mario Costa Cruz and Iuri Cordeiro Valadão from the Core Facility for Support Research (CEFAP, USP) for contributions on confocal protocol discussions.

References

1. Karizbodagh, M.P.; Rashidi, B.; Sahebkar, A.; Masoudifar, A.; Mirzaei, H. Implantation Window and Angiogenesis. *J. Cell. Biochem.* **2017**, *118*, 4141–4151. [CrossRef] [PubMed]
2. Aplin, J.D.; Ruane, P.T. Embryo-epithelium interactions during implantation at a glance. *J. Cell Sci.* **2017**, *130*, 15–22. [CrossRef] [PubMed]
3. Kim, S.-M.; Kim, J.-S. A Review of Mechanisms of Implantation. *Dev. Reprod.* **2017**, *21*, 351–359. [CrossRef] [PubMed]
4. Psychoyos, A. Uterine Receptivity for Nidation. *Ann. N. Y. Acad. Sci.* **1986**, *476*, 36–42. [CrossRef]
5. Matsumoto, H. Molecular and cellular events during blastocyst implantation in the receptive uterus: Clues from mouse models. *J. Reprod. Dev.* **2017**, *63*, 445–454. [CrossRef]
6. Su, R.W.; Fazleabas, A.T. Implantation and establishment of pregnancy in human and nonhuman primates. In *Advances in Anatomy Embryology and Cell Biology*; Springer: Berlin/Heidelberg, Germany, 2015; Volume 216, pp. 189–213.
7. Grund, S.; Grümmer, R. Direct cell–cell interactions in the endometrium and in endometrial pathophysiology. *Int. J. Mol. Sci.* **2018**, *19*, 2227. [CrossRef]
8. Machado, I.D.; Spatti, M.; Hastreiter, A.; Santin, J.R.; Fock, R.A.; Gil, C.D.; Oliani, S.M.; Perretti, M.; Farsky, S.H.P. Annexin A1 Is a Physiological Modulator of Neutrophil Maturation and Recirculation Acting on the CXCR4/CXCL12 Pathway. *J. Cell. Physiol.* **2016**, *231*, 2418–2427. [CrossRef]
9. Barbosa, C.M.V.; Fock, R.A.; Hastreiter, A.A.; Reutelingsperger, C.; Perretti, M.; Paredes-Gamero, E.J.; Farsky, S.H.P. Extracellular annexin-A1 promotes myeloid/granulocytic differentiation of hematopoietic stem/progenitor cells via the Ca2+/MAPK signalling transduction pathway. *Cell Death Discov.* **2019**, *5*, 1–11. [CrossRef]
10. Moraes, L.A.; Ampomah, P.B.; Lim, L.H.K. Annexin A1 in inflammation and breast cancer: A new axis in the tumor microenvironment. *Cell Adhes. Migr.* **2018**, *12*, 417–423. [CrossRef]
11. Fu, Z.; Zhang, S.; Wang, B.; Huang, W.; Zheng, L.; Cheng, A. Annexin A1: A double-edged sword as novel cancer biomarker. *Clin. Chim. Acta* **2020**, *504*, 36–42. [CrossRef]

12. Perretti, M.; D'Acquisto, F. Annexin A1 and glucocorticoids as effectors of the resolution of inflammation. *Nat. Rev. Immunol.* **2009**, *9*, 62–70. [CrossRef] [PubMed]
13. Purvis, G.S.D.; Solito, E.; Thiemermann, C. Annexin-A1: Therapeutic potential in microvascular disease. *Front. Immunol.* **2019**, *10*, 938. [CrossRef] [PubMed]
14. Bürli, R.W.; Xu, H.; Zou, X.; Muller, K.; Golden, J.; Frohn, M.; Adlam, M.; Plant, M.H.; Wong, M.; McElvain, M.; et al. Potent hFPRL1 (ALXR) agonists as potential anti-inflammatory agents. *Bioorg. Med. Chem. Lett.* **2006**, *16*, 3713–3718. [CrossRef] [PubMed]
15. Maderna, P.; Cottell, D.C.; Toivonen, T.; Dufton, N.; Dalli, J.; Perretti, M.; Godson, C. FPR2/ALX receptor expression and internalization are critical for lipoxin A 4 and annexin-derived peptide-stimulated phagocytosis. *FASEB J.* **2010**, *24*, 4240–4249. [CrossRef] [PubMed]
16. Corminboeuf, O.; Leroy, X. FPR2/ALXR agonists and the resolution of inflammation. *J. Med. Chem.* **2015**, *58*, 537–559. [CrossRef] [PubMed]
17. Hughes, E.L.; Becker, F.; Flower, R.J.; Buckingham, J.C.; Gavins, F.N.E. Mast cells mediate early neutrophil recruitment and exhibit anti-inflammatory properties via the formyl peptide receptor 2/lipoxin A4 receptor. *Br. J. Pharmacol.* **2017**, *174*, 2393–2408. [CrossRef]
18. Migeotte, I.; Communi, D.; Parmentier, M. Formyl peptide receptors: A promiscuous subfamily of G protein-coupled receptors controlling immune responses. *Cytokine Growth Factor Rev.* **2006**, *17*, 501–519. [CrossRef]
19. Bena, S.; Brancaleone, V.; Wang, J.M.; Perretti, M.; Flower, R.J. Annexin A1 interaction with the FPR2/ALX receptor: Identification of distinct domains and downstream associated signaling. *J. Biol. Chem.* **2012**, *287*, 24690–24697. [CrossRef]
20. Romisch, J.; Schuler, E.; Bastian, B.; Burger, T.; Dunkel, F.G.; Schwinn, A.; Hartmann, A.A.; Paques, E.P. Annexins I to VI: Quantitative determination in different human cell types and in plasma after myocardial infarction. *Blood Coagul. Fibrinolysis* **1992**, *3*, 11–17. [CrossRef]
21. Sun, M.; Liu, Y.; Gibb, W. Distribution of annexin I and II in term human fetal membranes, decidua and placenta. *Placenta* **1996**, *17*, 181–184. [CrossRef]
22. Aalberts, M.; van Dissel-Emiliani, F.M.; van Adrichem, N.P.; van Wijnen, M.; Wauben, M.H.; Stout, T.A.; Stoorvogel, W. Identification of Distinct Populations of Prostasomes That Differentially Express Prostate Stem Cell Antigen, Annexin A1, and GLIPR2 in Humans1. *Biol. Reprod.* **2012**, *86*, 82. [CrossRef] [PubMed]
23. Bennett, P.; Slater, D.; Berger, L.; Moore, G. The expression of phospholipase A2 and lipocortins (annexins) I, II and V in human fetal membranes and placenta in association with labour. *Prostaglandins* **1994**, *48*, 81–90. [CrossRef]
24. Gavins, F.N.E.; Hickey, M.J. Annexin A1 and the regulation of innate and adaptive immunity. *Front. Immunol.* **2012**, *3*, 3. [CrossRef] [PubMed]
25. Lynch-Salamon, D.I.; Everson, W.V.; Myatt, L. Decrease in annexin I messenger ribonucleic acid expression in human amnion with labor. *Am. J. Obstet. Gynecol.* **1992**, *167*, 1657–1663. [CrossRef]
26. Myatt, L.; Hirth, J.; Everson, W.V. Changes in annexin (lipocortin) content in human amnion and chorion at parturition. *J. Cell. Biochem.* **1992**, *50*, 363–373. [CrossRef] [PubMed]
27. Hebeda, C.B.; Machado, I.D.; Reif-Silva, I.; Moreli, J.B.; Oliani, S.M.; Nadkarni, S.; Perretti, M.; Bevilacqua, E.; Farsky, S.H.P. Endogenous annexin A1 (AnxA1) modulates early-phase gestation and offspring sex-ratio skewing. *J. Cell. Physiol.* **2018**, *233*, 6591–6603. [CrossRef]
28. Prins, J.R.; Gomez-Lopez, N.; Robertson, S.A. Interleukin-6 in pregnancy and gestational disorders. *J. Reprod. Immunol.* **2012**, *95*, 1–14. [CrossRef]
29. Huang, X.; Liu, H.; Li, R. Prostaglandin E2 promotes BeWo spheroids implantation in RL95-2 cell monolayers. *Gynecol. Endocrinol.* **2017**, *33*, 548–552. [CrossRef]
30. Miyazaki, Y.; Horie, A.; Tani, H.; Ueda, M.; Okunomiya, A.; Suginami, K.; Kondoh, E.; Baba, T.; Konishi, I.; Shinomura, T.; et al. Versican V1 in human endometrial epithelial cells promotes BeWo spheroid adhesion in vitro. *Reproduction* **2019**, *157*, 53–64. [CrossRef]
31. Cheng, X.; Liu, J.; Shan, H.; Sun, L.; Huang, C.; Yan, Q.; Jiang, R.; Ding, L.; Jiang, Y.; Zhou, J.; et al. Activating transcription factor 3 promotes embryo attachment via up-regulation of leukemia inhibitory factor in vitro. *Reprod. Biol. Endocrinol.* **2017**, *15*. [CrossRef]
32. Andrade, F.D.O.; Furtado, K.S.; Heidor, R.; Sandri, S.; Hebeda, C.B.; Miranda, M.L.P.; Fernandes, L.H.G.; Yamamoto, R.C.; Horst, M.A.; Farsky, S.H.P.; et al. Antiangiogenic effects of the chemopreventive agent

tributyrin, a butyric acid prodrug, during the promotion phase of hepatocarcinogenesis. *Carcinogenesis* **2019**, *40*, 979–988. [CrossRef] [PubMed]

33. DeCicco-Skinner, K.L.; Henry, G.H.; Cataisson, C.; Tabib, T.; Curtis Gwilliam, J.; Watson, N.J.; Bullwinkle, E.M.; Falkenburg, L.; O'Neill, R.C.; Morin, A.; et al. Endothelial cell tube formation assay for the in vitro study of angiogenesis. *J. Vis. Exp.* **2014**. [CrossRef] [PubMed]
34. Linden, S.K.; Sutton, P.; Karlsson, N.G.; Korolik, V.; McGuckin, M.A. Mucins in the mucosal barrier to infection. *Mucosal Immunol.* **2008**, *1*, 183–197. [CrossRef] [PubMed]
35. Genbacev, O.D.; Prakobphol, A.; Foulk, R.A.; Krtolica, A.R.; Ilic, D.; Singer, M.S.; Yang, Z.Q.; Kiessling, L.L.; Rosen, S.D.; Fisher, S.J. Trophoblast L-selectin-mediated adhesion at the maternal-fetal interface. *Science* **2003**, *299*, 405–408. [CrossRef] [PubMed]
36. Carson, D.D.; Julian, J.A.; Lessey, B.A.; Prakobphol, A.; Fisher, S.J. MUC1 is a scaffold for selectin ligands in the human uterus. *Front. Biosci.* **2006**, *11*, 2903–2908. [CrossRef]
37. Singh, M.; Chaudhry, P.; Asselin, E. Bridging endometrial receptivity and implantation: Network of hormones, cytokines, and growth factors. *J. Endocrinol.* **2011**, *210*, 5–14. [CrossRef]
38. Li, Y.Y.; Hsieh, L.L.; Tang, R.P.; Liao, S.K.; Yeh, K.Y. Macrophage-derived interleukin-6 up-regulates MUC1, but down-regulates MUC2 expression in the human colon cancer HT-29 cell line. *Cell. Immunol.* **2009**, *256*, 19–26. [CrossRef]
39. Brayman, M.J.; Dharmaraj, N.; Lagow, E.; Carson, D.D. MUC1 expression is repressed by protein inhibitor of activated signal transducer and activator of transcription-y. *Mol. Endocrinol.* **2007**, *21*, 2725–2737. [CrossRef]
40. Theodoropoulos, G.; Carraway, K.L. Molecular signaling in the regulation of mucins. *J. Cell. Biochem.* **2007**, *102*, 1103–1116. [CrossRef]
41. Shigetomi, K.; Ikenouchi, J. Regulation of the epithelial barrier by post-translational modifications of tight junction membrane proteins. *J. Biochem.* **2018**, *163*, 265–272. [CrossRef]
42. Pupjalis, D.; Goetsch, J.; Kottas, D.J.; Gerke, V.; Rescher, U. Annexin A1 released from apoptotic cells acts through formyl peptide receptors to dampen inflammatory monocyte activation via JAK/STAT/SOCS signalling. *EMBO Mol. Med.* **2011**, *3*, 102–114. [CrossRef]
43. Bist, P.; Phua, Q.H.; Shu, S.; Yi, Y.; Anbalagan, D.; Lee, L.H.; Sethi, G.; Low, B.C.; Lim, L.H.K. Annexin-A1 controls an ERK-RhoA-NFκB activation loop in breast cancer cells. *Biochem. Biophys. Res. Commun.* **2015**, *461*, 47–53. [CrossRef] [PubMed]
44. Odenwald, M.A.; Choi, W.; Buckley, A.; Shashikanth, N.; Joseph, N.E.; Wang, Y.; Warren, M.H.; Buschmann, M.M.; Pavlyuk, R.; Hildebrand, J.; et al. ZO-1 interactions with F-actin and occludin direct epithelial polarization and single lumen specification in 3D culture. *J. Cell Sci.* **2017**, *130*, 243–259. [CrossRef] [PubMed]
45. Hartsock, A.; Nelson, W.J. Adherens and tight junctions: Structure, function and connections to the actin cytoskeleton. *Biochim. Biophys. Acta - Biomembr.* **2008**, *1778*, 660–669. [CrossRef] [PubMed]
46. Torry, D.S.; Leavenworth, J.; Chang, M.; Maheshwari, V.; Groesch, K.; Ball, E.R.; Torry, R.J. Angiogenesis in implantation. *J. Assist. Reprod. Genet.* **2007**, *24*, 303–315. [CrossRef]
47. Sharkey, A.M.; Macklon, N.S. The science of implantation emerges blinking into the light. *Reprod. Biomed. Online* **2013**, *27*, 453–460. [CrossRef]
48. Modi, D.; Godbole, G. HOXA10 signals on the highway through pregnancy. *J. Reprod. Immunol.* **2009**, *83*, 72–78. [CrossRef]
49. Aikawa, S.; Deng, W.; Liang, X.; Yuan, J.; Bartos, A.; Sun, X.; Dey, S.K. Uterine deficiency of high-mobility group box-1 (HMGB1) protein causes implantation defects and adverse pregnancy outcomes. *Cell Death Differ.* **2019**, *27*, 1489–1504. [CrossRef]
50. Mazur, E.C.; Vasquez, Y.M.; Li, X.; Kommagani, R.; Jiang, L.; Chen, R.; Lanz, R.B.; Kovanci, E.; Gibbons, W.E.; DeMayo, F.J. Progesterone receptor transcriptome and cistrome in decidualized human endometrial stromal cells. *Endocrinology* **2015**, *156*, 2239–2253. [CrossRef]
51. Taylor, H.S.; Arici, A.; Olive, D.; Igarashi, P. HOXA10 is expressed in response to sex steroids at the time of implantation in the human endometrium. *J. Clin. Investig.* **1998**, *101*, 1379–1384. [CrossRef]
52. Daftary, G.S.; Troy, P.J.; Bagot, C.N.; Young, S.L.; Taylor, H.S. Direct Regulation of β 3 -Integrin Subunit Gene Expression by HOXA10 in Endometrial Cells. *Mol. Endocrinol.* **2002**, *16*, 571–579. [PubMed]
53. Chen, G.; Xin, A.; Liu, Y.; Shi, C.; Chen, J.; Tang, X.; Chen, Y.; Yu, M.; Peng, X.; Li, L.; et al. Integrins β1 and β3 are biomarkers of uterine condition for embryo transfer. *J. Transl. Med.* **2016**, *14*, 303. [CrossRef] [PubMed]

54. Aplin, J.D.; Meseguer, M.; Simón, C.; Ortíz, M.E.; Croxatto, H.; Jones, C.J.P. MUC1, glycans and the cell-surface barrier to embryo implantation. *Proc. Biochem. Soc. Trans.* **2001**, *29*, 153–156. [CrossRef]
55. Surveyor, G.A.; Gendler, S.J.; Pemberton, L.; Das, S.K.; Chakraborty, I.; Julian, J.; Pimental, R.A.; Wegner, C.C.; Dey, S.K.; Carson, D.D. Expression and steroid hormonal control of muc-1 in the mouse uterus. *Endocrinology* **1995**, *136*, 3639–3647. [CrossRef] [PubMed]
56. Gipson, I.K.; Blalock, T.; Tisdale, A.; Spurr-Michaud, S.; Allcorn, S.; Stavreus-Evers, A.; Gemzell, K. MUC16 Is Lost from the Uterodome (Pinopode) Surface of the Receptive Human Endometrium: In Vitro Evidence That MUC16 Is a Barrier to Trophoblast Adherence1. *Biol. Reprod.* **2008**, *78*, 134–142. [CrossRef] [PubMed]
57. Bastu, E.; Mutlu, M.F.; Yasa, C.; Dural, O.; Aytan, A.N.; Celik, C.; Buyru, F.; Yeh, J. Role of Mucin 1 and Glycodelin A in recurrent implantation failure. *Fertil. Steril.* **2015**, *103*, 1059–1064.e2. [CrossRef]
58. Horne, A.W.; Lalani, E.N.; Margara, R.A.; White, J.O. The effects of sex steroid hormones and interleukin-1-beta on MUC1 expression in endometrial epithelial cell lines. *Reproduction* **2006**, *131*, 733–742. [CrossRef]
59. Meseguer, M.; Aplin, J.D.; Caballero-Campo, P.; O'Connor, J.E.; Martín, J.C.; Remohí, J.; Pellicer, A.; Simón, C. Human Endometrial Mucin MUC1 Is Up-Regulated by Progesterone and Down-Regulated In Vitro by the Human Blastocyst1. *Biol. Reprod.* **2001**, *64*, 590–601. [CrossRef]
60. Ballester, B.; Milara, J.; Cortijo, J. Mucins as a New Frontier in Pulmonary Fibrosis. *J. Clin. Med.* **2019**, *8*, 1447. [CrossRef]
61. Hajishengallis, G.; Lambris, J.D. Microbial manipulation of receptor crosstalk in innate immunity. *Nat. Rev. Immunol.* **2011**, *11*, 187–200. [CrossRef]
62. Beutel, O.; Maraspini, R.; Pombo-García, K.; Martin-Lemaitre, C.; Honigmann, A. Phase Separation of Zonula Occludens Proteins Drives Formation of Tight Junctions. *Cell* **2019**, *179*, 923–936.e11. [CrossRef] [PubMed]
63. Heiskala, M.; Peterson, P.A.; Yang, Y. The roles of claudin superfamily proteins in paracellular transport. *Traffic* **2001**, *2*, 92–98. [CrossRef] [PubMed]
64. Itoh, M.; Furuse, M.; Morita, K.; Kubota, K.; Saitou, M.; Tsukita, S. Direct binding of three tight junction-associated MAGUKs, ZO-1, ZO-2, and ZO-3, with the COOH termini of claudins. *J. Cell Biol.* **1999**, *147*, 1351–1363. [CrossRef] [PubMed]
65. Kirschner, N.; Rosenthal, R.; Furuse, M.; Moll, I.; Fromm, M.; Brandner, J.M. Contribution of tight junction proteins to ion, macromolecule, and water barrier in keratinocytes. *J. Investig. Dermatol.* **2013**, *133*, 1161–1169. [CrossRef] [PubMed]
66. Bhat, A.A.; Syed, N.; Therachiyil, L.; Nisar, S.; Hashem, S.; Macha, M.A.; Yadav, S.K.; Krishnankutty, R.; Muralitharan, S.; Al-Naemi, H.; et al. Claudin-1, a double-edged sword in cancer. *Int. J. Mol. Sci.* **2020**, *21*, 569. [CrossRef] [PubMed]
67. Pope, J.L.; Bhat, A.A.; Sharma, A.; Ahmad, R.; Krishnan, M.; Washington, M.K.; Beauchamp, R.D.; Singh, A.B.; Dhawan, P. Claudin-1 regulates intestinal epithelial homeostasis through the modulation of Notch-signalling. *Gut* **2014**, *63*, 622–634. [CrossRef] [PubMed]
68. Fujita, H.; Chalubinski, M.; Rhyner, C.; Indermitte, P.; Meyer, N.; Ferstl, R.; Treis, A.; Gomez, E.; Akkaya, A.; O'Mahony, L.; et al. Claudin-1 expression in airway smooth muscle exacerbates airway remodeling in asthmatic subjects. *J. Allergy Clin. Immunol.* **2011**, *127*, 1612–1621.e8. [CrossRef]
69. Timmons, B.C.; Mitchell, S.M.; Gilpin, C.; Mahendroo, M.S. Dynamic changes in the cervical epithelial tight junction complex and differentiation occur during cervical ripening and parturition. *Endocrinology* **2007**, *148*, 1278–1287. [CrossRef]
70. Buddle, A.L.; Thompson, M.B.; Lindsay, L.A.; Murphy, C.R.; Whittington, C.M.; McAllan, B.M. Dynamic changes to claudins in the uterine epithelial cells of the marsupial Sminthopsis crassicaudata (Dasyuridae) during pregnancy. *Mol. Reprod. Dev.* **2019**, *86*, 639–649. [CrossRef] [PubMed]
71. Murphy, C.R. Uterine receptivity and the plasma membrane transformation. *Cell Res.* **2004**, *14*, 259–267. [CrossRef]
72. Khan, N.; Asif, A.R. Transcriptional regulators of claudins in epithelial tight junctions. *Mediators Inflamm.* **2015**, *2015*, 1–6. [CrossRef] [PubMed]
73. Bhat, A.A.; Uppada, S.; Achkar, I.W.; Hashem, S.; Yadav, S.K.; Shanmugakonar, M.; Al-Naemi, H.A.; Haris, M.; Uddin, S. Tight junction proteins and signaling pathways in cancer and inflammation: A functional crosstalk. *Front. Physiol.* **2019**, *9*, 10. [CrossRef]
74. Li, N.; Li, Q.; Zhou, X.D.; Kolosov, V.P.; Perelman, J.M. The effect of quercetin on human neutrophil elastase-induced mucin5AC expression in human airway epithelial cells. *Int. Immunopharmacol.* **2012**, *14*, 195–201. [CrossRef] [PubMed]

75. Lai, T.; Li, Y.; Mai, Z.; Wen, X.; Lv, Y.; Xie, Z.; Lv, Q.; Chen, M.; Wu, D.; Wu, B. Annexin a1 is elevated in patients with COPD and affects lung fibroblast function. *Int. J. COPD* **2018**, *13*, 473–486. [CrossRef] [PubMed]
76. Yang, Y.H.; Toh, M.-L.; Clyne, C.D.; Leech, M.; Aeberli, D.; Xue, J.; Dacumos, A.; Sharma, L.; Morand, E.F. Annexin 1 Negatively Regulates IL-6 Expression via Effects on p38 MAPK and MAPK Phosphatase-1. *J. Immunol.* **2006**, *177*, 8148–8153. [CrossRef] [PubMed]
77. Van Itallie, C.M.; Tietgens, A.J.; Anderson, J.M. Visualizing the dynamic coupling of claudin strands to the actin cytoskeleton through ZO-1. *Mol. Biol. Cell* **2017**, *28*, 524–534. [CrossRef]
78. McArthur, S.; Yazid, S.; Christian, H.; Sirha, R.; Flower, R.; Buckingham, J.; Solito, E. Annexin A1 regulates hormone exocytosis through a mechanism involving actin reorganization. *FASEB J.* **2009**, *23*, 4000–4010. [CrossRef]
79. Cristante, E.; McArthur, S.; Mauro, C.; Maggioli, E.; Romero, I.A.; Wylezinska-Arridge, M.; Couraud, P.O.; Lopez-Tremoleda, J.; Christian, H.C.; Weksler, B.B.; et al. Identification of an essential endogenous regulator of blood-brain barrier integrity, and its pathological and therapeutic implications. *Proc. Natl. Acad. Sci. USA* **2013**, *110*, 832–841. [CrossRef]
80. Yi, B.; Zeng, J.; Wang, G.; Qian, G.; Lu, K. Annexin A1 protein regulates the expression of PMVEC cytoskeletal proteins in CBDL rat serum-induced pulmonary microvascular remodeling. *J. Transl. Med.* **2013**, *11*, 98. [CrossRef]
81. Wang, Z.; Chen, Z.; Yang, J.; Yang, Z.; Yin, J.; Zuo, G.; Duan, X.; Shen, H.; Li, H.; Chen, G. Identification of two phosphorylation sites essential for annexin A1 in blood–brain barrier protection after experimental intracerebral hemorrhage in rats. *J. Cereb. Blood Flow Metab.* **2017**, *37*, 2509–2525. [CrossRef]
82. Park, J.C.; Baik, S.H.; Han, S.H.; Cho, H.J.; Choi, H.; Kim, H.J.; Choi, H.; Lee, W.; Kim, D.K.; Mook-Jung, I. Annexin A1 restores Aβ1-42-induced blood–brain barrier disruption through the inhibition of RhoA-ROCK signaling pathway. *Aging Cell* **2017**, *16*, 149–161. [CrossRef] [PubMed]
83. Lacerda, J.Z.; Drewes, C.C.; Mimura, K.K.O.; Zanon, C.d.F.; Ansari, T.; Gil, C.D.; Greco, K.V.; Farsky, S.H.P.; Oliani, S.M. Annexin A12-26 treatment improves skin heterologous transplantation by modulating inflammation and angiogenesis processes. *Front. Pharmacol.* **2018**, *9*, 9. [CrossRef] [PubMed]
84. Demir, R.; Yaba, A.; Huppertz, B. Vasculogenesis and angiogenesis in the endometrium during menstrual cycle and implantation. *Acta Histochem.* **2010**, *112*, 203–214. [CrossRef] [PubMed]
85. Okano, M.; Oshi, M.; Butash, A.L.; Katsuta, E.; Tachibana, K.; Saito, K.; Okayama, H.; Peng, X.; Yan, L.; Kono, K.; et al. Triple-negative breast cancer with high levels of annexin a1 expression is associated with mast cell infiltration, inflammation, and angiogenesis. *Int. J. Mol. Sci.* **2019**, *20*, 4197. [CrossRef] [PubMed]
86. Yi, M.; Schnitzer, J.E. Impaired tumor growth, metastasis, angiogenesis and wound healing in annexin A1-null mice. *Proc. Natl. Acad. Sci. USA* **2009**, *106*, 17886–17891. [CrossRef]
87. Ferraro, B.; Leoni, G.; Hinkel, R.; Ormanns, S.; Paulin, N.; Ortega-Gomez, A.; Viola, J.R.; de Jong, R.; Bongiovanni, D.; Bozoglu, T.; et al. Pro-Angiogenic Macrophage Phenotype to Promote Myocardial Repair. *J. Am. Coll. Cardiol.* **2019**, *73*, 2990–3002. [CrossRef]

6

Expression of Annexin A2 Promotes Cancer Progression in Estrogen Receptor Negative Breast Cancers

Amira F. Mahdi [1,2], Beatrice Malacrida [1,2], Joanne Nolan [1,2], Mary E. McCumiskey [3,4], Anne B. Merrigan [4], Ashish Lal [4], Shona Tormey [4], Aoife J. Lowery [5], Kieran McGourty [2,6] and Patrick A. Kiely [1,2,*]

1 Graduate Entry Medical School, University of Limerick, V94 T9PX Limerick, Ireland; amira.mahdi@ul.ie (A.F.M.); b.malacrida@qmul.ac.uk (B.M.); joanne.nolan@ul.ie (J.N.)
2 Health Research Institute, University of Limerick and Bernal Institute, University of Limerick, V94 T9PX Limerick, Ireland; kieran.mcgourty@ul.ie
3 Stokes Laboratories, Bernal Institute, University of Limerick, V94 T9PX Limerick, Ireland; mary.mccumiskey@ul.ie
4 Department of Surgery, University Hospital Limerick, V94 F858 Limerick, Ireland; anne.merrigan@hse.ie (A.B.M.); ashish.lal@hse.ie (A.L.); shona.tormey@hse.ie (S.T.)
5 Lambe Institute for Translational Research, National University of Ireland Galway, H91 TK33 Galway, Ireland; aoife.lowery@nuigalway.ie
6 Department of Chemical Sciences, University of Limerick, V94 T9PX Limerick, Ireland
* Correspondence: patrick.kiely@ul.ie

Abstract: When breast cancer progresses to a metastatic stage, survival rates decline rapidly and it is considered incurable. Thus, deciphering the critical mechanisms of metastasis is of vital importance to develop new treatment options. We hypothesize that studying the proteins that are newly synthesized during the metastatic processes of migration and invasion will greatly enhance our understanding of breast cancer progression. We conducted a mass spectrometry screen following bioorthogonal noncanonical amino acid tagging to elucidate changes in the nascent proteome that occur during epidermal growth factor stimulation in migrating and invading cells. Annexin A2 was identified in this screen and subsequent examination of breast cancer cell lines revealed that Annexin A2 is specifically upregulated in estrogen receptor negative (ER-) cell lines. Furthermore, siRNA knockdown showed that Annexin A2 expression promotes the proliferation, wound healing and directional migration of breast cancer cells. In patients, Annexin A2 expression is increased in ER- breast cancer subtypes. Additionally, high Annexin A2 expression confers a higher probability of distant metastasis specifically for ER- patients. This work establishes a pivotal role of Annexin A2 in breast cancer progression and identifies Annexin A2 as a potential therapeutic target for the more aggressive and harder to treat ER- subtype.

Keywords: breast cancer; estrogen receptor negative; Annexin A2; mass spectrometry; metastasis

1. Introduction

All cells, including tumor cells, exist not in a vacuum but in a continuous, interconnected and unbounded network of relationships. This includes relationships with other cells and with the extracellular or stromal environment in which the cells reside. In the case of tumor cells, these surroundings are collectively known as the tumor microenvironment (TME) [1]. Understanding how the TME can affect cellular behavior is of particular interest in attempts to elucidate the cellular mechanisms that enable a metastatic phenotype. This is because two of the cellular behaviors essential to the process of metastasis (migration and invasion) rely on interactions between the tumor cell and its surrounding TME [2,3]. In migration, cells gain the ability to become aberrantly motile. This motility is given directionality via chemotaxis in which a receptor expressed by the cancer cells interacts with a chemoattractant ligand present within the TME [4]. Invasion, referred to as the defining feature of malignancy, is the ability of cancer cells to change their morphology and alter their surroundings in order to penetrate the basement membrane, degrade encompassing stroma and escape usual tissue boundaries [5]. These pivotal mechanisms of cancer disease progression can be modelled in vitro using cell lines and well-defined transwell migration and invasion assays [6].

Deciphering the steps involved in the metastatic process is of vital importance in breast cancer. Breast cancer begins as a disease localized to the breast tissue. Advancements in multi-modal treatments such as surgery and radiation, have increased the chances for cure for over 70% of patients [7]. However, the mortality associated with the disease of breast cancer does not arise from these localized breast tumors but rather, from when the cancer undergoes metastasis and spreads to distant sites and colonizes vital organs of the body. At present, metastatic breast cancer is considered incurable. Once this stage of disease is reached, treatment aims to prolong survival and control symptoms. For this reason, there is an urgent need to decipher the specific environmental cues and resultant mechanisms involved in the metastatic process. This is of particular importance in estrogen receptor (ER) negative breast cancer subtypes (including both the triple negative (TNBC) and HER2+ groupings) as these tumors have a more aggressive phenotype and higher rate of distant metastasis than ER positive tumors [8]. This understanding will guide the development of interventions aimed at preventing the metastasis and associated mortality seen in those diagnosed with ER negative breast cancer.

One method to assess the effect of environmental cues is to study the resulting influence on the cell's proteome [9]. There is strong evidence that proteomic analysis offers a truer physiological insight over transcriptomic methods. Measurements of mRNA transcript abundance alone have been reported to be a poor predictor of protein synthesis due to the fact it is subject to an independent layer of translational control that controls protein expression [10]. Transcriptomics also fails to take into account protein turnover and degradation as well as activation state through post-translational modification. Proteomic analysis of breast cancers has revealed prognostic markers [11], novel subtype classifications [12] and targetable protein abundance differences as cancer progresses through clinical stages [13]. Analysis of proteins that are newly translated and synthesized by cells allows for an accurate snapshot of cellular activities during a specific process or duration. In fact, the execution of cancer hallmarks—such as migration and invasion—which drive progression are ultimately achieved by dynamic alterations in protein expression and post-translational modifications of those proteins [14].

With this in mind, we have developed a model to investigate changes in the nascent proteome of aggressive, ER negative breast cancer cells as they undergo migration and invasion, stimulated through modulation of the extra-cellular environment. In this study, we utilized these in vitro transwell models in combination with bioorthogonal noncanonical amino acid tagging (BONCAT) to investigate proteomic changes in MDA-MB-231 breast cancer cells as they undergo epidermal growth factor (EGF) mediated migration and invasion. BONCAT is a widely used method to assess the primary nascent proteomic response to a specific stimulus within a specific short time period [9]. This method allowed us to isolate and identify newly synthesized proteins that were translated in the two hours following directional EGF stimulation. Metastasis is a dynamic process of changes in cellular expression and behaviors and thus, must be investigated via a dynamic and temporal model. Isolating the nascent proteome in this manner facilitates the precise examination of the cancer cell's translational response to environmental cues, which trigger a metastatic phenotype. This knowledge may elucidate novel therapeutic strategies or treatments that prevent the metastatic spread of cancer cells.

2. Materials and Methods

2.1. Cell Culture

Human breast cancer cell lines (MDA-MB-231, MCF-7, ZR-75-1) were purchased from the ECACC culture collection (Sigma Aldrich, Wicklow, Ireland). Cell lines were routinely tested upon freezing and thawing for mycoplasma contamination using PCR. Cells were maintained at 37 °C, 5% CO_2 humidified incubator in DMEM-high glucose (MDA-MB-231 & MCF-7) or RPMI (ZR-75-1) supplemented with 10% fetal bovine serum, 5% l-glutamine and 5% penicillin/streptomycin (all obtained from Sigma Aldrich). ZR-75-1 media was additionally supplemented with 5% sodium pyruvate (Sigma Aldrich). For 3-dimensional culture, the on-top method was used where cells were seeded into wells containing Growth-Factor Reduced Matrigel™ (Biosciences, Dublin, Ireland) then covered with DMEM supplemented with 2% Matrigel. For analysis of cell signaling, cells were washed with PBS before being covered with serum free medium for 4 h prior to EGF stimulation (Peprotech, London, UK).

2.2. Click-iT Reaction

Cell cultures plated in 2-D (24 h) and 3-D (6 day) on transwell membranes were starved of serum and methionine for 4 h. After this, serum free, methionine free DMEM supplemented with 50 ng/mL EGF was placed into the lower chamber. At the same time, the upper chamber media was replaced with serum free, methionine free DMEM supplemented with 50 μM Azido Homo Alaine (AHA) (Thermo Fisher, Dublin, Ireland), a methionine analogue that is incorporated into newly synthesized proteins. Cells were then allowed to migrate and invade for 2 h. Cells were harvested from Matrigel or scraped from the membrane then lysed. The Click-iT reaction and incorporation of the tetramethylrhodamine (TAMRA) tag was then carried out according to Click-iT TAMRA Protein Analysis Detection Kit protocol (Thermo Fisher) to isolate newly synthesized proteins which had incorporated the AHA labelling. A control replicate omitting the EGF supplementation was carried out for both 2-D and 3-D cultures.

2.3. Immunoprecipitation

Newly synthesized proteins that had incorporated the TAMRA residue were then isolated from total cell lysate using anti-TAMRA immunoprecipitation. Lysates were precleared with protein G agarose beads (Roche, via Sigma Aldrich). The supernatant was then incubated with Anti-TAMRA antibody (Thermo Fisher) and protein G agarose beads at 4 °C overnight. The beads were washed with cell lysis buffer before being boiled with SDS loading buffer for 5 min. Immunoprecipitated samples were resolved on a 12% SDS gel and stained with Coomassie Instant Blue (Expedeon, Cambridge, UK). A control containing only beads was also performed to account for non-specific binding and contaminant exclusion.

2.4. Mass Spectrometry and Data Analysis

Following SDS-PAGE, gel lanes containing protein bands stained with Coomassie were excised from the gel and placed in 20% glycerol for transport to Mass Spectrometry and Proteomics Facility, University of St Andrews, Fife. Gel chunks were subjected to trypsin digestion. Next, proteolytic peptide mixtures were separated by nano-LC utilizing an Eksigent two-dimensional LC NanoLC system (Eksigent/Applied Biosystems Sciex, MA, USA) interfaced with a QStar XL mass spectrometer (Applied Biosystems Sciex, MA, USA). Data sets were searched against the NCBInr 20160830 database using MASCOT software (Matrix Science, MA, USA) under the following parameters: maximum one missed cleavage of trypsin digestion, carbamidomethyl (C) as a fixed modification, oxidation (M) as a variable modification, a peptide mass tolerance of ± 20 ppm and a fragment mass tolerance of ±0.05 Da. Only scores higher than the significance threshold ($p < 0.05$) were reported. Proteins identified in bead only and non-EGF supplemented controls were subtracted from the protein lists.

Proteins were then characterized functionally based on the available literature associated with their NCBI entry. Gene ontology (Molecular Function–Slim) enrichment analysis was carried out using the Panther overrepresentation test (Version 14.1) [15] with the Fisher's exact test to calculate *p*-value and Benjamini–Hochberg procedure to calculate false discovery rate (FDR). FDR cut-off was <0.05. Protein-protein interaction networks were probed using STRING v11 [16] and visualized using Cytoscape StringApp [17].

2.5. Cell Lysis, SDS PAGE and Western Blotting

The cells were lysed in 1% NP-40 lysis buffer then quantified and denatured with SDS loading buffer and boiled for 5 min. Lysates were separated on 12% SDS acrylamide gels and subsequently transferred to nitrocellulose membrane. Membranes were blocked using 5% BSA for 1 h at room temperature then probed with corresponding primary antibodies at 4 °C overnight. Dilutions of antibodies were as follows: Annexin A2 1:1000 (Abcam, Cambridge UK), Annexin A2 1:1000 (BD Bioscience), B-Actin 1:1000 (Sigma), phospo-Tyr24-Annexin A2 1:250 (Santa Cruz Biotechnology, Heidelberg, Germany) E-Cadherin 1:1000 (Abcam). IRdye700- or IRdye800-conjugated secondary antibodies, were then coupled to the primary antibody for 1 h at room temperature. Protein bands were detected using the Odyssey Sc (LI-COR, Cambridge, UK).and quantified using Image Studio 5.2 (LI-COR)).

2.6. siRNA Transfection

To reduce Annexin A2 expression, the Neon Transfection System (Thermo Fisher) was used to transiently transfect MDA-MB-231 cells with siRNA oligonucleotides targeting ANXA2 mRNA according to the protocol supplied by Neon. Two different siRNA oligonucleotides were obtained from Qiagen (Manchester, UK) and used at 10 nM for 72 h (Sequences reported in Table S3). MISSION® siRNA Universal Negative Control (Sigma Aldrich) was transfected as negative control.

2.7. Tissue Collection

Ethical approval was granted by the University Hospital Limerick's Ethics Committee and was allocated the ethical approval numbers 22/14 and 141/12. From the resected surgical specimens, core biopsies from the tumor mass were extracted by the operating surgeon. Where possible, pathologically normal breast tissue samples were also collected. The samples were stored in Allprotect Tissue Reagent (Qiagen) at −80 °C until extraction.

2.8. RNA Extraction

RNA from breast cancer cell lines was extracted using RNeasy Plus Mini kit (Qiagen) according to manufacturer's instructions. For breast tissue samples, RNA was extracted by submerging tissue into liquid nitrogen, grinding into a fine powder then using the RNeasy Lipid Tissue Mini Kit (Qiagen) in combination with Maxtract tubes (Qiagen) according to manufacturer's instructions. RNA was quantified using Nanodrop (Thermo Fisher) and quality was assessed by agarose gel separation. cDNA was then synthesized from 1 µg or 500 ng of total RNA using a Vilo cDNA kit (Invitrogen, via Biosciences, Dublin, Ireland).

2.9. RT-qPCR

Applied Biosystems QuantStudio 7 Flex Real-Time PCR was used to measure gene expression using Taqman Gene Expression Assay Kits (Thermo Fisher). Seven different housekeeping genes were tested using Normfinder and the three most stable genes between the three cell lines were used to normalise gene expression across samples as previously described [18]. Relative fold change in gene expression was calculated using the $2^{-\Delta\Delta CT}$ method and represented relative to a control sample set to 1 for each experiment. Data presented is from three independent experiments displayed as mean +/− SEM.

2.10. Cell Migration Assays

For wound healing assays, transfected cells were seeded into cell culture wells containing ibidi inserts (ibidi GmbH, Gräfelfing, Germany). After 72 h incubation, the insert was carefully removed and cells imaged using Olympus cellSens Dimension 1.12. Cells were incubated for 24 h, after which images of the wound closure were taken. Six fields of view were imaged for each condition and percent wound closure was measured using Image J (version 1.52n). Three individual experiments were carried out to verify results. Data is displayed as mean across three individual experiments +/− SEM.

For directional migration assays, the underside of 8 μm transwell membranes were coated with fibronectin before 6×10^5 cells in serum free DMEM media were plated onto the upper chamber. Cells were allowed to migrate towards a lower chamber containing 50 ng/mL EGF in serum free media for 24 h. Cells were removed from the upper chamber of the membrane. Migrated cells were fixed and stained with crystal violet and imaged. To quantify the cell migration, crystal violet stain was dissolved 10% acetic acid and intensity measured as absorbance at 595 nm in triplicate. Three individual experiments were carried out to verify results.

2.11. Real-Time Proliferation Assay

The rate of proliferation was monitored in real time using the xCELLigence RTCA DP E-plate system (ACEA, CA, USA). 1.5×10^4 MDA-MB-231 cells transfected either with ANXA2 siRNA or negative control RNA were seeded into each well. The impedance value of each well was measured by the xCELLigence system every 15 min for 96 h and expressed as a cell index value (CI). CI was normalized at 4 h post seeding. Wells were analyzed in duplicate. Three individual experiments were carried out to verify results. Data is displayed as mean cell index +/− SEM.

2.12. Use of Publicly Available Gene & Protein Expression Datasets

Protein expression data of breast cancer cell lines was provided by D P Nusinow et al. [19]. Protein expression was measured using quantitative mass spectrometry and is expressed as normalized, relative values of protein abundance. Full details of normalization is available at [20]. Gene expression data of breast cancer cell lines was downloaded from the Cancer Cell Line Encyclopedia (CCLE) (Available at: https://portals.broadinstitute.org/ccle; GEO Series GSE36139). This data set was collected using GPL15308 Affymetrix Human Genome U133 Plus 2.0 Array. Microarray data was evaluated using the robust multichip average method (RMA) and quantile normalized before being made available through the CCLE [21]. Breast cancer cell lines were then classified by ER expression (positive or negative) according to the available literature [22,23] and AnxA2 mRNA expression was compared between groups. Dataset GSE42568 [24] was downloaded from the GEO data repository (http://www.ncbi.nlm.nih.gov/geo) using the GEOR function. This data set was collected using GPL570 [HG-U133_Plus_2] Affymetrix Human Genome U133 Plus 2.0 Array. Microarray data was evaluated using the GC robust multichip average (GCRMA) method and quantile normalized before being made available through GEO. Protein expression data using SILAC quantitative mass spectrometry of 40 breast cancer tissue samples was obtained from Tyanova et al. (2016) [25]. Normalized protein expression depicted as log2 (ratio of L/H). Distant metastasis free survival (interval from surgery to date of metastasis diagnosis) was investigated by downloading data from http://kmplot.com, a manually curated database which collates gene expression and survival information from GEO, EGA and TCGA datasets (35 breast cancer datasets, using Affymetrix HGU133A and HGU133 Plus 2.0 microarrays) [26]. The median ANXA2 expression value for each cohort was chosen as the high/low cut off point. Graphpad Prism 8 was then used to produce Kaplan–Meier graphs and to calculate log-rank test result and hazard ratios.

2.13. Statistical Analysis

Statistical analyses were performed using GraphPad Prism version 8.0.2. The Shapiro–Wilk test was used to assess normality of sample distribution. Differences between two groups were calculated using Student's T-test (normally distributed samples) or Mann Whitney U test (non-normally distributed samples). Differences between three or more groups was calculated using ANOVA and multiple comparisons were performed using Bonferroni Correction. Data was reported as mean +/− standard deviation for technical replicates or mean +/− standard error of mean for biological replicates. A p value of <0.05 was considered statistically significant. Within figures, asterisks denote significance levels as such: * $p \leq 0.05$; ** $p \leq 0.01$; *** $p \leq 0.001$.

3. Results

3.1. Isolation and Identification of Newly Synthesised Proteins Involved in Breast Cancer Cell Metastasis

In order to develop a model to investigate breast cancer metastasis in vitro, we chose MDA-MB-231s due to their aggressive, epithelial to mesenchymal phenotype and well-evidenced ability to migrate and invade in vitro. Additionally, MDA-MB-231 belong to the basal/TNBC sub-type and are thus estrogen receptor negative and known to express the EGF receptor [23]. To determine whether EGF would be an effective chemoattractant in our models of migration and invasion, a series of transwell migration experiments were carried out. The highest level of migration of MDA-MB-231 cells was observed when EGF was present only in the lower chamber, as evidenced in Figure 1A; demonstrating that EGF elicits a specific increase in directional migration. Our own observations are in accordance with those published in similar studies [27–30] and gives evidence to the selection of EGF as a trigger for the processes of migration and invasion in MDA-MB-23 cells.

Having determined 50 ng/mL EGF to be an appropriate chemoattractant for our breast cancer cell line, it was selected as the stimulus in transwell migration and invasion assays in our model. The workflow of this model is summarized in Figure 1B. To identify the newly synthesized proteins while breast cancer cells undergo migration and invasion, MDA-MB-231 breast cancer cells were grown in 2-D and 3-D culture as described in Materials and Methods. Cell cultures were serum and methionine starved then stimulated with EGF and supplied with AHA. After stimulation, a fluorescent TAMRA-alkyne was added which binds to the azide moiety of the AHA-tagged newly synthesized proteins. This allowed for isolation of the newly synthesized proteins using anti-TAMRA in an immunoprecipitation reaction, with the resultant product being proteins newly synthesized following stimulation to migrate and invade towards an EGF chemoattractant. Proteins in these samples were then identified using mass spectrometry analysis. To specifically identify proteins that were newly translated during EGF stimulation, a control experiment without the addition of EGF was conducted for both the 2-D and 3-D models. Proteins identified in the EGF omitted controls were then subtracted from their respective EGF stimulated protein lists.

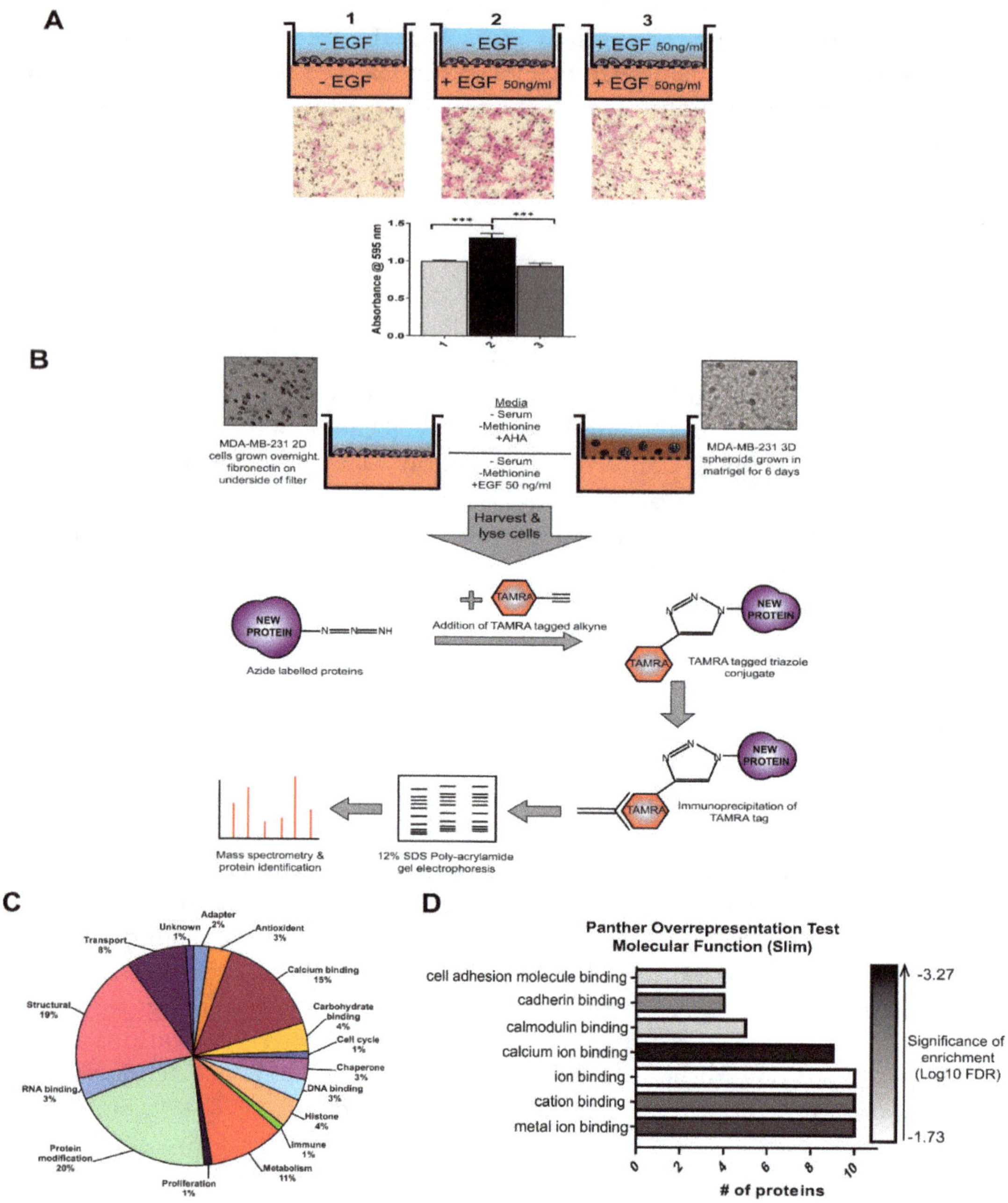

Figure 1. Isolation, identification, and functional characterization of newly synthesized proteins in MDA-MB-231 migration and invasion. (**A**) MDA-MB-231 cells were plated in the upper chamber of transwell plates and media was supplemented (+/− EGF) according to diagram. Cell migration after 4 h was measured by crystal violet (CV) staining of cells which moved through the well and adhered to the underside of the membrane. Non-migrated cells were removed prior to staining. Membranes were then imaged using inverted microscope and migration of cells was quantified by dissolving of CV stain and measuring absorbance at 595 nm. Data displayed as mean ± SEM, of 3 independent experiments, CV absorbance normalized to average value for well 1. Statistical analysis by one way ANOVA, $p = < 0.0001$. (**B**) Flow chart of model set-up, isolation of newly synthesized proteins, and mass spectrographic analysis carried out in this study. (**C**) Characterization of identified newly synthesized proteins according to the NCBI database entries of each protein and displayed as pie chart. (**D**) Molecular function enrichment analysis was carried out using the PANTHER overrepresentation test. The numbers of proteins annotated with each molecular function was plotted as a bar chart with the color scale representing the significance of the enrichment of molecular function within the list.

We identified a total of 95 newly synthesized proteins potentially involved in the migration and invasion of MDA-MB-231 breast cancer cells towards EGF (Table S1). Characterization of the protein list according to NCBI Protein [31] database entries found that the list spans a wide array of functions within the cell, with the highest proportions being: protein modification (e.g., kinases, phosphatases, proteases; 20%), structural (19%), and calcium binding (19%) as summarized in Figure 1C. We then carried out molecular function enrichment analysis using the PANTHER overrepresentation test which revealed the most significantly overrepresented category was that of proteins involved in calcium ion binding (GO:0045296, expected:0.97, input: 9, fold enrichment: 9.31, p-value: 7.56×10^{-7} false discovery rate: 3.82×10^{-4}, Figure 1D). This functional characterization showed that proteins newly synthesized in the processes of cell migration and invasion encompass a wide variety of functions and numerous essential processes in cell signaling and behavior, many of which are implicated in cancer progression.

3.2. Identification and Verification of Annexin A2 as A Newly Synthesized Protein in EGF Stimulated Migration and Invasion

Of the total 95 proteins identified as newly synthesized, 41 proteins were identified in the 2-dimensional culture migration experiment only, and 40 unique proteins were identified in the 3-dimensional culture invasion experiment only. When displayed in the Venn diagram in Figure 2A, 14 proteins were found to be common to both lists, meaning that they had been identified in both the 2-D migration and 3-D invasion models. Protein-protein interaction analysis of these 14 proteins using STRING [16] revealed an extensively linked network within the group, (with the exception of NCCRP1) as seen in Figure 2B. To identify key proteins within the identified network, each node was assessed for its degree of centrality, edge connectivity and whether it has been previously reported as important in breast cancer metastasis. Annexin A2, having satisfied these criteria and was chosen as a candidate protein to investigate how perturbing this elucidated network would affect breast cancer cell migration and invasion. Annexin A2 was also chosen as it was identified in both experiments, was relatively highly scored and due to its molecular function as a calcium ion binding protein [32], a gene ontology term found to be most significantly enriched (FDR = 3.82×10^{-4}) following analysis of the total protein list. Annexin A2 is a member of the 12 protein annexin family, a group of proteins which have the ability to bind negatively charged membrane phospholipids in a calcium dependent manner [33,34]. In addition to the interactions illustrated in Figure 2B, Annexin A2 has proven links with a number of other identified proteins such as beta- actin [35] and tubulin [36]. Thus, it is highly probable that Annexin A2 is playing a central role in influencing this network, for example through its roles in actin polymerization [37] and cytoskeletal rearrangement [38,39]. In addition, due to the integral role and embryonic lethality of essential cytoskeletal proteins like beta actin [40], we believe there is more potential therapeutic value in investigating proteins that have a more regulatory, non-essential role, such as Annexin A2 [41]. In addition, Annexin A2 has a known clinical association with cancer [42] and has a role downstream of growth factor signaling pathways such as the IGF-IR [43] and the EGFR [44,45] pathways. This association was first established in hepatocellular carcinoma in 1990 [46] but to date, it has been found to be commonly overexpressed in multiple cancer types such as colorectal, breast, lung and pancreatic [42]. The amino acid sequence of Annexin A2 with the matched peptide sequences is shown in Figure S1 (Score: 169, Monoisotopic mass (Mr): 38808, Calculated pI: 7.57, Matches: 5, Sequences: 4, Coverage: 12%). The reported mass agrees with Annexin A2's predicted molecular weight of 38 kDa. A representative MS/MS spectrum of one Annexin A2 peptide is also shown. Apart from Annexin A2 several other proteins of potential future interest were identified (Figure 2A,B). Intriguingly, a number of metabolism associated proteins, including GAPDH, LDHA, and GSTP1 were identified. Altered metabolism is a fundamental hallmark of cancer [47] and it has even been suggested that specific deregulations in the metabolism of metastasizing cells could be a specific therapeutic target [48].

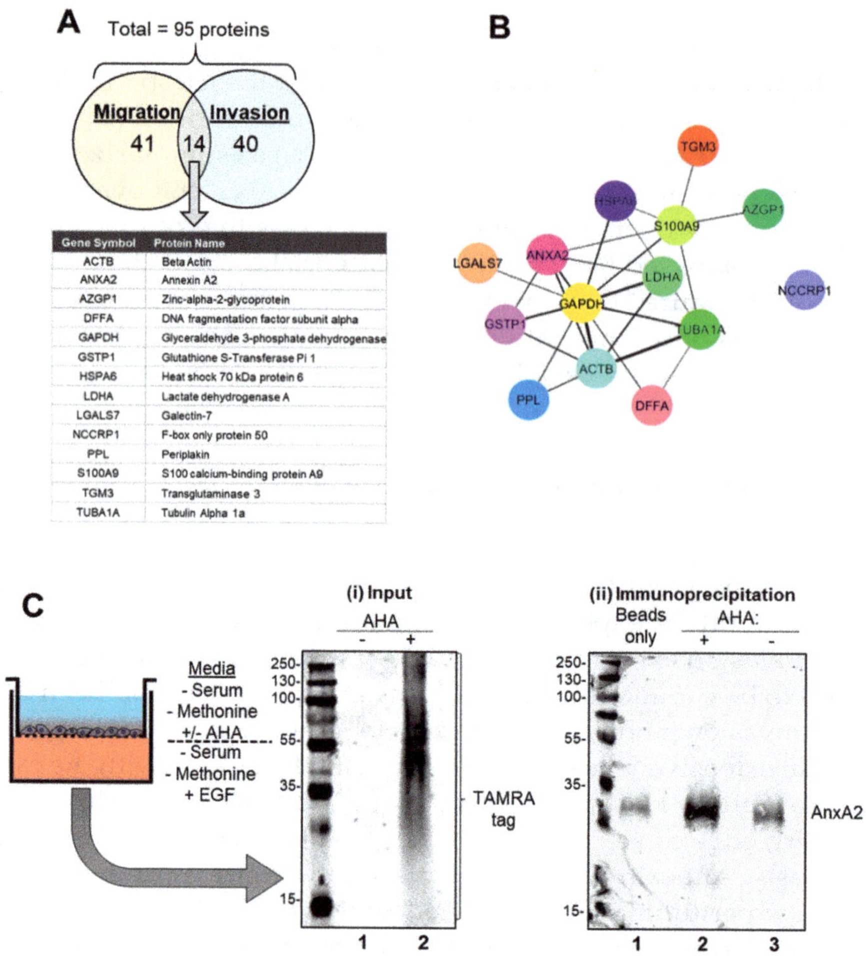

Figure 2. Annexin A2 is newly synthesized during the EGF stimulated migration and invasion of MDA-MB-231 cells. (**A**) 95 proteins in total were identified following mass spectrometry analysis and MASCOT peptide identification. Of these, 41 were found in the migration experiment only, 40 were found in the invasion experiment only and 14 were found to be common to both, listed here. (**B**) Protein-protein interactions between the 14 proteins found to be common to both the migration and invasion lists were assessed with String (http://string-db.org) [16] and visualized using the StringApp in Cytoscape set at 0.2 confidence level [17]. Edge line thickness denotes the confidence of association between protein nodes. (**C**) (**i**) Input: Western blot analysis of Click-It reaction lysate shows successful incorporation of the TAMRA tag into newly synthesized proteins when AHA was added. (Lane 1: Cell lysate of EGF migrating cells without AHA, Lane 2: Cell lysate of EGF migrating cells with AHA). (**ii**) Immunoprecipitation: Consequent TAMRA immunoprecipitation assay showed Annexin A2 pulldown following EGF induced cell migration as measured by SDS-PAGE and western blotting with anti-Annexin A2 antibody. (Lane 1: Beads only control, Lane 2: AHA+ lysate from lane 2 of (**i**), Lane 3 AHA- lysate from lane 1 of (**i**)).

Following identification of Annexin A2, we proceeded to validate it as a newly synthesized protein by replicating the initial migration experiment. This was done by performing a TAMRA immunoprecipitation (IP) of the Click-iT reaction lysate to isolate proteins that had incorporated the TAMRA tag then western blot analysis to show successful Annexin A2 pulldown (Figure 2C). Pull down of Annexin A2 is clearly increased in the AHA+ lysate sample (Panel (ii) lane 2) in comparison to the beads only (no antibody) control (Panel (ii), lane 1) and AHA- (non-TAMRA labeled) lysate control (Panel (ii), lane 3).This pulldown validates our approach and verifies our identification of Annexin A2 as a newly synthesized protein via mass spectrometry analysis.

3.3. *Annexin A2 Expression is Increased in Estrogen Receptor Negative Breast Cancer Cells*

Following identification and verification of Annexin A2 as a protein newly synthesized in EGF mediated MDA-MB-231 cell migration and invasion, we next examined its expression profile across a number of breast cancer cell lines. Western blotting analysis of cell protein content shows an increased expression of Annexin A2 in the estrogen receptor negative (ER-), basal/TNBC cell line MDA-MB-231 when compared to the estrogen receptor positive (ER+), luminal cell lines MCF-7 and ZR-75-1 (both ER+/PR+) (Figure 3A). Expression of Annexin A2 appears to inversely correlate with expression of E-Cadherin, the loss of which is a known marker of epithelial to mesenchymal transition. To further support these findings, we analyzed data from D. P. Nusinow et al. (2020) [19] to generate a protein expression heatmap (Figure 3B) showing how the expression of Annexin A2, EGFR, and E-Cadherin vary across a larger sample of cell lines, according to breast cancer subtype classification. From this it is evident that there is an increase in both Annexin A2 and EGFR expression in the ER- (HER2 and TNBC) cell lines when compared to ER+ cell lines (further depicted in Figure S2). This supports and is in agreement with our conclusions drawn from our western blot data (Figure 3A). Interestingly, we do not see an association between the loss of E-Cadherin and elevated Annexin A2 when we query this larger cohort of cell lines, thus the trend observed in Figure 3A may be attributed to individual characteristics of the three cell lines tested. When we consider the potential relationship between Annexin A2 and EGFR expression, our analysis shows that in ER- cell lines, there is a strong positive and significant correlation between the expression of these two proteins (Pearson's correlation, $r = 0.6026$, $p = 0.0023$). Importantly, and in support of our hypothesis, we do not see this significant correlation in ER+ cell lines ($r = 0.5952$, $p = 0.1196$).

Next, we examined Annexin A2 mRNA expression using RT-qPCR in breast cancer cell lines, under normal, serum starved and EGF stimulated conditions. Again, we observed Annexin A2 expression to be increased in ER- cells in comparison to ER+ cell lines under all culture conditions (Figure 3C). In particular, we found that following 4 h of 50 ng/mL EGF stimulation ANXA2 is even more highly expressed in MDA-MB-231s when compared to the other ER+ cell lines. This suggests that the increased expression of ANXA2 in response to EGF is specific to the ER-, EGFR expressing MDA-MB-231 cell line. We then utilized the available data within the Cancer Cell Line Encyclopedia (CCLE) [21] to further probe the relationship between ER status and AnxA2 expression. The CCLE is a large, publicly available data repository which compiles gene expression, chromosomal copy number and sequencing data from 947 human cancer cell lines, spanning 36 tumor types [21]. Breast cancer cell lines with ANXA2 expression data available ($n = 49$) were first categorized according to positive or negative ER status according to previously published molecular characterization studies [22,23]. This stratification revealed 18 of the available cell lines to be ER+ and 31 to be ER-. Comparing levels of ANXA2 mRNA expression between ER+ and ER- cell lines, as measured by Affymetrix U133 plus 2.0 arrays, showed ER- lines exhibit significantly higher expression of ANXA2 ($p = 0.0008$) as shown in Figure 3D. The cell lines analyzed and their respective ANXA2 expression values are outlined in Figure S2. These results show that expression of Annexin A2 in breast cancer is strongly influenced by estrogen receptor status and that its overexpression in breast cancer may be restricted to the ER- subtype. ER- negative breast cancers include both TNBC and HER-2 over expressing subtypes, both of which are more aggressive cancers with a poorer prognosis [49].

As the synthesis of Annexin A2 identified in our initial experiment was triggered by stimulation with EGF, we investigated the effect of EGF on the regulation of Annexin A2 (Figure 3E). To control our experiment, we examined the p-Akt response, which was used as a read-out of our serum starvation and EGF stimulation. We found that stimulation of MDA-MB-231 cells with EGF resulted in an increase in phosphorylation of Annexin A2 on Tyrosine 24 (Tyr24) which reached a peak phosphorylation at 30 min, before decreasing towards, initial, pre-stimulation levels over 120 min (Figure 3E). This post-translational modification allows Annexin A2 to bind to S100A10 (also known as p11) to form the AnxA2/p11 heterotetramer which then is translocated to the extracellular cell surface [50,51]. This modulation is of particular relevance in cancer phenotype as once on the cell surface, AnxA2/p11 promotes the conversion of plasminogen to active plasmin, a protease which plays a role in extracellular matrix degradation and activation of matrix metalloproteases, critical steps in cancer cell metastasis [34,52]. In fact, Tyr24 phosphorylation has been shown to be critical for the invasive potential of multi-drug resistant breast cancer cells [53] and is proposed to modulate epithelial to mesenchymal transition in a number of cancer types, including pancreatic [54], cervical [55], colon [56] and lung [57].

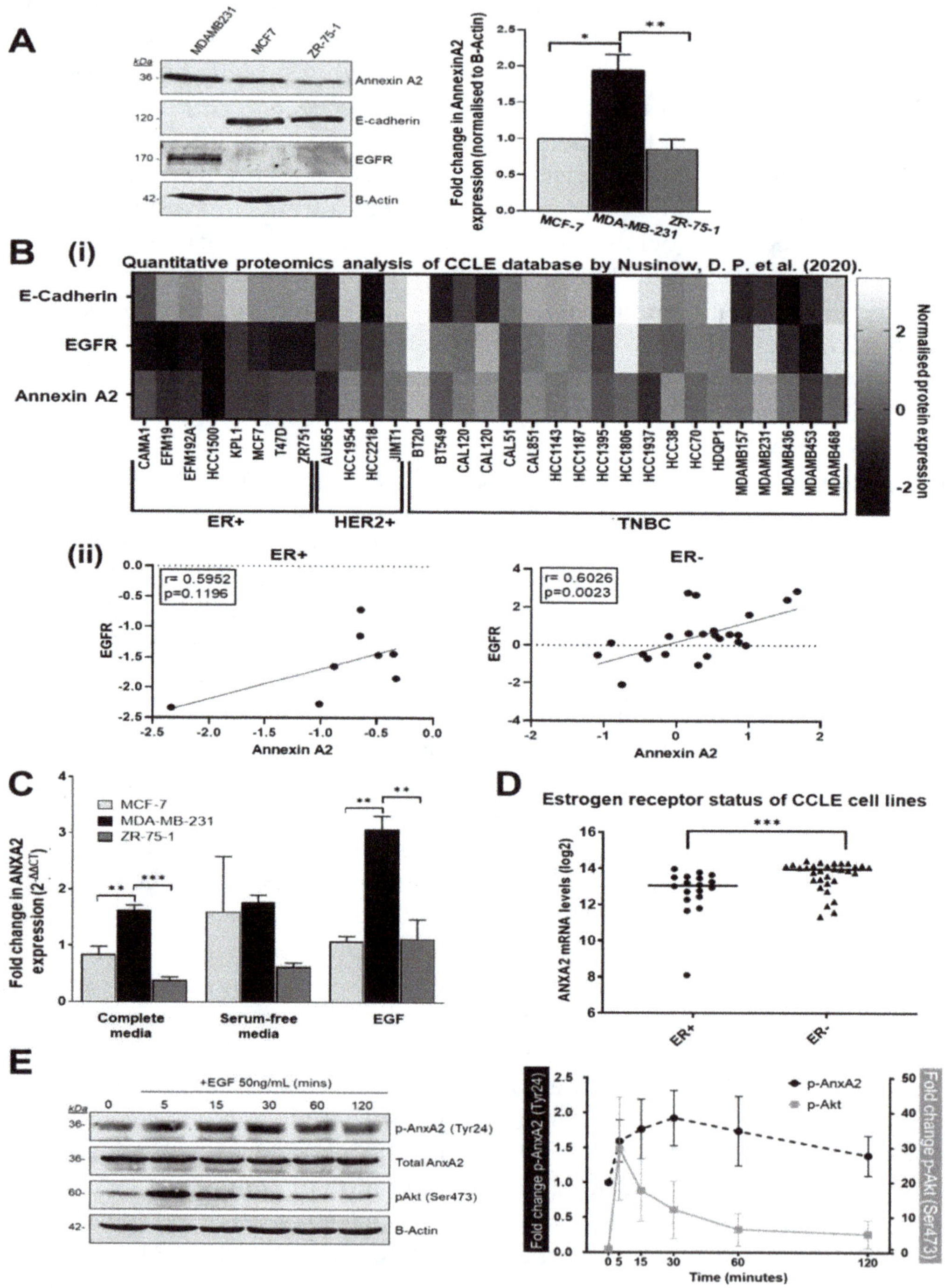

Figure 3. Annexin A2 expression is increased in ER negative breast cancer cells and undergoes transient phosphorylation following EGF stimulation. (**A**) Comparison of Annexin A2 expression in three breast

cancer cell lines: MDA-MB-23, MCF-7 and ZR-75-1 as measured by SDS-PAGE and western blot. Protein expression was quantified by densitometric signal analysis on Image Studio software (LI-COR). Annexin A2 signal was normalized to beta-actin signal, then expressed in comparison to MCF-7 expression as 1. Data shown is mean +/− SEM for 3 individual experiments, measured by ANOVA with Bonferroni, $p = 0.0034$. (**B**) (**i**) Heatmap of Annexin A2, EGFR and E-Cadherin protein expression data for breast cancer cell lines from Nusinow, D. P. et al. (2020), stratified according to breast cancer subtype. Protein expression represented as normalized relative protein expression, as measured by quantitative proteomics. (ii) Scatterplots of the same data indicating the relationship between Annexin A2 and EGFR protein expression. A significant and strong positive correlation between Annexin A2 and EGFR expression is seen for ER- cell lines (Pearson's correlation, $r = 0.6026$, $p = 0.0023$) (**C**) Gene expression of ANXA2 in three breast cancer cell lines: MDA-MB-231, MCF-7 and ZR-75-1. Expression was calculated using delta CT method and displayed as fold change (corrected RQ) in comparison to MCF-7 expression for each condition. (i.e., MCF-7 expression normalized to 1) differences between cell lines analyzed by ANOVA with Bonferroni. (**D**) Publicly available Affymetrix gene expression data for human breast cancer cell lines was obtained from the CCLE. Cell lines were categorized according to ER expression status and difference in ANXA2 expression between the 2 groups was measured using Mann–Whitney U test, $p = 0.0008$ (**E**) Western blot showing the effect of EGF stimulation on AnxA2 expression and phosphorylation in MDA-MB-231 cells. After 4 h of serum starvation, cells were treated with 50 ng/mL EGF. Representative blot and corresponding time course graph showing the increase in AnxA2 phosphorylation at Tyr24 followed by decrease after 30 min. Densitometry signal normalized as phosho-AnxA2/total AnxA2 and phospho-Akt/total beta-actin. Data points displayed as mean of 3 individual experiments ±SEM.

3.4. Annexin A2 is Required for Cell Proliferation, Wound Healing and EGF Directed Cell Migration of ER Negative Breast Cancer Cells

Following evidence of Annexin A2's involvement in the migration and invasion in our ER negative cell model, we next investigated its influence on the cancer phenotype with a series of functional assays. AnxA2 expression was suppressed in MDA-MB-231 cells using siRNA specific to AnxA2 mRNA transcript (Table S2). Figure 4A shows a representative blot of knockdown efficiency, showing reduced level of Annexin A2 protein in cells 72 h following transfection. Firstly, using a real time cell analysis platform (xCELLigence) the effect on the growth and proliferation of MDA-MB-231 cells was monitored by measuring cell index every 15 min over the course of 96 h. As shown in Figure 4B, the depletion of AnxA2 significantly reduced the proliferation of MDA-MB-231 cells when compared to control cells. Next, the effect of AnxA2 knockdown on cell migration was then assessed using two independent assays. We first used wound healing assays which revealed knockdown cells to have a significantly decreased ability to close the wound after 24 h (Figure 4C).

This was followed by transwell chamber experiments, which specifically tested the effect on the directional migration and the response to an EGF chemoattractant. In this assay, MDA-MB-231 cells were plated in serum-free media on the upper chamber of the transwell and allowed to migrate towards the bottom chamber, containing serum free media supplemented with 50 ng/mL EGF. We found that ANXA2 knockdown cells were significantly reduced in their ability to migrate towards an EGF gradient (Figure 4D). These results are consistent with previous reports of Annexin A2 being a positive regulator of wound healing in MDA-MB-231 cells [44,58]. However, they also point towards the role of Annexin A2 in promoting specific EGF directed migration. In addition, we have shown that Annexin A2 knockdown inhibits the proliferative ability of the ER- MDA-MB-231 cells. These functional results were confirmed by replication of each assay using another clone targeting ANXA2 (Figure S4).

Taken together, this data strongly supports the hypothesis that Annexin A2 promotes cell proliferation, wound healing and cell migration, all critical processes which support the malignant phenotype of ER- breast cancer cells in vitro.

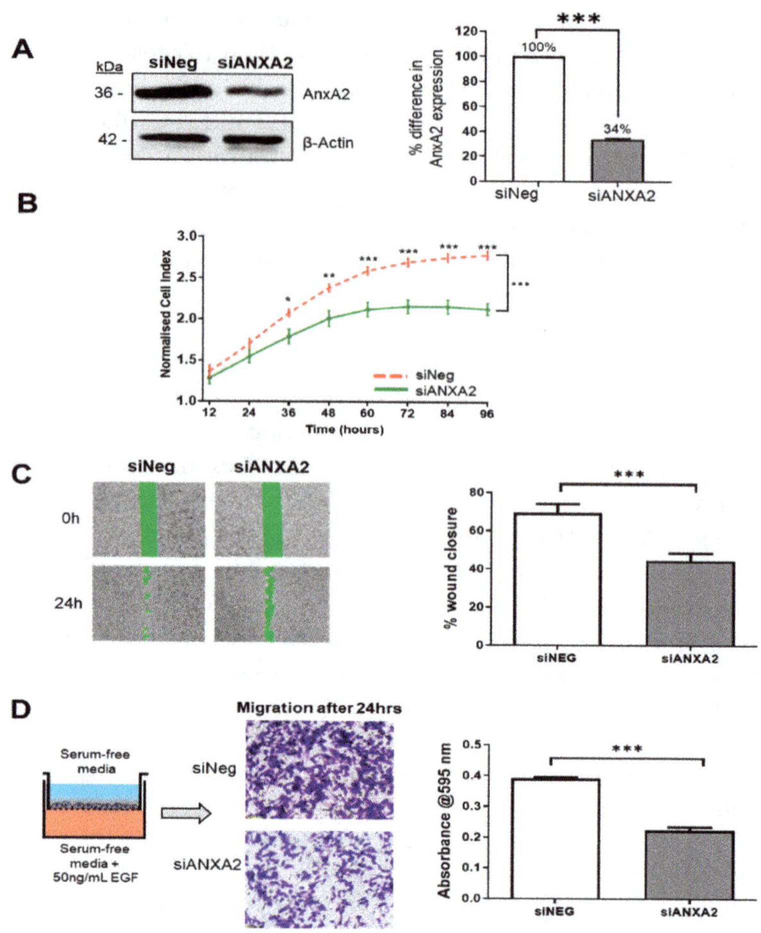

Figure 4. siRNA mediated knockdown of AnxA2 inhibits the proliferation and migration of MDA-MB-231 cells. (**A**) MDA-MB-231 cells were transfected with a small interference oligonucleotide against ANXA2 [10 nM] for 72 h, using the Neon Transfection system. Representative blot showing the efficiency of ANXA2 knockdown (KD) after 72 h and densitometry analysis of Annexin A2 protein bands normalized against beta-actin loading control and expressed as a percentage of AnxA2 level in negative control, measured using Image Studio software. Data displayed as mean +/− SEM for three individual experiments. (**B**) Reduction of Annexin A2 expression in MDA-MB-231 cells significantly inhibits cell proliferation as measured by real time change in cell impedance. KD and negative control cells were seeded in duplicate into E-plate wells and the rate of proliferation was measured in real-time using the xCELLigence system. Proliferation of KD cells (green) was compared to the proliferation of negative control wells (red) over the course of 96 h. Data displayed as mean cell index +/− SEM for three individual experiments. (**C**) Reduction of Annexin A2 expression in MDA-MB-231 cells significantly inhibits cell migration as measured by wound healing assay. Knockdown and negative control cells were plated into dishes containing ibidi Culture Inserts. After 72 h, the inserts were removed to generate a cell free wound. The green zone indicates the wound size as measured at 0 h and 24 h post-removal of insert. Data displayed as percent wound closure for 3 individual experiments (6 fields of view each) +/− SEM, $p = 0.0002$. (**D**) Reduction of Annexin A2 expression significantly inhibits EGF directed cell migration as measured with transwell migration chambers. Cells were plated onto the upper chamber of transwell membrane and allowed to migrate towards 50 ng/mL EGF chemoattractant for 24 h. Cells were removed from the upper chamber, leaving only those that had migrated to the lower chamber. Migrated cells were stained with crystal violet and imaged using an inverted light microscope. Crystal violet stain was dissolved, and intensity measured as absorbance at 595 nm in triplicate. Data displayed as mean absorbance values for 3 individual experiments +/− SEM, $p < 0.0001$.

3.5. Gene Expression Analysis of Breast Cancer Tissue Shows ANXA2 is Specifically Upregulated in ER Negative Breast Cancer and Correlates with Rates of Metastasis

Due to our accumulating evidence of Annexin A2's importance in breast cancer progression in our in vitro cell models, we next investigated its expression in breast cancer patient tissue. RT-qPCR was used to analyze the level of ANXA2 mRNA in core biopsy tissue samples (normal = 5, cancer = 30). Surprisingly, despite several reports in the literature that ANXA2 is upregulated in breast cancer [42,59,60], we observed that expression of ANXA2 is significantly down regulated ($p = 0.0002$, Mann Whitney U test) in breast cancer tissue in comparison to pathologically normal breast tissue (Figure 5A). When ANXA2 expression levels were stratified according to nodal involvement—an indicator of breast cancer disease progression—we found a trend of increased ANXA2 expression in those who had nodal positive disease, but this was not statistically significant ($p = 0.8674$, Mann Whitney U test). Due to the sample size of this local cohort, the expression of ANXA2 according to receptor status could not be carried out reliably (ER+ = 24, ER- = 2, see Table S3). Thus, to add power to our investigation and to further interrogate the relationship between ER status and ANXA2 expression we then extended our study to publicly available gene expression datasets.

The NCBI GEO database is a public functional genomics data repository through which anonymized gene expression and corresponding clinical attributes can be accessed and analyzed. The dataset GSE42568 was chosen to closely reflect our local experimental cohort as it also contained gene expression data from normal ($n = 17$) and breast cancer ($n = 104$) tissue from Irish patients. Fold change of ANXA2 expression was measured as log2 using Affymetrix microarrays and probe set 201590_x_at and differences between groups was measured using Student's t-test. Analysis of this data set (Figure 5B) also showed a significant down regulation in ANXA2 in breast tumor tissue compared to normal breast tissue ($p < 0.0001$, Mann–Whitney U). Further, by categorizing the GSE42568 cohort according to ER expression, we see that ANXA2 expression is significantly upregulated in those with ER negative breast cancer tissue ($p = 0.0094$, Student's t-test), which is consistent with our cell line data. Reflecting the results seen in our RT-qPCR local cohort, ANXA2 appears to be slightly increased in those with nodal progression in GSE42568 ($p = 0.1121$, Student's t-test). The expression status of other receptors such as progesterone receptor (PR) and HER-2 were not recorded for this dataset (Table S4). To further verify the relationship between Annexin A2 and ER- breast cancer tissue, we interrogated available quantitative proteomic data from S. Tyanova et al. (2016) [25]. Through this we found, in support of our argument, that Annexin A2 expression at the level of the protein is significantly higher in ER- breast cancer tissue ($p = 0.0185$, Student's t-test) (Figure S5).

These trends are further evidenced by querying the METABRIC (Molecular Taxonomy of Breast Cancer International Consortium) cohort for ANXA2 expression. METABRIC is a publicly available database containing genomic and clinical data for over 2000 breast cancer patients [61]. As illustrated in Figure 5C, again, ER negative patients appear to have a higher expression of ANXA2 ($p < 0.0001$, Mann–Whitney U). When considering expression of other receptors, it can be seen that PR negative patients have a slight increase in ANXA2 expression overall ($p = 0.0037$, Mann–Whitney U), whereas HER-2 positive patients have increased ANXA2 expression compared to those who do not have HER-2 overexpression ($p < 0.0001$, Mann–Whitney U). The 3-gene classifier postulated by Haibe-Kains B et al. [62] allows for the stratification of tumor samples into four main clinical subtypes: ER+ high proliferation, ER+ low proliferation, TNBC (denoted as ER-/HER2- in METABRIC) and HER2+. Comparing ANXA2 expression between these subtypes reveals both ER positive subtypes to have a significantly lower expression of ANXA2 than their ER negative counterparts. (TNBC vs. ER+hi prolif, $p < 0.0001$; TNBC- vs. HER2+, $p = 0.2737$; TNBC vs. ER+ lo prolif, $p < 0.0001$; ER+ hi prolif vs. HER2+, $p < 0.0001$; ER+ hi prolif vs. ER+ lo prolif, $p = 0.4909$; HER2+ vs. ER+lo prolif, $p < 0.0001$, Kruskal-Wallis with Dunns adjustment). Once more, analysis of this large breast cancer cohort appears to link ANXA2 expression with ER negative breast cancer, encompassing both the aggressive triple negative and HER-2 overexpressing subtypes. In addition, those with a positive nodal status also have an increased expression of ANXA2 when compared to those without nodal involvement, regardless of receptor status ($p = 0.0050$, Mann–Whitney U).

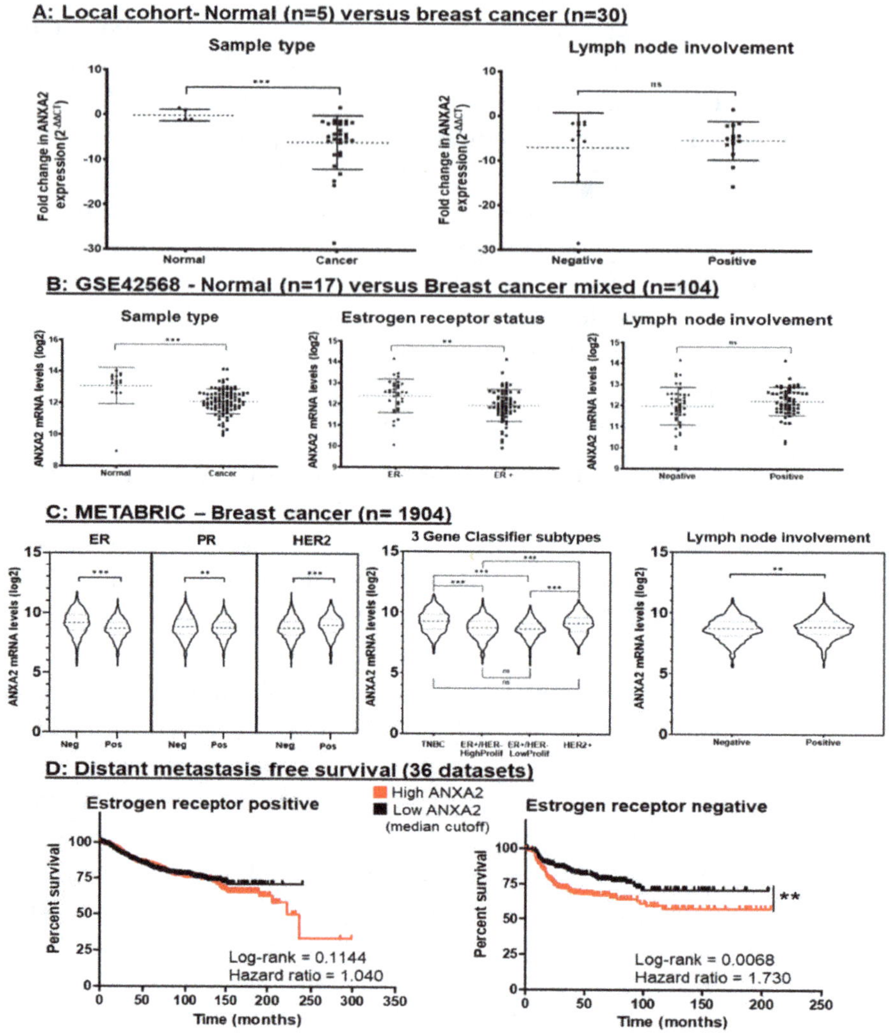

Figure 5. ANXA2 expression is associated with and is a prognostic indicator in ER negative breast cancer. (**A**) Analysis of ANXA2 expression using qRT-PCR of normal ($n = 5$) and breast cancer ($n = 30$) tissue samples. Analysis showed ANXA2 is significantly upregulated in normal tissue versus cancer tissue ($p = 0.0002$). Gene expression represented as fold change from normal closest to mean. Horizontal bars represent the mean and SD of fold change. Statistical difference between groups tested using Mann Whitney U test. (**B**) Analysis of ANXA2 mRNA expression in GSE42568 showing ANXA2 is upregulated in normal tissue versus cancer tissue, ($p < 0.0001$), in ER- breast cancer versus ER+ breast cancer ($p = 0.0094$) and in patients with nodal positive disease ($p = 0.1121$). Horizontal bars represent the mean and SD gene expression levels. Statistical difference between groups was measured using Student's t-test. (**C**) Analysis of ANXA2 mRNA expression of the METABRIC cohort showing upregulation of ANXA2 in ER negative ($p < 0.0001$), PR positive ($p = 0.0013$) HER2 positive ($p < 0.0001$) and in nodal positive ($p = 0.0050$) breast cancer patients. Using the 3 gene classifier (42) to stratify patients shows TNBC and HER2+ subtypes to have significantly higher expression of ANXA2 than both ER+/HER2- subtypes. Statistical difference between groups was calculated using Mann–Whitney U test. (**D**) Kaplan–Meier curves comparing the DMFS of patients with high ANXA2 expression (red) versus low ANXA2 expression (black) as determined by median expression value cut-off. Curves were analyzed using univariate log-rank tests. Expression of ANXA2 was found to not have a significant effect in ER+ patients whereas it did significantly increase probability of distant metastasis for ER- patients ($p = 0.0068$).

Taking into account the specificity we see of ANXA2 overexpression in ER-negative breast cancer and the potential influence ANXA2 has on disease progression in terms of lymph node involvement, we then wanted to ascertain the effect ANXA2 expression has on prognosis for breast cancer patients. Gene expression and survival data sets were downloaded from KMPlot.com and categorized according to ER expression (derived from gene expression data) [26]. Kaplan–Meier plots and log-rank tests were used to compare the distant metastasis free survival (DMFS) rates of patients stratified according to high or low ANXA2 expression, illustrated in Figure 5D. Using a median value as high/low cut off for ANXA2 expression, we found that for ER positive patients (n = 1395), ANXA2 expression had no significant effect on DMFS. Contrastingly, in patients with ER negative breast cancer (n = 351), high ANXA2 expression had a significant detrimental effect on DMFS, with a hazard ratio showing that those with high expression were almost 2 times more likely to develop distant metastasis (p = 0.0068, HR = 1.730). These results indicate that ANXA2 expression is correlated with poor clinical outcome, specifically distant metastasis for only the ER negative grouping of breast cancer patients.

4. Discussion

In this study, we utilized well-established transwell chamber assays, in combination with the BONCAT method and mass spectrometry to identify the nascent proteome that is synthesized when breast cancer cells undergo migration and invasion towards an EGF chemoattractant. This is an innovative method to assess the global proteomic changes that occur during these processes of cancer progression using an in vitro model. This experimental set-up allowed for a highly specific and controlled insight into the early translational events involved in the metastatic process in a model of aggressive ER negative breast cancer. Functional analysis of the proteins identified in this procedure showed that a wide range of cellular functions were regulated by the chemotactic EGF stimulation. Of these functions, gene ontology enrichment analysis revealed calcium ion binding to be the most significantly overrepresented GO term within our list. Calcium signaling has well-established links to processes of cancer progression such as proliferation, apoptosis and migration and invasion [63,64]. Furthermore, calcium flux has an intrinsic role in the breast due to lactation and the expression of several calcium pumps, involved in milk production, have been implicated in breast cancer [65]. Altered calcium homeostasis has been observed within breast cancer, however, it is currently unclear whether the dysregulation of calcium binding proteins like Annexin A2 are a cause or a consequence of this imbalance [64]. Despite this, it is known that Ca^{2+} level within the cell regulates Annexin A2's affinity for its binding partners and cellular components such as the plasma membrane and the actin cytoskeleton [51,66,67]. Apart from Annexin A2, calcium binding proteins such as the S100 family (several of which were identified in our screen, Table S1) have been thoroughly investigated in the context of cancer, as reviewed by Bresnick et al. [68] and these, along with Annexin A2, have the potential to be utilized as prognostic or therapeutic targets in breast cancer [69]. As well as successfully identifying Annexin A2 as a mediator of breast cancer cell migration and invasion, our screen has provided an extensive list of proteins potentially involved in breast cancer metastasis which require further investigation (Table S1). This valuable information can guide our and others' future research into deciphering the mechanics of cancer progression.

As migration and invasion constitute integral mechanisms in the development of cancer metastasis, the proteins identified by our model and mass spectrometry screen potentially have a role in cancer progression. We thus tested the value of our experimental design by conducting a focused study on Annexin A2—a protein identified in both our migration and invasion experiments. Annexin A2 is a calcium ion regulated membrane binding protein [67] found to play a role in a wide range of cellular processes, from endo- and exo- cytosis to proliferation and apoptosis [51]. One of its best-known functions is its hand in the proteolytic cascade, triggered by the activation of plasmin. The plasminogen activation system is vital for the tissue remodeling required for wound healing. This process however, has also been associated with cancer progression [70]. AnxA2 promotes the conversion of the inactive pre-protein plasminogen into its active version, the serine protease called plasmin. AnxA2 does this by binding both tissue plasminogen activator (tPA) and plasminogen thus bringing enzyme and substrate together spatially and significantly increasing the rate of conversion [71]. This fibrinolytic process produces proteases with the ability to degrade extracellular matrix (ECM) proteins and further activate matrix metallo-proteases (MMPs). When dysregulated, this degradation ability greatly accelerates the progression of cancer by giving the tumor a greater chance to invade through the ECM and metastasize to other locations in the body [60]. Annexin A2 has been proposed as a potential biomarker and therapeutic target for aggressive and metastatic cancers [72,73]. A recent study by Shen et al. even postulated the use of phosphorylated Annexin A2 for the selective imaging of solid tumors within the clinic [74]. Studies by Sharma et al. [58] and Yeatman et al. [75] have actually shown that the expression of Annexin A2 is higher in metastatic cancer cell lines versus non-metastatic cancer cells. The recent authoritative review by Sharma discusses these studies and similar findings in a number of cancer types and concluded by proposing Annexin A2 to be a universal signature of aggressive and metastatic cancer [72]. Our approach has added to the consensus that Annexin A2 plays an important role in the progression of breast cancer, suggesting that stimulus to migrate and invade with an EGF chemoattractant actually induces the nascent translation of Annexin A2.

The patterns of higher Annexin A2 expression at both a protein and gene level, as demonstrated using both western blotting and RT-qPCR methods (Figure 3A,C), as well as through interrogating independent genomic and proteomic data repositories (Figure 3B,C) suggested the importance of Annexin A2 expression in ER negative breast cancer. Thus, we next assessed the functional consequences of decreasing Annexin A2 expression using RNA interference techniques within an ER negative breast cancer model. Knockdown of Annexin A2 attenuated many of the traditional hallmarks of cancer exhibited by MDA-MB-231 cells. Cell growth and proliferation was diminished, adding to the evidence of Annexin A2's role in cell division, as reported in a number of experimental models [76,77]. Cell migration, as measured by both wound healing assay and EGF-induced directional migration, was significantly diminished upon reduction of Annexin A2 expression. As discussed earlier, cancer cell motility and chemotactic response to factors found within the tumor microenvironment are pillars of metastatic progression. Annexin A2 has been implicated in cell motility due to its role in binding actin filaments, regulating filament polymerization, and cytoskeletal organization [35–37]. In addition, it has been shown to recruit Rho GTPases, cofilin and other motility associated effector proteins [38,39,45,78,79] This role of Annexin A2 in promoting breast cancer cell migration equates with results from previous research [44,80,81] but to our knowledge is the first instance to show the specific reaction to an EGF gradient. To further validate the clinically relevant role of Annexin A2 in metastasis, future investigations are needed. Studies using Matrigel plug assays [82] and xenograft models [83] have indicated the potential of targeting Annexin A2 in angiogenesis and tumor growth in vivo. However, various measurements of invasion such as 3-D spheroid formation and gelatin zymography, in combination with physiologically relevant, spontaneous metastatic mouse models (such as those used in [84]) are essential for a fully comprehensive insight into Annexin A2's role in the metastatic cascade.

A link with EGF signaling could be contributed to the interaction between Annexin A2 and the EGF receptor (EGFR). Our results (Figure 3B(ii)) indicate a strong correlation between the expression of EGFR and Annexin A2 in ER negative cell line models. However, within the literature, this relationship is riddled with conflicting reports with some data showing that reduction of Annexin A2 dampens the downstream activation of the EGFR pathway [44]; while others report that inhibition of Annexin A2 expression enhanced both EGF induced cell migration and downstream activation of JNK and Akt in mouse models [45]. Indeed, our results (Figure 3D) and others [85], show that EGF stimulation appears to transiently increase Annexin A2 phosphorylation at Tyr24. This phosphorylation is likely caused by EGFR mediated activation of Src kinase, of which Annexin A2 is a known substrate [44]. Current literature suggest that this phosphorylation promotes Annexin A2 translocation to cell and exosomal surfaces where it can enact its pro-metastatic functions by bridging proteins together to activate ECM remodeling enzymes and even attract pro-tumorigenic macrophages into the TME [54,58,81,86–88].

Recent work by Y Fan et al., showed that the reduction in migration and invasion attributed to Annexin A2 suppression can be rescued by the exogenous expression of a phospho-mimic-ANXA2 but not by a phospho-null-ANXA2, highlighting the importance of the phosphorylation we have observed [53]. Furthermore, a number of studies have linked growth factor mediated phosphorylation of Annexin A2 to the promotion of EMT in a variety cancer types [55–57,81]. Although the methods used for assessing EMT vary, many come to the consensus that Annexin A2 plays a role in triggering the mesenchymal and migratory phenotype in cancer cells via the transcriptional program of EMT. This agrees with our observations of Annexin A2 suppression diminishing the migratory capacity of breast cancer cells (Figure 4C,D) and potentially correlating with loss of EMT marker, E-Cadherin, in specific cases (Figure 3A). Despite the increase in Annexin A2 phosphorylation and the nascent translation proposed by our mass spectrometry identification (Figure 3D), the level of total Annexin A2 within the cell remains constant when measured by western blot. This is similar to results seen by Maji et al. in which exosomal Annexin A2 was seen to increase in a progressively metastatic breast cancer cell line model, despite seeing no change when the whole cell lysate is analyzed [87]. This could suggest that the total cellular pool outweighs the newly synthesized pool of Annexin A2 that is translated upon directional EGF stimulation, thus masking any potential increase in protein levels. This is supported by reports of Annexin A2 having a long half-life (40–50 h) [89–91] and by our own findings that suggest that it is only 72 h post transfection with siRNA against ANXA2 that we see the knockdown and maintenance of its knockdown beyond 35% (data not shown). Further examination is ongoing to ascertain whether Annexin A2 is newly synthesized during this process to replace the protein lost to cell surface translocation. Exosomal or secreted Annexin A2 has been reported to promote breast cancer metastasis and angiogenesis in a number of studies [58,82,87] and has even been proposed as a serum based biomarker for breast cancer detection [82,92]. Taken together, this suggests that Annexin A2 is involved in processes that promote breast cancer disease progression and this could be particularly true for tumors driven by EGFR overexpression. This is further compacted by our findings of a significant correlation between Annexin A2 and EGFR expression and by the observation that breast cancers found to have EGFR overexpression are more commonly ER negative than ER positive [93].

Having established the importance of Annexin A2 expression in maintaining proliferation, wound healing, and migration within our in vitro model (Figure 3), we next moved to investigate its expression in patient tissue. Interestingly, we found that when unstratified breast cancer samples were compared to normal breast tissue, Annexin A2 expression was significantly lower in the cancer tissue. This finding was validated in the larger GSE4825 cohort. This is in contrast to immunohistochemical analysis by Sharma et al. [58] which showed Annexin A2 to be undetectable in normal breast epithelia but strongly and consistently expressed in invasive breast cancer tissue. This discrepancy could be explained by a differential level of translational control between mRNA to protein. However, this requires further investigation as our result poses a challenge to the current general consensus that Annexin A2 is upregulated in breast cancer compared to pathologically normal tissue [42,59,60].

The stratification of breast cancer patients according to their ER expression status (GSE42568, METABRIC and S. Tyanova, et al. (2016) cohorts) revealed that the trend seen in cell lines also applies to human breast cancer tissue samples. ER negative breast cancer encompasses both the triple negative and HER-2 overexpressing subtypes [49]. These are often more aggressive and have fewer treatment strategies in comparison with the more common ER positive breast cancers. Women diagnosed with ER negative tumors are usually diagnosed at an earlier age and experience a higher mortality rate [94]. With the exception of trastuzumab's unmatched beneficial effect for HER-2 positive patients [95], traditionally, toxic and unspecific chemotherapy is used to treat ER negative tumors. In addition, a caveat to the success of trastuzumab for HER-2 positive breast cancer is that a large majority of those treated develop resistance over time [96]. This highlights the need for targeted therapies which can mimic the wide reaching success of anti-estrogens such as tamoxifen for ER positive cancers [97]. The large sample size of the METABRIC analysis also revealed Annexin A2 to be upregulated in HER2 positive breast cancer, contradicting what has been previously published [44] and suggesting the value of investigating Annexin A2 in breast cancer is not limited to the triple negative subtypes. The potential of Annexin A2 as a therapeutic target or marker of aggressiveness in ER negative breast cancer is further solidified by its links to disease progression and metastasis. Analysis of our local tissue cohort (RT-qPCR), GSE42568, and the METABRIC (Affymetrix) cohort all show that Annexin A2 expression is higher in patients with lymph node involvement (Figure 5A–C). Due to the lymphatic drainage of the breast tissue, the lymph nodes are often the first location of metastatic spread. Presence of metastasized tumor cells in the axillary lymph nodes is one of the strongest prognostic indicators in breast cancer [98] and as a result, lymph node status is a parameter used to stage the disease [99]. The trend of higher ANXA2 expression in patients with lymph node involvement in our local cohort and the larger GSE42568, is not statistically significant, however, this could be attributed to the fact that only a small proportion of patients in these cohorts are ER negative (Tables S3 and S4), thus masking the possible effect of ANXA2 expression on disease progression. Annexin A2 is again associated with metastatic progression of disease when we show that risk of distant metastasis is significantly higher in ER negative breast cancer patients with high expression of ANXA2. This statistically significant result does not hold true for ER positive patients, providing evidence that the potential of targeting Annexin A2 holds promise for the more aggressive and harder to treat ER negative breast cancers.

5. Conclusions

Collectively, we have developed a novel approach to assess the changes that occur in the nascent proteome when cancer cells undergo stimulation to carry out the pro-metastatic functions of invasion and migration. We have proven the validity and usefulness of this approach by demonstrating how Annexin A2, one of the elucidated proteins, plays a pivotal role in maintaining the malignant phenotype of estrogen receptor negative breast cancer. The variety of experimental techniques and independent transcriptomic and proteomic datasets encompassing both cell line models and patient tissue in this study provides an emphatic argument that Annexin A2 plays a key role in ER negative breast cancer. Our work has used innovative models to add substantial weight to the findings by others, linking Annexin A2 to breast cancer. This study provides novel insight into the potential mechanisms of metastasis in ER negative breast cancer. Further validation is needed to build upon this study and truly assess the viability of targeting Annexin A2 to prevent cancer metastasis in vivo. Ultimately, this knowledge suggests that targeting Annexin A2 merits investigation as a novel therapeutic strategy in treating the pharmacologically challenging ER negative subtype, and this may help prevent breast cancer progression to the later metastatic stages of disease.

Supplementary Materials
Figure S1: (A) Score and sequence coverage of Annexin A2 from mass spectrometry and MASCOT server results (B) representative mass spectrum of AnxA2 peptide Figure S2: Proteomic expression data for CCLE cell lines from Nusinow, D. P. et al. (2020) Figure S3: Publicly available Affymetrix gene expression data for human breast cancer cell lines. Figure S4: Additional functional assays with second siRNA clone on MDA-MB-231 cells. Figure S5: Proteomic expression data for breast cancer tissue samples from S. Tyanova, et al. (2016). Figure S6, S7, and S8: full, uncropped images of western blots in manuscript. Table S1: List of newly synthesised proteins identified in 2D migration experiment. Table S2: List of newly synthesised proteins identified in 3D invasion experiment. Table S3: Sequences of the small interfering RNA used for ANXA2 knockdown. Table S4: Clinical information of local breast cancer cohort (n = 30) Table S5. Clinical information of GSE42568 breast cancer cohort (n = 104).

Author Contributions: A.F.M. and P.A.K. designed the experiments and wrote the manuscript. A.B.M. performed the experiments and interpreted results. B.M., J.N. & K.M. assisted with experimental design and collection, interpretation and visualization of data. B.M., J.N., A.J.L. & K.M. also performed manuscript editing. M.E.M., A.B.M., A.L. and S.T. collected informed consent, tissue, and pathological data from patients. P.A.K., K.M. and A.J.L. conceived and supervised the study. All authors have read and agreed to the published version of the manuscript.

Acknowledgments: We are grateful to our colleagues in the Laboratory of Cellular and Molecular Biology for helpful discussions and critical review. Additionally, we are grateful to all patients and their families from University Hospital Limerick who generously donated tissue to the Breast Biobank.

References

1. Balkwill, F.R.; Capasso, M.; Hagemann, T. The tumor microenvironment at a glance. *J. Cell Sci.* **2012**, *125*, 5591–5596. [CrossRef]
2. Sahai, E. Illuminating the metastatic process. *Nat. Rev. Cancer* **2007**, *7*, 737–749. [CrossRef] [PubMed]
3. Chambers, A.F.; Groom, A.C.; MacDonald, I.C. Dissemination and growth of cancer cells in metastatic sites. *Nat. Rev. Cancer* **2002**, *2*, 563–572. [CrossRef] [PubMed]
4. Stuelten, C.H.; Parent, C.A.; Montell, D.J. Cell motility in cancer invasion and metastasis: Insights from simple model organisms. *Nat. Rev. Cancer* **2018**, *18*, 296–312. [CrossRef] [PubMed]
5. Welch, D.R.; Hurst, D.R. Defining the Hallmarks of Metastasis. *Cancer Res.* **2019**, *79*, 3011–3027. [CrossRef] [PubMed]
6. Albini, A.; Benelli, R.; Noonan, D.M.; Brigati, C. The "chemoinvasion assay": A tool to study tumor and endothelial cell invasion of basement membranes. *Int. J. Dev. Biol.* **2004**, *48*, 563–571. [CrossRef]
7. Harbeck, N.; Penault-Llorca, F.; Cortes, J.; Gnant, M.; Houssami, N.; Poortmans, P.; Ruddy, K.; Tsang, J.; Cardoso, F. Breast cancer. *Nat. Rev. Dis. Primers* **2019**, *5*, 66. [CrossRef]
8. Kennecke, H.; Yerushalmi, R.; Woods, R.; Cheang, M.C.; Voduc, D.; Speers, C.H.; Nielsen, T.O.; Gelmon, K. Metastatic behavior of breast cancer subtypes. *J. Clin. Oncol.* **2010**, *28*, 3271–3277. [CrossRef]
9. Dieterich, D.C.; Link, A.J.; Graumann, J.; Tirrell, D.A.; Schuman, E.M. Selective identification of newly synthesized proteins in mammalian cells using bioorthogonal noncanonical amino acid tagging (BONCAT). *Proc. Natl. Acad. Sci. USA* **2006**, *103*, 9482–9487. [CrossRef]
10. Schwanhausser, B.; Busse, D.; Li, N.; Dittmar, G.; Schuchhardt, J.; Wolf, J.; Chen, W.; Selbach, M. Global quantification of mammalian gene expression control. *Nature* **2011**, *473*, 337–342. [CrossRef]
11. Geiger, T.; Madden, S.F.; Gallagher, W.M.; Cox, J.; Mann, M. Proteomic portrait of human breast cancer progression identifies novel prognostic markers. *Cancer Res.* **2012**, *72*, 2428–2439. [CrossRef] [PubMed]
12. Yanovich, G.; Agmon, H.; Harel, M.; Sonnenblick, A.; Peretz, T.; Geiger, T. Clinical Proteomics of Breast Cancer Reveals a Novel Layer of Breast Cancer Classification. *Cancer Res.* **2018**, *78*, 6001–6010. [CrossRef] [PubMed]
13. Al-Wajeeh, A.S.; Salhimi, S.M.; Al-Mansoub, M.A.; Khalid, I.A.; Harvey, T.M.; Latiff, A.; Ismail, M.N. Comparative proteomic analysis of different stages of breast cancer tissues using ultra high performance liquid chromatography tandem mass spectrometer. *PLoS ONE* **2020**, *15*, e0227404. [CrossRef] [PubMed]
14. Chakraborty, S.; Hosen, M.I.; Ahmed, M.; Shekhar, H.U. Onco-Multi-OMICS Approach: A New Frontier in Cancer Research. *Biomed. Res. Int.* **2018**, *2018*, 9836256. [CrossRef] [PubMed]
15. Mi, H.; Muruganujan, A.; Ebert, D.; Huang, X.; Thomas, P.D. PANTHER version 14: More genomes, a new PANTHER GO-slim and improvements in enrichment analysis tools. *Nucleic Acids Res.* **2019**, *47*, D419–D426. [CrossRef]
16. Szklarczyk, D.; Gable, A.L.; Lyon, D.; Junge, A.; Wyder, S.; Huerta-Cepas, J.; Simonovic, M.; Doncheva, N.T.; Morris, J.H.; Bork, P.; et al. STRING v11: Protein-protein association networks with increased coverage, supporting functional discovery in genome-wide experimental datasets. *Nucleic Acids Res.* **2019**, *47*, D607–D613. [CrossRef]
17. Doncheva, N.T.; Morris, J.H.; Gorodkin, J.; Jensen, L.J. Cytoscape StringApp: Network Analysis and Visualization of Proteomics Data. *J. Proteome Res.* **2019**, *18*, 623–632. [CrossRef]
18. Dowling, C.M.; Walsh, D.; Coffey, J.C.; Kiely, P.A. The importance of selecting the appropriate reference genes for quantitative real time PCR as illustrated using colon cancer cells and tissue. *F1000Research* **2016**, *5*, 99. [CrossRef]
19. Nusinow, D.P.; Szpyt, J.; Ghandi, M.; Rose, C.M.; McDonald, E.R.; Kalocsay, M.; Jane-Valbuena, J.; Gelfand, E.; Schweppe, D.K.; Jedrychowski, M.; et al. Quantitative Proteomics of the Cancer Cell Line Encyclopedia. *Cell* **2020**, *180*, 387–402.e16. [CrossRef]
20. Nusinow, D.P.; Gygi, S.P. A Guide to the Quantitative Proteomic Profiles of the Cancer Cell Line Encyclopedia. *bioRxiv* **2020**. [CrossRef]
21. Barretina, J.; Caponigro, G.; Stransky, N.; Venkatesan, K.; Margolin, A.A.; Kim, S.; Wilson, C.J.; Lehar, J.; Kryukov, G.V.; Sonkin, D.; et al. The Cancer Cell Line Encyclopedia enables predictive modelling of anticancer drug sensitivity. *Nature* **2012**, *483*, 603–607. [CrossRef] [PubMed]

22. Smith, S.E.; Mellor, P.; Ward, A.K.; Kendall, S.; McDonald, M.; Vizeacoumar, F.S.; Vizeacoumar, F.J.; Napper, S.; Anderson, D.H. Molecular characterization of breast cancer cell lines through multiple omic approaches. *Breast Cancer Res.* **2017**, *19*, 65. [CrossRef]
23. Dai, X.; Cheng, H.; Bai, Z.; Li, J. Breast Cancer Cell Line Classification and Its Relevance with Breast Tumor Subtyping. *J. Cancer* **2017**, *8*, 3131–3141. [CrossRef] [PubMed]
24. Clarke, C.; Madden, S.F.; Doolan, P.; Aherne, S.T.; Joyce, H.; O'Driscoll, L.; Gallagher, W.M.; Hennessy, B.T.; Moriarty, M.; Crown, J.; et al. Correlating transcriptional networks to breast cancer survival: A large-scale coexpression analysis. *Carcinogenesis* **2013**, *34*, 2300–2308. [CrossRef] [PubMed]
25. Tyanova, S.; Albrechtsen, R.; Kronqvist, P.; Cox, J.; Mann, M.; Geiger, T. Proteomic maps of breast cancer subtypes. *Nat. Commun.* **2016**, *7*, 10259. [CrossRef] [PubMed]
26. Gyorffy, B.; Lanczky, A.; Eklund, A.C.; Denkert, C.; Budczies, J.; Li, Q.; Szallasi, Z. An online survival analysis tool to rapidly assess the effect of 22,277 genes on breast cancer prognosis using microarray data of 1809 patients. *Breast Cancer Res. Treat.* **2010**, *123*, 725–731. [CrossRef] [PubMed]
27. Price, J.T.; Tiganis, T.; Agarwal, A.; Djakiew, D.; Thompson, E.W. Epidermal growth factor promotes MDA-MB-231 breast cancer cell migration through a phosphatidylinositol 3′-kinase and phospholipase C-dependent mechanism. *Cancer Res.* **1999**, *59*, 5475–5478.
28. Wang, S.J.; Saadi, W.; Lin, F.; Minh-Canh Nguyen, C.; Li Jeon, N. Differential effects of EGF gradient profiles on MDA-MB-231 breast cancer cell chemotaxis. *Exp. Cell Res.* **2004**, *300*, 180–189. [CrossRef]
29. Lu, Z.; Jiang, G.; Blume-Jensen, P.; Hunter, T. Epidermal growth factor-induced tumor cell invasion and metastasis initiated by dephosphorylation and downregulation of focal adhesion kinase. *Mol. Cell. Biol.* **2001**, *21*, 4016–4031. [CrossRef]
30. Biswenger, V.; Baumann, N.; Jurschick, J.; Hackl, M.; Battle, C.; Schwarz, J.; Horn, E.; Zantl, R. Characterization of EGF-guided MDA-MB-231 cell chemotaxis in vitro using a physiological and highly sensitive assay system. *PLoS ONE* **2018**, *13*, e0203040. [CrossRef]
31. Sayers, E.W.; Agarwala, R.; Bolton, E.E.; Brister, J.R.; Canese, K.; Clark, K.; Connor, R.; Fiorini, N.; Funk, K.; Hefferon, T.; et al. Database resources of the National Center for Biotechnology Information. *Nucleic Acids Res.* **2019**, *47*, D23–D28. [CrossRef] [PubMed]
32. Gerke, V.; Moss, S.E. Annexins: From structure to function. *Physiol. Rev.* **2002**, *82*, 331–371. [CrossRef] [PubMed]
33. Boye, T.L.; Nylandsted, J. Annexins in plasma membrane repair. *Biol. Chem.* **2016**, *397*, 961–969. [CrossRef]
34. Hajjar, K.A. The Biology of Annexin A2: From Vascular Fibrinolysis to Innate Immunity. *Trans. Am. Clin. Clim. Assoc.* **2015**, *126*, 144–155.
35. Grieve, A.G.; Moss, S.E.; Hayes, M.J. Annexin A2 at the interface of actin and membrane dynamics: A focus on its roles in endocytosis and cell polarization. *Int. J. Cell Biol.* **2012**, *2012*, 852430. [CrossRef] [PubMed]
36. Xiao, L.; Jin, H.; Duan, W.; Hou, Y. Roles of N-terminal Annexin A2 phosphorylation sites and miR-206 in colonic adenocarcinoma. *Life Sci.* **2020**, *253*, 117740. [CrossRef]
37. Morel, E.; Parton, R.G.; Gruenberg, J. Annexin A2-dependent polymerization of actin mediates endosome biogenesis. *Dev. Cell* **2009**, *16*, 445–457. [CrossRef]
38. Rescher, U.; Ludwig, C.; Konietzko, V.; Kharitonenkov, A.; Gerke, V. Tyrosine phosphorylation of annexin A2 regulates Rho-mediated actin rearrangement and cell adhesion. *J. Cell Sci.* **2008**, *121*, 2177–2185. [CrossRef]
39. de Graauw, M.; Tijdens, I.; Smeets, M.B.; Hensbergen, P.J.; Deelder, A.M.; van de Water, B. Annexin A2 phosphorylation mediates cell scattering and branching morphogenesis via cofilin Activation. *Mol. Cell. Biol.* **2008**, *28*, 1029–1040. [CrossRef]
40. Bunnell, T.M.; Burbach, B.J.; Shimizu, Y.; Ervasti, J.M. beta-Actin specifically controls cell growth, migration, and the G-actin pool. *Mol. Biol. Cell* **2011**, *22*, 4047–4058. [CrossRef]
41. Grewal, T.; Wason, S.J.; Enrich, C.; Rentero, C. Annexins—Insights from knockout mice. *Biol. Chem.* **2016**, *397*, 1031–1053. [CrossRef] [PubMed]
42. Christensen, M.V.; Hogdall, C.K.; Jochumsen, K.M.; Hogdall, E.V.S. Annexin A2 and cancer: A systematic review. *Int. J. Oncol.* **2018**, *52*, 5–18. [CrossRef] [PubMed]
43. Zhao, W.Q.; Chen, G.H.; Chen, H.; Pascale, A.; Ravindranath, L.; Quon, M.J.; Alkon, D.L. Secretion of Annexin II via activation of insulin receptor and insulin-like growth factor receptor. *J. Biol. Chem.* **2003**, *278*, 4205–4215. [CrossRef]

44. Shetty, P.K.; Thamake, S.I.; Biswas, S.; Johansson, S.L.; Vishwanatha, J.K. Reciprocal regulation of annexin A2 and EGFR with Her-2 in Her-2 negative and herceptin-resistant breast cancer. *PLoS ONE* **2012**, *7*, e44299. [CrossRef]
45. de Graauw, M.; Cao, L.; Winkel, L.; van Miltenburg, M.H.; le Devedec, S.E.; Klop, M.; Yan, K.; Pont, C.; Rogkoti, V.M.; Tijsma, A.; et al. Annexin A2 depletion delays EGFR endocytic trafficking via cofilin activation and enhances EGFR signaling and metastasis formation. *Oncogene* **2014**, *33*, 2610–2619. [CrossRef] [PubMed]
46. Frohlich, M.; Motte, P.; Galvin, K.; Takahashi, H.; Wands, J.; Ozturk, M. Enhanced expression of the protein kinase substrate p36 in human hepatocellular carcinoma. *Mol. Cell. Biol.* **1990**, *10*, 3216–3223. [CrossRef] [PubMed]
47. Hanahan, D.; Weinberg, R.A. Hallmarks of cancer: The next generation. *Cell* **2011**, *144*, 646–674. [CrossRef]
48. Fendt, S.M. Metabolic vulnerabilities of metastasizing cancer cells. *BMC Biol.* **2019**, *17*, 54. [CrossRef]
49. Sorlie, T.; Perou, C.M.; Tibshirani, R.; Aas, T.; Geisler, S.; Johnsen, H.; Hastie, T.; Eisen, M.B.; van de Rijn, M.; Jeffrey, S.S.; et al. Gene expression patterns of breast carcinomas distinguish tumor subclasses with clinical implications. *Proc. Natl. Acad. Sci. USA* **2001**, *98*, 10869–10874. [CrossRef]
50. Grindheim, A.K.; Saraste, J.; Vedeler, A. Protein phosphorylation and its role in the regulation of Annexin A2 function. *Biochim. Biophys. Acta Gen. Subj.* **2017**, *1861*, 2515–2529. [CrossRef]
51. Bharadwaj, A.; Bydoun, M.; Holloway, R.; Waisman, D. Annexin A2 heterotetramer: Structure and function. *Int. J. Mol. Sci.* **2013**, *14*, 6259–6305. [CrossRef]
52. Lokman, N.A.; Ween, M.P.; Oehler, M.K.; Ricciardelli, C. The role of annexin A2 in tumorigenesis and cancer progression. *Cancer Microenviron.* **2011**, *4*, 199–208. [CrossRef] [PubMed]
53. Fan, Y.; Si, W.; Ji, W.; Wang, Z.; Gao, Z.; Tian, R.; Song, W.; Zhang, H.; Niu, R.; Zhang, F. Rack1 mediates tyrosine phosphorylation of Anxa2 by Src and promotes invasion and metastasis in drug-resistant breast cancer cells. *Breast Cancer Res.* **2019**, *21*, 66. [CrossRef] [PubMed]
54. Zheng, L.; Foley, K.; Huang, L.; Leubner, A.; Mo, G.; Olino, K.; Edil, B.H.; Mizuma, M.; Sharma, R.; Le, D.T.; et al. Tyrosine 23 phosphorylation-dependent cell-surface localization of annexin A2 is required for invasion and metastases of pancreatic cancer. *PLoS ONE* **2011**, *6*, e19390. [CrossRef] [PubMed]
55. Cui, L.; Song, J.; Wu, L.; Cheng, L.; Chen, A.; Wang, Y.; Huang, Y.; Huang, L. Role of Annexin A2 in the EGF-induced epithelial-mesenchymal transition in human CaSki cells. *Oncol. Lett.* **2017**, *13*, 377–383. [CrossRef] [PubMed]
56. Rocha, M.R.; Barcellos-de-Souza, P.; Sousa-Squiavinato, A.C.M.; Fernandes, P.V.; de Oliveira, I.M.; Boroni, M.; Morgado-Diaz, J.A. Annexin A2 overexpression associates with colorectal cancer invasiveness and TGF-ss induced epithelial mesenchymal transition via Src/ANXA2/STAT3. *Sci. Rep.* **2018**, *8*, 11285. [CrossRef]
57. Wu, M.; Sun, Y.; Xu, F.; Liang, Y.; Liu, H.; Yi, Y. Annexin A2 Silencing Inhibits Proliferation and Epithelial-to-mesenchymal Transition through p53-Dependent Pathway in NSCLCs. *J. Cancer* **2019**, *10*, 1077–1085. [CrossRef]
58. Sharma, M.R.; Koltowski, L.; Ownbey, R.T.; Tuszynski, G.P.; Sharma, M.C. Angiogenesis-associated protein annexin II in breast cancer: Selective expression in invasive breast cancer and contribution to tumor invasion and progression. *Exp. Mol. Pathol.* **2006**, *81*, 146–156. [CrossRef]
59. Wang, C.Y.; Lin, C.F. Annexin A2: Its molecular regulation and cellular expression in cancer development. *Dis. Markers* **2014**, *2014*, 308976. [CrossRef]
60. Xu, X.H.; Pan, W.; Kang, L.H.; Feng, H.; Song, Y.Q. Association of annexin A2 with cancer development (Review). *Oncol. Rep.* **2015**, *33*, 2121–2128. [CrossRef]
61. Pereira, B.; Chin, S.F.; Rueda, O.M.; Vollan, H.K.; Provenzano, E.; Bardwell, H.A.; Pugh, M.; Jones, L.; Russell, R.; Sammut, S.J.; et al. The somatic mutation profiles of 2,433 breast cancers refines their genomic and transcriptomic landscapes. *Nat. Commun.* **2016**, *7*, 11479. [CrossRef] [PubMed]
62. Haibe-Kains, B.; Desmedt, C.; Loi, S.; Culhane, A.C.; Bontempi, G.; Quackenbush, J.; Sotiriou, C. A three-gene model to robustly identify breast cancer molecular subtypes. *J. Natl. Cancer Inst.* **2012**, *104*, 311–325. [CrossRef] [PubMed]
63. Prevarskaya, N.; Skryma, R.; Shuba, Y. Calcium in tumour metastasis: New roles for known actors. *Nat. Rev. Cancer* **2011**, *11*, 609–618. [CrossRef]
64. Azimi, I.; Roberts-Thomson, S.J.; Monteith, G.R. Calcium influx pathways in breast cancer: Opportunities for pharmacological intervention. *Br. J. Pharm.* **2014**, *171*, 945–960. [CrossRef] [PubMed]
65. VanHouten, J.; Sullivan, C.; Bazinet, C.; Ryoo, T.; Camp, R.; Rimm, D.L.; Chung, G.; Wysolmerski, J. PMCA2

regulates apoptosis during mammary gland involution and predicts outcome in breast cancer. *Proc. Natl. Acad. Sci. USA* **2010**, *107*, 11405–11410. [CrossRef]

66. Ikebuchi, N.W.; Waisman, D.M. Calcium-dependent regulation of actin filament bundling by lipocortin-85. *J. Biol. Chem.* **1990**, *265*, 3392–3400.
67. Grill, D.; Matos, A.L.L.; de Vries, W.C.; Kudruk, S.; Heflik, M.; Dorner, W.; Mootz, H.D.; Jan Ravoo, B.; Galla, H.J.; Gerke, V. Bridging of membrane surfaces by annexin A2. *Sci. Rep.* **2018**, *8*, 14662. [CrossRef]
68. Bresnick, A.R.; Weber, D.J.; Zimmer, D.B. S100 proteins in cancer. *Nat. Rev. Cancer* **2015**, *15*, 96–109. [CrossRef]
69. Cancemi, P.; Buttacavoli, M.; Di Cara, G.; Albanese, N.N.; Bivona, S.; Pucci-Minafra, I.; Feo, S. A multiomics analysis of S100 protein family in breast cancer. *Oncotarget* **2018**, *9*, 29064–29081. [CrossRef]
70. Ranson, M.; Andronicos, N.M. Plasminogen binding and cancer: Promises and pitfalls. *Front. Biosci.* **2003**, *8*, s294–s304. [CrossRef]
71. Hajjar, K.A.; Krishnan, S. Annexin II: A mediator of the plasmin/plasminogen activator system. *Trends Cardiovasc. Med.* **1999**, *9*, 128–138. [CrossRef]
72. Sharma, M.C. Annexin A2 (ANX A2): An emerging biomarker and potential therapeutic target for aggressive cancers. *Int. J. Cancer* **2019**, *144*, 2074–2081. [CrossRef] [PubMed]
73. Jaiswal, J.K.; Nylandsted, J. S100 and annexin proteins identify cell membrane damage as the Achilles heel of metastatic cancer cells. *Cell Cycle* **2015**, *14*, 502–509. [CrossRef] [PubMed]
74. Shen, D.; Xu, B.; Liang, K.; Tang, R.; Sudlow, G.P.; Egbulefu, C.; Guo, K.; Som, A.; Gilson, R.; Maji, D.; et al. Selective imaging of solid tumours via the calcium-dependent high-affinity binding of a cyclic octapeptide to phosphorylated Annexin A2. *Nat. Biomed. Eng.* **2020**, *4*, 298–313. [CrossRef]
75. Yeatman, T.J.; Updyke, T.V.; Kaetzel, M.A.; Dedman, J.R.; Nicolson, G.L. Expression of annexins on the surfaces of non-metastatic and metastatic human and rodent tumor cells. *Clin. Exp. Metastasis* **1993**, *11*, 37–44. [CrossRef]
76. Benaud, C.; Le Dez, G.; Mironov, S.; Galli, F.; Reboutier, D.; Prigent, C. Annexin A2 is required for the early steps of cytokinesis. *EMBO Rep.* **2015**, *16*, 481–489. [CrossRef]
77. Vishwanatha, J.K.; Kumble, S. Involvement of annexin II in DNA replication: Evidence from cell-free extracts of Xenopus eggs. *J. Cell Sci.* **1993**, *105 Pt 2*, 533–540.
78. Yang, S.F.; Hsu, H.L.; Chao, T.K.; Hsiao, C.J.; Lin, Y.F.; Cheng, C.W. Annexin A2 in renal cell carcinoma: Expression, function, and prognostic significance. *Urol. Oncol.* **2015**, *33*, 22.e11–22.e21. [CrossRef]
79. Zhao, P.; Zhang, W.; Wang, S.J.; Yu, X.L.; Tang, J.; Huang, W.; Li, Y.; Cui, H.Y.; Guo, Y.S.; Tavernier, J.; et al. HAb18G/CD147 promotes cell motility by regulating annexin II-activated RhoA and Rac1 signaling pathways in hepatocellular carcinoma cells. *Hepatology* **2011**, *54*, 2012–2024. [CrossRef]
80. Sharma, M.; Ownbey, R.T.; Sharma, M.C. Breast cancer cell surface annexin II induces cell migration and neoangiogenesis via tPA dependent plasmin generation. *Exp. Mol. Pathol.* **2010**, *88*, 278–286. [CrossRef]
81. Wang, T.; Yuan, J.; Zhang, J.; Tian, R.; Ji, W.; Zhou, Y.; Yang, Y.; Song, W.; Zhang, F.; Niu, R. Anxa2 binds to STAT3 and promotes epithelial to mesenchymal transition in breast cancer cells. *Oncotarget* **2015**, *6*, 30975–30992. [CrossRef] [PubMed]
82. Chaudhary, P.; Gibbs, L.D.; Maji, S.; Lewis, C.M.; Suzuki, S.; Vishwanatha, J.K. Serum exosomal-annexin A2 is associated with African-American triple-negative breast cancer and promotes angiogenesis. *Breast Cancer Res.* **2020**, *22*, 11. [CrossRef]
83. Sharma, M.C.; Tuszynski, G.P.; Blackman, M.R.; Sharma, M. Long-term efficacy and downstream mechanism of anti-annexinA2 monoclonal antibody (anti-ANX A2 mAb) in a pre-clinical model of aggressive human breast cancer. *Cancer Lett.* **2016**, *373*, 27–35. [CrossRef] [PubMed]
84. Kijewska, M.; Viski, C.; Turrell, F.; Fitzpatrick, A.; van Weverwijk, A.; Gao, Q.; Iravani, M.; Isacke, C.M. Using an in-vivo syngeneic spontaneous metastasis model identifies ID2 as a promoter of breast cancer colonisation in the brain. *Breast Cancer Res.* **2019**, *21*, 4. [CrossRef] [PubMed]
85. Sawyer, S.T.; Cohen, S. Epidermal growth factor stimulates the phosphorylation of the calcium-dependent 35,000-dalton substrate in intact A-431 cells. *J. Biol. Chem.* **1985**, *260*, 8233–8236. [PubMed]
86. Valapala, M.; Maji, S.; Borejdo, J.; Vishwanatha, J.K. Cell surface translocation of annexin A2 facilitates glutamate-induced extracellular proteolysis. *J. Biol. Chem.* **2014**, *289*, 15915–15926. [CrossRef] [PubMed]
87. Maji, S.; Chaudhary, P.; Akopova, I.; Nguyen, P.M.; Hare, R.J.; Gryczynski, I.; Vishwanatha, J.K. Exosomal Annexin II Promotes Angiogenesis and Breast Cancer Metastasis. *Mol. Cancer Res.* **2017**, *15*, 93–105. [CrossRef]

88. Yuan, J.; Yang, Y.; Gao, Z.; Wang, Z.; Ji, W.; Song, W.; Zhang, F.; Niu, R. Tyr23 phosphorylation of Anxa2 enhances STAT3 activation and promotes proliferation and invasion of breast cancer cells. *Breast Cancer Res. Treat.* **2017**, *164*, 327–340. [CrossRef]
89. Rescher, U.; Gerke, V. Annexins—Unique membrane binding proteins with diverse functions. *J. Cell Sci.* **2004**, *117*, 2631–2639. [CrossRef]
90. Cuervo, A.M.; Gomes, A.V.; Barnes, J.A.; Dice, J.F. Selective degradation of annexins by chaperone-mediated autophagy. *J. Biol. Chem.* **2000**, *275*, 33329–33335. [CrossRef]
91. Aukrust, I.; Rosenberg, L.A.; Ankerud, M.M.; Bertelsen, V.; Hollas, H.; Saraste, J.; Grindheim, A.K.; Vedeler, A. Post-translational modifications of Annexin A2 are linked to its association with perinuclear nonpolysomal mRNP complexes. *FEBS Open Bio* **2017**, *7*, 160–173. [CrossRef] [PubMed]
92. Jeon, Y.R.; Kim, S.Y.; Lee, E.J.; Kim, Y.N.; Noh, D.Y.; Park, S.Y.; Moon, A. Identification of annexin II as a novel secretory biomarker for breast cancer. *Proteomics* **2013**, *13*, 3145–3156. [CrossRef] [PubMed]
93. Rimawi, M.F.; Shetty, P.B.; Weiss, H.L.; Schiff, R.; Osborne, C.K.; Chamness, G.C.; Elledge, R.M. Epidermal growth factor receptor expression in breast cancer association with biologic phenotype and clinical outcomes. *Cancer* **2010**, *116*, 1234–1242. [CrossRef] [PubMed]
94. Putti, T.C.; El-Rehim, D.M.; Rakha, E.A.; Paish, C.E.; Lee, A.H.; Pinder, S.E.; Ellis, I.O. Estrogen receptor-negative breast carcinomas: A review of morphology and immunophenotypical analysis. *Mod. Pathol.* **2005**, *18*, 26–35. [CrossRef]
95. Pinto, A.C.; Ades, F.; de Azambuja, E.; Piccart-Gebhart, M. Trastuzumab for patients with HER2 positive breast cancer: Delivery, duration and combination therapies. *Breast* **2013**, *22* (Suppl 2), S152–S155. [CrossRef]
96. Derakhshani, A.; Rezaei, Z.; Safarpour, H.; Sabri, M.; Mir, A.; Sanati, M.A.; Vahidian, F.; Gholamiyan Moghadam, A.; Aghadoukht, A.; Hajiasgharzadeh, K.; et al. Overcoming trastuzumab resistance in HER2-positive breast cancer using combination therapy. *J. Cell Physiol.* **2020**, *235*, 3142–3156. [CrossRef]
97. Speers, C.; Tsimelzon, A.; Sexton, K.; Herrick, A.M.; Gutierrez, C.; Culhane, A.; Quackenbush, J.; Hilsenbeck, S.; Chang, J.; Brown, P. Identification of novel kinase targets for the treatment of estrogen receptor-negative breast cancer. *Clin. Cancer Res.* **2009**, *15*, 6327–6340. [CrossRef]
98. Verheuvel, N.C.; Ooms, H.W.; Tjan-Heijnen, V.C.; Roumen, R.M.; Voogd, A.C. Predictors for extensive nodal involvement in breast cancer patients with axillary lymph node metastases. *Breast* **2016**, *27*, 175–181. [CrossRef]
99. Amin, M.B.; Greene, F.L.; Edge, S.B.; Compton, C.C.; Gershenwald, J.E.; Brookland, R.K.; Meyer, L.; Gress, D.M.; Byrd, D.R.; Winchester, D.P. The Eighth Edition AJCC Cancer Staging Manual: Continuing to build a bridge from a population-based to a more "personalized" approach to cancer staging. *CA Cancer J. Clin.* **2017**, *67*, 93–99. [CrossRef]

7

Annexin A2 Egress during Calcium-Regulated Exocytosis in Neuroendocrine Cells

Marion Gabel [1], Cathy Royer [2], Tamou Thahouly [1], Valérie Calco [1], Stéphane Gasman [1], Marie-France Bader [1], Nicolas Vitale [1] and Sylvette Chasserot-Golaz [1,2,*]

1 Centre National de la Recherche Scientifique, Université de Strasbourg, Institut des Neurosciences Cellulaires et Intégratives, F-67000 Strasbourg, France; m.gabel@wanadoo.fr (M.G.); tam@inci-cnrs.unistra.fr (T.T.); calco@unistra.fr (V.C.); gasman@inci-cnrs.unistra.fr (S.G.); badermf@inci-cnrs.unistra.fr (M.-F.B.); vitalen@inci-cnrs.unistra.fr (N.V.)

2 Plateforme Imagerie In Vitro, Neuropôle, Université de Strasbourg, F-67000 Strasbourg, France; cathy.royer@unistra.fr

* Correspondence: chasserot@inci-cnrs.unistra.fr

Abstract: Annexin A2 (AnxA2) is a calcium- and lipid-binding protein involved in neuroendocrine secretion where it participates in the formation and/or stabilization of lipid micro-domains required for structural and spatial organization of the exocytotic machinery. We have recently described that phosphorylation of AnxA2 on Tyr^{23} is critical for exocytosis. Considering that Tyr^{23} phosphorylation is known to promote AnxA2 externalization to the outer face of the plasma membrane in different cell types, we examined whether this phenomenon occurred in neurosecretory chromaffin cells. Using immunolabeling and biochemical approaches, we observed that nicotine stimulation triggered the egress of AnxA2 to the external leaflets of the plasma membrane in the vicinity of exocytotic sites. AnxA2 was found co-localized with tissue plasminogen activator, previously described on the surface of chromaffin cells following secretory granule release. We propose that AnxA2 might be a cell surface tissue plasminogen activator receptor for chromaffin cells, thus playing a role in autocrine or paracrine regulation of exocytosis.

Keywords: annexinA2 egress; exocytosis; chromaffin cells

---

1. Introduction

Molecules such as neurotransmitters and hormones are secreted by calcium-regulated exocytosis [1,2]. In neuroendocrine cells, exocytosis implies the recruitment and subsequent fusion of secretory granules at specific sites of the plasma membrane. The scaffolding protein annexin A2 (AnxA2) is a promoter of these sites of exocytosis in cells activated for secretion. AnxA2 binds two major actors of exocytosis, actin and phospholipids. In chromaffin cells, electron tomography has revealed that actin filaments bundled by AnxA2 contribute to the formation of lipid micro-domains at the plasma membrane required for the spatial and functional organization of the exocytotic machinery [3–5]. More recently, we have found that AnxA2 needs to be phosphorylated on Tyr^{23} to stabilize the lipid platform determining the exocytotic site, and then dephosphorylated to bundle actin filaments for stably anchoring granules, implying that the phosphorylation cycle of AnxA2 on Tyr^{23} is critical for neuroendocrine secretion [6].

In some cell types, one of the consequences of the AnxA2 phosphorylation is the egress of AnxA2, i.e., its passage through the plasma membrane. For instance, AnxA2 appeared on the cell surface of endothelial cells following a thermal shock [7], calcium-stimulated fibroblasts [8] or depolarized cortical neurons [9]. In neurons, cell surface AnxA2 was found to interact with tissue plasminogen activator (t-PA), which is involved in synaptic plasticity and memorization [10] as well as in neuronal death via plasmin proteolysis activity [11]. In chromaffin cells, one study reported that AnxA2 is "secreted" unconventionally in the extracellular medium following nicotine stimulation [12]. This release of AnxA2 was correlated with catecholamine secretion and found to be independent of cell death.

Having recently shown that the AnxA2 is phosphorylated on Tyr^{23} during exocytosis [6], we examined whether Tyr^{23} phosphorylation of AnxA2 could lead to its externalization in chromaffin cells. Using immunolabeling and biochemical approaches, we show here that nicotine stimulation triggers the egress of AnxA2 to the external leaflets of the plasma membrane and identify tissue plasminogen activator (t-PA) as a potential partner at the chromaffin cell surface.

2. Materials and Methods

2.1. Antibodies and Reagents

Rabbit polyclonal antibodies directed against AnxA2 purified from bovine aorta were a generous gift from J.C. Cavadore (Inserm U-249) [13]. Monoclonal antibodies anti-AnxA2 and anti-$pTyr^{23}$-AnxA2 (85.Tyr24) were purchased respectively from BD Transduction Laboratories (Pont de Claix, France) and Santa Cruz Biotechnology Inc. (Dallas, TX, USA). Mouse monoclonal antibodies directed against dopamine ß-hydroxylase (EC.1.14.17.1: DBH), used to specifically label secretory granules in chromaffin cells [14], were purchased from Sigma-Aldrich (St. Louis, MO, USA) and the rabbit polyclonal anti-chromogranin A (CgA) from Abcam (Amsterdam, Netherlands). Rabbit polyclonal anti-mouse tissue plasminogen activator (t-PA) antibodies were from Molecular Innovations (Novi, MI, USA). Mouse monoclonal antibodies against S100A10 were from Transduction Laboratories (Lexington, KY, USA). Sheep polyclonal antibodies against bovine phenyl ethanolamine *N*-transferase (PNMT) were purchased from Chemicon International Inc., (Temecula, CA, USA) and TRITC- or Atto-647 *N*-phalloidin from Sigma-Aldrich (St. Louis). Rabbit polyclonal anti-CD63 antibodies were from Santa Cruz Biotechnology Inc (Dallas, TX, USA). Secondary antibodies coupled to Alexa Fluor® conjugates (488, 561 or 647) or gold particles were from Molecular Probes (Invitrogen, Cergy Pontoise, France) and Aurion (Wageningen, Netherlands) respectively.

The construct allowing for the simultaneous expression of the catalytic subunit of the tetanus toxin (TTx) and the GFP was generated by subcloning both the eGFP and the TTx (residues 1 to 457, a generous gift from Thomas Binz [15]) into a bidirectional expression vector. The eGFP was PCR amplified as a BglII/PstI fragment using 5′-TATAGATCTCGCCACCATGGTGAGCAAGGGCGA-3′ and 5′-CGCCTGCAGTTACTTGTACAG CTCGTCCATGC-3′ primers and cloned into the MCS2 of the pBI-CMV1 vector (TAKARA Bio USA, Mountain View, CA, USA). Then, the TTx was PCR amplified as an MluI/NotI fragment using 5′-TATACGCGTGCCACCATGCCGATCACCATCAAC-3′ and 5′-TATGCGGCCGCTTAAGCGGT ACGGTTGTACAG-3′ primers and cloned into the MCS1 of the pBI-GFP vector. The botulinum C toxin (Bot C Tx) plasmid expressing the light chain of the toxin [16] was co-transfected with pmaxGFP to identify cells expressing the toxin.

2.2. Chromaffin Cell Culture and Transfection

Chromaffin cells were isolated from fresh bovine adrenal glands by perfusion with collagenase A, purified on self-generating Percoll gradients and maintained in culture as previously described [17]. To induce exocytosis, chromaffin cells were washed twice with Locke's solution (140 mM NaCl, 4.7 mM KCl, 2.5 mM $CaCl_2$, 1.2 mM KH_2PO_4, 1.2 mM $MgSO_4$, 11 mM glucose, 0.56 mM ascorbic acid, 0.01 mM ethylene diamine tetraacetic acid (EDTA) and 15 mM Hepes, pH 7.5), and then stimulated either with Locke's solution containing 20 μM nicotine or high K^+ solution (86,9 mM NaCl, 59 mM KCl, 2.5 mM $CaCl_2$, 1.2 mM KH_2PO_4, 1.2 mM $MgSO_4$, 11 mM glucose, 0.56 mM ascorbic acid, 0.01 mM EDTA and 15 mM Hepes, pH 7.2).

TTx-GFP was transfected into chromaffin cells (5×10^6 cells) by electroporation (Amaxa Nucleofactor systems, Lonza, Levallois, France) according to the manufacturer's instructions. Electroporated cells were immediately recovered in warm culture medium and plated onto fibronectin-coated glass coverslips. Experiments were performed 48 h after transfection.

2.3. Cell Stimulation and Extracellular AnxA2 and t-PA Measurements

Two days after plating in 5 cm Petri dishes, 10×10^6 chromaffin cells were washed for 5 min with Locke's solution, then 5 min with Locke's solution without Ca^{2+} (140 mM NaCl, 4.7 mM KCl, 1.2 mM KH_2PO_4, 1.2 mM $MgSO_4$, 0.25 mM ethylene glycol-tetraacetic acid (EGTA), 1 mM glucose, 125 μM ascorbate and 15 mM HEPES, pH 7.5) to discard cellular debris. Cells were then stimulated for 5 min with 1 mL of either 20 μM nicotine in Locke's solution or of high K^+ solution. Similarly, control cells were treated with Locke's solution. The media of stimulated cells were recovered (fraction "secreted material") and the cells were further treated for 5 min with calcium-free Locke's solution containing 20 mM EGTA (fraction "EGTA eluate"). Finally, cells were scraped into 500 μL lysis buffer (Cell Extraction Buffer, Novex®, Fisher Scientific, Illkirch-Graffenstaden, France) supplemented with protease inhibitor cocktail (P8340, Sigma-Aldrich, St. Louis). The lysates were sonicated and centrifuged and the supernatants were recovered. Secreted material and the EGTA eluate fractions were concentrated 10 times to obtain 50 μL (Spin-X UF, Corning, Wiesbaden, Germany) for AnxA2 and t-PA detection. Aliquots of supernatants were used for the lactate dehydrogenase (LDH) activity assay (QuantiChromTM LDH, D2DH-100, BioAssay Systems, Hayward, CA, USA). To isolate extracellular vesicles, the secreted material was centrifuged at 300× *g* for 10 min at 4 °C to discard cellular debris, then the supernatant was centrifuged at 100,000× *g* for 2 h at 4 °C. The pellet was suspended in 15 μL of lane marker reducing sample buffer (Thermo Scientific, Illkirch Graffenstaden, Franch). Then the samples were separated on a 4–12% SDS-PAGE gel (Novex, Thermo Scientific), blotted to nitrocellulose with the Trans-blot Turbo System (Biorad, Marnes-la Coquette, Franch) and revealed with Substrat chemiluminescent SuperSignal™ (West Femto, Thermo Scientific).

2.4. Immunofluorescence and Confocal Microscopy

For immunocytochemistry, chromaffin cells, grown on fibronectin-coated glass coverslips, were fixed and labeled as described previously [14]. The transient accessibility of DBH to the plasma membrane of chromaffin cells was tested by incubating cells for 5 min in Locke's solution containing 20 μM nicotine and anti-DBH antibodies diluted to 1:100. F-actin was stained with TRITC-phalloidin (0.5 μg/mL) for 15 min in the dark at room temperature. Labeled cells were visualized using a Leica SP5II confocal microscope. Nonspecific fluorescence was assessed by incubating cells with the secondary fluorescent-conjugated antibodies. To compare the labeling of cells from different conditions within the same experiment, images were acquired at the equatorial plane of the nucleus with the same parameters of the lasers and photomultipliers. The amount of AnxA2 or t-PA labeling associated with the plasma membrane was measured with ICY software [18] and expressed as the average fluorescence intensity normalized to the labeling surface, and divided by the total area of each cell. This allowed a quantitative cell-to-cell comparison of the fluorescence detected in cells.

2.5. Plasma Membrane Sheet Preparation and Transmission Electron Microscopy Observation

Cytoplasmic face-up membrane sheets were prepared and processed as previously described [19]. Briefly, carbon-coated Formvar films on nickel electron grids were inverted onto unstimulated or nicotine-stimulated chromaffin cells incubated with antibodies. To prepare membrane sheets, pressure was applied to the grids for 20 s, then the grids were lifted so that the fragments of the upper cell surface adhered to the grid. These membrane portions were fixed in 2% paraformaldehyde for 10 min at 4 °C. After blocking in PBS with 1% BSA and 1% acetylated BSA, the immune labeling was performed and revealed with gold particle-conjugated secondary antibodies. These membrane portions were fixed in 2.5% glutaraldehyde in PBS, postfixed with 0.5% OsO4, dehydrated in a graded ethanol series, treated with hexamethyldisilazane (Sigma-Aldrich, St. Louis), air-dried and observed using a Hitachi 7500 transmission electron microscope.

2.6. Statistical Analysis

As specified in figure legends, groups of data are presented as mean (±SEM) or median and were analyzed using a Mann-Whitney test. Asterisks in each box and whisker plot indicate statistical significance.

3. Results

3.1. AnxA2 Crosses the Plasma Membrane in Stimulated Chromaffin Cells in a Ca^{2+}-Dependent Manner and Accumulates on the Extracellular Membrane Leaflet

Bovine chromaffin cells represent a good model to study regulated exocytosis [20]. They express various nicotinic receptors and accordingly exocytosis is triggered by nicotinic agonists [21,22]. To examine whether AnxA2 could be found on the cell surface of stimulated chromaffin cells, living chromaffin cells maintained under resting conditions or stimulated with nicotine were incubated in the presence of anti-AnxA2 antibodies to specifically label AnxA2 potentially present on the cell surface. Cells were then fixed and labeled with phalloidin-TRITC to reveal the actin cytoskeleton. Resting cells characterized by a typical F-actin ring at the cell periphery displayed only faint cell surface AnxA2 labeling (Figure 1a). In contrast, while F-actin labeling decreased in stimulated cells in line with the actin depolymerization occurring during exocytosis, the labeling of AnxA2 on the cell surface increased (Figure 1a). In some experiments, cells stimulated with nicotine in the presence of the AnxA2 antibodies were further washed with a calcium-free Locke's solution containing 20 mM EGTA to collect proteins bound in a calcium-dependent manner to the cell surface (S + EGTA). Washing cells with EGTA led to a diminution of the cell surface AnxA2 labeling. Semi-quantitative analysis of the confocal images (Figure 1b) indicated that cell stimulation increased by approximately three times the amount of AnxA2 detected on the surface of chromaffin cells and confirmed the calcium sensitivity of AnxA2 binding to the plasma membrane.

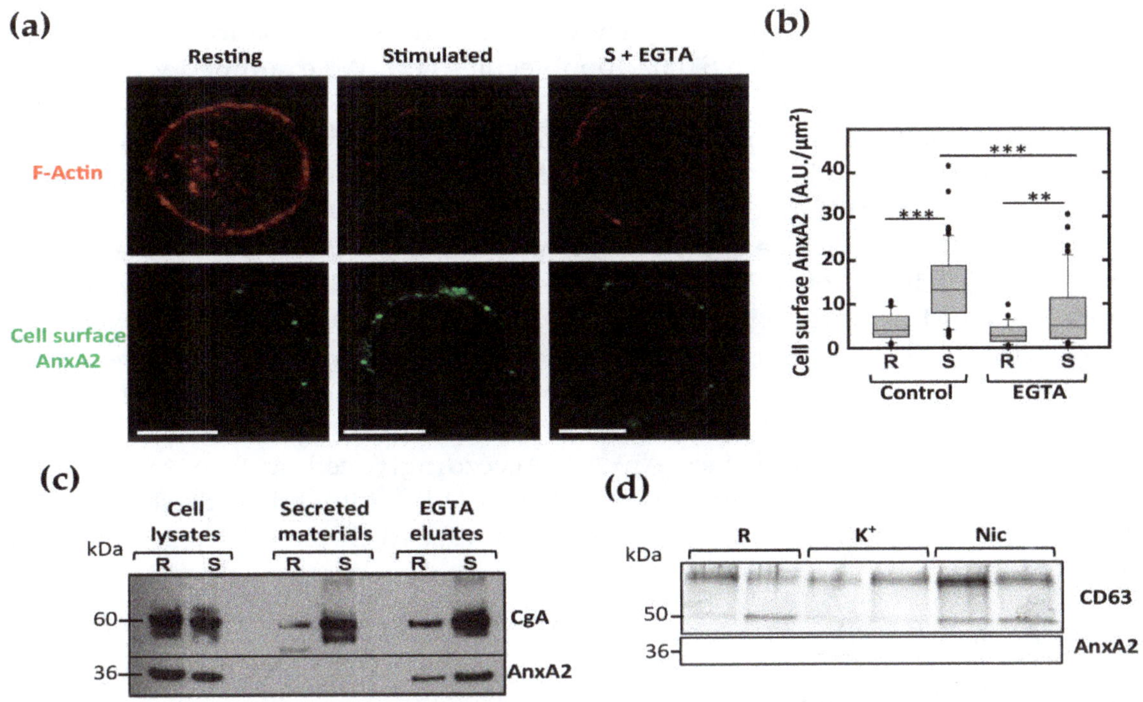

Figure 1. Annexin A2 (AnxA2) is present on the outer face of the plasma membrane of stimulated chromaffin cells. (**a**) AnxA2 labeling at the surface of chromaffin cells in the resting condition (Resting), stimulated with 20 µM nicotine without (Stimulated) or after washing with calcium-free Locke's solution

containing 20 mM EGTA (Stimulated + EGTA). Anti-AnxA2 antibodies were revealed with Alexa Fluor®488-conjugated anti-rabbit antibodies and F-actin with TRITC-phalloidin. Confocal images were recorded in the same optical section by a dual exposure procedure with the same parameters of lasers and photomultipliers. Scale bar: 10 μM. (**b**) Semi-quantitative analysis of cell surface AnxA2 labeling in chromaffin cells in resting condition (R) or stimulated with 20 μM nicotine (S) without (Control) or after EGTA wash (EGTA). AnxA2 labeling is expressed in arbitrary units. Statistical significance for medians was determined using a Mann-Whitney test. Dotted lines indicate the mean and asterisks statistical significance (*** = $p < 0.001$, ** = $p < 0.01$). Three experiments were done on independent cell cultures and pooled ($n = 32$ and 63 control cells, 28 and 63 EGTA-treated cells for resting and stimulated conditions, respectively). (**c**) Lysate, secreted material and EGTA eluate from chromaffin cells in the resting condition (R), or stimulated with 20 μM nicotine (S) were analyzed by western blot and revealed with anti-CgA and anti-AnxA2 antibodies. Data correspond to a typical experiment representative of three independent experiments. (**d**) The 100,000 g pellet of secreted materials from chromaffin cells in the resting condition (R), or stimulated with high K^+ solution (K^+) or 20 μM nicotine (Nic) were analyzed by western blot and revealed with anti-CD63 and anti-AnxA2 antibodies. Samples analyzed were obtained from two independent experiments.

Next, a biochemical approach was performed using fractions collected from nicotine-stimulated chromaffin cells. As illustrated in Figure 1c, AnxA2 was not found in the secreted material containing chromogranin A (CgA), but it could be detected in the calcium-free EGTA solution (EGTA eluates) used to wash cells after stimulation and in the cell lysates. In line with the absence of AnxA2 in secretory granules, these data suggest that AnxA2 is not secreted via the conventional exocytotic pathway but it is found bound in a calcium-dependent manner on the extracellular face of the plasma membrane following cell stimulation. Moreover, we did not detect AnxA2 in 100,000 g pellets of secreted materials (Figure 1d) containing the specific marker of extracellular vesicle CD63 [23]. Thus, AnxA2 did not seem to be associated with the extracellular vesicles released after cell stimulation.

3.2. *Tyr ^{23} Phosphorylated AnxA2 Tetramer Binds the External Face of the Plasma Membrane*

In chromaffin cells, we have previously showed that AnxA2 can be found in monomeric and tetrameric forms, associated with two S100A10 molecules [19]. We examined whether cell surface AnxA2 was phosphorylated on Tyr^{23} and whether S100A10 was present on the cell surface upon cell stimulation. Live chromaffin cells were stimulated with nicotine in the presence of anti-$pTyr^{23}$ AnxA2 or anti-S100A10 antibodies. Both antibodies labeled the surface of stimulated cells, suggesting that cell surface AnxA2 is Tyr^{23} phosphorylated and associated with S100A10. Thus, AnxA2 tetramer could be present on the external face of the plasma membrane (Figure 2a,b). Bovine chromaffin cells are constituted of two populations: the adrenergic cells secreting adrenaline and noradrenergic cells secreting noradrenaline [24], and we have previously showed that S100A10 is selectively expressed in adrenergic cells [14]. Thus, we performed a staining experiment for cell surface AnxA2 together with S100A10 or phenylethanolamine *N*-methyltransferase (PNMT), selectively expressed in adrenergic cells [24]. Figure 2c shows a S100A10-positive cell close to two S100A10-negative cells and all cells display a similar staining of cell surface AnxA2. Accordingly, cell surface AnxA2 labeling was observed on PNMT-positive and -negative cells (Figure 2d). Altogether, these data suggest that AnxA2 translocated through the plasma membrane of stimulated chromaffin cells even in the absence of S100A10.

3.3. *The Egress of AnxA2 is Linked to Exocytosis*

To probe the idea that AnxA2 egress is related to exocytosis, we labeled in parallel cell surface AnxA2 and the exocytotic sites using anti-DBH antibodies [14]. In live cells stimulated for exocytosis, the granule-associated DBH becomes transiently accessible to the antibody only at sites of exocytosis, leading to the appearance of fluorescent patches at the cell surface [14].

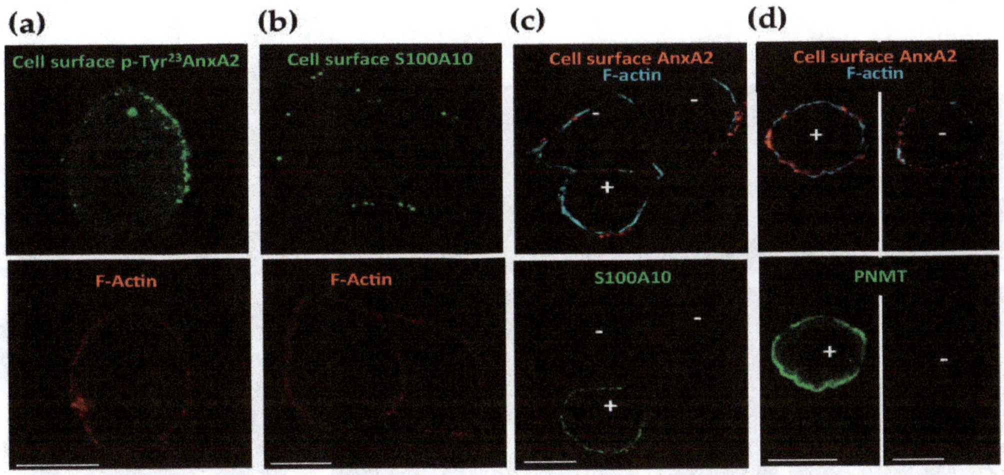

Figure 2. Tyr^{23}-phosphorylated AnxA2 and S100A10 are associated with the surface of stimulated chromaffin cells. (**a**) The p-Tyr^{23}AnxA2 labeling at the surface of stimulated chromaffin cells. Anti-p-Tyr^{23}AnxA2 antibodies were revealed with Alexa Fluor®488-conjugated anti-mouse antibodies and F-actin with TRITC-phalloidin to visualize the cell shape. (**b**) S100A10 labeling at the surface of stimulated chromaffin cells. Anti-S100A10 antibodies were revealed with Alexa Fluor®488-conjugated anti-mouse antibodies and F-actin with TRITC-phalloidin. (**c**) Confocal micrograph of the triple labeling of cell surface AnxA2, of intracellular S100A10 and of F-actin labeled with Atto-647 *N*-phalloidin in stimulated chromaffin cells. (**d**) Confocal micrographs of the triple labeling of cell surface AnxA2, of intracellular PNMT and of F-actin labeled with Atto-647 *N*-phalloidin in stimulated chromaffin cells. For (**a**–**d**), confocal images were recorded in the same optical section by a dual exposure procedure. Scale bar: 10 μM.

As illustrated in Figure 3a, resting chromaffin cells exhibited only a few DBH patches, confirming the low levels of baseline exocytotic activity in the absence of secretagogue, and displayed only a weak cell surface AnxA2 staining. Stimulation for 5 min with nicotine triggered the appearance of a patchy pattern of DBH surface staining and concomitantly increased cell surface AnxA2 labeling (Figure 3a). We also observed the co-localization between DBH and AnxA2 at the cell surface (Figure 3a, merge). Semi-quantitative analysis indicated that 65.5 ± 4.12% (±SEM, $n = 27$) of the cell surface AnxA2 labeling colocalized with DBH labeling. This indicated that the AnxA2 egress primarily takes place in the vicinity of the exocytotic sites. Next, we examined the time course of AnxA2 appearance on the cell surface of nicotine-stimulated chromaffin cells using a semi-quantitative analysis (Figure 3b). Egress of AnxA2 was observed after 30 s of stimulation and it peaked after 60 s. Thus, the kinetics of AnxA2 egress correlated well with the rapid kinetics of AnxA2 Tyr^{23} phosphorylation during exocytosis, but preceded the maximal exocytotic response usually observed after 180 s of stimulation in our experimental conditions [6]. To confirm the link between the AnxA2 egress and exocytosis, we examined whether the formation of the Soluble NSF Attachment Proteins REceptor (SNARE) complexes was necessary for the externalization of AnxA2. A bicistronic plasmid encoding for tetanus toxin (TTx) and GFP was expressed in chromaffin cells. TTx is known to specifically cleave VAMP2, a protein required for SNARE complex formation and exocytosis [25]. As illustrated in Figure 3c, cell surface AnxA2 labeling was reduced in stimulated cells expressing TTx as compared to electroporated control cells not expressing TTx. Semi-quantitative analysis confirmed that expression of TTx reduced AnxA2 egress in stimulated cells (Figure 3d). Thus, VAMP2 cleavage and the consequent inhibition of the SNARE complex formation reduced AnxA2 externalization and its appearance on the cell surface. Similar results were obtained in cells expressing botulinum toxin C, which cleaves SNAP-25 and syntaxin-1 (Figure 3c,d). These findings indicate that the egress of AnxA2 seems to depend on the formation of SNARE complexes and therefore on the completion of exocytosis in chromaffin cells.

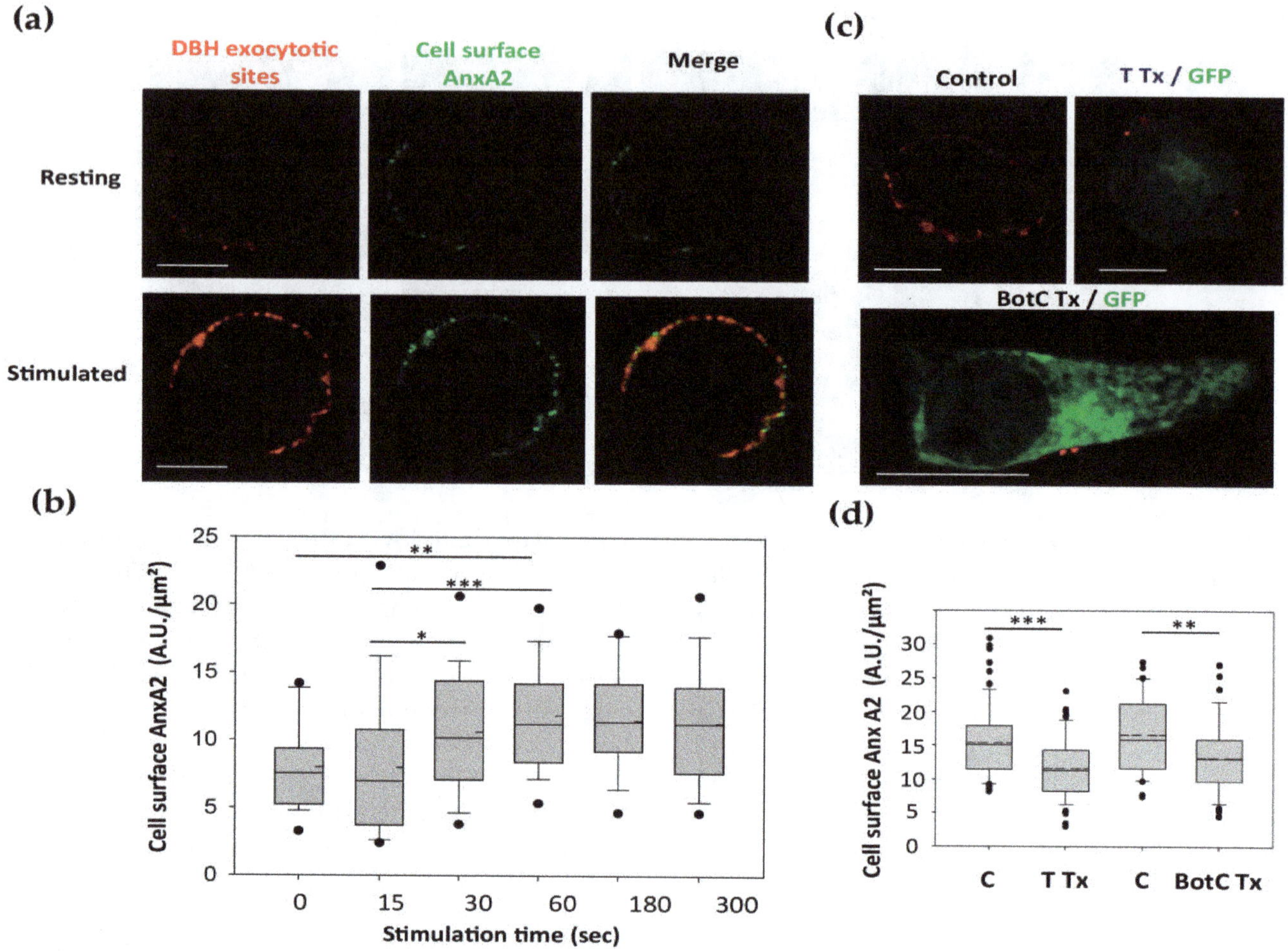

Figure 3. AnxA2 egress was correlated with exocytosis. (**a**) Dual labeling of cell surface AnxA2 and exocytotic sites. Cells were stimulated with nicotine 20 µM in the presence of anti-DBH and anti-AnxA2 antibodies. Cells were then fixed and incubated with secondary antibodies coupled to Alexa Fluor®561 and Alexa Fluor®488, respectively. Confocal images were recorded in the same optical section and with the same parameters of lasers and photomultipliers. Scale bar: 10 µM. (**b**) Time course of AnxA2 egress after cell stimulation. Chromaffin cells were stimulated with 20 µM nicotine during different times in the presence of AnxA2 antibodies. The cell surface AnxA2 labeling is expressed in arbitrary units. Statistical significance for medians was determined using a Mann-Whitney test. Dotted lines indicate the mean and asterisks statistical significance (*** = $p < 0.001$, ** = $p < 0.01$, * = $p < 0.05$). Two experiments were done on independent cell cultures and pooled. Number of cells analyzed were 19 (0 s), 25 (15 s), 29 (30 s), 23 (60 s), 21 (180 s). (**c**) Effect of tetanus and botulinum C toxins on the AnxA2 egress in chromaffin cells. Electroporated cells expressing TTx/GFP, BotC Tx/GFP or no toxin (Control, C) were stimulated with high K^+ solution in the presence of anti-AnxA2 antibodies, fixed and then incubated with secondary antibodies coupled to Alexa Fluor®561. Confocal images were recorded in the same optical section. Scale bar: 10 µM. (**d**) Semi-quantitative analysis of cell surface AnxA2 labeling in stimulated cells is expressed in arbitrary units (Control $n = 79$ cells, TTx/GFP $n = 69$ cells, Control $n = 39$ cells, BotC Tx/GFP $n = 39$ cells). Statistical significance for medians was determined using a Mann-Whitney test. Dotted lines indicate the mean and asterisks statistical significance (*** = $p < 0.001$, ** = $p < 0.01$). Three experiments were done on independent cell cultures and pooled.

3.4. Cell Surface Membrane-Associated t-PA is Present Close to Exocytotic Sites

Cell surface AnxA2 has been described as a co-receptor of t-PA [26]. Cell surface AnxA2 tetramer was shown to capture circulating plasminogen and t-PA to promote plasmin generation that participates in fibrinolysis [26]. Moreover, previous studies reported the presence of t-PA in secretory granules and its release upon chromaffin cell stimulation [27], and t-PA was found to bind to the cell surface by interacting with an unknown receptor [28]. We confirmed the presence of t-PA at the surface of chromaffin cells (Figure 4). Double labeling of living cells stimulated with nicotine in the presence of anti-t-PA and anti-DBH antibodies revealed the appearance of a patchy pattern of DBH surface staining and a concomitant increase in t-PA labeling (Figure 4a). We also observed the co-localization between t-PA and DBH at the cell surface (Figure 3a, merge). Semi-quantitative analysis indicated that 68 ± 2.7% (±SEM, n = 33) of the cell surface t-PA labeling colocalized with DBH labeling.

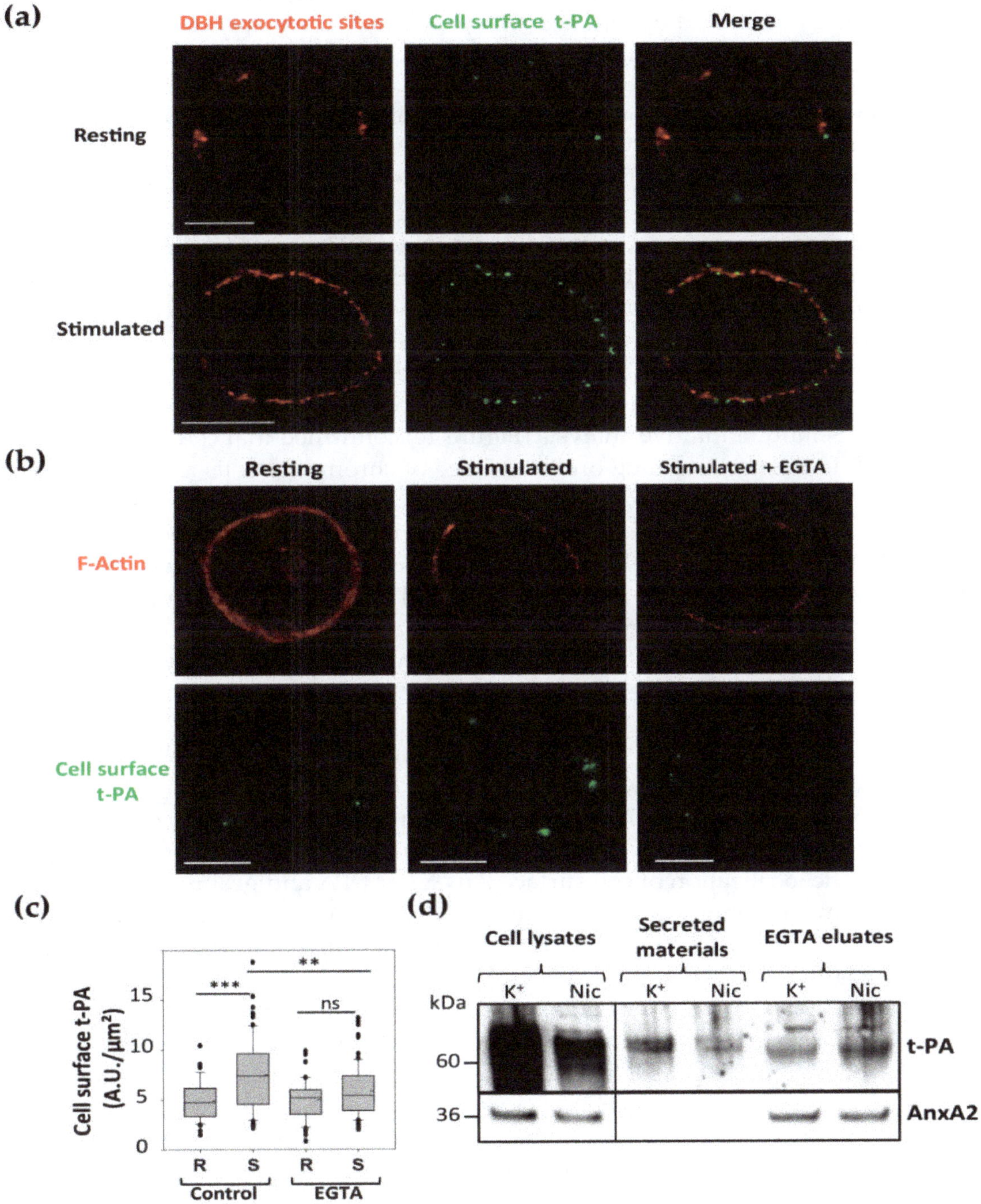

Figure 4. Tissue plasminogen activator (t-PA) is present on the outer leaflet of the plasma membrane of stimulated chromaffin cells. (**a**) Dual labeling of cell surface t-PA and exocytotic sites. Cells were stimulated with nicotine 20 µM in the presence of anti-t-PA and anti-DBH antibodies. Cells were then fixed and incubated with secondary antibodies coupled to Alexa Fluor®488 and Alexa Fluor®551,

respectively. Confocal images were recorded in the same optical section and with the same parameters of lasers and photomultipliers. Scale bar: 10 μM. (**b**) The t-PA labeling on the surface of chromaffin cells in the resting condition (R), stimulated with 20 μM nicotine without (S) or after EGTA wash (S + EGTA). Anti-t-PA antibodies were revealed with Alexa Fluor®488-conjugated anti-rabbit antibodies and F-actin with TRITC-phalloidin to visualize the cell shape. Confocal images were recorded in the same optical section by a dual exposure procedure. Scale bar: 10 μM. (**c**) Semi-quantitative analysis of t-PA labeling on the cell surface of chromaffin cells in the resting condition (R), stimulated with 20 μM nicotine (S) without (Control) or after EGTA wash (S + EGTA). The t-PA labeling is expressed in arbitrary units. Statistical significance for medians was determined using a Mann-Whitney test. Dotted lines indicate the mean and asterisks statistical significance (*** = $p < 0.001$, ** = $p < 0.01$). Three experiments were done on independent cell cultures and pooled (n = 36 and 65 control cells, 40 and 75 EGTA-treated cells for resting and stimulated conditions, respectively). (**d**) Lysate, secreted material and EGTA eluate from chromaffin cells in the resting condition (R) or stimulated with nicotine 20 μM (S) were analyzed by western blot and revealed with anti-t-PA and anti-AnxA2 antibodies. Data correspond to a typical experiment representative of three independent experiments.

We further characterized the binding of t-PA at the cell surface of stimulated chromaffin cells using a similar approach to that used to visualize cell surface AnxA2. Live cells were maintained at rest or stimulated with nicotine in the presence of anti-t-PA antibodies prior to fixation. In resting cells, t-PA labeling was barely visible, whereas in stimulated cells, a clear increase in cell surface labeling was observed (Figure 4b). Washing the nicotine-stimulated cells with calcium-free EGTA containing Locke's solution prior to incubation with the anti-t-PA antibodies resulted in a clear reduction in t-PA labeling (Figure 4b). Semi-quantitative analysis (Figure 4c) confirmed that cell stimulation increased by approximately 2.5-fold the t-PA staining on the surface of chromaffin cells and validated the calcium sensitivity of t-PA binding to the plasma membrane. Figure 4d illustrates a western blot analysis of the fractions collected from cells stimulated with nicotine or high K^+ solution and subsequently washed for 5 min with EGTA solution. In the cell lysates and secreted material, t-PA was found, in line with the presence of t-PA in secretory granules. It was also detected in the EGTA eluates, indicating that part of the t-PA released by exocytosis remained bound to the cell surface in a calcium-dependent manner. CgA, a major component stored in chromaffin granules and released by exocytosis, also bound to the cell surface following secretion (Figure 1c).

3.5. AnxA2 and t-PA are Side by Side at Cell Surface of Stimulated Chromaffin Cells

To further explore the spatial relationship between t-PA, AnxA2 and the sites of exocytosis, experiments were designed to analyze at the ultrastructural level the outer face of the chromaffin cell plasma membrane. The localization of cell surface AnxA2 or t-PA and granule membranes transiently inserted into the plasma membrane after exocytosis was examined on plasma membrane sheets from chromaffin cells stimulated with 20 μM nicotine in the presence of anti-DBH antibodies to reveal exocytotic granule membranes [29] and anti-AnxA2 or anti-t-PA antibodies (Figure 5). Cells were fixed and labeled with anti-mouse antibodies revealed with 10 nm gold particles and anti-rabbit antibodies revealed with 15 nm gold particles to label DBH/anti-DBH complexes and AnxA2/anti-AnxA2 or t-PA/anti-t-PA, respectively. Ultrastructural images obtained by transmission electron microscopy revealed the appearance of gold-labeled DBH clusters on the cell surface (Figure 5a), corresponding to the insertion of the granular membrane in the plasma membrane of nicotine-stimulated cells [29]. Cell surface AnxA2, as well as t-PA (Figure 5a,b), tended to localize at the periphery of the DBH-labeled areas. Double-labeling experiments with anti-t-PA and anti-AnxA2 antibodies using a combination of 10 and 15 nm gold particles indicated AnxA2 and t-PA formed mixed clusters at the cell surface (Figure 5c,d). Both types of beads were found in close proximity, suggesting a possible interaction of t-PA with AnxA2 at the surface of stimulated chromaffin cells, in line with the idea that AnxA2 could be a t-PA receptor at the surface of chromaffin cells.

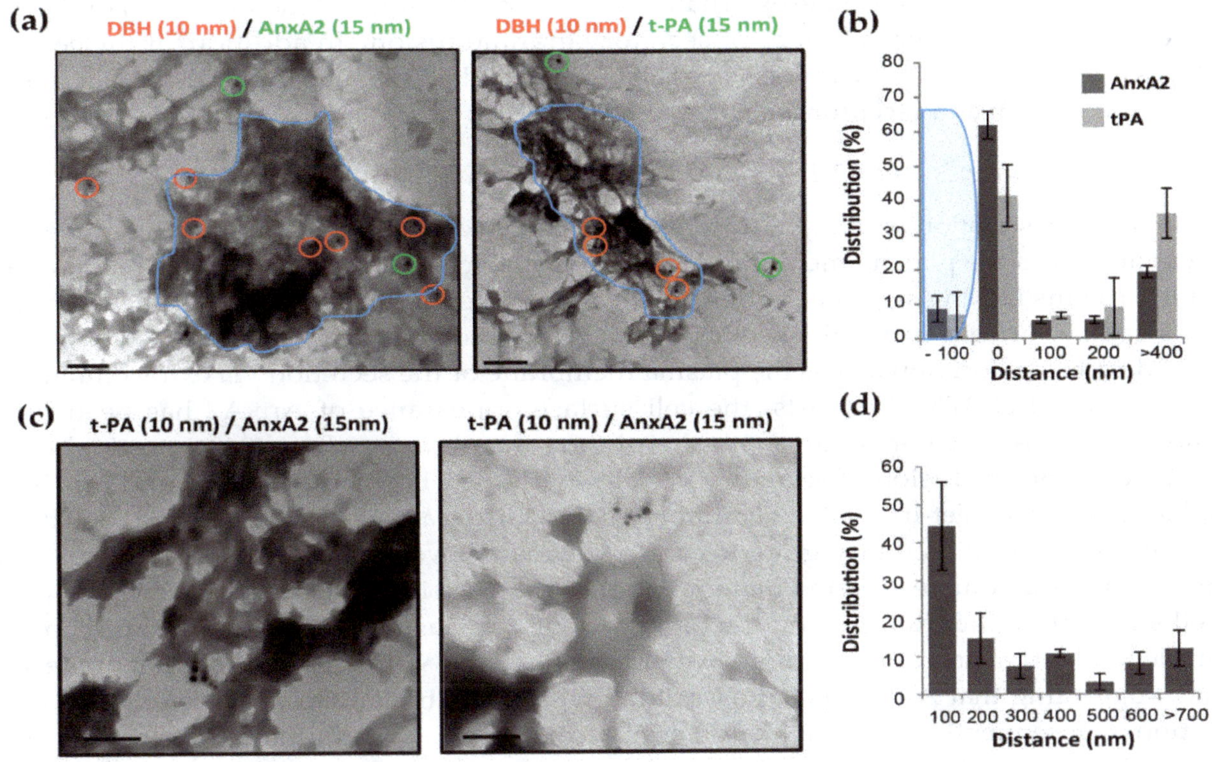

Figure 5. Membrane topography of AnxA2, t-PA and exocytotic sites after immunogold labeling of the outer face of the plasma membrane sheets prepared from stimulated chromaffin cells. (**a**) Plasma membrane sheets were prepared from bovine chromaffin cells stimulated by nicotine for 5 min. To label DBH, AnxA2 and t-PA exposed at the surface of cells undergoing exocytosis, anti-DBH, anti-AnxA2 and anti-t-PA antibodies were added during stimulation. Membrane sheets were labeled with anti-mouse antibodies coupled to 10 nm gold particles to detect DBH antibodies revealing exocytotic sites (red circle) and rabbit antibodies coupled to 15 nm gold particles to label AnxA2 or t-PA (green circle). (**b**) The histogram represents the relative distribution of 15 nm gold particles as a function of their distance from the granule membrane once inserted in the plasma membrane (blue line). The distance was measured and the number of particles was counted manually with Photoshop. Three experiments were done on independent cell cultures (**c**). Double staining experiment for t-PA (10 nm gold particles) and AnxA2 (15 nm gold particles) were performed with the same protocol. Scale bar: 100 nm. (**d**) The histogram represents the relative distribution of 10 nm gold particles (t-PA) as a function of their distance from 15 nm gold particles (AnxA2). The distance was measured and the number of particles was counted manually with Photoshop. Two experiments were done on two independent cell cultures.

4. Discussion

During neuroendocrine secretion, AnxA2 participates together with the actin cytoskeleton in the formation of lipid micro-domains required for the docking and fusion of secretory granules with the plasma membrane [5,13,19]. We have recently observed that AnxA2 needs to be phosphorylated on Tyr^{23} to form these lipid platforms supporting exocytosis [6]. Since Tyr^{23} phosphorylation of AnxA2 has been reported to induce its translocation from the inner to the outer side of the plasma membrane [7], we examined here whether AnxA2 might be externalized in neurosecretory chromaffin cells. In agreement with these findings, the present report favors a model in which AnxA2 is not conventionally secreted in the extracellular medium, but rather translocates through the plasma membrane and remains attached to the surface of stimulated chromaffin cells. Here, p-Tyr^{23}-AnxA2 was found associated with the outer face of the plasma membrane in a calcium-dependent interaction. Hence, our observations are in agreement with previous results obtained in endothelial cells [7] and keratinocytes [30], but also in neurosecretory GABAergic neurons [9]. AnxA2 egress is closely linked to the process of exocytosis since the externalization of AnxA2 required the formation of the SNARE complexes and occurred near the sites of secretory granule fusion. In addition, at the cell surface, AnxA2 was found to co-localize with t-PA in a calcium-dependent manner. We propose that cell surface AnxA2 could function as a receptor for secreted t-PA.

4.1. *By Which Mechanism is AnxA2 Translocated to the Cell Surface in Chromaffin Cells?*

Despite the description of many extracellular functions for AnxA2, the mechanism underlying its translocation across the plasma membrane to the cell surface remains unclear. There are a number of soluble proteins lacking signal peptides which are secreted in the extracellular medium through a process called "unconventional secretion" [31]. Two major pathways for this secretion involve either the direct translocation across the plasma membrane or the secretion via extracellular vesicles. For instance, in NIH 3T3 fibroblasts, the cell surface appearance of AnxA2 has been linked to the fusion of multi-vesicular bodies with the plasma membrane, whereas in intestinal epithelial cells, it depends on the fusion of secretory vesicles with the plasma membrane, and in stimulated macrophages or ultraviolet-irradiated keratinocytes it requires caspase-1 activation [32]. In chromaffin cells, the implication of secretory granules and extracellular vesicles is unlikely since AnxA2 was neither found in the soluble secreted material nor associated with the extracellular vesicles released after cell stimulation, but detected on the outer face of the plasma membrane where it bound in a calcium-dependent manner. Furthermore, we did not find AnxA2 in a 100,000 g pellet of the secreted material, suggesting that AnxA2 is not associated with the extracellular vesicles released after cell stimulation. Consequently, the most likely mechanism for AnxA2 membrane translocation involves direct insertion into the lipid bilayer, allowing the passage of AnxA2 through the plasma membrane.

The insertion of annexins into model membranes has been demonstrated in vitro for several members of this large family of proteins (AnxA1, A2, A4, A5 and A6) [33]. Within cells, AnxA2 was also shown to translocate across membranes [34]. AnxA2 membrane translocation requires calcium-dependent binding to negatively charged phospholipids such as phosphatidylserine (PS) and a lipid flipping activity. Accordingly, the phospholipid scramblase TMEM16F was reported to contribute to AnxA2 membrane crossing [35]. A similar mechanism has been described for the unconventional secretion of fibroblast growth factors (FGFs) [36], which also requires calcium and is linked to PS egress [37]. It is tempting to draw parallels with AnxA2 in chromaffin cells, since secretagogue-evoked stimulation of chromaffin cells triggers the appearance of PS at the cell surface, presumably at the periphery of the granule fusion sites [29,38]. This PS egress depends on the lipid scramblase PLSCR-1 [39]. The contribution of PLSCR-1 in AnxA2 egress needs now to be tested in chromaffin cells. Finally, an alternative mechanism has been proposed in enterocytes, which involves the extrusion of AnxA2 during hemifusion [40].

4.2. *What Might be the Role of Cell Surface AnxA2?*

The most well-documented role of cell surface AnxA2 is as co-receptor of t-PA and plasminogen [26]. Plasminogen and t-PA were previously described on the surface of chromaffin cells but their receptors remained unidentified [41]. Of note, t-PA was also detected in the secretory granules of a subpopulation of chromaffin cells [42]. In the present report, t-PA and AnxA2 were both found on the surface of chromaffin cells. Both t-PA and cell surface AnxA2 were eluted by EGTA, indicating that both proteins bind to the membrane in a calcium-dependent manner. Although the direct interaction of t-PA with chromaffin cell surface AnxA2 remains to be demonstrated, AnxA2 is able to interact with t-PA and its primary substrate, plasminogen [43]. Thus, by recruiting released t-PA and circulating plasminogen, cell surface AnxA2 might well serve as an extracellular proteolytic center that locally generates plasmin. Plasmin is known to cleave released CgA into a variety of biologically active peptides, some of which may significantly inhibit the nicotinic stimulation of catecholamine release from PC12 cells and primary bovine adrenal chromaffin cells [28]. For instance, the fragment $CgA_{360\text{-}373}$ is selectively generated by plasmin and the corresponding synthetic peptide markedly inhibited nicotine-induced catecholamine release [44]. Plasmin generated at the cell surface could also activate signals, leading to protein kinase C-mediated phosphorylation of intracellular AnxA2, thereby dissociating the AnxA2 complex and preventing further catecholamine release [32]. In neuronal tissues, plasminogen activators and cell surface AnxA2 have both been detected at high levels and implicated in processes like neuronal plasticity and synaptic remodeling within the hippocampus and cerebellum during memory [10].

Additionally, AnxA2, via its ability to sequester and laterally organize PS [45], could promote the formation or stabilization of PS-rich domains in the external leaflet of the plasma membrane. These AnxA2-mediated clusters of PS at the cell surface may represent a concentrated signal for endocytosis as has been shown for AnxA5 and phagocytosis [46]. Finally, we cannot exclude that the AnxA2 tetramer could potentially interact with neighboring phospholipid membranes and therefore serve as a bridge between two adjacent cells [47].

To summarize, we propose that, upon cell stimulation, AnxA2 is recruited to the plasma membrane to form a tetramer with S100A10. Once phosphorylated at the plasma membrane, AnxA2 phosphorylated on Tyr^{23} stabilizes lipid micro-domains required to recruit/organize the priming/docking machinery for exocytosis. Although AnxA2 Tyr^{23} dephosphorylation is required to promote the formation of actin bundles that strongly anchor secretory granules to the exocytotic sites, a fraction of Tyr^{23}-phosphorylated AnxA2 appears to cross the plasma membrane near exocytotic sites. At the cell surface, AnxA2 as a co-receptor of t-PA could thus participate in various autocrine and/or paracrine activities.

Author Contributions: Conceptualization, S.C.-G.; Data curation, M.G., C.R. and S.C.-G.; Formal analysis, S.C.-G.; Funding acquisition, S.G. and N.V.; Resources, M.G., C.R., T.T., V.C. and S.C.-G.; Supervision, S.G. and N.V.; Visualization, M.G. and S.C.-G.; Writing—original draft, M.G. and S.C.-G.; Writing—review and editing, M.G., M.-F.B., S.G., N.V. and S.C.-G. All authors have read and agreed to the published version of the manuscript.

Acknowledgments: We are grateful to Thomas Binz for generously providing us with tetanus toxin plasmid. We thank the municipal slaughterhouse of Haguenau (France) for providing bovine adrenal glands.

References

1. Tanguy, E.; Carmon, O.; Wang, Q.; Jeandel, L.; Chasserot-Golaz, S.; Montero-Hadjadje, M.; Vitale, N. Lipids implicated in the journey of a secretory granule: From biogenesis to fusion. *J. Neurochem.* **2016**, *137*, 904–912. [CrossRef]
2. Bombardier, J.P.; Munson, M. Three steps forward, two steps back: Mechanistic insights into the assembly and disassembly of the SNARE complex. *Curr. Opin. Chem. Biol.* **2015**, *29*, 66–71. [CrossRef] [PubMed]
3. Bharadwaj, A.; Bydoun, M.; Holloway, R.; Waisman, D. Annexin A2 heterotetramer: Structure and function. *Int. J. Mol. Sci.* **2013**, *14*, 6259–6305. [CrossRef] [PubMed]
4. Gabel, M.; Chasserot-Golaz, S. Annexin A2, an essential partner of the exocytotic process in chromaffin cells. *J. Neurochem.* **2016**, *137*, 890–896. [CrossRef] [PubMed]
5. Gabel, M.; Delavoie, F.; Demais, V.; Royer, C.; Bailly, Y.; Vitale, N.; Bader, M.F.; Chasserot-Golaz, S. Annexin A2-dependent actin bundling promotes secretory granule docking to the plasma membrane and exocytosis. *J. Cell Biol.* **2015**, *210*, 785–800. [CrossRef] [PubMed]
6. Gabel, M.; Delavoie, F.; Royer, C.; Tahouly, T.; Gasman, S.; Bader, M.F.; Vitale, N.; Chasserot-Golaz, S. Phosphorylation cycling of Annexin A2 Tyr23 is critical for calcium-regulated exocytosis in neuroendocrine cells. *Biochim. Biophys. Acta Mol. Cell Res.* **2019**, *1866*, 1207–1217. [CrossRef]
7. Deora, A.B.; Kreitzer, G.; Jacovina, A.T.; Hajjar, K.A. An annexin 2 phosphorylation switch mediates p11-dependent translocation of annexin 2 to the cell surface. *J. Biol. Chem.* **2004**, *279*, 43411–43418. [CrossRef]
8. Valapala, M.; Vishwanatha, J.K. Lipid raft endocytosis and exosomal transport facilitate extracellular trafficking of annexin A2. *J. Biol. Chem.* **2011**, *286*, 30911–30925. [CrossRef]
9. Zhao, W.Q.; Lu, B. Expression of annexin A2 in GABAergic interneurons in the normal rat brain. *J. Neurochem.* **2007**, *100*, 1211–1223. [CrossRef]

10. Calabresi, P.; Napolitano, M.; Centonze, D.; Marfia, G.A.; Gubellini, P.; Teule, M.A.; Berretta, N.; Bernardi, G.; Frati, L.; Tolu, M.; et al. Tissue plasminogen activator controls multiple forms of synaptic plasticity and memory. *Eur. J. Neurosci.* **2000**, *12*, 1002–1012. [CrossRef]
11. Tsirka, S.E. Tissue plasminogen activator as a modulator of neuronal survival and function. *Biochem. Soc. Trans.* **2002**, *30*, 222–225. [CrossRef] [PubMed]
12. Faure, A.V.; Migne, C.; Devilliers, G.; Ayala-Sanmartin, J. Annexin 2 "secretion" accompanying exocytosis of chromaffin cells: Possible mechanisms of annexin release. *Exp. Cell Res.* **2002**, *276*, 79–89. [CrossRef] [PubMed]
13. Chasserot-Golaz, S.; Vitale, N.; Umbrecht-Jenck, E.; Knight, D.; Gerke, V.; Bader, M.F. Annexin 2 promotes the formation of lipid microdomains required for calcium-regulated exocytosis of dense-core vesicles. *Mol. Biol. Cell* **2005**, *16*, 1108–1119. [CrossRef] [PubMed]
14. Chasserot-Golaz, S.; Vitale, N.; Sagot, I.; Delouche, B.; Dirrig, S.; Pradel, L.A.; Henry, J.P.; Aunis, D.; Bader, M.F. Annexin II in exocytosis: Catecholamine secretion requires the translocation of p36 to the subplasmalemmal region in chromaffin cells. *J. Cell Biol.* **1996**, *133*, 1217–1236. [CrossRef]
15. McMahon, H.T.; Ushkaryov, Y.A.; Edelmann, L.; Link, E.; Binz, T.; Niemann, H.; Jahn, R.; Sudhof, T.C. Cellubrevin is a ubiquitous tetanus-toxin substrate homologous to a putative synaptic vesicle fusion protein. *Nature* **1993**, *364*, 346–349. [CrossRef]
16. Lam, A.D.; Tryoen-Toth, P.; Tsai, B.; Vitale, N.; Stuenkel, E.L. SNARE-catalyzed fusion events are regulated by Syntaxin1A-lipid interactions. *Mol. Biol. Cell* **2008**, *19*, 485–497. [CrossRef]
17. Bader, M.F.; Thierse, D.; Aunis, D.; Ahnert-Hilger, G.; Gratzl, M. Characterization of hormone and protein release from alpha-toxin-permeabilized chromaffin cells in primary culture. *J. Biol. Chem.* **1986**, *261*, 5777–5783.
18. de Chaumont, F.; Dallongeville, S.; Chenouard, N.; Herve, N.; Pop, S.; Provoost, T.; Meas-Yedid, V.; Pankajakshan, P.; Lecomte, T.; Le Montagner, Y.; et al. Icy: An open bioimage informatics platform for extended reproducible research. *Nat. Methods* **2012**, *9*, 690–696. [CrossRef]
19. Umbrecht-Jenck, E.; Demais, V.; Calco, V.; Bailly, Y.; Bader, M.F.; Chasserot-Golaz, S. S100A10-mediated translocation of annexin-A2 to SNARE proteins in adrenergic chromaffin cells undergoing exocytosis. *Traffic* **2010**, *11*, 958–971. [CrossRef]
20. Bader, M.F.; Holz, R.W.; Kumakura, K.; Vitale, N. Exocytosis: The chromaffin cell as a model system. *Ann. N. Y. Acad. Sci.* **2002**, *971*, 178–183. [CrossRef]
21. Livett, B.G.; Kozousek, V.; Mizobe, F.; Dean, D.M. Substance P inhibits nicotinic activation of chromaffin cells. *Nature* **1979**, *278*, 256–257. [CrossRef]
22. Trifaro, J.M.; Lee, R.W. Morphological characteristics and stimulus-secretion coupling in bovine adcrenal chromaffin cell cultures. *Neuroscience* **1980**, *5*, 1533–1546. [CrossRef]
23. Kowal, J.; Arras, G.; Colombo, M.; Jouve, M.; Morath, J.P.; Primdal-Bengtson, B.; Dingli, F.; Loew, D.; Tkach, M.; Thery, C. Proteomic comparison defines novel markers to characterize heterogeneous populations of extracellular vesicle subtypes. *Proc. Natl. Acad. Sci. USA* **2016**, *113*, E968–E977. [CrossRef] [PubMed]
24. Verhofstad, A.A.; Coupland, R.E.; Parker, T.R.; Goldstein, M. Immunohistochemical and biochemical study on the development of the noradrenaline- and adrenaline-storing cells of the adrenal medulla of the rat. *Cell Tissue Res.* **1985**, *242*, 233–243. [CrossRef] [PubMed]
25. Ahnert-Hilger, G.; Bigalke, H. Molecular aspects of tetanus and botulinum neurotoxin poisoning. *Prog. Neurobiol.* **1995**, *46*, 83–96. [CrossRef]
26. Kim, J.; Hajjar, K.A. Annexin II: A plasminogen-plasminogen activator co-receptor. *Front. Biosci.* **2002**, *7*, d341–d348. [CrossRef] [PubMed]
27. Santell, L.; Marotti, K.R.; Levin, E.G. Targeting of tissue plasminogen activator into the regulated secretory pathway of neuroendocrine cells. *Brain Res.* **1999**, *816*, 258–265. [CrossRef]
28. Parmer, R.J.; Mahata, M.; Gong, Y.; Mahata, S.K.; Jiang, Q.; O'Connor, D.T.; Xi, X.P.; Miles, L.A. Processing of chromogranin A by plasmin provides a novel mechanism for regulating catecholamine secretion. *J. Clin. Investig.* **2000**, *106*, 907–915. [CrossRef]
29. Ceridono, M.; Ory, S.; Momboisse, F.; Chasserot-Golaz, S.; Houy, S.; Calco, V.; Haeberle, A.M.; Demais, V.; Bailly, Y.; Bader, M.F.; et al. Selective recapture of secretory granule components after full collapse exocytosis in neuroendocrine chromaffin cells. *Traffic* **2011**, *12*, 72–88. [CrossRef]
30. Dziduszko, A.; Ozbun, M.A. Annexin A2 and S100A10 regulate human papillomavirus type 16 entry and intracellular trafficking in human keratinocytes. *J. Virol.* **2013**, *87*, 7502–7515. [CrossRef]

31. Zhang, M.; Schekman, R. Cell biology. Unconventional secretion, unconventional solutions. *Science* **2013**, *340*, 559–561. [CrossRef] [PubMed]
32. Luo, M.; Hajjar, K.A. Annexin A2 system in human biology: Cell surface and beyond. *Semin. Thromb. Hemost.* **2013**, *39*, 338–346. [CrossRef]
33. Lizarbe, M.A.; Barrasa, J.I.; Olmo, N.; Gavilanes, F.; Turnay, J. Annexin-phospholipid interactions. Functional implications. *Int. J. Mol. Sci.* **2013**, *14*, 2652–2683. [CrossRef] [PubMed]
34. Popa, S.J.; Stewart, S.E.; Moreau, K. Unconventional secretion of annexins and galectins. *Semin. Cell Dev. Biol.* **2018**, *83*, 42–50. [CrossRef]
35. Stewart, S.E.; Ashkenazi, A.; Williamson, A.; Rubinsztein, D.C.; Moreau, K. Transbilayer phospholipid movement facilitates the translocation of annexin across membranes. *J. Cell Sci.* **2018**, *131*. [CrossRef] [PubMed]
36. Rabouille, C.; Malhotra, V.; Nickel, W. Diversity in unconventional protein secretion. *J. Cell Sci.* **2012**, *125*, 5251–5255. [CrossRef] [PubMed]
37. Kirov, A.; Al-Hashimi, H.; Solomon, P.; Mazur, C.; Thorpe, P.E.; Sims, P.J.; Tarantini, F.; Kumar, T.K.; Prudovsky, I. Phosphatidylserine externalization and membrane blebbing are involved in the nonclassical export of FGF1. *J. Cell Biochem.* **2012**, *113*, 956–966. [CrossRef]
38. Vitale, N.; Caumont, A.S.; Chasserot-Golaz, S.; Du, G.; Wu, S.; Sciorra, V.A.; Morris, A.J.; Frohman, M.A.; Bader, M.F. Phospholipase D1: A key factor for the exocytotic machinery in neuroendocrine cells. *EMBO J.* **2001**, *20*, 2424–2434. [CrossRef]
39. Ory, S.; Ceridono, M.; Momboisse, F.; Houy, S.; Chasserot-Golaz, S.; Heintz, D.; Calco, V.; Haeberle, A.M.; Espinoza, F.A.; Sims, P.J.; et al. Phospholipid scramblase-1-induced lipid reorganization regulates compensatory endocytosis in neuroendocrine cells. *J. Neurosci.* **2013**, *33*, 3545–3556. [CrossRef]
40. Danielsen, E.M.; van Deurs, B.; Hansen, G.H. "Nonclassical" secretion of annexin A2 to the lumenal side of the enterocyte brush border membrane. *Biochemistry* **2003**, *42*, 14670–14676. [CrossRef]
41. Parmer, R.J.; Mahata, S.K.; Jiang, Q.; Taupenot, L.; Gong, Y.; Mahata, M.; O'Connor, D.T.; Miles, L.A. Tissue plasminogen activator and chromaffin cell function. *Adv. Exp. Med. Biol.* **2000**, *482*, 179–192. [CrossRef] [PubMed]
42. Weiss, A.N.; Anantharam, A.; Bittner, M.A.; Axelrod, D.; Holz, R.W. Lumenal protein within secretory granules affects fusion pore expansion. *Biophys. J.* **2014**, *107*, 26–33. [CrossRef] [PubMed]
43. He, K.L.; Sui, G.; Xiong, H.; Broekman, M.J.; Huang, B.; Marcus, A.J.; Hajjar, K.A. Feedback regulation of endothelial cell surface plasmin generation by PKC dependent phosphorylation of annexin A2. *J. Biol. Chem.* **2010**. [CrossRef]
44. Jiang, Q.; Yasothornsrikul, S.; Taupenot, L.; Miles, L.A.; Parmer, R.J. The local chromaffin cell plasminogen/plasmin system and the regulation of catecholamine secretion. *Ann. N. Y. Acad. Sci.* **2002**, *971*, 445–449. [CrossRef] [PubMed]
45. Menke, M.; Gerke, V.; Steinem, C. Phosphatidylserine membrane domain clustering induced by annexin A2/S100A10 heterotetramer. *Biochemistry* **2005**, *44*, 15296–15303. [CrossRef] [PubMed]
46. van Genderen, H.O.; Kenis, H.; Hofstra, L.; Narula, J.; Reutelingsperger, C.P. Extracellular annexin A5: Functions of phosphatidylserine-binding and two-dimensional crystallization. *Biochim. Biophys. Acta* **2008**, *1783*, 953–963. [CrossRef]
47. Jung, Y.; Wang, J.; Song, J.; Shiozawa, Y.; Wang, J.; Havens, A.; Wang, Z.; Sun, Y.X.; Emerson, S.G.; Krebsbach, P.H.; et al. Annexin II expressed by osteoblasts and endothelial cells regulates stem cell adhesion, homing, and engraftment following transplantation. *Blood* **2007**, *110*, 82–90. [CrossRef]

8

Annexin A2 in Inflammation and Host Defense

Valentina Dallacasagrande and Katherine A. Hajjar *

Department of Pediatrics, Weill Cornell Medicine, 1300 York Avenue, New York, NY 10065, USA; vad2010@med.cornell.edu

* Correspondence: khajjar@med.cornell.edu

Abstract: Annexin A2 (AnxA2) is a multifunctional calcium^{2+} (Ca^{2+}) and phospholipid-binding protein that is expressed in a wide spectrum of cells, including those participating in the inflammatory response. In acute inflammation, the interaction of AnxA2 with actin and adherens junction VE-cadherins underlies its role in regulating vascular integrity. In addition, its contribution to endosomal membrane repair impacts several aspects of inflammatory regulation, including lysosome repair, which regulates inflammasome activation, and autophagosome biogenesis, which is essential for macroautophagy. On the other hand, AnxA2 may be co-opted to promote adhesion, entry, and propagation of bacteria or viruses into host cells. In the later stages of acute inflammation, AnxA2 contributes to the initiation of angiogenesis, which promotes tissue repair, but, when dysregulated, may also accompany chronic inflammation. AnxA2 is overexpressed in malignancies, such as breast cancer and glioblastoma, and likely contributes to cancer progression in the context of an inflammatory microenvironment. We conclude that annexin AnxA2 normally fulfills a spectrum of anti-inflammatory functions in the setting of both acute and chronic inflammation but may contribute to disease states in settings of disordered homeostasis.

Keywords: annexin A2; inflammation; infection; adherens junction; angiogenesis; macroautophagy

---

1. Introduction

Inflammation is defined as the local host response to injury caused by infectious or noninfectious agents; it results in elimination or compartmentalization of the inciting agent, clearance of necrotic cells, and repair of damaged tissue [1,2]. Initially, host cells recognize danger-associated molecular patterns (DAMPs) or pathogen-associated molecular patterns (PAMPs) through germline-encoded pattern-recognition receptors (PRRs) expressed mainly in monocytes, macrophages, neutrophils, and dendritic cells [1]. The four known classes of PRR molecules include transmembrane proteins, such as Toll-like (TLRs) and C-type lectin receptors (CLRs), as well as cytoplasmic proteins, such as retinoic acid-inducible gene (RIG)-like receptors (RLRs) and nucleotide-binding oligomerization domain-like receptors (NLRs) [3]. These pathways induce the release of proinflammatory cytokines, which increase vascular permeability, thus facilitating the extravasation of immune cells into the damaged tissue, and chemokines, which recruit supplemental immune cells that phagocytose and kill pathogens [1]. Inflammation that resolves within days is classified as acute, whereas chronic inflammation may persist for months to years. Chronic inflammation underlies the pathogenesis of an array of disease entities, including diabetes, asthma, arthritis, cancer, atherosclerosis, vasculitis, and inflammatory bowel disease.

The annexins are Ca^{2+}-regulated, phospholipid- and membrane-binding proteins that are named after the Greek word "*annex*", meaning to attach or bridge because of their ability to link membranes to other membranes or other structures [4]. All but one (A6) of the twelve annexin family members (A1–A11 and A13) identified in vertebrates have a highly homologous core domain (~30–35 kilodaltons) containing four multi-alpha helical repeats with potential Ca^{2+}-binding activity and an N-terminal domain (~3 kilodaltons), which is specific for each family member [5,6]. The annexins are expressed ubiquitously throughout the phylogenetic tree and are evolutionarily ancient.

Annexin A2 (AnxA2) is one of the most extensively studied members of the annexin superfamily [6]. AnxA2 is produced by a wide spectrum of cell types, including endothelial, trophoblast, epithelial, and tumor cells, as well as innate immune cells, such as macrophages, monocytes, and dendritic cells. AnxA2 may exist in either monomeric or heterotetrameric form. The heterotetramer $(A2\bullet S100A10)_2$ is composed of two copies each of AnxA2 and protein S100A10, also called p11. AnxA2 is present in the cytoplasm and on cell surfaces, and its functions are largely location specific. On the endothelial cell surface, the $(A2\bullet S100A10)_2$ complex binds components of the fibrinolytic system, plasminogen and tissue plasminogen activator (tPA), accelerating the activation of the serine protease plasmin [7,8]. As the primary fibrinolytic protease, plasmin enables fibrin breakdown and angiogenesis [9,10]. Intracellularly, AnxA2 seems to fulfill many functions, including organization of specialized membrane microdomains, facilitation of vesicle budding, and regulation of additional membrane dynamic events such as fusion, endocytosis, endosomal biogenesis, and membrane repair [4,5,11–21].

In view of this wide array of functions, we have developed a working model depicting the disparate anti- and pro-inflammatory roles of AnxA2 at the various stages of inflammation (Figure 1). In the initial phase, AnxA2 limits vascular permeability, thereby modulating recruitment of leukocytes and their subsequent release of inflammatory mediators. AnxA2 also supports internal membrane repair, thus modulating inflammasome activation, and participates in the biogenesis of the phagophore in autophagy, thereby facilitating the removal of pathogens and damaged organelles. Later, AnxA2 enables angiogenesis and tissue healing.

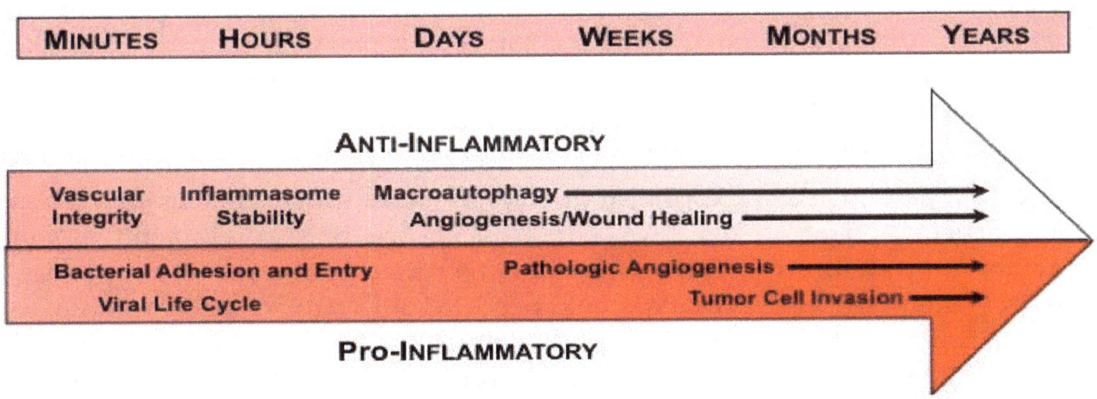

Figure 1. Annexin A2 in inflammatory responses—working model. Under homeostatic conditions (top arrow), AnxA2 plays an anti-inflammatory role in response to injury or infection. In the immediate response to injury, AnxA2 maintains vascular integrity, thereby preventing edema and extravasation of blood cells. Within minutes to hours, AnxA2 also protects internal membranes, such as those delimiting the lysosome, thus preventing inflammasome activation and tempering cytokine production. AnxA2 is also necessary for biogenesis of the phagophore, a double membrane structure that engulfs and destroys intracellular pathogens and allows the cell to adapt to environmental stresses in the process of macroautophagy. In the later stages of inflammation, AnxA2 likely promotes angiogenesis and wound healing by supporting cell surface fibrinolytic activity. In disease states, however (bottom arrow), AnxA2 may be co-opted to perform pro-inflammatory actions. On the plasma membrane, AnxA2 appears to serve as a site for bacterial adhesion and entry into cells, and may enable viral infection, replication, and release. In chronic inflammatory states, excessive angiogenesis that is sustained by AnxA2 may induce tissue damage, as in the diabetic retina, or may support the progression of cancer. In addition, some tumor cells may utilize cell surface AnxA2 to generate protease activity necessary for migration and metastasis within a pro-inflammatory micromilieu.

2. Annexin A2 and Infection

Infection results from the interaction of a pathogen (e.g., bacterium, fungus, or virus) with a host organism, leading to tissue invasion, proliferation of the pathogen, production of toxins, and cellular damage [22,23]. The first response of the mammalian host to infection is inflammation, mediated by the innate immune system, and is not specific to the inciting agent. The ensuing adaptive immune response is highly specific to the invading pathogen and typically provides long-lasting protection [23]. Annexin A2 fulfills a number of protective roles during pathogenic infection (Table 1).

In a model of *Klebsiella* pneumonia in mice, AnxA2 appeared to reduce infection-associated inflammation [24]. In *Anxa2*$^{-/-}$ mice infected intranasally with *Klebsiella* pneumonia, the level of pro-inflammatory cytokines was significantly elevated, and *Anxa2*$^{-/-}$ mice exhibited 100% mortality, versus 100% survival for *Anxa2*$^{+/+}$ mice at 50 h post injection. Additionally, peritoneal macrophages lacking AnxA2 showed worsening of the TLR4-triggered inflammatory response. These investigators concluded that AnxA2 has a role in limiting inflammation by promoting anti-inflammatory signals [24].

Annexin A2 also protects the host in a murine model of *Cryptococcal* infection. *Cryptococcus neoformans* is an encapsulated budding yeast that, unlike others, can replicate under acidic conditions; its virulence depends upon its interaction with host macrophages [25]. In this infection, *Anxa2*$^{-/-}$ bone-marrow-derived macrophages were less efficient than control macrophages at phagocytosing yeast cells, resulting in a lower frequency of non-lytic exocytosis. In addition, the *Cryptococcus* capsule was enlarged in *Anxa2*$^{-/-}$-deficient macrophages, possibly impacting nonlytic exocytosis, and potentially reflecting the ability of AnxA2 to promote intracellular membrane interactions and vesicle adhesion to the cell membrane [6,26–31]. In vivo, mice lacking AnxA2 showed a lower survival rate when infected with *Cryptococcus*, reflecting a dysregulated inflammatory response [25].

On the other hand, there is evidence that cell surface annexin AnxA2 may be co-opted to facilitate infection by invading bacteria such as *Pseudomonas aeruginosa* [32], *Escherichia coli* [33], *Salmonella typhimurium* [34], and *Rickettsial* species [35]. *P. aeruginosa*, an organism that causes chronic pulmonary infection in patients with cystic fibrosis, anchors to AnxA2 on respiratory epithelial cells, undergoes internalization, and causes apoptotic cell death and release of pro-inflammatory cytokines [32]. In this instance, AnxA2 enables the initiation of productive infection.

Similarly, *E. coli*, particularly the EspL2 strain, reorganizes the cytoskeleton of the host epithelial cells with the participation of annexin A2, which is known to regulate actin dynamics at the bacterium–membrane contact site [36,37]. AnxA2 is recruited to the membrane adhesion site by *E. coli*, where its F-actin bundling activity is enhanced [33,38]. AnxA2 also promotes cytoskeletal rearrangements during host cell invasion by *Salmonella*. *S. typhimurium*, a common cause of gastroenteritis, uses the contact-dependent type 3 secretion system of the host cell to extensively remodel actin. AnxA2 is engaged in various mechanisms driven by actin rearrangement (e.g., endocytosis, cell–cell adhesion, and membrane ruffling) [34]. AnxA2 participates in reorganization of the actin cytoskeleton during *Salmonella* invasion; depletion of annexin A2 and S100A10 reduced bacteria invasion, probably because of the ability of these proteins to bind the large phosphoprotein AHNAK at the bacteria entry site [34].

In addition, annexin A2 acts as an adhesion receptor in *Rickettsial* infections [35]. In these disorders, bacteria adhere to and invade vascular endothelial cells in a manner that must be strong enough to overcome the shear stress of flowing blood. In human umbilical vein endothelial cells, immunofluorescence–confocal microscopy revealed colocalization of AnxA2 and *Rickettsial* organisms on the external face of the plasma membrane. The numbers of *Rickettsia* adherent to wild-type mouse brain microvascular endothelial cells was significantly increased compared to those associated with cells from AnxA2-null mice. In-vivo findings corroborated the hypothesis that host endothelial cell AnxA2 serves as an adhesion receptor for *Rickettsial* species. In fact, using a plaque assay and with confocal imaging, He et al. showed an increase in *Rickettsia* in circulating blood and a concomitant decrease on the endothelial cell surface in AnxA2-null mice [35].

Finally, annexin A2 appears to be co-opted during the life cycle of at least 13 human viruses. Reported roles include attachment, penetration by receptor-mediated endocytosis or direct membrane fusion, replication, assembly, and release [39]. For example, the (A2•S100A10)$_2$ tetramer was identified as a central mediator in human papillomavirus (HPV) attachment and intracellular trafficking, leading to progression of genital cancers. Suppression of AnxA2 altered the entry of HPV into target cells, whereas antibodies against S100A10 did not have the same effect [40]. In addition, the infection was significantly reduced when the (A2•S100A10)$_2$ complex was eliminated, a result that was not seen when S100A10 alone was deleted [41,42].

Table 1. Role of annexin A2 in pathogen–host cell interactions in infection.

Pathogen	AnxA2 role	References
Bacteria		
	Anti-Inflammatory Actions	
Klebsiella pneumoniae	Reduced cytokine response	[24]
	Pro-Inflammatory Actions	
Pseudomonas aeruginosa	Plasma membrane adhesion	[32]
Escherichia coli	Plasma membrane adhesion	[33,38]
Salmonella typhimurium	Host cell invasion	[34]
Rickettsia australis	Plasma membrane adhesion	[35]
Fungus		
	Anti-Inflammatory Actions	
Cryptococcus neoformans	Phagocytosis and exocytosis	[25]
	Pro-Inflammatory Actions	
none reported	—	—
Virus		
	Anti-Inflammatory Actions	
Human Papillomavirus	Attachment and intracellular trafficking	[40–42]
	Pro-Inflammatory Actions	
none reported	—	—

3. Annexin A2 and Regulation of Vascular Integrity

One of the earliest responses to inflammation is a loss of vascular integrity, which facilitates the invasion of innate immune cells into the affected tissue. The maintenance of blood vessel integrity is an active process that, if interrupted, can cause hemorrhage, edema, and progressive inflammation. Vascular permeability is regulated by endothelial–endothelial cell junctions [43]. Homotypic interactions between vascular endothelial cadherin (VE-cad), a major constituent of the adherens-type junction, are regulated by Src-mediated tyrosine phosphorylation of VE-cad, which induces opening of the junction [44–46].

Annexin A2 interacts directly with VE-cad and is essential to maintaining VE-cad within cell junctions [45]. AnxA2 and S100 A10 interact independently with VE-cad via actin filaments, and vascular endothelial growth factor (VEGF) treatment disconnects AnxA2 and actin from the VE-cad complex, increasing vasculature permeability. Phosphorylation of the adherens junctions can be induced by either loss of Src homology phosphatase 2 (SHP2) or destabilization of cholesterol drafts, both of which implicate annexin A2 in vascular integrity [45]. Annexin A2 also regulates tyrosine phosphorylation of VE-cad in the pulmonary microvasculature. Under hypoxia, *Anxa2*$^{-/-}$ mice, unlike control *Anxa2*$^{+/+}$ animals, showed a significant increase in Src-related phosphorylation of VE-cad. *Anxa2*$^{-/-}$, but not *Anxa2*$^{+/+}$, mice also displayed an acute inflammatory response with infiltration of neutrophils into the lung parenchyma and development of pulmonary edema. With VE-cad and SHP2, AnxA2 forms a complex, which is disrupted in the absence of AnxA2, thus preventing dephosphorylation of VE-Cad and inducing vascular leak [47].

In addition, in-vitro and in-vivo studies support a role for annexin A2 in maintaining the blood–brain barrier. Analysis of *Anxa2*$^{-/-}$ mice showed fewer tight junctions containing zonulin-1 and claudin-5 and fewer adherens junctions containing VE-cad. In cultured primary human brain microvascular endothelial cells, AnxA2 seems to promote F-actin and VE-cad interactions by binding Robo4 and contributing to its complex formation with paxillin. This process appears to be essential for maintaining vasculature integrity in the central nervous system [48].

4. Annexin A2 and Recruitment of Inflammatory Cells

It is likely that annexin A2 is important for the recruitment of some classes of leukocytes to sites of inflammation. AnxA2 may interact with CD44 in the chemotaxis of neutrophil-like cells in response to complement factor 5a in vitro, and anti-AnxA2 appears to block this activity in human neutrophils [49]. At the same time, the level of expression of AnxA2 in freshly isolated human neutrophils appears to be low compared to other circulating leukocytes, especially human monocytes and monocyte-derived macrophages, where it is highly expressed [50]. In fact, anti-AnxA2 IgG impairs cytokine-directed monocyte migration through the extracellular matrix [50] This migratory activity also requires tPA-dependent activation of plasminogen to the serine protease plasmin. In addition, an in-vitro wound healing assay revealed that loss of AnxA2 expression in intestinal epithelial cells led to increased cell-matrix adhesion via a β integrin, and reduced cell migration [51], but whether this mechanism applies to inflammatory cells is unknown. Nevertheless, the full extent of the actions of AnxA2 in inflammatory cell recruitment in vivo remains to be determined.

5. Annexin A2 in Inflammasome Dynamics

As an adaptive response, inflammation must be tightly regulated; inadequate inflammation can lead to persistent infection, whereas excessive or prolonged inflammation can cause longstanding disease, such as chronic arthritis, neurodegeneration, inflammatory bowel disease, or metabolic syndrome (atherosclerosis, type 2 diabetes and obesity) [52,53]. Inflammasomes are large intracellular protein complexes that sense inflammatory triggers in macrophages, monocytes, dendritic cells, neutrophils, and epithelial cells. They activate caspase-1 in response to both pathogens and host-derived signals of cellular stress including leakage of lysosomal cathepsins into the cytosol, damage to mitochondria, and production of reactive oxygen species [52].

As an adaptive response, inflammation must be tightly regulated; inadequate inflammation can lead to persistent infection, whereas excessive or prolonged inflammation can cause longstanding disease, such as chronic arthritis, neurodegeneration, inflammatory bowel disease, or metabolic syndrome (atherosclerosis, type 2 diabetes and obesity) [52,53]. Inflammasomes are large intracellular protein complexes that sense inflammatory triggers in macrophages, monocytes, dendritic cells, neutrophils, and epithelial cells. They activate caspase-1 in response to both pathogens and host-derived signals of cellular stress including leakage of lysosomal cathepsins into the cytosol, damage to mitochondria, and production of reactive oxygen species [52].

In humans with joint replacement devices, artificial articular surfaces may generate non-biodegradable wear debris particles, which, upon phagocytosis by dendritic cells, damage the endolysosomal limiting membrane. Damage to membranes leads to discharge of lysosomal cathepsins and H^+ ions into the cytosol and activation of the NLRP3 inflammasome [54]. In the cytosol of dendritic cells exposed to wear debris particles, AnxA2 was profoundly downregulated, and both AnxA2 and S100A10 translocated to the endosome, to promote resealing of damaged membranes. These findings confirmed the hypothesis that AnxA2 is involved in endolysosomal membrane repair, thus limiting inflammasome activation [54].

On the other hand, upon infection with *Anaplasma phagocytophilum*, a *Rickettsial* organism, annexin AnxA2 null macrophages showed evidence of impaired inflammasome activation. Interleukin-1β secretion, caspase-1 activation, and NLRC4 oligomerization were all reduced compared to wild-type macrophages under the same conditions. Moreover, *Anxa2*$^{-/-}$-infected mice were significantly more susceptible to infection than wild-type mice, indicating that AnxA2 helps this pathogen evade the NLRC4 inflammasome-based host defense system [55].

6. Annexin A2 and Macroautophagy

Once the innate immune cells have identified the pathogens or damaged cells during the inflammatory response, the process of macroautophagy may be initiated. Macroautophagy is a cell survival mechanism, in which certain cytoplasmic constituents (damaged organelles, misfolded proteins, or bacteria) are degraded in a lysosome-derived, double-membrane-enclosed vacuole, and then recycled to maintain cellular homeostasis [56–58]. In inflammatory macroautophagy, the process begins when a pre-autophagosomal membrane elongates and forms an autophagosome by engulfing a portion of the cytosol. Autophagosomes initially fuse with endosomes to form amphisomes, and then ultimately fuse with lysosomes to degrade their contents [59]. Macroautophagy is associated with a number of human pathologies including cancer [60], myopathies [61], diabetes [62], neurodegenerative processes [63], and infectious disease [63,64].

Annexin A2 participates in macroautophagy by interacting with the autophagy-related protein Atg16, especially in the biogenesis of Atg16L-positive vesicles. In primary dendritic cells, which have the highest rate of plasma membrane turnover and are crucial for immunosurveillance, proteomic analyses identified a population of vesicles positive for both AnxA2 and Atg16L. Comparison of cells from *Anxa2*$^{+/+}$ and *Anxa2*$^{-/-}$ mice demonstrated that AnxA2 facilitates Atg16L-positive vesicle fusion, a fundamental step in phagophore formation and elongation. Thus, AnxA2 appears to link Atg16L to membrane vesicles [15].

Macroautophagy is upregulated in response to a wide range of chronic inflammatory disorders and other stress conditions. These include persistent infections, such as tuberculosis, neurodegenerative disorders, such as Parkinson's disease, and chronic inflammatory disorders, such as Crohn's disease [65]. Under starvation conditions, for example, macroautophagy correlates with increased expression of annexin A2 in cultured cells and in mouse brain through a transcriptional pathway involving Jun N-terminal kinase (JNK) and c-Jun. Biogenesis of autophagosomes was enhanced in cells fed exogenous AnxA2 and abrogated in AnxA2-null mice. Autophagosome formation may be regulated by the effect of AnxA2 on actin-mediated trafficking of the transmembrane protein Atg9a, as AnxA2 and Atg9a were found to colocalize with actin filaments, leading to the conclusion that AnxA2 anchors actin on Atg9a-positive vesicles [66].

In another study, activation of autophagy during oxygen–glucose deprivation was reported to be influenced by AnxA2 in human retinal endothelial cells. Knockdown of AnxA2 attenuated the initiation of autophagy, reduced cell viability, and fostered cell apoptosis [67]. In another study, *Anxa2*$^{-/-}$ mice sustained a more severe *Pseudomonas aeruginosa* infection, which spread more readily to the lung and other organs. This appeared to correlate with an inability to clear the infection when annexin A2 is silenced and with the role of AnxA2 in regulating autophagosome formation through the Akt1–mTOR–ULK1/2 signaling pathway [68].

Macroautophagy is also central to the pathogenesis of osteoarthritis, a form of chronic joint inflammation. Autophagy may be protective in normal cartilage, as expression of ULK1, Beclin1 and LC3, the primary genes regulating the autophagy pathway, is decreased in osteoarthritic cartilage and chondrocytes [69]. In an in-vivo study with green fluorescent protein (GFP)-conjugated LC3 (GFP-LC3)-transgenic reporter mice, which allow one to monitor autophagy, there was a decrease in autophagic activity associated with a size and number reduction of autophagosomes [70]. These studies confirmed that age-related osteoarthritis correlates with autophagy, underlining a possible role for AnxA2 in this chronic inflammatory disorder.

7. Annexin A2 in Angiogenesis

Angiogenesis is the process by which endothelial cells, derived from pre-existing vasculature, proliferate and migrate to create new blood vessels. It is distinct from vasculogenesis, in which endothelial cells or their precursors coalesce to form new vasculature [71]. Under physiologic conditions, blood vessels undergo constant turnover, in which the rate of vessel degeneration is counterbalanced by the rate of blood vessel renewal. Insufficient angiogenesis is associated with myocardial infarction, stroke, preeclampsia, and neurodegeneration, whereas excessive vascular growth supports inflammation, obesity-associated disorders, development and spread of malignant tumors, and vascular-based ocular disorders such as diabetic retinopathy [72]. During inflammation, secreted pro-angiogenic factors promote neovascularization. The formation of new blood vessels, in turn, facilitates infiltration of innate immune cells, thus perpetuating the inflammatory process [73]. Additionally, upregulation of the inflammatory response caused by the persistence of the instigating agent may also facilitate angiogenesis through endothelial cell proliferation and migration [74].

Early in angiogenesis, pro-angiogenic signals activate endothelial cells, which migrate into the extracellular matrix, proliferate, elongate and create a lumen. The $(A2 \bullet S100A10)_2$ complex, residing on the endothelial cell surface, converts plasminogen into plasmin, which can then activate a cascade of matrix metalloproteases. Matrix metalloproteases (MMP) can proteolyze components of the basement membrane, thus liberating endothelial cells and permitting their directed migration [72]. Based on these observations, it has been hypothesized that AnxA2 may be involved in excessive, pathological angiogenesis across a range of disease states [71].

Several in-vivo studies demonstrate a role for annexin A2 in postnatal angiogenesis. In the corneal pocket assay, for example, AnxA2-deficient mice displayed a decreased capability of generating new blood vessels in the cornea stimulated with fibroblast growth factor [75]. In wild-type mice, this response was blocked when mice were placed on a high-methionine diet; methionine is metabolized to homocysteine, which modifies AnxA2 by forming a covalent adduct with cysteine 8 in the N-terminal tail region. Interestingly, when hyperhomocysteinemic wild-type mice were treated with intravenous AnxA2, normal corneal angiogenesis ensued [76]. In the Matrigel implant assay, similarly, neovessel formation was significantly impaired in $Anxa2^{-/-}$ mice. Unlike $Anxa2^{-/-}$ mice, wild-type mice showed neovascularization of implanted Matrigel plugs. Addition of a peptide that mimics the N-terminal domain of annexin A2 and blocks tPA binding reduced the number of von Willebrand factor positive cells by 80%, whereas a scrambled control peptide had no effect [75].

In the oxygen-induced retinopathy (OIR) model of angiogenesis, 7-day-old neonatal mice are subjected to a 75% O_2 environment for 5 days and then returned to room air (21% O_2) for an additional 5-day recovery period. Immunohistologic analysis of retinas revealed a significant reduction of both neovascular tufts and tuft cell nuclei in $Anxa2^{-/-}$ mice. Under OIR conditions, the accumulation of fibrin in the perivasculature was extensive in $Anxa2^{-/-}$ retinas, but almost indiscernible in $Anxa2^{+/+}$ retinas. Treatment of pups with ancrod, a fibrinogen-depleting agent, upon return to room air almost completely eliminated fibrin in both groups of mice, while at the same time doubling the extent of neovascularization in $Anxa2^{-/-}$ but not $Anxa2^{+/+}$ mice. On the other hand, pups treated with tranexamic acid, an inhibitor of fibrinolysis, drastically increased fibrin deposition and lowered neovascularization in $Anxa2^{+/+}$ but not $Anxa2^{-/-}$ retinas. In addition, it was established that hypoxia inducible factor-1, a master hypoxia-responsive transcription factor, directly regulates AnxA2 gene expression and that AnxA2 binding with S100A10 promotes neovascularization by enhancing plasmin activation and fibrin remodeling. Together, these data confirmed that both fibrinolysis and angiogenesis were impaired under ablation of annexin A2 [77]. Similarly, AnxA2 appears to be upregulated by vascular endothelial growth factor, and blockade of AnxA2 in ischemic mice inhibits retinal neovascularization [78]. In choroidal neovascularization, moreover, the AnxA2-binding agent, TM601, a synthetic form of chlorotoxin, significantly suppressed neovascularization by inducing apoptosis in endothelial cells [79]. Together, these studies indicate that the annexin A2 fibrinolytic system is a key regulator of angiogenesis.

8. Annexin A2 and Tumor Progression

The correlation between inflammation, angiogenesis, and cancer is well known. In fact, hypoxia-induced release of VEGF, a feature of both cancer progression and tumor angiogenesis, increases cellular proliferation, invasion and metastasis [74]. The involvement of annexin A2 in tumor progression may be related to the overproduction of plasmin on the surface of aggressive cancer cells, on the surface of endothelial cells, or both [80]. In addition, AnxA2 may facilitate the recruitment of pro-angiogenic inflammatory cells into the tumor microenvironment, which may in turn promote tumor progression.

Overexpression of annexin A2 has been observed in many malignancies, including aggressive breast cancer. AnxA2 was found to be elevated in stromal cells and epithelial cells of invasive ductal carcinoma (MDA-MB231) but not in a less aggressive cell line (MCF-7). MDA-MB231 cells, unlike MCF-7 cells, had the ability to activate plasminogen, leading to the hypothesis that plasmin generating capacity correlates with breast cancer cell migration and aggressiveness, implicating AnxA2 as a key mediator in metastasis. They also reported that blocking AnxA2 with angiostatin, a competitor for plasminogen binding on endothelial cell surface, or an anti-AnxA2 monoclonal antibody hindered MDA-MB231 invasion and migration. Immunohistochemical analyses of human breast cancer tissues revealed AnxA2 and elevated tPA on the surface of cancer cells, but not normal cells, as well as evidence of inflammation within the tumor. Quantitative analysis of the microvascular density showed a correlation between new blood vessel formation and the annexin A2 expression pattern [81–83]. Furthermore, immunoneutralization of AnxA2 impaired plasmin-mediated activation of MMP-2/9 in the breast tumor microenvironment [84]. Together, these data suggest that AnxA2 plays a pivotal role in breast cancer invasion and angiogenesis.

Annexin A2 expression is significantly elevated in glioblastoma, a highly vascularized and exceedingly aggressive brain tumor [85,86]. Intracerebral inoculation of canine glioma cells into the brains of athymic rats showed that AnxA2 is present in clusters of tumor cells surrounding dilated tumor vessels. Overexpression of AnxA2 correlated with high expression of VEGF and platelet-derived growth factor (PDGF), suggesting that the level of AnxA2 reflected invasiveness [87]. In addition, the involvement of AnxA2 in glioma angiogenesis was highlighted by the demonstration that tumors in wild-type mice displayed significantly higher microvascular density and more dilated blood vessels than tumors in *Anxa2*$^{-/-}$ mice [88].

Increased expression of AnxA2 appears to correlate with tumor invasiveness in renal cell carcinoma [89,90], hepatocellular carcinoma [91], colorectal cancer [92] and lung cancer [93]. These data suggest that AnxA2 may be a potential target in cancer therapeutics. In these latter studies, expression of AnxA2 has not yet been shown to correlate with activation of plasmin, as it has in breast cancer and glioblastoma. In colorectal cancer, upregulation of AnxA2 is regulated by TGF-β, which induces epithelial–mesenchymal transition [92]. In addition, silencing of AnxA2 in lung cancer stem cells in mice induced a reduction in tumor weight, which correlated with the loss of both β-catenin and S100A100, suggesting that AnxA2 may directly or indirectly regulate metastasis [93]. The mechanism for these effects, however, remains poorly understood.

9. Conclusions

Annexin A2 appears to have a spectrum of functions across a multitude of inflammatory disorders. In early infection, according to our working model, AnxA2 may support infection by acting as an anchor protein that promotes adhesion and internalization of bacteria and viruses, regulating actin dynamics at adhesion sites, and enabling virus assembly. AnxA2 may also help recruit leukocytes to some sites of inflammation. AnxA2 also maintains the integrity of adherens junctions early in the inflammatory process by regulating phosphorylation of VE-cadherin. Numerous studies demonstrate that AnxA2 is required for optimal endosomal membrane stabilization and autophagosome biogenesis and promotes membrane repair during inflammation and infection. Cell-surface AnxA2 in complex with S100A10, finally, is a key factor in both malignant and non-malignant angiogenesis, and it contributes to tumor cell invasion and metastasis within an inflammatory microenvironment. We conclude, therefore, that annexin A2 has a primarily anti-inflammatory role, although it occasionally facilitates pathogen activity. Many of these actions have been elucidated through the use of the *Anxa2*$^{-/-}$ mouse, which has become a useful tool for understanding the role of AnxA2 in vivo (Table 2). Taken together, these activities identify annexin A2 as a robust biomarker and potential therapeutic target.

Table 2. Studies in *Anxa2-/-* mice revealing a role of annexin A2 in inflammation.

Action	Model System	Result in *Anxa2-/-*	References
Infection	*Klebsiella* pneumonia	Increased mortality	[24]
	Cryptococcal pneumonia	Increased mortality and enhanced inflammatory response	[25]
	Rickettsia australis	Increased bacteria in blood and reduced adhesion to vascular endothelial cells	[35]
Vascular integrity	Hypoxia	Increased vascular leak in lungs and skin	[47]
	Tracer injection in embryonic mice	Increased leakage of 10-kDA-dextran	[48]
Inflammasome Dynamics	*Anaplasma phagocytophilum* peritonitis	Increased bacterial load, splenomegaly, and cytopenias	[55]
Macroautophagy	Starvation-induced autophagy	Abrogation of autophagosome biogenesis	[66]
	Pseudomonas aeruginosa pneumonia	Reduced survival, increased bacterial dissemination, and blunted autophagy	[68]
Angiogenesis	Cornea pocket assay	Reduced growth factor induced neoangiogenesis	[75]
	Matrigel implant assay	Reduced growth factor induced neoangiogenesis	[78]
	Oxygen-induced retinopathy	Reduced angiogenesis and fibrinolysis with fibrin accumulation	[77]
Tumor progression	Intracerebral glioma cell implantation	Increased tumor size and reduced vascularity	[88]

Author Contributions: V.D. and K.A.H. conceptualized the manuscript. V.D. wrote the initial draft, and K.A.H. provided further editing. Both authors have read and agreed to the published version of the manuscript.

Acknowledgments: We thank Dena Almeida, Min Lucy Luo, Deyin Doreen Hsing, and Hana Lim, who provided valuable comments on the manuscript.

References

1. Netea, M.G.; Balkwill, F.; Chonchol, M.; Cominelli, F.; Donath, M.Y.; Giamarellos-Bourboulis, E.J.; Golenbock, D.; Gresnigt, M.S.; Heneka, M.T.; Hoffman, H.M.; et al. A guiding map for inflammation. *Nat. Immunol.* **2017**, *18*, 826–831. [CrossRef]
2. Chen, L.; Deng, H.; Cui, H.; Fang, J.; Zuo, Z.; Deng, J.; Li, Y.; Wang, X.; Zhao, L. Inflammatory responses and inflammation-associated diseases in organs. *Oncotarget* **2018**, *9*, 7204–7218. [CrossRef] [PubMed]
3. Takeuchi, O.; Akira, S. Pattern recognition receptors and inflammation. *Cell* **2010**, *140*, 805–820. [CrossRef] [PubMed]
4. Gerke, V.; Creutz, C.E.; Moss, S.E. Annexins: Linking Ca^{2+} signalling to membrane dynamics. *Nat. Rev. Mol. Cell Biol.* **2005**, *6*, 449–461. [CrossRef]

5. Hajjar, K.A. The Biology of Annexin A2: From Vascular Fibrinolysis to Innate Immunity. *Trans. Am. Clin. Climatol. Assoc.* **2015**, *126*, 144–155. [PubMed]
6. Rescher, U.; Gerke, V. Annexins unique membrane binding proteins with diverse functions. *J. Cell Sci.* **2004**, *117*, 2631–2639. [CrossRef]
7. Cesarman, G.M.; Guevara, C.A.; Hajjar, K.A. An endothelial cell receptor for plasminogen/tissue plasminogen activator (t-PA). II. Annexin II-mediated enhancement of t-PA-dependent plasminogen activation. *J. Biol. Chem.* **1994**, *269*, 21198–21203.
8. Hajjar, K.A.; Jacovina, A.T.; Chacko, J. An endothelial cell receptor for plasminogen/tissue plasminogen activator. I. Identity with annexin II. *J. Biol. Chem.* **1994**, *269*, 21191–21197.
9. Dassah, M.; Deora, A.B.; He, K.; Hajjar, K.A. The Endothelial Cell Annexin A2 System and Vascular Fibrinolysis. *Gen. Physiol. Biophys.* **2009**, *28*, F20–F28.
10. Hajjar, K.A.; Acharya, S.S. Annexin II and Regulation of Cell Surface Fibrinolysis. *Ann. N. Y. Acad. Sci.* **2000**, *902*, 265–271. [CrossRef]
11. Donnelly, S.R.; Moss, S.E. Annexins in the secretory pathway. *Cell. Mol. Life Sci.* **1997**, *53*, 533–538. [CrossRef]
12. Gerke, V.; Moss, S.E. Annexins: From structure to function. *Physiol. Rev.* **2002**, *82*, 331–371. [CrossRef] [PubMed]
13. Hedhli, N.; Falcone, D.J.; Huang, B.; Cesarman-Maus, G.; Kraemer, R.; Zhai, H.; Tsirka, S.E.; Santambrogio, L.; Hajjar, K.A. The annexin A2/S100A10 system in health and disease: Emerging paradigms. *J. Biomed. Biotechnol.* **2012**, *2012*, 406273. [CrossRef] [PubMed]
14. Law, A.L.; Ling, Q.; Hajjar, K.A.; Futter, C.E.; Greenwood, J.; Adamson, P.; Wavre-Shapton, S.T.; Moss, S.E.; Hayes, M.J. Annexin A2 regulates phagocytosis of photoreceptor outer segments in the mouse retina. *Mol. Biol. Cell* **2009**, *20*, 3896–3904. [CrossRef]
15. Morozova, K.; Sridhar, S.; Sidhar, S.; Zolla, V.; Clement, C.C.; Scharf, B.; Verzani, Z.; Diaz, A.; Larocca, J.N.; Hajjar, K.A.; et al. Annexin A2 promotes phagophore assembly by enhancing Atg16L$^+$ vesicle biogenesis and homotypic fusion. *Nat. Commun.* **2015**, *6*, 5856. [CrossRef] [PubMed]
16. Moss, S.E.; Morgan, R.O. The annexins. *Genome Biol.* **2004**, *5*, 219. [CrossRef] [PubMed]
17. Jaiswal, J.K.; Lauritzen, S.P.; Scheffer, L.; Sakaguchi, M.; Bunkenborg, J.; Simon, S.M.; Kallunki, T.; Jäättelä, M.; Nylandsted, J. S100A11 is required for efficient plasma membrane repair and survival of invasive cancer cells. *Nat. Commun.* **2014**, *5*, 3795. [CrossRef] [PubMed]
18. Defour, A.; Medikayala, S.; Van der Meulen, J.H.; Hogarth, M.W.; Holdreith, N.; Malatras, A.; Duddy, W.; Boehler, J.; Nagaraju, K.; Jaiswal, J.K. Annexin A2 links poor myofiber repair with inflammation and adipogenic replacement of the injured muscle. *Hum. Mol. Genet.* **2017**, *26*, 1979–1991. [CrossRef]
19. Demonbreun, A.R.; Quattrocelli, M.; Barefield, D.Y.; Allen, M.V.; Swanson, K.E.; McNally, E.M. An actin-dependent annexin complex mediates plasma membrane repair in muscle. *J. Cell Biol.* **2016**, *213*, 705–718. [CrossRef]
20. Koerdt, S.N.; Gerke, V. Annexin A2 is involved in Ca^{2+}-dependent plasma membrane repair in primary human endothelial cells. *Biochim. Biophys. Acta* **2017**, *1864*, 1046–1053. [CrossRef]
21. Jaiswal, J.K.; Nylandsted, J. S100 and annexin proteins identify cell membrane damage as the Achilles heel of metastatic cancer cells. *Cell Cycle* **2015**, *14*, 502–509. [CrossRef] [PubMed]
22. Durmus Tekir, S.; Cakir, T.; Ulgen, K. Infection Strategies of Bacterial and Viral Pathogens through Pathogen–Human Protein–Protein Interactions. *Front. Microbiol.* **2012**, *3*. [CrossRef] [PubMed]
23. Signore, A. About inflammation and infection. *EJNMMI Res.* **2013**, *3*, 8. [CrossRef] [PubMed]
24. Zhang, S.; Yu, M.; Guo, Q.; Li, R.; Li, G.; Tan, S.; Li, X.; Wei, Y.; Wu, M. Annexin A2 binds to endosomes and negatively regulates TLR4-triggered inflammatory responses via the TRAM-TRIF pathway. *Sci. Rep.* **2015**, *5*, 15859. [CrossRef]
25. Stukes, S.; Coelho, C.; Rivera, J.; Jedlicka, A.E.; Hajjar, K.A.; Casadevall, A. The Membrane Phospholipid Binding Protein Annexin A2 Promotes Phagocytosis and Nonlytic Exocytosis of Cryptococcus neoformans and Impacts Survival in Fungal Infection. *J. Immunol.* **2016**, *197*, 1252–1261. [CrossRef]
26. Creutz, C.E. The annexins and exocytosis. *Science* **1992**, *258*, 924–931. [CrossRef]
27. Gruenberg, J.; Emans, N. Annexins in membrane traffic. *Trends Cell Biol.* **1993**, *3*, 224–227. [CrossRef]
28. Gabel, M.; Chasserot-Golaz, S. Annexin A2, an essential partner of the exocytotic process in chromaffin cells. *J. Neurochem.* **2016**, *137*, 890–896. [CrossRef]

29. Gerke, V. Annexins A2 and A8 in endothelial cell exocytosis and the control of vascular homeostasis. *Biol. Chem.* **2016**, *397*, 995–1003. [CrossRef]
30. Rentero, C.; Blanco-Muñoz, P.; Meneses-Salas, E.; Grewal, T.; Enrich, C. Annexins—Coordinators of Cholesterol Homeostasis in Endocytic Pathways. *Int. J. Mol. Sci.* **2018**, *19*, 1444. [CrossRef]
31. Enrich, C.; Rentero, C.; Meneses-Salas, E.; Tebar, F.; Grewal, T. Annexins: Ca^{2+} Effectors Determining Membrane Trafficking in the Late Endocytic Compartment. *Adv. Exp. Med. Biol.* **2017**, *981*, 351–385. [CrossRef] [PubMed]
32. Kirschnek, S.; Adams, C.; Gulbins, E. Annexin II is a novel receptor for Pseudomonas aeruginosa. *Biochem. Biophys. Res. Commun.* **2005**, *327*, 900–906. [CrossRef] [PubMed]
33. Miyahara, A.; Nakanishi, N.; Ooka, T.; Hayashi, T.; Sugimoto, N.; Tobe, T. Enterohemorrhagic Escherichia coli effector EspL2 induces actin microfilament aggregation through annexin 2 activation. *Cell. Microbiol.* **2009**, *11*, 337–350. [CrossRef] [PubMed]
34. Jolly, C.; Winfree, S.; Hansen, B.; Steele-Mortimer, O. The Annexin A2/p11 complex is required for efficient invasion of Salmonella Typhimurium in epithelial cells. *Cell. Microbiol.* **2014**, *16*, 64–77. [CrossRef]
35. He, X.; Zhang, W.; Chang, Q.; Su, Z.; Gong, D.; Zhou, Y.; Xiao, J.; Drelich, A.; Liu, Y.; Popov, V.; et al. A new role for host annexin A2 in establishing bacterial adhesion to vascular endothelial cells: Lines of evidence from atomic force microscopy and an in vivo study. *Lab. Investig.* **2019**, *99*, 1650–1660. [CrossRef]
36. Hayes, M.J.; Rescher, U.; Gerke, V.; Moss, S.E. Annexin–Actin Interactions. *Traffic* **2004**, *5*, 571–576. [CrossRef]
37. Hayes, M.J.; Shao, D.; Bailly, M.; Moss, S.E. Regulation of actin dynamics by annexin 2. *EMBO J.* **2006**, *25*, 1816–1826. [CrossRef]
38. Tobe, T. Cytoskeleton-modulating effectors of enteropathogenic and enterohemorrhagic *Escherichia coli*: Role of EspL2 in adherence and an alternative pathway for modulating cytoskeleton through Annexin A2 function. *FEBS J.* **2010**, *277*, 2403–2408. [CrossRef]
39. Taylor, J.R.; Skeate, J.G.; Kast, W.M. Annexin A2 in Virus Infection. *Front. Microbiol.* **2018**, *9*. [CrossRef]
40. Dziduszko, A.; Ozbun, M.A. Annexin A2 and S100A10 Regulate Human Papillomavirus Type 16 Entry and Intracellular Trafficking in Human Keratinocytes. *J. Virol.* **2013**, *87*, 7502–7515. [CrossRef]
41. Taylor, J.R.; Fernandez, D.J.; Thornton, S.M.; Skeate, J.G.; Lühen, K.P.; Da Silva, D.M.; Langen, R.; Kast, W.M. Heterotetrameric annexin A2/S100A10 (A2t) is essential for oncogenic human papillomavirus trafficking and capsid disassembly, and protects virions from lysosomal degradation. *Sci. Rep.* **2018**, *8*, 1–15. [CrossRef] [PubMed]
42. Woodham, A.W.; Da Silva, D.M.; Skeate, J.G.; Raff, A.B.; Ambroso, M.R.; Brand, H.E.; Isas, J.M.; Langen, R.; Kast, W.M. The S100A10 subunit of the annexin A2 heterotetramer facilitates L2-mediated human papillomavirus infection. *PLoS ONE* **2012**, *7*, e43519. [CrossRef] [PubMed]
43. Murakami, M.; Simons, M. Regulation of vascular integrity. *J. Mol. Med.* **2009**, *87*, 571–582. [CrossRef] [PubMed]
44. Giannotta, M.; Trani, M.; Dejana, E. VE-cadherin and endothelial adherens junctions: Active guardians of vascular integrity. *Dev. Cell* **2013**, *26*, 441–454. [CrossRef]
45. Heyraud, S.; Jaquinod, M.; Durmort, C.; Dambroise, E.; Concord, E.; Schaal, J.P.; Huber, P.; Gulino-Debrac, D. Contribution of annexin 2 to the architecture of mature endothelial adherens junctions. *Mol. Cell. Biol.* **2008**, *28*, 1657–1668. [CrossRef]
46. Dejana, E.; Orsenigo, F.; Lampugnani, M.G. The role of adherens junctions and VE-cadherin in the control of vascular permeability. *J. Cell Sci.* **2008**, *121*, 2115–2122. [CrossRef]
47. Luo, M.; Flood, E.C.; Almeida, D.; Yan, L.; Berlin, D.A.; Heerdt, P.M.; Hajjar, K.A. Annexin A2 supports pulmonary microvascular integrity by linking vascular endothelial cadherin and protein tyrosine phosphatases. *J. Exp. Med.* **2017**, *214*, 2535–2545. [CrossRef]
48. Li, W.; Chen, Z.; Yuan, J.; Yu, Z.; Cheng, C.; Zhao, Q.; Huang, L.; Hajjar, K.A.; Chen, Z.; Lo, E.H.; et al. Annexin A2 is a Robo4 ligand that modulates ARF6 activation-associated cerebral trans-endothelial permeability. *J. Cereb. Blood Flow Metab. Off. J. Int. Soc. Cereb. Blood Flow Metab.* **2019**, *39*, 2048–2060. [CrossRef]
49. McVoy, L.A.; Kew, R.R. CD44 and Annexin A2 Mediate the C5a Chemotactic Cofactor Function of the Vitamin D Binding Protein. *J. Immunol.* **2005**, *175*, 4754–4760. [CrossRef]
50. Brownstein, C.; Deora, A.B.; Jacovina, A.T.; Weintraub, R.; Gertler, M.; Khan, K.M.F.; Falcone, D.J.; Hajjar, K.A. Annexin II mediates plasminogen-dependent matrix invasion by human monocytes: Enhanced expression by macrophages. *Blood* **2004**, *103*, 317–324. [CrossRef]

51. Rankin, C.R.; Hilgarth, R.S.; Leoni, G.; Kwon, M.; Beste, K.A.D.; Parkos, C.A.; Nusrat, A. Annexin A2 Regulates β1 Integrin Internalization and Intestinal Epithelial Cell Migration. *J. Biol. Chem.* **2013**, *288*, 15229–15239. [CrossRef] [PubMed]
52. Guo, H.; Callaway, J.B.; Ting, J.P.Y. Inflammasomes: Mechanism of action, role in disease, and therapeutics. *Nat. Med.* **2015**, *21*, 677–687. [CrossRef] [PubMed]
53. Strowig, T.; Henao-Mejia, J.; Elinav, E.; Flavell, R. Inflammasomes in health and disease. *Nature* **2012**, *481*, 278–286. [CrossRef] [PubMed]
54. Scharf, B.; Clement, C.C.; Wu, X.X.; Morozova, K.; Zanolini, D.; Follenzi, A.; Larocca, J.N.; Levon, K.; Sutterwala, F.S.; Rand, J.; et al. Annexin A2 binds to endosomes following organelle destabilization by particulate wear debris. *Nat. Commun.* **2012**, *3*, 755. [CrossRef]
55. Wang, X.; Shaw, D.K.; Sakhon, O.S.; Snyder, G.A.; Sundberg, E.J.; Santambrogio, L.; Sutterwala, F.S.; Dumler, J.S.; Shirey, K.A.; Perkins, D.J.; et al. The Tick Protein Sialostatin L2 Binds to Annexin A2 and Inhibits NLRC4-Mediated Inflammasome Activation. *Infect. Immun.* **2016**, *84*, 1796–1805. [CrossRef]
56. Feng, Y.; He, D.; Yao, Z.; Klionsky, D.J. The machinery of macroautophagy. *Cell Res.* **2014**, *24*, 24–41. [CrossRef]
57. Ravikumar, B.; Sarkar, S.; Davies, J.E.; Futter, M.; Garcia-Arencibia, M.; Green-Thompson, Z.W.; Jimenez-Sanchez, M.; Korolchuk, V.I.; Lichtenberg, M.; Luo, S.; et al. Regulation of mammalian autophagy in physiology and pathophysiology. *Physiol. Rev.* **2010**, *90*, 1383–1435. [CrossRef]
58. Xie, Z.; Klionsky, D.J. Autophagosome formation: Core machinery and adaptations. *Nat. Cell Biol.* **2007**, *9*, 1102–1109. [CrossRef]
59. Ravikumar, B.; Futter, M.; Jahreiss, L.; Korolchuk, V.I.; Lichtenberg, M.; Luo, S.; Massey, D.C.O.; Menzies, F.M.; Narayanan, U.; Renna, M.; et al. Mammalian macroautophagy at a glance. *J. Cell Sci.* **2009**, *122*, 1707–1711. [CrossRef]
60. Chen, N.; Karantza-Wadsworth, V. Role and regulation of autophagy in cancer. *Biochim. Biophys. Acta* **2009**, *1793*, 1516–1523. [CrossRef]
61. Margeta, M. Autophagy Defects in Skeletal Myopathies. *Annu. Rev. Pathol.* **2020**, *15*, 261–285. [CrossRef] [PubMed]
62. Gonzalez, C.D.; Lee, M.S.; Marchetti, P.; Pietropaolo, M.; Towns, R.; Vaccaro, M.I.; Watada, H.; Wiley, J.W. The emerging role of autophagy in the pathophysiology of diabetes mellitus. *Autophagy* **2011**, *7*, 2–11. [CrossRef] [PubMed]
63. Rubinsztein, D.C. The roles of intracellular protein-degradation pathways in neurodegeneration. *Nature* **2006**, *443*, 780–786. [CrossRef]
64. Rubinsztein, D.C.; Bento, C.F.; Deretic, V. Therapeutic targeting of autophagy in neurodegenerative and infectious diseases. *J. Exp. Med.* **2015**, *212*, 979–990. [CrossRef] [PubMed]
65. Mizushima, N.; Komatsu, M. Autophagy: Renovation of Cells and Tissues. *Cell* **2011**, *147*, 728–741. [CrossRef] [PubMed]
66. Moreau, K.; Ghislat, G.; Hochfeld, W.; Renna, M.; Zavodszky, E.; Runwal, G.; Puri, C.; Lee, S.; Siddiqi, F.; Menzies, F.M.; et al. Transcriptional regulation of Annexin A2 promotes starvation-induced autophagy. *Nat. Commun.* **2015**, *6*, 8045. [CrossRef] [PubMed]
67. Jiang, S.; Xu, Y. Annexin A2 upregulation protects human retinal endothelial cells from oxygen-glucose deprivation injury by activating autophagy. *Exp. Ther. Med.* **2019**, *18*, 2901–2908. [CrossRef]
68. Li, R.; Tan, S.; Yu, M.; Jundt, M.C.; Zhang, S.; Wu, M. Annexin A2 Regulates Autophagy in Pseudomonas aeruginosa Infection through the Akt1-mTOR-ULK1/2 Signaling Pathway. *J. Immunol.* **2015**, *195*, 3901–3911. [CrossRef]
69. Caramés, B.; Taniguchi, N.; Otsuki, S.; Blanco, F.J.; Lotz, M. Autophagy is a Protective Mechanism in Normal Cartilage and its Aging-related Loss is Linked with Cell Death and Osteoarthritis. *Arthritis Rheum.* **2010**, *62*, 791–801. [CrossRef]
70. Caramés, B.; Olmer, M.; Kiosses, W.B.; Lotz, M.K. The Relationship of Autophagy Defects to Cartilage Damage during Joint Aging in a Mouse Model. *Arthritis Rheumatol.* **2015**, *67*, 1568–1576. [CrossRef]
71. Liu, W.; Hajjar, K.A. The annexin A2 system and angiogenesis. *Biol. Chem.* **2016**, *397*, 1005–1016. [CrossRef]
72. Potente, M.; Gerhardt, H.; Carmeliet, P. Basic and therapeutic aspects of angiogenesis. *Cell* **2011**, *146*, 873–887. [CrossRef]

73. Ribatti, D. Inflammation and Angiogenesis. In *Inflammation and Angiogenesis*; Ribatti, D., Ed.; Springer International Publishing: Cham, Switzerland, 2017; pp. 25–26. ISBN 978-3-319-68448-2.
74. Aguilar-Cazares, D.; Chavez-Dominguez, R.; Carlos-Reyes, A.; Lopez-Camarillo, C.; Hernadez de la Cruz, O.N.; Lopez-Gonzalez, J.S. Contribution of Angiogenesis to Inflammation and Cancer. *Front. Oncol.* **2019**, *9*, 1399. [CrossRef] [PubMed]
75. Ling, Q.; Jacovina, A.T.; Deora, A.; Febbraio, M.; Simantov, R.; Silverstein, R.L.; Hempstead, B.; Mark, W.H.; Hajjar, K.A. Annexin II regulates fibrin homeostasis and neoangiogenesis in vivo. *J. Clin. Investig.* **2004**, *113*, 12. [CrossRef]
76. Jacovina, A.T.; Deora, A.B.; Ling, Q.; Broekman, M.J.; Almeida, D.; Greenberg, C.B.; Marcus, A.J.; Smith, J.D.; Hajjar, K.A. Homocysteine inhibits neoangiogenesis in mice through blockade of annexin A2-dependent fibrinolysis. *J. Clin. Investig.* **2009**, *119*, 3384–3394. [CrossRef] [PubMed]
77. Huang, B.; Deora, A.B.; He, K.L.; Chen, K.; Sui, G.; Jacovina, A.T.; Almeida, D.; Hong, P.; Burgman, P.; Hajjar, K.A. Hypoxia-inducible factor-1 drives annexin A2 system-mediated perivascular fibrin clearance in oxygen-induced retinopathy in mice. *Blood* **2011**, *118*, 2918–2929. [CrossRef] [PubMed]
78. Zhao, S.; Huang, L.; Wu, J.; Zhang, Y.; Pan, D.; Liu, X. Vascular endothelial growth factor upregulates expression of annexin A2 in vitro and in a mouse model of ischemic retinopathy. *Mol. Vis.* **2009**, *15*, 1231–1242. [PubMed]
79. Silva, R.L.E.; Shen, J.; Gong, Y.Y.; Seidel, C.P.; Hackett, S.F.; Kevasan, K.; Jacoby, D.B.; Campochiaro, P.A. Agents That Bind Annexin A2 Suppress Ocular Neovascularization. *J. Cell. Physiol.* **2010**, *225*, 855–864. [CrossRef]
80. Andreasen, P.A.; Egelund, R.; Petersen, H.H. The plasminogen activation system in tumor growth, invasion, and metastasis. *Cell. Mol. Life Sci.* **2000**, *57*, 25–40. [CrossRef]
81. Sharma, M.; Ownbey, R.T.; Sharma, M.C. Breast cancer cell surface annexin II induces cell migration and neoangiogenesis via tPA dependent plasmin generation. *Exp. Mol. Pathol.* **2010**, *88*, 278–286. [CrossRef]
82. Sharma, M.R.; Koltowski, L.; Ownbey, R.T.; Tuszynski, G.P.; Sharma, M.C. Angiogenesis-associated protein annexin II in breast cancer: Selective expression in invasive breast cancer and contribution to tumor invasion and progression. *Exp. Mol. Pathol.* **2006**, *81*, 146–156. [CrossRef] [PubMed]
83. Sharma, M.; Blackman, M.R.; Sharma, M.C. Antibody-directed neutralization of annexin II (ANX II) inhibits neoangiogenesis and human breast tumor growth in a xenograft model. *Exp. Mol. Pathol.* **2012**, *92*, 175–184. [CrossRef] [PubMed]
84. Sharma, M.C.; Tuszynski, G.P.; Blackman, M.R.; Sharma, M. Long-term efficacy and downstream mechanism of anti-annexinA2 monoclonal antibody (anti-ANX A2 mAb) in a pre-clinical model of aggressive human breast cancer. *Cancer Lett.* **2016**, *373*, 27–35. [CrossRef] [PubMed]
85. Nygaard, S.J.; Haugland, H.K.; Kristoffersen, E.K.; Lund-Johansen, M.; Laerum, O.D.; Tysnes, O.B. Expression of annexin II in glioma cell lines and in brain tumor biopsies. *J. Neurooncol.* **1998**, *38*, 11–18. [CrossRef] [PubMed]
86. Reeves, S.A.; Chavez-Kappel, C.; Davis, R.; Rosenblum, M.; Israel, M.A. Developmental regulation of annexin II (Lipocortin 2) in human brain and expression in high grade glioma. *Cancer Res.* **1992**, *52*, 6871–6876.
87. Onishi, M.; Ichikawa, T.; Kurozumi, K.; Inoue, S.; Maruo, T.; Otani, Y.; Fujii, K.; Ishida, J.; Shimazu, Y.; Yoshida, K.; et al. Annexin A2 regulates angiogenesis and invasion phenotypes of malignant glioma. *Brain Tumor Pathol.* **2015**, *32*, 184–194. [CrossRef]
88. Zhai, H.; Acharya, S.; Gravanis, I.; Mehmood, S.; Seidman, R.J.; Shroyer, K.R.; Hajjar, K.A.; Tsirka, S.E. Annexin A2 Promotes Glioma Cell Invasion and Tumor Progression. *J. Neurosci.* **2011**, *31*, 14346–14360. [CrossRef]
89. Ohno, Y.; Izumi, M.; Kawamura, T.; Nishimura, T.; Mukai, K.; Tachibana, M. Annexin II represents metastatic potential in clear-cell renal cell carcinoma. *Br. J. Cancer* **2009**, *101*, 287–294. [CrossRef]
90. Zimmermann, U.; Woenckhaus, C.; Pietschmann, S.; Junker, H.; Maile, S.; Schultz, K.; Protzel, C.; Giebel, J. Expression of annexin II in conventional renal cell carcinoma is correlated with Fuhrman grade and clinical outcome. *Virchows Arch. Int. J. Pathol.* **2004**, *445*, 368–374. [CrossRef]
91. Sun, Y.; Gao, G.; Cai, J.; Wang, Y.; Qu, X.; He, L.; Liu, F.; Zhang, Y.; Lin, K.; Ma, S.; et al. Annexin A2 is a discriminative serological candidate in early hepatocellular carcinoma. *Carcinogenesis* **2013**, *34*, 595–604. [CrossRef]

92. Rocha, M.R.; Barcellos-de-Souza, P.; Sousa-Squiavinato, A.C.M.; Fernandes, P.V.; de Oliveira, I.M.; Boroni, M.; Morgado-Diaz, J.A. Annexin A2 overexpression associates with colorectal cancer invasiveness and TGF-ß induced epithelial mesenchymal transition via Src/ANXA2/STAT3. *Sci. Rep.* **2018**, *8*, 11285. [CrossRef] [PubMed]
93. Andey, T.; Marepally, S.; Patel, A.; Jackson, T.; Sarkar, S.; O'Connell, M.; Reddy, R.C.; Chellappan, S.; Singh, P.; Singh, M. Cationic lipid guided short-hairpin RNA interference of annexin A2 attenuates tumor growth and metastasis in a mouse lung cancer stem cell model. *J. Control. Release Off. J. Control. Release Soc.* **2014**, *184*, 67–78. [CrossRef] [PubMed]

Diverse Roles of Annexin A6 in Triple-Negative Breast Cancer Diagnosis, Prognosis and EGFR-Targeted Therapies

Olga Y. Korolkova [1], Sarrah E. Widatalla [1], Stephen D. Williams [1], Diva S. Whalen [1], Heather K. Beasley [1], Josiah Ochieng [1], Thomas Grewal [2] and Amos M. Sakwe [1,*]

1 Department of Biochemistry and Cancer Biology, School of Graduate Studies and Research, Meharry Medical College, Nashville, TN 37208, USA; okorolkova@mmc.edu (O.Y.K.); swidatalla13@email.mmc.edu (S.E.W.); swilliams17@email.mmc.edu (S.D.W.); dwhalen15@email.mmc.edu (D.S.W.); hbeasley17@email.mmc.edu (H.K.B.); jochieng@mmc.edu (J.O.)

2 School of Pharmacy, Faculty of Medicine and Health, University of Sydney, Sydney, NSW 2006, Australia; thomas.grewal@sydney.edu.au

* Correspondence: asakwe@mmc.edu

Abstract: The calcium (Ca^{2+})-dependent membrane-binding Annexin A6 (AnxA6), is a multifunctional, predominantly intracellular scaffolding protein, now known to play relevant roles in different cancer types through diverse, often cell-type-specific mechanisms. AnxA6 is differentially expressed in various stages/subtypes of several cancers, and its expression in certain tumor cells is also induced by a variety of pharmacological drugs. Together with the secretion of AnxA6 as a component of extracellular vesicles, this suggests that AnxA6 mediates distinct tumor progression patterns via extracellular and/or intracellular activities. Although it lacks enzymatic activity, some of the AnxA6-mediated functions involving membrane, nucleotide and cholesterol binding as well as the scaffolding of specific proteins or multifactorial protein complexes, suggest its potential utility in the diagnosis, prognosis and therapeutic strategies for various cancers. In breast cancer, the low AnxA6 expression levels in the more aggressive basal-like triple-negative breast cancer (TNBC) subtype correlate with its tumor suppressor activity and the poor overall survival of basal-like TNBC patients. In this review, we highlight the potential tumor suppressor function of AnxA6 in TNBC progression and metastasis, the relevance of AnxA6 in the diagnosis and prognosis of several cancers and discuss the concept of therapy-induced expression of AnxA6 as a novel mechanism for acquired resistance of TNBC to tyrosine kinase inhibitors.

Keywords: breast cancer; annexin A6; RasGRF2; EGFR; cholesterol; cell growth; cell motility; acquired resistance; tyrosine kinase inhibitors

1. Introduction

Breast cancer is the most common cancer among women in the USA, with an incidence of 63,410 cases of in situ disease, 268,600 new cases of invasive disease, and 41,760 deaths estimated in 2019 [1]. In addition to classification into intrinsic subtypes such as luminal A, luminal B, HER2-enriched, basal-like, claudin-low and normal-like [2,3], breast cancer and triple-negative breast cancer (TNBC) in particular are known to be molecularly heterogeneous diseases. Basal-like breast cancers which are mostly TNBCs, lack or express low levels of the estrogen receptor (ER), progesterone receptor (PR) and human epidermal growth factor receptor-2 (HER2) [4,5]. Based on gene expression profiling of bulk tumors, TNBC tumors are now known to belong to at least four molecular subtypes. These include the immune active basal-like 1 (BL1/BLIA), the immunosuppressed basal-like

2 (BL2/BLIS), the mesenchymal-like (MES) and the luminal androgen-receptor-expressing (LAR) TNBC subtypes [6–8]. These mostly high-grade tumors with poor prognosis are particularly prevalent in younger patients, with frequent relapses and metastases to distant organs [9]. About 60–80% of these cancers express variable levels of the epidermal growth factor receptor (EGFR) [10,11], which for many years was considered to be a major oncogene and a promising therapeutic target in these tumors.

The discovery of EGFR as a major oncogene in TNBC sparked intense research on its therapeutic potential and several tyrosine kinase inhibitors (TKIs) and therapeutic monoclonal antibodies (mAbs) targeting this receptor have been developed. Therapeutic monoclonal antibodies against EGFR such as cetuximab bind to the ligand-binding site in the extracellular domain of the receptor. By competing with the receptor ligands, these drugs provoke receptor internalization and degradation, which is accompanied by cell cycle arrest and cell death [12]. Other studies have shown that cetuximab and perhaps other therapeutic monoclonal antibodies induce apoptosis by stimulating the expression of the cell cycle inhibitor p27Kip1 [13]. On the other hand, TKIs such as lapatinib, erlotinib, gefitinib, as well as the more recent generations of these drugs, block the kinase activity of the receptor by competing with ATP binding to the ATP binding pocket in the cytosolic tyrosine kinase domain of the receptor [14]. Some of these TKIs have been approved for the treatment of TNBC, while others are approved for other cancer types, and inhibit tumor growth by promoting cell cycle arrest and apoptosis [15]. However, the use of these EGFR-targeted therapies in the treatment of TNBC and other cancer types have led to dismal outcomes with rapid disease recurrence and metastasis (reviewed in [16]). Although the mechanisms for the frequently acquired resistance to these drugs are continually being unraveled, the failure of these drugs in the treatment of TNBC remains a major challenge. As the recurrence and subsequent disease progression are sustained by residual therapy-resistant tumor cells, remedial approaches will require a better understanding of the mechanisms underlying the ability of the therapy-resistant tumor cells to grow aggressively and/or to become invasive.

Annexin A6 (AnxA6), the largest member (with eight rather than four core domains) of the annexin family of calcium (Ca^{2+})-dependent membrane-binding proteins, is a multifunctional, predominantly intracellular scaffolding protein. In addition, AnxA6 is frequently detected in extracellular vesicles (EVs, ExoCarta exosome database: www.exocarta.org), suggesting that AnxA6 also functions extracellularly. AnxA6 is known to bind to negatively-charged phospholipids, cholesterol, nucleotides as well as a plethora of proteins in a Ca^{2+}-dependent manner, and these properties underlie, at least in part, its diverse cellular functions [17,18]. It is increasingly becoming clear that the AnxA6 expression status varies greatly in breast cancer cells as tumor cells with mesenchymal-like phenotypes express higher levels of the protein compared to those with basal-like morphology [19,20]. Although considered to be constitutively expressed in most cell types, AnxA6 expression is also inducible by treatment of tumor cells with a variety of pharmacological drugs [21], and it is differentially expressed in various stages/subtypes of several cancer types including breast cancer [22]. Additionally, it is increasingly becoming evident that disease and/or therapy-associated changes in the expression status of AnxA6 may be useful in the diagnosis, prognosis, as well as in the prediction of patient response to chemotherapy and certain targeted therapeutic options. Here, we highlight the current developments on the potential tumor suppressor and proinvasive roles of AnxA6 in TNBC and other cancers, and how this may be relevant for TNBC diagnosis and prognosis. Finally, we will discuss the novel concept of therapy-induced upregulation of AnxA6 especially in basal-like TNBC cells with low AnxA6 levels (AnxA6-low TNBC cells) as another mechanism for acquired resistance of this hard-to-treat breast cancer subtype to TKIs.

2. Molecular Characteristics and AnxA6-Mediated Functions

Based on ample evidence provided over the years, it is now well established that AnxA6 is not only a Ca^{2+}-dependent phospholipid-binding protein, but AnxA6 also binds to cholesterol and nucleotides, and serves as a scaffolding protein for several proteins to regulate and/or establish the dynamic association of multifactorial complexes in specialized membrane domains [22–24]. These molecular features appear to be critical in the multiple functions of AnxA6 in various cell types and pathological conditions including breast cancer.

2.1. Ca^{2+}-Dependent Interaction with Cellular Membranes

Elevated concentrations of extracellular Ca^{2+} or activation of Ca^{2+}-mobilizing G protein-coupled receptors (GPCRs) and receptor tyrosine kinases (RTKs) lead to an increase in cytosolic Ca^{2+} levels. This transient elevation of intracellular Ca^{2+} levels promotes the translocation of AnxA6 to the plasma membrane and endosomal membranes predominantly via its core domains, as well as to membrane-associated subcellular structures such as microtubules, the actin cytoskeleton, intermediate filaments, membrane lipid rafts, focal adhesions and cell–cell contacts [22–24]. The targeting of AnxA6 to these structures, at least in part, underlies the various membrane-associated functions of AnxA6, including regulatory roles in membrane transport along endo-/exocytic and secretory pathways. The Ca^{2+}-dependent recruitment of AnxA6 to the plasma membrane has also been shown to contribute to the inactivation of RTKs such as EGFR in A431 epidermal carcinoma cells, HeLa and head and neck cancer cell lines (Fadu, Detroit), by acting as a scaffold for protein kinase C-α (PKC-α) [25,26].

In addition, the upregulation of AnxA6 in a variety of cell lines, including EGFR-overexpressing A431 cells, results in increased association of AnxA6 with late endosomes [19,21,25], which inhibits both cholesterol and endo-/exocytic vesicle trafficking [27,28]. On the other hand, elevated AnxA6 levels have been shown to stabilize activated EGFR and potentially other activated RTKs on the surface of TNBC cells to sustain Ras/MAP kinase signaling and cell proliferation [29]. While these effects may require Ca^{2+}-dependent interaction of AnxA6 with distinct effectors followed by the translocation of the complex to the plasma membrane and/or endosomal compartment, the AnxA6-mediated inactivation or sustained activation of RTKs seems to significantly alter the proliferation and motility of tumor cells in a cell type-specific manner. In addition to the regulation of oncogenic receptor function, membrane-localized AnxA6 has also been shown to modulate store-operated Ca^{2+} entry [23], which is a major component of GPCR and RTK function and consequently affects the critical roles of these receptors in cell proliferation, motility and differentiation. Finally, membrane-translocated AnxA6 may also facilitate membrane repair of tumor cells especially following extracellular insults [30–33]. Therefore, the Ca^{2+}-dependent membrane-binding property of AnxA6 triggers numerous Ca^{2+}-modulated cellular processes, providing multiple links to the modulation of cell growth, cell motility, differentiation, apoptosis and, presumably, the resistance of TNBC cells to drugs targeting these receptors.

2.2. Cholesterol Binding and Subcellular Localization

The abundance and subcellular distribution of cholesterol, an essential component of biological membranes, is strongly influenced by the expression levels and subcellular localization of AnxA6 [32,34]. The possibility that this constitutes a direct interaction of AnxA6 with cholesterol is supported by a proteome-wide mapping of cholesterol interacting proteins that confirmed that several annexins including AnxA6, directly bind cholesterol [35]. Some studies have revealed that the core domain of most annexins is responsible for the cholesterol-mediated effects [36], while others have shown that W343 in the linker region is important for the interaction between AnxA6 and cholesterol [37]. Most strikingly, other cell-based studies revealed Ca^{2+}-insensitive, but cholesterol-dependent interaction of cytosolic AnxA6 with late endosomal membranes [38]. Interestingly, a recent study in Niemann–Pick-type C1 mutant cells identified a feedback loop involving selective lysosomal degradation of AnxA6 in the regulation of AnxA6 and cholesterol levels in this subcellular site [39]. On the other hand, upregulation of AnxA6 levels in A431, Chinese hamster ovary and other cell models triggered cholesterol accumulation in the late endosomal compartment [21,27,28]. Overexpression of AnxA6 in TNBC cells has also been linked to the accumulation of cholesterol, especially in late endosomes [21,27]. Together, these studies suggest that AnxA6 contributes to the cellular distribution of cholesterol via this feedback mechanism and presumably, other yet to be discovered mechanisms.

Overall, the association and/or cellular dynamics of AnxA6 and cholesterol appear to underlie the involvement of AnxA6 in not only cell migration [28], vesicle trafficking, exocytosis and endocytosis [27,40], but also in viral uptake, propagation and release [41,42]. The AnxA6/cholesterol paradigm may also be important in membrane repair, cell survival and protection of cells from extracellular insults [43,44]. This includes protection from a cytotoxic surge in intracellular Ca^{2+} induced by Ca^{2+} ionophores or oncogene addiction, which leads to persistent physiological activation of Ca^{2+} mobilizing receptors such as EGFR in TNBC cells [20].

2.3. *Nucleotide-Binding Characteristic of Annexins*

Annexins and AnxA6 in particular, have been shown to bind to nucleotides including ATP and GTP at least in vitro [45–47]. This appears to occur via a nucleotide-binding domain located in the N-terminus of nucleotide-sensitive annexins [46]. While the molecular details still remain sparse, the interaction may occur directly [48] or in the case of AnxA7, via bona fide nucleotide-binding proteins such as guanine nucleotide-binding protein subunit beta-2-like 1 (also known as the receptor for activated C kinase 1 (RACK1)) [49]. Earlier studies on the interaction of AnxA6 with nucleotides suggested the existence of two AnxA6 domains within residues 293-301 and 641-649 that potentially bind the phosphate groups of GTP [48]. However, in a follow-up study, the expression of a W343S AnxA6 mutant led to not only diminished GTP binding and GTP-induced ion channel activity, but also the formation of AnxA6 trimers in the presence of GTP [50]. This notwithstanding and given that nucleotides are required for the activation of protein kinases and small GTPases, it is plausible to suggest that the nucleotide-binding characteristic of AnxA6 contributes to cellular functions related to tumorigenesis, endocytosis, exocytosis, vesicular transport and signal transduction pathways. However, whether the interaction of AnxA6 with GTP or other nucleotides is relevant for TNBC progression remains to be fully elucidated.

2.4. *Scaffolding Functions of AnxA6*

Accumulating evidence suggests that the multifunctional role of AnxA6 in cancer may largely depend on its multiple and diverse scaffolding functions. Over the years, it has been demonstrated that in addition to its Ca^{2+}, phospholipid, nucleotide and cholesterol-binding properties, AnxA6 also interacts with a vast and diverse number of proteins or protein complexes which to some extent justify its multifunctional role in TNBC and other cell types. The interaction of AnxA6 with F-actin and the F-actin cross-linking protein alpha-actinin [23,51] suggests a role in the remodeling of the actin cytoskeleton, especially during cell adhesion, spreading and motility.

Several members of the S100 family of Ca^{2+} binding proteins are known to associate with cytoskeletal structures including the actin cytoskeleton, intermediate filaments and microtubules. These interactions enable S100 proteins to influence multiple cellular functions, including cell motility [52,53]. The interaction of AnxA6 with some members of the S100 proteins is not only a critical requirement to facilitate the secretion of these S100 proteins [54], but also links the plasma membrane to the cytoskeleton and/or enhances the formation of signaling complexes on biological membranes [55]. In addition, S100/AnxA6 interaction facilitates the bridging of adjacent intracellular vesicles (via annexin molecules) during membrane fusion events [56]. These multiple cellular activities driven by the interaction of AnxA6 with S100 proteins could be critical for the organization of membrane microdomains including lipid rafts, focal adhesions and cell–cell contacts, which independently contribute to enhanced cell motility.

The binding of AnxA6 to the microtubule-associated protein Tau modulates the distribution of Tau in pathologic conditions [57]. Besides its well-established role in tubulin polymerization and stabilization of microtubules, Tau protein has also been shown to be enriched in metastatic breast tumors [58] and upregulation of Tau protein is associated with resistance to paclitaxel [59] and potentially other taxane-based chemotherapeutic drugs. Although these noncanonical functions of

Tau protein are similar to some of the functions listed for AnxA6, whether the Tau/AnxA6 interaction is critical in TNBC metastasis and resistance to taxane-based therapies requires further investigations.

The interaction of AnxA6 with glycosaminoglycans such as chondroitin sulfate has also been shown to influence cell motility and invasion [60], while AnxA6 expression and its interaction with influenza A virus protein M2 strongly impaired virus release [42,61]. AnxA6 also interacts with the mu subunits of the clathrin assembly protein complex, which further supports its role in endocytosis and vesicular transport [62]. Although AnxA6 may interact with several Ca^{2+} channels to regulate Ca^{2+} influx via its interaction with L-type Ca^{2+} channels, Na^+/Ca^{2+} exchangers and store-operated Ca^{2+} entry channels, AnxA6 also influences the release of Ca^{2+} from internal stores via its interaction with sarcoplasmic reticulum Ca^{2+}-release (SERCA) channels [63–65]. Other studies have shown that AnxA6 interacts with extracellular signal effector proteins such as PKC-α [25,66], the Ras GTPase Activating Protein p120RasGAP [67,68], H-Ras [19], Raf-1 [69]; transcription factors such as the p65 subunit of nuclear factor-κB (NF-κB) [70] as well as members of the TBC family of GTPase activating proteins that regulate Rab GTPases [71]. These interactions suggest the involvement of AnxA6 in cellular signaling mechanisms that ultimately lead to tumor cell growth, motility and differentiation.

Based on evidence from previous studies, the putative binding domains of cholesterol, nucleotides and some of the known AnxA6-interacting proteins are schematically represented in Figure 1. The annexin core domains have been shown to mediate the interaction of AnxA6 with membrane phospholipids [72] and Tau [57], while the C and N terminal halves of AnxA6 are required for its interaction with S100 proteins [73] and chondroitin sulfate [60]. Actin [74], α-actinin [51] and the mu subunit of clathrin assembly protein [62] have all been shown to interact with the N-terminal of AnxA6. Meanwhile, the minimal interaction segment(s) of some AnxA6 interactors have been mapped. This includes residues 325-363 for p120GAP [75], residues 44 to 147 for influenza A virus M2 protein [61], residues 629–673 for PKC-α [76], residues 157–163, 241–247 and 590–596 for S100A11 [55] and residues 293–301 and 641–647 for nucleotides [48].

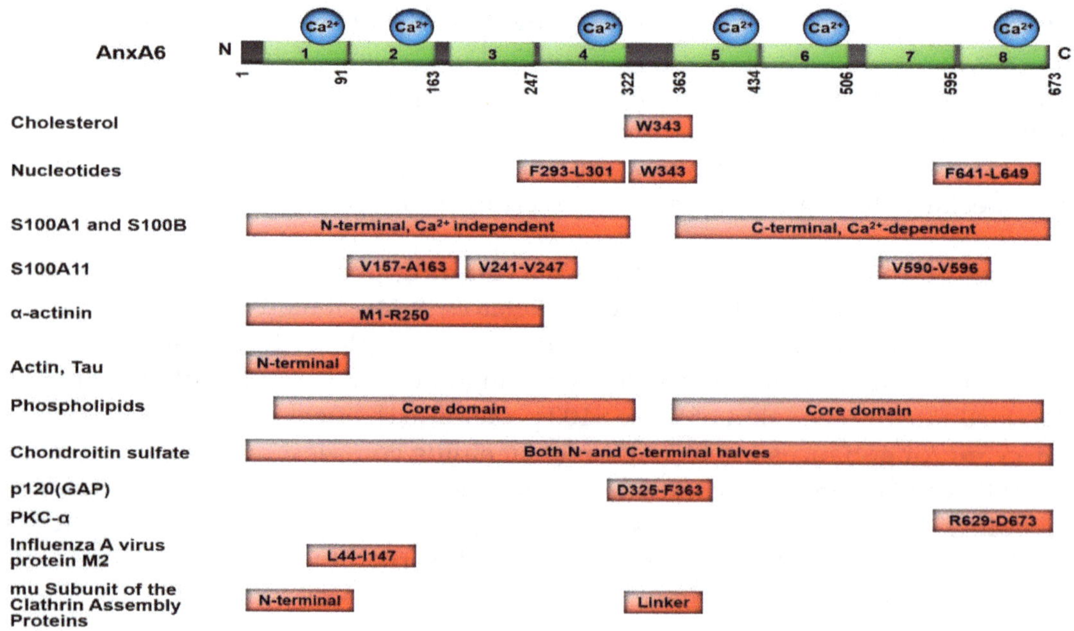

Figure 1. Putative binding domains of some AnxA6 interacting partners. The primary structure of AnxA6 showing the eight annexin repeats and the six potential Ca^{2+} binding sites. The putative binding domains of known interactors including phospholipids, cholesterol, nucleotides and other proteins relative to the primary structure of AnxA6 isoform 1 (NM_001155.5; NP_001146.2; 673 amino acid residues) are indicated (not to scale). See text for further details.

Interestingly, a single-residue W343 has been shown to be critical in the binding of both cholesterol [37] and nucleotides [50]. While the list of AnxA6 interacting proteins and/or protein complexes is not yet exhaustive, it is possible that the discovery of novel interacting proteins and the relationship with AnxA6-mediated functions will provide a further premise for its multiple functions in cancer.

3. The Multiple and Diverse Roles of AnxA6 in Tumor Cell Growth and Motility

Over the years, several studies have examined the effects of AnxA6 on hallmarks of cancer such as growth, motility and differentiation in several tumor cell models. However, our understanding of how AnxA6 promotes the progression of TNBC as well as a very diverse set of other cancers [77] remains largely unknown. An emerging concept is that differences in the expression status of intracellular and/or extracellular pools of AnxA6 underlie, at least in part, the distinct phenotypic characteristics of cancer cells and consequently, their propensity to grow rapidly or to become invasive. Here, we review our current knowledge on the role of AnxA6 in cancer progression.

3.1. Altered Expression of AnxA6 in Tumor Cell Proliferation

As a Ca^{2+}-dependent membrane-binding protein, AnxA6 has been shown to be involved in several cellular functions that define tumor cell growth. However, this function often appears to be cell- and/or cancer-type specific. In 3T3-L1 preadipocytes, siRNA-mediated AnxA6 knock-down impaired proliferation and differentiation [34]. In vivo proliferation of $CD4^+$ T cells, but not $CD8^+$ T cells, has also been shown to be impaired in $AnxA6^{-/-}$ relative to WT mice [78]. In human squamous A431 epithelial carcinoma cells, which overexpress EGFR but lack endogenous AnxA6, stable expression of AnxA6 was associated with reduced cell growth [79]. The ectopic AnxA6 expression in these cells led to growth arrest in the G1 phase of the cell cycle, longer doubling times, contact inhibition upon confluence and reduced proliferation when cultured in low serum media [80]. Furthermore, stable AnxA6 expression in A431 xenografts also reduced tumor growth in vivo [79]. Meanwhile, in breast, gastric cancer, hepatocellular and several other cancer types, elevated AnxA6 expression has also been reported to inhibit tumor cell proliferation [22,29,77]. In these tumors and in basal-like TNBCs in particular, AnxA6 appears to act as a tumor suppressor, as reduced expression of AnxA6 led to the rapid growth of xenograft tumors and was associated with poor overall survival of basal-like TNBC patients [20,29]. Interestingly, AnxA6 expression levels are relatively lower in the more aggressive basal-like breast cancer TNBC cell lines than in the more invasive mesenchymal-like TNBC cell lines [20].

The mechanisms underlying the role of AnxA6 in cell proliferation are gradually being unraveled and may include modulation of cell cycle progression [79], regulation of plasma membrane permeability to extracellular Ca^{2+} [23,31,81], inhibition of EGFR and Ras/MAP kinase signaling [19,25,68], interference with cholesterol homeostasis [27,28,71] and protection of cells via facilitation of membrane repair [31–33]. Accumulating evidence now suggests that AnxA6 may also affect the growth of tumor cells by modulating glucose and lipid metabolism, with consequences in cellular energy status and consumption. In support of this notion, AnxA6 deficiency compromised alanine-dependent gluconeogenesis during liver regeneration [82]. AnxA6 expression is also associated with fatty-acid-induced lipid droplet formation, with increased lipid droplet numbers and size in a cytoplasmic phospholipase A2α-dependent manner [83]. However, in 3T3-L1 adipocytes, AnxA6 expression decreased cellular triglycerides and adiponectin secretion suggesting that AnxA6 expression in adipose tissues contributes to impaired triglyceride storage and adiponectin release [34]. In TNBC cells, downregulation of AnxA6, which is associated with increased cell proliferation, was accompanied by the upregulation of fatty-acid-binding protein 4 (FAPB4), a protein critical for the import of fatty acids into mitochondria for degradation [20]. Finally, the involvement of AnxA6 in the sensing of intracellular pH during hypoxia [84], mitochondrial morphogenesis [85] and in cellular differentiation [76,86,87], may also contribute to cell viability and growth. While these studies suggest a potential role of AnxA6 in energy metabolism especially in rapidly growing tumor cells, the molecular details and underlying mechanism(s) remain to be fully elucidated.

3.2. Tumor Cell Motility and Invasiveness Mediated by AnxA6

Contrary to its role in the proliferation of tumor cells, the expression status of AnxA6 appears to differentially influence the motility of tumor cells in a cell-type- or cancer-type-dependent manner. Overexpression of AnxA6 in human squamous A431 epithelial carcinoma cells is associated with decreased migration in wound healing, as well as reduced invasion in matrigel and organotypic matrices [28]. These anti-invasive properties of ectopically expressed AnxA6 in human squamous A431 epithelial carcinoma cells can be attributed to mislocalization of several SNARE proteins, including SNAP23, syntaxin 4 (Stx4) and Stx6, responsible for the secretion of fibronectin (FN) and the recycling of $\alpha V\beta 3$ and $\alpha 5\beta 1$ integrins, respectively [28,40]. The SNARE dysfunction was due to the AnxA6-induced accumulation of cholesterol in late endosomes, leading to obstruction of cholesterol-sensitive exocytic and recycling pathways that deliver FN and integrins to the cell surface [28,40]. These anti-invasive activities of AnxA6 in A431 cells are, however, opposite to the proinvasive effects of AnxA6 in several other cancer cell types including invasive breast cancer cells [22], and migratory neural crest cells [88]. While this reiterates cell-type-specific roles of intracellular AnxA6 pools in cell motility, it is possible that the cell context-dependent requirement of AnxA6 in cell motility is mediated by distinct factors and/or mechanisms. Overexpression of AnxA6 in TNBC cells and presumably other solid tumor types promotes cell motility via activation of the small GTPase Cdc42 [20], which together with other Rho and Rac GTPases are known to promote several different types of cell migration and invasiveness [89]. Other studies have further suggested that the proinvasive roles of AnxA6 in TNBC cells may be linked to the functional status of focal adhesions since the loss of AnxA6 is associated with mislocalization and dysfunction of focal contacts and consequently, loss of invasiveness [22]. Whether this mechanism can be replicated in other cancer cells require further investigations.

Remarkably, several other cell-based studies have suggested a critical role of extracellular AnxA6 in tumor cell motility and invasiveness, as well as cancer metastasis in vivo [20,22,29]. Although a predominantly intracellular protein, AnxA6 is also secreted as a component of extracellular vesicles (EVs, exosomes). These small vesicles are secreted by most cells to facilitate cell–cell communications, and to deliver bioactive molecules that strongly alter the behavior of recipient cells [90,91]. The existence of an extracellular pool of AnxA6 is supported by the detection of AnxA6 in serum exosomes [92] and confirmed in TNBC cell-derived exosomes [90,92]. That the extracellular pool of AnxA6 plays a critical role in cellular adhesion, spreading and motility is also implicated in studies showing that AnxA6 is a cell surface receptor for the abundant liver-derived serum protein Fetuin-A [93], and for chondroitin sulfate [60], both of which important in cell adhesion and motility as components of the extracellular matrix. The proinvasive activity of AnxA6 is further reinforced by the identification of a monoclonal antibody (9E1) against AnxA6 with anti-invasive properties on several aggressive cancer cells including pancreatic, lung squamous and breast cancer cells [94]. These studies presumably suggest that in breast and other cancer types, the proinvasive properties of AnxA6 are mediated by receptor-like properties of AnxA6, as well as by AnxA6-enriched EVs and that the mechanisms may not be limited to the activation of NF-κB and Wnt signaling as recently reported [95].

Despite its Ca^{2+}-dependent membrane-binding activity, and the implication of AnxA6 in several membrane-associated events, direct evidence for the involvement of AnxA6 in EV biogenesis and secretion is still lacking. Exosomes are generated from intraluminal membranes (ILVs) in the late endosomal compartment and recent studies identified substantial amounts of AnxA6 to be associated with ILVs [39]. Hence, one could envisage a direct physical association of AnxA6 with the final steps of exosome release at the cell surface, which needs further investigation. However, the notion that tumor cell motility and invasiveness may be mediated by AnxA6-enriched EVs has now been independently demonstrated in breast and pancreatic cancers. These include the findings that AnxA6-enriched EVs from cancer-associated fibroblasts elicited proinvasive properties when taken up by pancreatic and breast cancer cells [96]. Additionally, chemotherapy-stimulated EVs were found to be enriched with AnxA6, and that these EVs facilitated the establishment of breast metastatic lesions in the lungs [97]. While these studies are in strong support for a critical role of AnxA6 in cancer metastasis, this also suggests that AnxA6 enrichment in EVs can be targeted for therapeutic purposes.

3.3. *Modulation of the Effector Functions of Ca^{2+}-Activated Ras Guanine Nucleotide Releasing Factor 2 (RasGRF2) by AnxA6*

The demonstration that AnxA6 interacts with PKC-α and p120RasGAP and that these interactions attenuated the activity of EGFR and Ras signaling [19], precludes the role of other components of the Ras signaling pathway such as Ras protein-specific guanine nucleotide exchange factors (RasGEFs). It is possible that other intracellular effectors sensitive to AnxA6 up- or downregulation with roles in cell growth and motility may provide the missing link between AnxA6 expression status and cell growth and motility. Indeed, in a quest to provide a comprehensive picture on the molecular mechanisms underlying the role of AnxA6 in breast cancer cell motility and proliferation, Whalen et al. identified RasGRF2, a Ras protein-specific guanine nucleotide exchange factor (RasGEF), as a major effector of AnxA6-mediated cell growth and motility [20]. In these studies, the expression levels of RasGRF2 and AnxA6 were inversely related in TNBC cells. This is further supported by the relatively high levels of RasGRF2 following downregulation of AnxA6, and the reduced RasGRF2 cellular levels upon AnxA6 overexpression in TNBC cells [20].

As discussed in the preceding sections, although relatively high AnxA6 expression is proinvasive, it has been shown to be antiproliferative in TNBC cells. RasGRF2, on the other hand, promotes cell proliferation via activation of Ras proteins, but inhibits cell motility and invasiveness via inhibition of Rho GTPases, e.g., Cdc42 and Rac1 [20,98,99]. The reciprocal expression of these two proteins is further supported by studies showing that RasGRF2 is activated by a surge in intracellular Ca^{2+} [20,100] and that this is accompanied by its degradation in proteasomes [101]. The activity of RasGRF2 is closely linked to Ca^{2+} homeostasis, in that elevated Ca^{2+} levels stimulate calmodulin-mediated RasGRF2 activation via its IQ domain, with subsequent activation of Ras and/or Rac1 mediated MAP kinase signaling [102]. Overexpression of AnxA6 in TNBC cells not only blocked the EGF/EGFR or Ca^{2+} ionophore stimulated a surge in intracellular Ca^{2+} levels, but also the degradation of RasGRF2 [20,100]. This suggests that the previously demonstrated inhibition of Ca^{2+} influx by AnxA6 [23] underlies the reciprocal expression patterns of AnxA6 and RasGRF2 in TNBC cells. Recent studies from our laboratories revealed that the gene expression profiles of AnxA6-low/RasGRF2-high rapidly growing TNBC cells are distinct from the expression patterns obtained from AnxA6-high/RasGRF2-low invasive TNBC cells [103]. Together, this not only links these proteins to Ca^{2+} mobilization by oncogenic cell surface receptors but also implicates the reciprocal expression of these proteins in TNBC progression and metastasis.

As shown in the model depicted in Figure 2, modulation of Ca^{2+} influx by AnxA6 in the EGFR/Ca^{2+} influx/RasGRF2 axis represents a critical molecular link between the expression status of AnxA6 in TNBC cells and differences in the invasiveness and proliferation of these cells. In AnxA6-high/RasGRF2-low TNBC cells such as BT-549 cells (Figure 2, left panel), activated EGFR is sustained on the cell surface and the accompanying persistent store-operated Ca^{2+} entry (SOCE) promotes the degradation of RasGRF2. Since RasGRF2 has been shown to interact with RhoGEFs to prevent these GEFs from activating their targets [98], reduced RasGRF2 cellular levels will, therefore, release its inhibition on Rho GTPase GEFs (e.g., Cdc42 GEF), which subsequently activate Cdc42. Cdc42 is also an endogenous inhibitor of the EGFR ubiquitin ligase c-Cbl [104,105], which will lead to inhibition of EGFR ubiquitination, reduced endocytosis and sustained cell surface expression of activated EGFR [29]. Hence, enhanced EGFR activity, together with increased activity of Cdc42 or related Rho GTPases [106] may drive the invasiveness of AnxA6-high/ RasGRF2 low TNBC cells. On the contrary, and as demonstrated by Koumangoye et al., activated EGFR in AnxA6-low/RasGRF2-high TNBC cells, such as MDA-MB-468, is rapidly internalized and degraded [29]. This leads to reduced SOCE, and relatively higher cellular levels of RasGRF2 [20], which not only inhibit the activation of Cdc42 and/or related GTPases, but also activates Ras proteins to drive cell proliferation (Figure 2, right panel). Although this model explains the regulation of the poorly studied receptor tyrosine kinase activated Ca^{2+} influx/RasGRF2 axis by AnxA6 in TNBC cell proliferation and migration/invasiveness, whether this can be exploited for therapeutic purposes remains to be fully investigated.

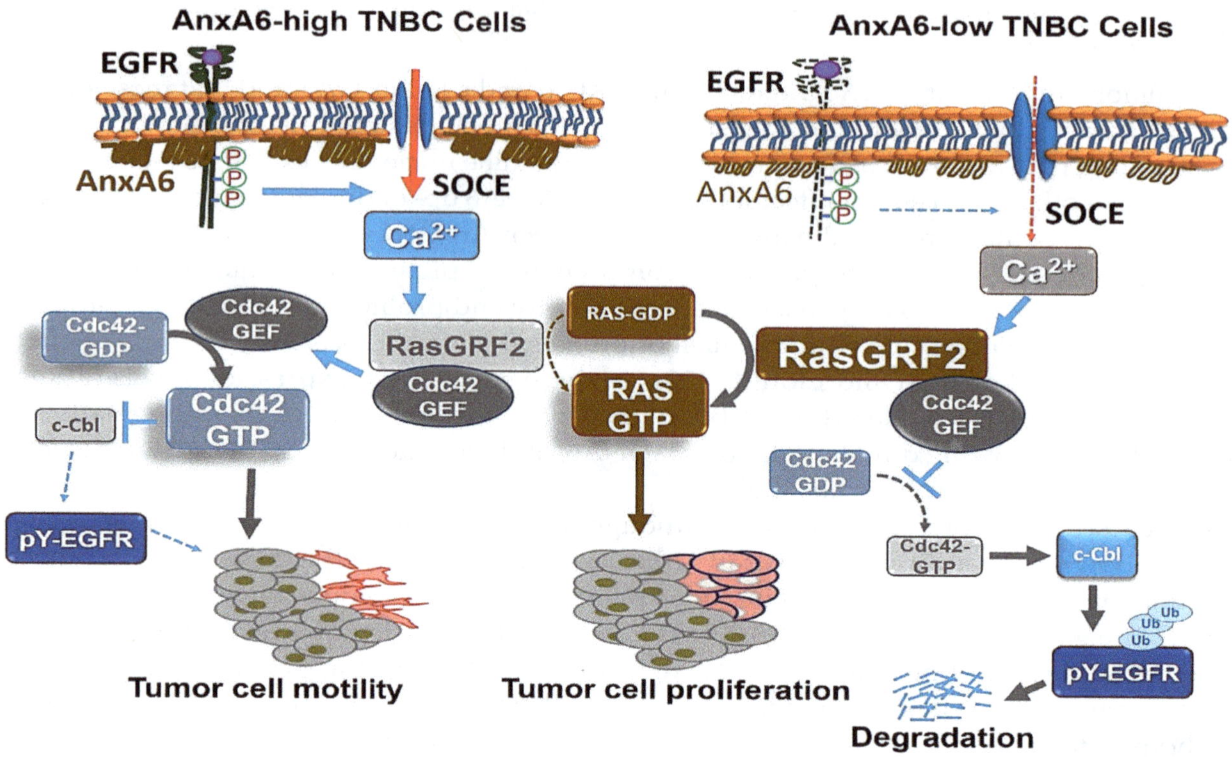

Figure 2. RasGRF2 as an effector of AnxA6-mediated TNBC cell growth and motility. AnxA6 is a predominantly cytosolic protein but upon an increase in intracellular Ca^{2+} levels, it translocates to the plasma membrane where it influences the stability of cell surface receptors, e.g., EGFR and the activity of certain Ca^{2+} channels presumably in membrane lipid rafts. In AnxA6 expressing TNBC cells (left panel), prolonged activation of EGFR leads to a sustained increase in store-operated Ca^{2+} entry (SOCE) and activation, followed by subsequent degradation of RasGRF2. The resulting decrease in cellular RasGRF2 levels enhances the activation of Cdc42 which inhibits the ubiquitin ligase c-Cbl and leads to the stabilization of activated EGFR on the cell surface. The activation of Cdc42 also sustains persistent cell motility. In AnxA6-low TNBC cells (right panel), activation of EGFR is not affected but the activated EGFR (pY-EGFR) is short-lived on the cell surface and therefore, the EGFR signal output is transient. This leads to reduced SOCE, stabilization of RasGRF2 levels and its interaction of Cdc42 GEFs, thereby inhibiting the activity of Cdc42. Reduced activity of Cdc42 enhances the activity of c-Cbl, subsequent ubiquitination of activated EGFR, internalization and degradation by proteasomes. The stabilization of RasGRF2 promotes the activation of the Ras/MAP kinase pathway and enhances cell growth. Abbreviations: AnxA6, annexin A6; EGFR, epidermal growth factor receptor; GEF, guanine nucleotide exchange factor; pY-EGFR, activated EGFR (Tyr-1068); Ras guanine nucleotide releasing factor 2, RasGRF2, SOCE, store-operated Ca^{2+} entry; TNBC, triple-negative breast cancer; Ras/MAP, Ras/mitogen-activated protein kinase.

4. Relevance of Annexin A6 in Diagnosis, Prognosis and Therapeutic Interventions

In addition to clinical and pathological characteristics, the most commonly used biomarker panel for intrinsic breast cancer classification is the expression status of ER, PR and HER2. Evaluation of these biomarkers remains the standard method for not only the evaluation of disease prognosis but also current treatment decisions. However, the classification of breast cancer solely on these biomarkers does not accurately represent the complexity of the disease, including very distinct patterns of disease progression, and significant challenges associated with the selection of patients for specific therapies. Serious challenges also remain in identifying relevant therapeutic targets and diagnostic biomarkers for certain breast cancer subtypes including TNBC. Here, we review our current understanding of the potential of AnxA6 as a biomarker for cancer diagnosis, the poor response of TNBC to EGFR-targeted therapies, and the prospects for the detection of AnxA6 as a predictor of the response of TNBC to EGFR-targeted therapies.

4.1. AnxA6 as a Biomarker for Cancer Progression

It has been amply reported that AnxA6 plays a role in the progression of TNBC and other cancer types based on changes in the expression of the protein in various neoplasms (reviewed in [77]). The potential usefulness of AnxA6 as a biomarker for the severity of certain cancer types has also been extensively reported. As indicated in Table 1, this includes studies showing that the upregulation of AnxA6 is an indicator of the progression of ovarian carcinomas [107], women's thyroid cancer [108], polycystic ovarian syndrome [109], pancreatic cancer [96] and esophageal adenocarcinoma [110]. Other studies have unambiguously demonstrated that AnxA6 may be useful to detect minimal residual disease in B-lineage acute lymphoblastic leukemia [111], the progression of melanomas [112] and squamous cervical cancer carcinogenesis [113]. Meanwhile, some studies have shown that AnxA6 is downregulated in the highly malignant forms of gastric cancer [114], hepatocellular carcinomas [115], cervical cancer [116] and breast cancer [22]. Although these studies emphasize the notion that detection of AnxA6 in several cancers might have diagnostic value, the effects of AxA6 suggest that it may act as either a tumor suppressor or a tumor promoter, depending on the type of cancer and stage of the disease [77,117]. In breast cancer, reduced expression of AnxA6 and its tumor suppressor function is more relevant in TNBC than in non-TNBC subtypes [103], an observation that is consistent with the differences in the malignancy of these breast tumors. In most of these studies, the detection of AnxA6 was carried out by either reverse transcriptase-PCR, immunohistochemistry (IHC) and/or Western blotting. Even though these assays are far from being reliable, it seems feasible that detection of AnxA6 in normal or benign tissues versus malignant tumors may be a reliable indicator of tumor progression and/or malignancy.

Table 1. Diverse diagnostic, prognostic and therapeutic significance of AnxA6 in cancer progression.

Cancer Type	AnxA6 Expression Status	Diagnostic, Prognostic or Therapeutic Value	Citation
Ovarian carcinoma	Markedly increased in advanced-stage tumors vs. benign controls	Diagnosis of advanced ovarian cancer stages	[107]
Pancreatic cancer	High expression in pancreatic cancer and lung squamous cancer vs. normal tissues	Monoclonal antibody 9E1 as a therapeutic option for invasive cancers	[94]
Pancreatic ductal adenocarcinoma	AnxA6 enriched in EVs from cancer-associated fibroblasts and following chemotherapy	Biomarker and therapeutic target	[96,97]
Esophageal adenocarcinoma	AnxA6 is a component of a 4-protein serum biomarker panel	Noninvasive detection of early tumor stages in patient serum	[110]
Squamous cervical cancer	Expression is increased in cervical intraepithelial neoplasia and microinvasive cervical cancer vs. squamous cervical cancer precursor lesions.	Diagnosis of cervical cancer progression	[113,116]
Acute lymphoblastic leukemia	Highly expressed in B-lineage acute lymphoblastic leukemia vs. normal B-cell progenitors	Diagnosis of B-lineage acute lymphoblastic leukemia	[118]
Breast cancer	Downregulated in EGFR-overexpressing and estrogen receptor (ER)-negative breast cancer cells	Biomarker for EGFR-overexpressing, ER-negative breast cancer	[19]
	Reduced expression in breast cancer tissues, but elevated in invasive breast cancer phenotypes	Biomarker for invasive breast cancer phenotypes	[22]
	Expression status significantly associated with the survival of patients with basal-like breast cancer	Predictive biomarker for basal-like breast cancer patient survival	[29]
	Elevated expression associated with acquired resistance to lapatinib in TNBC.	Predictive biomarker for response to EGFR-targeted therapies	[21]
	Loss of AnxA6 associated with the early onset and rapid growth of xenograft TNBC tumors in mice	Biomaker for TNBC progression	[20]

Table 1. *Cont.*

HER-2/neu-driven mammary tumor	Associated with tumor progression	Biomarker for rapidly growing breast cancer	[119]
Melanoma	Decrease or loss of expression as melanomas progress from benign to malignant phenotypes	Detection of melanoma progression	[112]
Gastric cancer	Downregulated in gastric cancer cells and primary gastric carcinomas	Diagnosis of gastric cancer	[114]

4.2. Lack of Efficacy of EGFR-Targeted Therapies in the Treatment of TNBC

Although TNBCs lack ER, PR and HER2, 60–80% of these cancers express variable levels of EGFR [5,10,11]. Unlike other cancer types, such as nonsmall cell lung cancer (NSCLC) that express oncogenic EGFR mutants, EGFR in TNBC is rarely mutated, but it is frequently overexpressed (reviewed in [16]). As a potential therapeutic target in TNBC and other cancers, several monoclonal antibodies (mAbs) and TKIs against the EGFR have been developed and tested in breast and other cancer types. These include therapeutic mAbs, such as cetuximab [120,121], and TKIs including lapatinib indicated for metastatic or advanced-stage TNBC [15,122]. Unfortunately, clinical trials to test the effectiveness of these and other EGFR-targeted drugs in TNBC patients have led to modest, poor or incredibly disappointing outcomes with relatively short progression-free survival [120,123].

Yet, EGFR-TKIs have shown promising results in the treatment of other cancers. For example, gefitinib is approved as a first-line treatment for metastatic NSCLC with EGFR exon 19 deletions or the L858R mutant EGFR with or without disease progression [124,125]. While these drugs are often initially effective against cancers with mutated EGFR, some patients acquire resistance due to the development of secondary, often activating mutations in exon 20 (e.g., T790M) [126,127]. Erlotinib has also shown encouraging results in the treatment of pancreatic cancer with EGFR mutations [128,129]. Cetuximab, on the other hand, has been demonstrated to be more effective in the treatment of head and neck squamous cell carcinoma (HNSCC) [130]. Other studies have tested combinations of therapeutic mAbs against EGFR and either TKIs or chemotherapeutic agents including the TBCRC-001 clinical trial of cetuximab and carboplatin [120]. Despite the disappointing results from clinical trials of EGFR-targeted therapies, the possibility that a subset of patients may effectively respond to these drugs remains to be fully explored. However, the challenge is to identify patients who can respond to these drugs with pathological complete response, as well as the need for appropriate biomarkers to monitor disease progression and/or drug efficacy.

4.3. AnxA6 as a Predictor of Breast Cancer Recurrence and Response to Therapy

Initial evidence suggesting that AnxA6 levels may influence drug sensitivity and possibly the development of drug resistance is based on the concept that AnxA6 exhibits tumor suppressor activity. In TNBC, high AnxA6 expression is associated with reduced cell growth, while reduced AnxA6 levels promote rapid cell growth. Thus, tumor cells with low AnxA6 expression levels are expected to respond more rapidly to therapeutic interventions, while cells with higher AnxA6 levels may be more refractory to treatment. In support of this concept, Koumangoye et al. demonstrated that reduced AnxA6 expression was associated with poor overall and distant metastasis-free survival of basal-like breast cancer patients and moreover, sensitized TNBC cells to TKIs [29]. This suggests that differential expression of AnxA6 may be useful for the prediction of not only the survival, but also the likelihood of basal-like breast cancer patients to respond to EGFR-targeted therapies. The sensitivity of AnxA6-low TNBC cells to EGFR-TKIs is more likely due to the rapid internalization and degradation of activated EGFR as demonstrated by AnxA6 depletion in TNBC cells [29]. Another recent report revealed that the upregulation of AnxA6 following prolonged treatment of TNBC cells with lapatinib or other EGFR-targeted TKIs was accompanied by accumulation of cholesterol in late endosomes and the development of acquired resistance [21], and this may be predictive of stable and/or progressive disease.

The abundance of AnxA6 in serum and cell-derived EVs also provides an opportunity to predict cancer metastasis in breast and other cancer types. Exosomes originate from internal vesicles of late endosomes/prelysosomes to then fuse with the plasma membrane for subsequent extracellular release. Hence, although the fate of late endosomal AnxA6 and cholesterol in TNBC cells following prolonged treatment with EGFR-TKIs remains to be clarified, it appears plausible to suggest that a proportion of the drug-induced AnxA6 as well as the late endosomal cholesterol may be secreted in EVs. This scenario may support recent studies showing that AnxA6-containing EVs are predictive of metastatic progression in pancreatic cancer [96,97]. Tumor-derived and AnxA6-enriched EVs have also been shown to be prometastatic in mouse mammary tumor virus-polyoma middle tumor-antigen (MMTV-PyMT) and 4T1 mouse models of breast cancer [97]. This suggests that AnxA6 levels in EVs from molecularly distinct breast cancer subtypes and/or chemotherapy-treated patients may be useful to predict the risk of metastasis [96,97] especially in patients who do not achieve a complete pathological response.

4.4. *Upregulation of AnxA6 Expression and the Development of Acquired Resistance*

The mechanisms underlying the poor efficacy of EGFR-targeted therapies are varied and remain poorly understood. As mentioned above, the development of acquired resistance against EGFR-TKIs is common [126,127]. This is often accompanied by overexpression and activation of other receptor tyrosine kinases such as c-MET (also called hepatocyte growth factor receptor), HER2, fibroblast growth factor receptor (FGFR) and AXL and is also associated with resistance to EGFR-TKIs of tumors with EGFR activating mutations (e.g., in NSCLC) [131]. EGFR mAbs are often more effective than TKIs, but resistance to these drugs in various carcinomas is also frequent and is mostly attributed to mutations of effectors downstream of RTKs such as Ras [132], phosphatidylinositol 3-kinase (PI3KCA) and/or loss of phosphatase and tensin homolog (PTEN) [133], as well as activation of BRAF [134]. Since activating mutations of EGFR, Ras and other effectors in the Ras/MAP kinase pathway are rare in TNBC tumors [135], it is possible that the failure of clinical trials and the development of acquired resistance of TNBCs to EGFR-targeted therapies may require distinct and/or unique molecular mechanisms.

Several lines of evidence from our studies and other reports suggest that AnxA6 expression levels are also regulated following the treatment of tumor cells with a variety of pharmacological drugs. This includes studies identifying that exposure of female mice to the suspected endocrine-disrupting xenobiotic plastic precursor bisphenol A, led to AnxA6 upregulation and that this may potentially contribute to the etiology of thyroid cancer in women [108]. Treatment of bone-marrow-derived mesenchymal stem cells with the polyaromatic hydrocarbon fluoranthene also led to AnxA6 upregulation [136]. Furthermore, the hapten challenge of a nonatopic asthma mouse model sensitive to the organofluorine dinitrofluorobenzene also led to increased AnxA6 levels [137]. Although the upregulation of AnxA6 in these studies was detected by proteomic profiling, this nevertheless suggests that detection of AnxA6 together with other coexpressed genes by RT-PCR or IHC, could be useful as novel biomarkers for exposure to these drugs.

We and others have also demonstrated that treatment of breast tumor cells with TKIs [21], nonselective Ca^{2+} channel blockers [103], and the DNA methyltransferase (DNMT) inhibitors 5-aza-2′-deoxycytidine or 5-aza-cytidine [114] led to AnxA6 upregulation. Given that the AnxA6 promoter is heavily methylated, for example in EGFR-overexpressing A431 and ER-negative MDA-MB-468 breast cancer cells, both of which with relatively low AnxA6 levels [19], these effects may be mediated via epigenetic mechanisms including inhibition of DNA methyltransferases and/or histone deacetylases. Within this context, it should also be noted that the human AnxA6 gene is located on chromosome 5q32–q34, with several studies identifying a statistically significant loss of 5q31–q35 in ER-negative tumors and its association with ErbB2 amplification [138–140]. It is not clear whether there are ER-negative cell lines with this chromosomal aberration, but it will be interesting to compare the response to therapy of such cell lines to those in which reduced AnxA6 expression could be due to epigenetic mechanisms.

Until recently, it has been unclear whether AnxA6 expression status could be associated with the response of TNBC cells to cytotoxic and/or EGFR-targeted therapies. Strikingly, analysis of AnxA6 expression in stage IV TNBC clinical samples from patients treated with cetuximab and/or carboplatin [120] revealed that treatment of patients with this combination regimen was associated with AnxA6 upregulation. Further analysis showed that the therapy-induced AnxA6 expression was associated with EGFR inhibition rather than chemotherapy in AnxA6-low TNBC cell lines [21]. This finding suggests that unlike treatment with cytotoxic chemotherapeutic drugs such as paclitaxel and carboplatin, treatment of TNBC cells with EGFR-targeted therapies is accompanied by the upregulation of AnxA6, which coincides with an accumulation of late endosomal cholesterol [21]. As indicated above and extensively reviewed elsewhere, cholesterol homeostasis is commonly dysregulated in cancer [141,142] and often accompanied by anticancer drug resistance [143–145]. Although AnxA6 strongly influences the intracellular distribution of cholesterol in most cell types [21,146], it is still feasible that AnxA6 expression and high cellular cholesterol levels are independently associated with drug resistance [29]. The latter study suggests that the lapatinib-induced AnxA6 upregulation and concomitant cholesterol accumulation constitutes a novel adaptive mechanism for EGFR-expressing TNBC cells to overcome prolonged treatment with EGFR-targeted TKIs [21]. As further evidence in support of this notion, the withdrawal of lapatinib from lapatinib-resistant cells reversed the expression of AnxA6 to basal levels. Similarly, stable expression of shRNAs targeting AnxA6 in lapatinib-resistant TNBC cells prevented the lapatinib-induced AnxA6 upregulation, suggesting that chronic treatment of TNBC cells with EGFR-TKIs affects AnxA6 mRNA expression rather than the protein stability. Thus far, it remains unclear whether the fate of AnxA6 and cholesterol in the late endosomal compartment of chronic EGFR-TKI treated TNBC cells includes subsequent secretion in EVs. However, these findings provide a strong rationale for further studies to validate the detection of AnxA6 by RT-PCR or IHC along with cholesterol or other markers found in cholesterol-rich and specialized membrane domains (e.g., lipid rafts), as biomarkers for acquired resistance of TNBC to EGFR-targeted therapies and/or other RTK antagonists.

5. Conclusions and Future Perspectives

For more than a decade, AnxA6 has drawn considerable interest as a potential suppressor or promoter of cancer initiation and progression in a variety of cancers. This coincides with increasing evidence for AnxA6 being a drug-inducible factor that primarily functions as a scaffolding protein. As high AnxA6 expression levels promote cell motility but attenuate cell growth in TNBCs, the potential for AnxA6 to influence both tumor progression and metastasis has generated great interest. It is also becoming evident that AnxA6 expression levels in various tumor settings most likely reflects the ability of AnxA6 to interact with or indirectly influence the activity of a specific protein or protein complex in a location- and cell type-specific manner; and that this is highly relevant in molecular events that drive cell growth, migration/invasion and differentiation that define carcinogenesis.

The precise role of AnxA6 in drug resistance still remains incomplete. However, the increase in the expression/localization of AnxA6 in late endosomes and the retention of cholesterol in this compartment following prolonged treatment of tumor cells with EGFR-targeting drugs that interfere with Ca^{2+} entry/signaling, provide an exciting first insight implicating high/low AnxA6 levels in drug resistance. Moreover, the upregulation of AnxA6 following chronic treatment of AnxA6-low basal-like TNBC cells with EGFR inhibitors constitutes a novel mechanism for the development of acquired resistance to these and presumably similar drugs targeting other RTKs.

The recent finding that the cellular levels of AnxA6 are inversely related to the levels of RasGRF2 suggests that the reciprocal expression of these proteins in distinct TNBC molecular subtypes underlies, at least in part, the difference in their propensity for growth and/or motility and consequently, tumor malignancy. Together, this provides a rationale for the use of TKIs targeting EGFR/HER2 in combination with inhibition of RasGRF2 to block hyperactive and wild type Ras/MAP kinase pathway-driven triple-negative breast tumors.

Finally, the finding that reduced AnxA6 levels are more relevant in TNBC than in non-TNBC tumors suggests that detection of AnxA6 may not only be useful as a potential biomarker for specific breast cancer subtypes, but also provides promise as a predictor of the response of especially basal-like TNBC to targeted therapeutic interventions. Further studies to exploit these characteristics of AnxA6 are warranted to clearly delineate its usefulness as a diagnostic biomarker and a predictive/prognostic factor in breast and other cancer types.

Author Contributions: This manuscript was written with the contributions of all authors. All authors have read and agreed to the published version of the manuscript.

Acknowledgments: We would like to thank all members of our laboratories, past and present, for their invaluable contributions and apologize to all those collaborators whose work could not be discussed in this review owing to space limitations.

References

1. DeSantis, C.E.; Ma, J.; Bryan, L.; Jemal, A. Breast cancer statistics. *CA A Cancer J. Clin.* **2013**, *64*, 52–62. [CrossRef] [PubMed]
2. Perou, C.M. Molecular Stratification of Triple-Negative Breast Cancers. *Oncologist* **2011**, *16*, 61–70. [CrossRef] [PubMed]
3. Prat, A.; Adamo, B.; Cheang, M.C.U.; Anders, C.K.; Carey, L.A.; Perou, C.M. Molecular Characterization of Basal-Like and Non-Basal-Like Triple-Negative Breast Cancer. *Oncologist* **2013**, *18*, 123–133. [CrossRef] [PubMed]
4. Cancer Genome Atlas Network. Comprehensive molecular portraits of human breast tumours. *Nature* **2012**, *490*, 61–70. [CrossRef]
5. Lehmann, B.D.; Bauer, J.A.; Chen, X.; Sanders, M.E.; Chakravarthy, A.B.; Shyr, Y.; Pietenpol, J.A. Identification of human triple-negative breast cancer subtypes and preclinical models for selection of targeted therapies. *J. Clin. Investig.* **2011**, *121*, 2750–2767. [CrossRef]
6. Burstein, M.D.; Tsimelzon, A.; Poage, G.M.; Covington, K.R.; Contreras, A.; Fuqua, S.A.; Savage, M.I.; Osborne, C.K.; Hilsenbeck, S.G.; Chang, J.C.; et al. Comprehensive genomic analysis identifies novel subtypes and targets of triple-negative breast cancer. *Clin. Cancer Res.* **2014**, *21*, 1688–1698. [CrossRef]
7. Lehmann, B.D.; Jovanović, B.; Chen, X.; Estrada, M.V.; Johnson, K.N.; Shyr, Y.; Moses, H.L.; Sanders, M.E.; Pietenpol, J.A. Refinement of Triple-Negative Breast Cancer Molecular Subtypes: Implications for Neoadjuvant Chemotherapy Selection. *PLoS ONE* **2016**, *11*, e0157368. [CrossRef]
8. Liu, Y.-R.; Jiang, Y.-Z.; Xu, X.-E.; Yu, K.-D.; Jin, X.; Hu, X.; Zuo, W.-J.; Hao, S.; Wu, J.; Liu, G.-Y.; et al. Comprehensive transcriptome analysis identifies novel molecular subtypes and subtype-specific RNAs of triple-negative breast cancer. *Breast Cancer Res.* **2016**, *18*, 33. [CrossRef]
9. Carey, L.A.; Perou, C.M.; Livasy, C.A.; Dressler, L.G.; Cowan, D.; Conway, K.; Karaca, G.; Troester, M.A.; Tse, C.K.; Edmiston, S.; et al. Race, Breast Cancer Subtypes, and Survival in the Carolina Breast Cancer Study. *JAMA* **2006**, *295*, 2492–2502. [CrossRef]
10. Cheang, M.C.U.; Voduc, D.; Bajdik, C.; Leung, S.; McKinney, S.; Chia, S.K.; Perou, C.M.; Nielsen, T.O. Basal-Like Breast Cancer Defined by Five Biomarkers Has Superior Prognostic Value than Triple-Negative Phenotype. *Clin. Cancer Res.* **2008**, *14*, 1368–1376. [CrossRef]
11. Rakha, E.A.; Reis-Filho, J.S. Basal-like breast carcinoma: From expression profiling to routine practice. *Arch. Pathol. Lab. Med.* **2009**, *133*, 1041–1063.
12. Doody, J.F.; Wang, Y.; Patel, S.N.; Joynes, C.; Lee, S.P.; Gerlak, J.; Rolser, R.L.; Li, Y.; Steiner, P.; Bassi, R.; et al. Inhibitory activity of cetuximab on epidermal growth factor receptor mutations in non small cell lung cancers. *Mol. Cancer Ther.* **2007**, *6*, 2642–2651. [CrossRef] [PubMed]
13. Baselga, J.; Albanell, J. Epithelial growth factor receptor interacting agents. *Hematol. Clin. N. Am.* **2002**, *16*, 1041–1063. [CrossRef]
14. Carey, K.D.; Garton, A.J.; Romero, M.S.; Kahler, J.; Thomson, S.; Ross, S.; Park, F.; Haley, J.D.; Gibson, N.; Sliwkowski, M.X. Kinetic analysis of epidermal growth factor receptor somatic mutant proteins shows increased sensitivity to the epidermal growth factor receptor tyrosine kinase inhibitor, erlotinib. *Cancer Res.* **2006**, *66*, 8163–8171. [CrossRef] [PubMed]

15. Arteaga, C.L. EGF Receptor As a Therapeutic Target: Patient Selection and Mechanisms of Resistance to Receptor-Targeted Drugs. *J. Clin. Oncol.* **2003**, *21*, 289–291. [CrossRef] [PubMed]
16. Nakai, K.; Hung, M.C.; Yamaguchi, H. A perspective on anti-EGFR therapies targeting triple-negative breast cancer. *Am. J. Cancer Res.* **2016**, *6*, 1609–1623.
17. Enrich, C.; Rentero, C.; De Muga, S.V.; Reverter, M.; Mulay, V.; Wood, P.; Köse, M.; Grewal, T. Annexin A6—Linking Ca2+ signaling with cholesterol transport. *Biochim. Biophys. Acta.* **2011**, *1813*, 935–947. [CrossRef]
18. Enrich, C.; Rentero, C.; Grewal, T. Annexin A6 in the liver: From the endocytic compartment to cellular physiology. *Biochim. Biophys. Acta.* **2017**, *1864*, 933–946. [CrossRef]
19. Vila de Muga, S.; Timpson, P.; Cubells, L.; Evans, R.; Hayes, T.E.; Rentero, C.; Hegemann, A.; Reverter, M.; Leschner, J.; Pol, A.; et al. Annexin A6 inhibits Ras signalling in breast cancer cells. *Oncogene* **2009**, *28*, 363–377. [CrossRef]
20. Whalen, D.S.; Widatalla, S.E.; Korolkova, O.Y.; Nangami, G.S.; Beasley, H.K.; Williams, S.D.; Virgous, C.; Lehmann, B.D.; Ochieng, J.; Sakwe, A.M. Implication of calcium activated RasGRF2 in Annexin A6-mediated breast tumor cell growth and motility. *Oncotarget* **2019**, *10*, 133–151. [CrossRef]
21. Widatalla, S.E.; Korolkova, O.Y.; Whalen, D.S.; Goodwin, J.S.; Williams, K.P.; Ochieng, J.; Sakwe, A.M. Lapatinib-induced annexin A6 upregulation as an adaptive response of triple-negative breast cancer cells to EGFR tyrosine kinase inhibitors. *Carcinogenesis* **2019**, *40*, 998–1009. [CrossRef]
22. Sakwe, A.M.; Koumangoye, R.; Guillory, B.; Ochieng, J. Annexin A6 contributes to the invasiveness of breast carcinoma cells by influencing the organization and localization of functional focal adhesions. *Exp. Cell Res.* **2011**, *317*, 823–837. [CrossRef]
23. Monastyrskaya, K.; Babiychuk, E.B.; Hostettler, A.; Wood, P.; Grewal, T.; Draeger, A. Plasma Membrane-associated Annexin A6 Reduces Ca2+Entry by Stabilizing the Cortical Actin Cytoskeleton. *J. Boil. Chem.* **2009**, *284*, 17227–17242. [CrossRef]
24. Strzelecka-Kiliszek, A.; Buszewska, M.E.; Podszywalow-Bartnicka, P.; Pikula, S.; Otulak, K.; Buchet, R.; Bandorowicz-Pikula, J. Calcium- and pH-dependent localization of annexin A6 isoforms in Balb/3T3 fibroblasts reflecting their potential participation in vesicular transport. *J. Cell Biochem.* **2008**, *104*, 418–434. [CrossRef]
25. Koese, M.; Rentero, C.; Kota, B.P.; Hoque, M.; Cairns, R.; Wood, P.; De Muga, S.V.; Reverter, M.; Alvarez-Guaita, A.; Monastyrskaya, K.; et al. Annexin A6 is a scaffold for PKCα to promote EGFR inactivation. *Oncogene* **2012**, *32*, 2858–2872. [CrossRef]
26. Seedorf, K.; Sherman, M.; Ullrich, A. Protein kinase C mediates short- and long-term effects on receptor tyrosine kinases. Regulation of tyrosine phosphorylation and degradation. *Ann. N. Y. Acad. Sci.* **1995**, *11*, 18953–18960. [CrossRef]
27. Cubells, L.; De Muga, S.V.; Tebar, F.; Wood, P.; Evans, R.; Ingelmo-Torres, M.; Calvo, M.; Gaus, K.; Pol, A.; Grewal, T.; et al. Annexin A6-Induced Alterations in Cholesterol Transport and Caveolin Export from the Golgi Complex. *Traffic* **2007**, *8*, 1568–1589. [CrossRef]
28. García-Melero, A.; Reverter, M.; Hoque, M.; Meneses-Salas, E.; Köse, M.; Conway, J.R.W.; Johnsen, C.H.; Alvarez-Guaita, A.; Morales-Paytuví, F.; Elmaghrabi, Y.A.; et al. Annexin A6 and Late Endosomal Cholesterol Modulate Integrin Recycling and Cell Migration. *J. Boil. Chem.* **2015**, *291*, 1320–1335. [CrossRef]
29. Koumangoye, R.B.; Nangami, G.; Thompson, P.D.; Agboto, V.; Ochieng, J.; Sakwe, A.M. Reduced annexin A6 expression promotes the degradation of activated epidermal growth factor receptor and sensitizes invasive breast cancer cells to EGFR-targeted tyrosine kinase inhibitors. *Mol. Cancer* **2013**, *12*, 167. [CrossRef]
30. Alvarez-Guaita, A.; De Muga, S.V.; Owen, D.M.; Williamson, D.J.; Magenau, A.; García-Melero, A.; Reverter, M.; Hoque, M.; Cairns, R.; Cornely, R.; et al. Evidence for annexin A6-dependent plasma membrane remodelling of lipid domains. *Br. J. Pharmacol.* **2015**, *172*, 1677–1690. [CrossRef]
31. Babiychuk, E.B.; Monastyrskaya, K.; Potez, S.; Draeger, A. Intracellular Ca2+ operates a switch between repair and lysis of streptolysin O-perforated cells. *Cell Death Differ.* **2009**, *16*, 1126–1134. [CrossRef] [PubMed]
32. Boye, T.L.; Maeda, K.; Pezeshkian, W.; Sønder, S.L.; Haeger, S.C.; Gerke, V.; Simonsen, A.C.; Nylandsted, J. Annexin A4 and A6 induce membrane curvature and constriction during cell membrane repair. *Nat. Commun.* **2017**, *8*, 1623. [CrossRef] [PubMed]
33. Demonbreun, A.R.; Fallon, K.S.; Oosterbaan, C.C.; Bogdanovic, E.; Warner, J.L.; Sell, J.J.; Page, P.G.; Quattrocelli, M.; Barefield, D.Y.; McNally, E.M. Recombinant annexin A6 promotes membrane repair

and protects against muscle injury. *J. Clin. Investig.* **2019**, *129*, 4657–4670. [CrossRef] [PubMed]

34. Krautbauer, S.; Haberl, E.M.; Eisinger, K.; Pohl, R.; Rein-Fischboeck, L.; Rentero, C.; Alvarez-Guaita, A.; Enrich, C.; Grewal, T.; Buechler, C.; et al. Annexin A6 regulates adipocyte lipid storage and adiponectin release. *Mol. Cell. Endocrinol.* **2017**, *439*, 419–430. [CrossRef] [PubMed]
35. Hulce, J.J.; Cognetta, A.B.; Niphakis, M.J.; Tully, S.E.; Cravatt, B.F. Proteome-wide mapping of cholesterol-interacting proteins in mammalian cells. *Nat. Methods* **2013**, *10*, 259–264. [CrossRef] [PubMed]
36. Ayala-Sanmartin, J. Cholesterol enhances phospholipid binding and aggregation of annexins by their core domain. *Biochem. Biophys. Res.* **2001**, *283*, 72–79. [CrossRef]
37. Domon, M.M.; Matar, G.; Strzelecka-Kiliszek, A.; Bandorowicz-Pikula, J.; Pikula, S.; Besson, F. Interaction of annexin A6 with cholesterol rich membranes is pH-dependent and mediated by the sterol OH. *J. Colloid Interface Sci.* **2010**, *346*, 436–441. [CrossRef]
38. De Diego, I.; Schwartz, F.; Siegfried, H.; Dauterstedt, P.; Heeren, J.; Beisiegel, U.; Enrich, C.; Grewal, T. Cholesterol Modulates the Membrane Binding and Intracellular Distribution of Annexin 6. *J. Boil. Chem.* **2002**, *277*, 32187–32194. [CrossRef]
39. Meneses-Salas, E.; García-Melero, A.; Blanco-Muñoz, P.; Jose, J.; Brenner, M.-S.; Lu, A.; Tebar, F.; Rentero, C.; Rentero, C.; Enrich, C. Selective Degradation Permits a Feedback Loop Controlling Annexin A6 and Cholesterol Levels in Endolysosomes of NPC1 Mutant Cells. *Cells* **2020**, *9*, 1152. [CrossRef]
40. Reverter, M.; Rentero, C.; de Muga, S.V.; Alvarez-Guaita, A.; Mulay, V.; Cairns, R.; Wood, P.; Monastyrskaya, K.; Pol, A.; Tebar, F.; et al. Cholesterol transport from late endosomes to the Golgi regulates t-SNARE trafficking, assembly, and function. *Mol. Biol Cell* **2011**, *22*, 4108–4123. [CrossRef]
41. Kuhnl, A.; Musiol, A.; Heitzig, N.; Johnson, D.E.; Ehrhardt, C.; Grewal, T.; Gerke, V.; Ludwig, S.; Rescher, U. Late Endosomal/Lysosomal Cholesterol Accumulation Is a Host Cell-Protective Mechanism Inhibiting Endosomal Escape of Influenza A Virus. *mBio* **2018**, *24*. [CrossRef]
42. Musiol, A.; Gran, S.; Ehrhardt, C.; Ludwig, S.; Grewal, T.; Gerke, V.; Rescher, U. Annexin A6-Balanced Late Endosomal Cholesterol Controls Influenza A Replication and Propagation. *mBio* **2013**, *4*, e00608–e00613. [CrossRef]
43. Creutz, C.E.; Hira, J.K.; Gee, V.E.; Eaton, J.M. Protection of the membrane permeability barrier by annexins. *Biochemistry* **2012**, *51*, 9966–9983.
44. Potez, S.; Luginbühl, M.; Monastyrskaya, K.; Hostettler, A.; Draeger, A.; Babiychuk, E.B. Tailored Protection against Plasmalemmal Injury by Annexins with Different Ca2+ Sensitivities. *J. Boil. Chem.* **2011**, *286*, 17982–17991. [CrossRef] [PubMed]
45. Bandorowicz-Pikula, J.; Danieluk, M.; Wrzosek, A.; Bus, R.; Buchet, R.; Pikula, S. Annexin VI: An intracellular target for ATP. *Acta Biochim Pol.* **1999**, *46*, 801–812. [CrossRef] [PubMed]
46. Danieluk, M.; Pikula, S.; Bandorowicz-Pikula, J. Annexin VI interacts with adenine nucleotides and their analogs. *Biochimie* **1999**, *81*, 717–726. [CrossRef]
47. Kirilenko, A.; Golczak, M.; Pikuła, S.; Bandorowicz-Pikuła, J. GTP-binding properties of the membrane-bound form of porcine liver annexin VI. *Acta Biochim. Pol.* **2001**, *48*, 851–865. [CrossRef] [PubMed]
48. Bandorowicz-Pikula, J.; Kirilenko, A.; van Deursen, R.; Golczak., M.; Kuhnel, M.; Lancelin, J.M.; Pikula, S.; Buchet, R. A putative consensus sequence for the nucleotide-binding site of annexin A6. *Biochemistry* **2003**, *42*, 9137–9146. [CrossRef] [PubMed]
49. Du, Y.; Meng, J.; Huang, Y.; Wu, J.; Wang, B.; Ibrahim, M.M.; Tang, J. Guanine nucleotide-binding protein subunit beta-2-like 1, a new Annexin A7 interacting protein. *Biochem. Biophys. Res. Commun.* **2014**, *445*, 58–63. [CrossRef]
50. Kirilenko, A.; Pikula, S.; Bandorowicz-Pikula, J. Effects of Mutagenesis of W343 in Human Annexin A6 Isoform 1 on Its Interaction with GTP: Nucleotide-Induced Oligomer Formation and Ion Channel Activity. *Biochemistry* **2006**, *45*, 4965–4973. [CrossRef]
51. Mishra, S.; Chander, V.; Banerjee, P.; Oh, J.G.; Lifirsu, E.; Park, W.J.; Kim, D.H.; Bandyopadhyay, A. Interaction of annexin A6 with alpha actinin in cardiomyocytes. *BMC Cell Biol.* **2011**, *12*, 7. [CrossRef] [PubMed]
52. Jaiswal, J.K.; Nylandsted, J. S100 and annexin proteins identify cell membrane damage as the Achilles heel of metastatic cancer cells. *Cell Cycle* **2015**, *14*, 502–509. [CrossRef] [PubMed]
53. Nedjadi, T.; Kitteringham, N.; Campbell, F.; Jenkins, R.E.; Park, B.K.; Navarro, P.; Ashcroft, F.; Tepikin, A.V.; Neoptolemos, J.; Costello, E. S100A6 binds to annexin 2 in pancreatic cancer cells and promotes pancreatic cancer cell motility. *Br. J. Cancer* **2009**, *101*, 1145–1154. [CrossRef] [PubMed]

54. Bode, G.; Luken, A.; Kerkhoff, C.; Roth, J.; Ludwig, S.; Nacken, W. Interaction between S100A8/A9 and annexin A6 is involved in the calcium-induced cell surface exposition of S100A8/A9. *J. Biol. Chem.* **2008**, *283*, 31776–31784. [CrossRef]
55. Chang, N.; Sutherland, C.; Hesse, E.; Winkfein, R.; Wiehler, W.B.; Pho, M.; Veillette, C.; Li, S.; Wilson, D.P.; Kiss, E.; et al. Identification of a novel interaction between the Ca2+-binding protein S100A11 and the Ca2+- and phospholipid-binding protein annexin A6. *Am. J. Physiol. Physiol.* **2007**, *292*, C1417–C1430. [CrossRef]
56. Garbuglia, M.; Verzini, M.; Donato, R. Annexin VI binds S100A1 and S100B and blocks the ability of S100A1 and S100B to inhibit desmin and GFAP assemblies into intermediate filaments. *Cell Calcium* **1998**, *24*, 177–191. [CrossRef]
57. Gauthier-Kemper, A.; Alonso, M.S.; Sündermann, F.; Niewidok, B.; Fernández, M.-P.; Bakota, L.; Heinisch, J.J.; Brandt, R. Annexins A2 and A6 interact with the extreme N terminus of tau and thereby contribute to tau's axonal localization. *J. Boil. Chem.* **2018**, *293*, 8065–8076. [CrossRef]
58. Matrone, M.A.; Whipple, R.A.; Thompson, K.; Cho, E.; Vitolo, M.I.; Balzer, E.M.; Yoon, J.R.; Ioffe, O.B.; Tuttle, K.C.; Tan, M.; et al. Metastatic breast tumors express increased tau, which promotes microtentacle formation and the reattachment of detached breast tumor cells. *Oncogene* **2010**, *29*, 3217–3227. [CrossRef]
59. Tanaka, S.; Nohara, T.; Iwamoto, M.; Sumiyoshi, K.; Kimura, K.; Takahashi, Y.; Tanigawa, N. Tau Expression and Efficacy of Paclitaxel Treatment in Metastatic Breast Cancer. *Poster Sess. Abstr.* **2009**, *69*, 1138. [CrossRef]
60. Takagi, H.; Asano, Y.; Yamakawa, N.; Matsumoto, I.; Kimata, K. Annexin 6 is a putative cell surface receptor for chondroitin sulfate chains. *J. Cell Sci.* **2002**, 115.
61. Ma, H.; Kien, F.; Manière, M.; Zhang, Y.; Lagarde, N.; Tse, K.S.; Poon, L.L.M.; Nal, B. Human Annexin A6 Interacts with Influenza A Virus Protein M2 and Negatively Modulates Infection. *J. Virol.* **2011**, *86*, 1789–1801. [CrossRef] [PubMed]
62. Creutz, C.E.; Snyder, S.L. Interactions of annexins with the mu subunits of the clathrin assembly proteins. *Biochemistry* **2005**, *44*, 13795–13806. [PubMed]
63. Gunteski-Hamblin, A.M.; Song, G.; Walsh, R.A.; Frenzke, M.; Boivin, G.P.; Dorn, G.W.; Kaetzel, M.A.; Horseman, N.D.; Dedman, J.R. Annexin VI overexpression targeted to heart alters cardiomyocyte function in transgenic mice. *Am. J. Physiol. Circ. Physiol.* **1996**, *270*, H1091–H1100. [CrossRef] [PubMed]
64. Hazarika, P.; Sheldon, A.; Kaetzel, M.A.; Diaz-Munoz, M.; Hamilton, S.L.; Dedman, J.R. Regulation of the sarcoplasmic reticulum Ca2+-release channel requires intact annexin VI. *J. Cell. Biochem.* **1991**, *46*, 86–93. [CrossRef]
65. Song, G.; Harding, S.E.; Duchen, M.R.; Tunwell, R.; O'Gara, P.; Hawkins, T.E.; Moss, S.E. Altered mechanical properties and intracellular calcium signaling in cardiomyocytes from annexin 6 null-mutant mice. *FASEB J.* **2002**, *16*, 622–624. [CrossRef]
66. Schmitz-Peiffer, C.; Browne, C.L.; Walker, J.H.; Biden, T.J. Activated protein kinase C alpha associates with annexin VI from skeletal muscle. *Biochem. J.* **1998**, *330*, 675–681. [CrossRef]
67. Davis, A.J.; Butt, J.T.; Walker, J.H.; Moss, S.E.; Gawler, D. The Ca2+-dependent lipid binding domain of P120GAP mediates protein-protein interactions with Ca2+-dependent membrane-binding proteins. Evidence for a direct interaction between annexin VI and P120GAP. *J. Boil. Chem.* **1996**, *271*, 24333–24336. [CrossRef]
68. Grewal, T.; Evans, R.; Rentero, C.; Tebar, F.; Cubells, L.; De Diego, I.; Kirchhoff, M.F.; Hughes, W.E.; Heeren, J.; Rye, K.-A.; et al. Annexin A6 stimulates the membrane recruitment of p120GAP to modulate Ras and Raf-1 activity. *Oncogene* **2005**, *24*, 5809–5820. [CrossRef]
69. Pons, M.; Grewal, T.; Rius, E.; Schnitgerhans, T.; Jäckle, S.; Enrich, C. Evidence for the Involvement of Annexin 6 in the Trafficking between the Endocytic Compartment and Lysosomes. *Exp. Cell Res.* **2001**, *269*, 13–22. [CrossRef]
70. Campbell, K.A.; Minashima, T.; Zhang, Y.; Hadley, S.; Lee, Y.J.; Giovinazzo, J.; Quirno, M.; Kirsch, T. Annexin A6 interacts with p65 and stimulates NF-kappaB activity and catabolic events in articular chondrocytes. *Arthritis Rheum.* **2013**, *65*, 3120–3129. [CrossRef]
71. Meneses-Salas, E.; Garcia-Melero, A.; Kanerva, K.; Blanco-Munoz, P.; Morales-Paytuvi, F.; Bonjoch, J.; Casas, J.; Egert, A.; Beevi, S.S.; Jose, J.; et al. Annexin A6 modulates TBC1D15/Rab7/StARD3 axis to control endosomal cholesterol export in NPC1 cells. *Cell Mol. Life Sci.* **2019**, *77*, 2839–2857. [CrossRef] [PubMed]
72. Huber, R.; Römisch, J.; Paques, E.P. The crystal and molecular structure of human annexin V, an anticoagulant protein that binds to calcium and membranes. *EMBO J.* **1990**, *9*, 3867–3874. [CrossRef] [PubMed]

73. Santamaria-Kisiel, L.; Rintala-Dempsey, A.C.; Shaw, G.S. Calcium-dependent and -independent interactions of the S100 protein family. *Biochem. J.* **2006**, *396*, 201–214. [CrossRef] [PubMed]
74. Locate, S.; Colyer, J.; Gawler, D.J.; Walker, J.H. Annexin A6 at the cardiac myocyte sarcolemma – Evidence for self-association and binding to actin. *Cell Boil. Int.* **2008**, *32*, 1388–1396. [CrossRef]
75. Chow, A.; Gawler, D.J. Mapping the site of interaction between annexin VI and the p120GAP C2 domain. *FEBS Lett.* **1999**, *460*, 166–172. [CrossRef]
76. Minashima, T.; Small, W.; Moss, S.E.; Kirsch, T. Intracellular Modulation of Signaling Pathways by Annexin A6 Regulates Terminal Differentiation of Chondrocytes. *J. Boil. Chem.* **2012**, *287*, 14803–14815. [CrossRef]
77. Qi, H.; Liu, S.; Guo, C.; Wang, J.; Greenaway, F.T.; Sun, M.-Z. Role of annexin A6 in cancer. *Oncol. Lett.* **2015**, *10*, 1947–1952. [CrossRef]
78. Cornely, R.; Pollock, A.H.; Rentero, C.; Norris, S.E.; Alvarez-Guaita, A.; Grewal, T.; Mitchell, T.; Enrich, C.; Moss, S.E.; Parton, R.G.; et al. Annexin A6 regulates interleukin-2-mediated T-cell proliferation. *Immunol. Cell Boil.* **2016**, *94*, 543–553. [CrossRef]
79. Theobald, J.; Smith, P.D.; Jacob, S.M.; Moss, S.E. Expression of annexin VI in A431 carcinoma cells suppresses proliferation: A possible role for annexin VI in cell growth regulation. *Biochim. Biophys. Acta* **1994**, *1223*, 383–390. [CrossRef]
80. Hoque, M.; Elmaghrabi, Y.A.; Köse, M.; Beevi, S.S.; Jose, J.; Meneses-Salas, E.; Blanco-Muñoz, P.; Conway, J.R.W.; Swarbrick, A.; Timpson, P.; et al. Annexin A6 improves anti-migratory and anti-invasive properties of tyrosine kinase inhibitors in EGFR overexpressing human squamous epithelial cells. *FEBS J.* **2019**. [CrossRef]
81. Fleet, A.; Ashworth, R.; Kubista, H.; Edwards, H.; Bolsover, S.; Mobbs, P.; Moss, S.E. Inhibition of EGF-Dependent Calcium Influx by Annexin VI is Splice Form-Specific. *Biochem. Biophys. Res. Commun.* **1999**, *260*, 540–546. [CrossRef]
82. Alvarez-Guaita, A.; Blanco-Muñoz, P.; Meneses-Salas, E.; Wahba, M.; Pollock, A.H.; Bosch, M.; Gaus, K.; Lu, A.; Pol, A.; Tebar, F.; et al. Annexin A6 is critical to maintain glucose homeostasis and survival during liver regeneration. *Hepatology* **2020**. [CrossRef]
83. Cairns, R.; Alvarez-Guaita, A.; Martínez-Saludes, I.; Wason, S.J.; Hanh, J.; Nagarajan, S.; Hosseini-Beheshti, E.; Monastyrskaya, K.; Hoy, A.J.; Buechler, C.; et al. Role of hepatic Annexin A6 in fatty acid-induced lipid droplet formation. *Exp. Cell Res.* **2017**, *358*, 397–410. [CrossRef]
84. Monastyrskaya, K.; Tschumi, F.; Babiychuk, E.B.; Stroka, D.; Draeger, A. Annexins sense changes in intracellular pH during hypoxia. *Biochem. J.* **2007**, *409*, 65–75. [CrossRef]
85. Chlystun, M.; Campanella, M.; Law, A.-L.; Duchen, M.; Fatimathas, L.; Levine, T.P.; Gerke, V.; Moss, S.E. Regulation of Mitochondrial Morphogenesis by Annexin A6. *PLoS ONE* **2013**, *8*, e53774. [CrossRef]
86. Pfander, D.; Swoboda, B.; Kirsch, T. Expression of Early and Late Differentiation Markers (Proliferating Cell Nuclear Antigen, Syndecan-3, Annexin VI, and Alkaline Phosphatase) by Human Osteoarthritic Chondrocytes. *Am. J. Pathol.* **2001**, *159*, 1777–1783. [CrossRef]
87. Shah, A.; Taneyhill, L.A. Differential expression pattern of Annexin A6 in chick neural crest and placode cells during cranial gangliogenesis. *Gene Expr. Patterns* **2015**, *18*, 21–28. [CrossRef]
88. Wu, C.Y.; Taneyhill, L.A. Annexin a6 modulates chick cranial neural crest cell emigration. *PLoS ONE* **2012**, *7*, e44903. [CrossRef]
89. Ridley, A.J. Rho GTPases and cell migration. *J. Cell Sci.* **2001**, *114*, 2713–2722.
90. Koumangoye, R.B.; Sakwe, A.M.; Goodwin, J.S.; Patel, T.; Ochieng, J. Detachment of Breast Tumor Cells Induces Rapid Secretion of Exosomes Which Subsequently Mediate Cellular Adhesion and Spreading. *PLoS ONE* **2011**, *6*, e24234. [CrossRef]
91. Sato, S.; Weaver, A.M. Extracellular vesicles: Important collaborators in cancer progression. *Essays Biochem.* **2018**, *62*, 149–163.
92. Ochieng, J.; Pratap, S.; Khatua, A.K.; Sakwe, A.M. Anchorage-independent growth of breast carcinoma cells is mediated by serum exosomes. *Exp. Cell Res.* **2009**, *315*, 1875–1888. [CrossRef]
93. Kundranda, M.N.; Ray, S.; Saria, M.; Friedman, D.; Matrisian, L.M.; Lukyanov, P.; Ochieng, J. Annexins expressed on the cell surface serve as receptors for adhesion to immobilized fetuin-A. *Biochim. Biophys. Acta (BBA) Bioenerg.* **2004**, *1693*, 111–123. [CrossRef]
94. O'Sullivan, D.; Dowling, P.; Joyce, H.; McAuley, E.; McCann, A.; Henry, M.; McGovern, B.; Barham, P.; Kelleher, F.C.; Murphy, J.; et al. A novel inhibitory anti-invasive MAb isolated using phenotypic screening

highlights AnxA6 as a functionally relevant target protein in pancreatic cancer. *Br. J. Cancer* **2017**, *117*, 1326–1335. [CrossRef]
95. Minashima, T.; Kirsch, T. Annexin A6 regulates catabolic events in articular chondrocytes via the modulation of NF-kappaB and Wnt/ss-catenin signaling. *PLoS ONE* **2018**, *13*, e0197690. [CrossRef]
96. Leca, J.; Martinez, S.; Lac, S.; Nigri, J.; Secq, V.; Rubis, M.; Bressy, C.; Sergé, A.; Lavaut, M.-N.; Dusetti, N.; et al. Cancer-associated fibroblast-derived annexin A6+ extracellular vesicles support pancreatic cancer aggressiveness. *J. Clin. Investig.* **2016**, *126*, 4140–4156. [CrossRef]
97. Keklikoglou, I.; Cianciaruso, C.; Güç, E.; Squadrito, M.L.; Spring, L.M.; Tazzyman, S.; Lambein, L.; Poissonnier, A.; Ferraro, G.B.; Baer, C.; et al. Chemotherapy elicits pro-metastatic extracellular vesicles in breast cancer models. *Nat. Cell Biol.* **2018**, *21*, 190–202. [CrossRef]
98. Calvo, F.; Sanz-Moreno, V.; Agudo-Ibáñez, L.; Wallberg, F.; Sahai, E.; Marshall, C.J.; Crespo, P. RasGRF suppresses Cdc42-mediated tumour cell movement, cytoskeletal dynamics and transformation. *Nat. Cell Biol.* **2011**, *13*, 819–826. [CrossRef]
99. Ma, X.; Espana-Serrano, L.; Kim, W.J.; Thayele Purayil, H.; Nie, Z.; Daaka, Y. betaArrestin1 regulates the guanine nucleotide exchange factor RasGRF2 expression and the small GTPase Rac-mediated formation of membrane protrusion and cell motility. *J. Biol. Chem.* **2014**, *289*, 13638–13650. [CrossRef]
100. Anborgh, P.H.; Qian, X.; Papageorge, A.G.; Vass, W.C.; DeClue, J.E.; Lowy, D.R. Ras-Specific Exchange Factor GRF: Oligomerization through Its Dbl Homology Domain and Calcium-Dependent Activation of Raf. *Mol. Cell. Boil.* **1999**, *19*, 4611–4622. [CrossRef]
101. De Hoog, C.L.; Koehler, J.A.; Goldstein, M.D.; Taylor, P.; Figeys, D.; Moran, M.F. Ras Binding Triggers Ubiquitination of the Ras Exchange Factor Ras-GRF2. *Mol. Cell. Boil.* **2001**, *21*, 2107–2117. [CrossRef] [PubMed]
102. Tebar, F.; Chavero, A.; Agell, N.; Lu, A.; Rentero, C.; Enrich, C.; Grewal, T. Pleiotropic Roles of Calmodulin in the Regulation of KRas and Rac1 GTPases: Functional Diversity in Health and Disease. *Int. J. Mol. Sci.* **2020**, *21*, 3680. [CrossRef] [PubMed]
103. Korolkova, O.Y.; Widatalla, S.E.; Whalen, D.S.; Nangami, G.N.; Abimbola, A.; Williams, S.D.; Beasley, H.K.; Reisenbichler, E.; Washington, M.K.; Ochieng, J.; et al. Reciprocal expression of Annexin A6 and RasGRF2 discriminates rapidly growing from invasive triple negative breast cancer subsets. *PLoS ONE* **2020**, *15*, e0231711. [CrossRef] [PubMed]
104. Schmidt, M.H.H.; Husnjak, K.; Szymkiewicz, I.; Haglund, K.; Dikic, I. Cbl escapes Cdc42-mediated inhibition by downregulation of the adaptor molecule betaPix. *Oncogene* **2006**, *25*, 3071–3078. [CrossRef]
105. Wu, W.J.; Tu, S.; Cerione, R.A. Activated Cdc42 Sequesters c-Cbl and Prevents EGF Receptor Degradation. *Cell* **2003**, *114*, 715–725. [CrossRef]
106. Hirsch, D.S.; Shen, Y.; Wu, W.J. Growth and Motility Inhibition of Breast Cancer Cells by Epidermal Growth Factor Receptor Degradation Is Correlated with Inactivation of Cdc42. *Cancer Res.* **2006**, *66*, 3523–3530. [CrossRef]
107. Noreen, S.; Gardener, Q.A.; Fatima, I.; Sadaf, S.; Akhtar, M.W. Up-Regulated Expression of Calcium Dependent Annexin A6: A Potential Biomarker of Ovarian Carcinoma. *Proteom. Clin. Appl.* **2019**, e1900078.
108. Lee, H.-S.; Kang, Y.; Tae, K.; Bae, G.-U.; Park, J.Y.; Cho, Y.H.; Yang, M. Proteomic Biomarkers for Bisphenol A–Early Exposure and Women's Thyroid Cancer. *Cancer Res. Treat.* **2018**, *50*, 111–117. [CrossRef]
109. Li, L.; Zhang, J.; Deng, Q.; Li, J.; Li, Z.; Xiao, Y.; Hu, S.; Li, T.; Tan, Q.; Li, X.; et al. Proteomic Profiling for Identification of Novel Biomarkers Differentially Expressed in Human Ovaries from Polycystic Ovary Syndrome Patients. *PLoS ONE* **2016**, *11*, e0164538. [CrossRef] [PubMed]
110. Zaidi, A.H.; Gopalakrishnan, V.; Kasi, P.M.; Zeng, X.; Malhotra, U.; Balasubramanian, J.; Visweswaran, S.; Sun, M.; Flint, M.; Davison, J.M.; et al. Evaluation of a 4-protein serum biomarker panel-biglycan, annexin-A6, myeloperoxidase, and protein S100-A9 (B-AMP)-for the detection of esophageal adenocarcinoma. *Cancer* **2014**, *120*, 3902–3913. [CrossRef] [PubMed]
111. Chen, X.; Pan, Q.; Stow, P.; Behm, F.G.; Goorha, R.; Pui, C.H.; Neale, G. Quantification of minimal residual disease in T-lineage acute lymphoblastic leukemia with the TAL-1 deletion using a standardized real-time PCR assay. *Leukemia* **2001**, *15*, 166–170. [CrossRef] [PubMed]
112. Francia, G.; Mitchell, S.D.; Moss, S.E.; Hanby, A.M.; Marshall, J.F.; Hart, I.R. Identification by differential display of annexin-VI, a gene differentially expressed during melanoma progression. *Cancer Res.* **1996**, *56*, 3855–3858. [PubMed]

113. Lomnytska, M.; Becker, S.; Bodin, I.; Olsson, A.; Hellman, K.; Hellström, A.-C.; Mints, M.; Hellman, U.; Auer, G.; Andersson, S. Differential expression of ANXA6, HSP27, PRDX2, NCF2, and TPM4 during uterine cervix carcinogenesis: Diagnostic and prognostic value. *Br. J. Cancer* **2010**, *104*, 110–119. [CrossRef] [PubMed]
114. Wang, X.; Zhang, S.; Zhang, J.; Lam, E.; Liu, X.; Sun, J.; Feng, L.; Lu, H.; Yu, J.; Jin, H. Annexin A6 is down-regulated through promoter methylation in gastric cancer. *Am. J. Transl. Res.* **2013**, *5*, 555–562.
115. Meier, E.M.; Rein-Fischboeck, L.; Pohl, R.; Wanninger, J.; Hoy, A.J.; Grewal, T.; Eisinger, K.; Krautbauer, S.; Liebisch, G.; Weiss, T.; et al. Annexin A6 protein is downregulated in human hepatocellular carcinoma. *Mol. Cell. Biochem.* **2016**, *418*, 81–90. [CrossRef]
116. Lomnytska, M.I.; Becker, S.; Hellman, K.; Hellstrom, A.C.; Souchelnytskyi, S.; Mints, M.; Hellman, U.; Andersson, S.; Auer, G. Diagnostic protein marker patterns in squamous cervical cancer. *Proteomics Clin. Appl* **2010**, *4*, 17–31. [CrossRef]
117. Grewal, T.; Hoque, M.; Conway, J.R.W.; Reverter, M.; Wahba, M.; Beevi, S.S.; Timpson, P.; Enrich, C.; Rentero, C. Annexin A6—A multifunctional scaffold in cell motility. *Cell Adhes. Migr.* **2017**, *11*, 288–304. [CrossRef]
118. Chen, J.-S.; Coustan-Smith, E.; Suzuki, T.; Neale, G.A.; Mihara, K.; Pui, C.-H.; Campana, D. Identification of novel markers for monitoring minimal residual disease in acute lymphoblastic leukemia. *Blood* **2001**, *97*, 2115–2120. [CrossRef]
119. Croci, S.; Recktenwald, C.V.; Lichtenfels, R.; Nicoletti, G.; Dressler, S.P.; De Giovanni, C.; Astolfi, A.; Palladini, A.; Shin-ya, K.; Landuzzi, L.; et al. Proteomic and PROTEOMEX profiling of mammary cancer progression in a HER-2/neu oncogene-driven animal model system. *Proteomics* **2010**, *10*, 3835–3853. [CrossRef]
120. Carey, L.A.; Rugo, H.S.; Marcom, P.K.; Mayer, E.L.; Esteva, F.J.; Ma, C.X.; Liu, M.C.; Storniolo, A.M.; Rimawi, M.F.; Forero-Torres, A.; et al. TBCRC 001: Randomized phase II study of cetuximab in combination with carboplatin in stage IV triple-negative breast cancer. *J Clin Oncol. Off. J. Am. Soc. Clin. Oncol.* **2012**, *30*, 2615–2623. [CrossRef]
121. Ferraro, D.A.; Gaborit, N.; Maron, R.; Cohen-Dvashi, H.; Porat, Z.; Pareja, F.; Lavi, S.; Lindzen, M.; Ben-Chetrit, N.; Sela, M.; et al. Inhibition of triple-negative breast cancer models by combinations of antibodies to EGFR. *Proc. Natl. Acad. Sci. USA* **2013**, *110*, 1815–1820. [CrossRef]
122. Burness, M.L.; Grushko, T.A.; Olopade, O.I. Epidermal growth factor receptor in triple-negative and basal-like breast cancer: Promising clinical target or only a marker? *Cancer J.* **2010**, *16*, 23–32. [CrossRef]
123. Baselga, J.; Gomez, P.; Greil, R.; Braga, S.; Climent, M.A.; Wardley, A.M.; Kaufman, B.; Stemmer, S.M.; Pego, A.; Chan, A.; et al. Randomized phase II study of the anti-epidermal growth factor receptor monoclonal antibody cetuximab with cisplatin versus cisplatin alone in patients with metastatic triple-negative breast cancer. *J. Clin. Oncol.* **2013**, *31*, 2586–2592. [CrossRef]
124. Inomata, M.; Shukuya, T.; Takahashi, T.; Ono, A.; Nakamura, Y.; Tsuya, A.; Tanigawara, Y.; Naito, T.; Murakami, H.; Harada, H.; et al. Continuous administration of EGFR-TKIs following radiotherapy after disease progression in bone lesions for non-small cell lung cancer. *Anticancer. Res.* **2011**, *31*.
125. Shukuya, T.; Takahashi, T.; Naito, T.; Kaira, R.; Ono, A.; Nakamura, Y.; Tsuya, A.; Tanigawara, Y.; Murakami, H.; Harada, H.; et al. Continuous EGFR-TKI administration following radiotherapy for non-small cell lung cancer patients with isolated CNS failure. *Lung Cancer* **2011**, *74*, 457–461. [CrossRef] [PubMed]
126. Soh, J.; Toyooka, S.; Ichihara, S.; Suehisa, H.; Kobayashi, N.; Ito, S.; Yamane, M.; Aoe, M.; Sano, Y.; Kiura, K.; et al. EGFR mutation status in pleural fluid predicts tumor responsiveness and resistance to gefitinib. *Lung Cancer* **2007**, *56*, 445–448. [CrossRef]
127. Zhang, Q.; Ke, E.; Niu, F.; Deng, W.; Chen, Z.; Xu, C.; Zhang, X.-C.; Zhao, N.; Su, J.; Yang, J.; et al. The role of T790M mutation in EGFR-TKI re-challenge for patients with EGFR-mutant advanced lung adenocarcinoma. *Oncotarget* **2017**, *8*, 4994–5002. [CrossRef] [PubMed]
128. Fountzilas, C.; Chhatrala, R.; Khushalani, N.; Hutson, A.; Tucker, C.; Ma, W.W.; Warren, G.; Boland, P.; Tan, W.; LeVea, C.; et al. A phase II trial of erlotinib monotherapy in advanced pancreatic cancer as a first- or second-line agent. *Cancer Chemother. Pharmacol.* **2017**, *80*, 497–505. [CrossRef]
129. Moore, M.J.; Goldstein, D.; Hamm, J.; Figer, A.; Hecht, J.R.; Gallinger, S.; Au, H.J.; Murawa, P.; Walde, D.; Wolff, R.A.; et al. Erlotinib Plus Gemcitabine Compared with Gemcitabine Alone in Patients With Advanced Pancreatic Cancer: A Phase III Trial of the National Cancer Institute of Canada Clinical Trials Group. *J. Clin. Oncol.* **2007** *25*, 1960–1966. [CrossRef]

130. Bernier, J. Drug Insight: Cetuximab in the treatment of recurrent and metastatic squamous cell carcinoma of the head and neck. *Nat. Clin. Pract. Oncol.* **2008**, *5*, 705–713. [CrossRef]
131. Yu, H.; Riely, G.J.; Lovly, C.M. Therapeutic strategies utilized in the setting of acquired resistance to EGFR tyrosine kinase inhibitors. *Clin. Cancer Res.* **2014**, *20*, 5898–5907. [CrossRef] [PubMed]
132. Misale, S.; Yaeger, R.; Hobor, S.; Scala, E.; Janakiraman, M.; Liska, D.; Valtorta, E.; Schiavo, R.; Buscarino, M.; Siravegna, G.; et al. Emergence of KRAS mutations and acquired resistance to anti-EGFR therapy in colorectal cancer. *Nature* **2012**, *486*, 532–536. [CrossRef] [PubMed]
133. Jhawer, M.; Goel, S.; Wilson, A.J.; Montagna, C.; Ling, Y.-H.; Byun, -S.; Nasser, S.; Arango, D.; Shin, J.; Klampfer, L.; et al. PIK3CA mutation/PTEN expression status predicts response of colon cancer cells to the epidermal growth factor receptor inhibitor cetuximab. *Cancer Res.* **2008**, *68*, 1953–1961. [CrossRef] [PubMed]
134. Berg, M.; Søreide, K. EGFR and downstream genetic alterations in KRAS/BRAF and PI3K/AKT pathways in colorectal cancer: Implications for targeted therapy. *Discov. Med.* **2012**, *14*.
135. Grob, T.J.; Heilenkötter, U.; Geist, S.; Paluchowski, P.; Wilke, C.; Jaenicke, F.; Quaas, A.; Wilczak, W.; Choschzick, M.; Sauter, G.; et al. Rare oncogenic mutations of predictive markers for targeted therapy in triple-negative breast cancer. *Breast Cancer Res. Treat.* **2012**, *134*, 561–567. [CrossRef]
136. Hoque, M.; Lee, Y.E.; Kim, H.R.; Shin, M.-G. Potential biomarkers and antagonists for fluoranthene-induced cellular toxicity of bone marrow derived mesenchymal stem cells. *Blood Res.* **2019**, *54*, 253–261. [CrossRef]
137. Houtman, R.; Krijgsveld, J.; Kool, M.; Romijn, E.P.; Redegeld, F.A.; Nijkamp, F.P.; Heck, A.J.; Humphery-Smith, I. Lung proteome alterations in a mouse model for nonallergic asthma. *Proteomics* **2003**, *3*, 2008–2018. [CrossRef]
138. Johannsdottir, H.K.; Jonsson, G.; Johannesdottir, G.; Agnarsson, B.A.; Eerola, H.; Arason, A.; Heikkila, P.; Egilsson, V.; Olsson, H.; Johannsson, O.T.; et al. Chromosome 5 imbalance mapping in breast tumors from BRCA1 and BRCA2 mutation carriers and sporadic breast tumors. *Int. J. Cancer* **2006**, *119*, 1052–1060. [CrossRef]
139. Loo, L.W.M. Array Comparative Genomic Hybridization Analysis of Genomic Alterations in Breast Cancer Subtypes. *Cancer Res.* **2004**, *64*, 8541–8549. [CrossRef]
140. Pierga, J.-Y.; Reis-Filho, J.S.; Cleator, S.J.; Dexter, T.; Mackay, A.; Simpson, P.T.; Fenwick, K.; Iravani, M.; Salter, J.; Hills, M.; et al. Microarray-based comparative genomic hybridisation of breast cancer patients receiving neoadjuvant chemotherapy. *Br. J. Cancer* **2006**, *96*, 341–351. [CrossRef]
141. Cruz, P.M.; Mo, H.; McConathy, W.J.; Sabnis, N.A.; Lacko, A.G. The role of cholesterol metabolism and cholesterol transport in carcinogenesis: A review of scientific findings, relevant to future cancer therapeutics. *Front. Pharmacol.* **2013**, *4*. [CrossRef] [PubMed]
142. Poirot, M.; Silvente-Poirot, S.; Weichselbaum, R.R. Cholesterol metabolism and resistance to tamoxifen. *Curr. Opin. Pharmacol.* **2012**, *12*, 683–689. [CrossRef] [PubMed]
143. Gu, L.; Saha, S.T.; Thomas, J.; Kaur, M. Targeting cellular cholesterol for anticancer therapy. *FEBS J.* **2019**, *286*, 4192–4208. [CrossRef]
144. Guillaumond, F.; Bidaut, G.; Ouaissi, M.; Servais, S.; Gouirand, V.; Olivares, O.; Lac, S.; Borge, L.; Roques, J.; Gayet, O.; et al. Cholesterol uptake disruption, in association with chemotherapy, is a promising combined metabolic therapy for pancreatic adenocarcinoma. *Proc. Natl. Acad. Sci. USA* **2015**, *112*, 2473–2478. [CrossRef]
145. Vasseur, S.; Guillaumond, F. LDL Receptor: An open route to feed pancreatic tumor cells. *Mol. Cell. Oncol.* **2015**, *3*, e1033586. [CrossRef]
146. Rentero, C.; Blanco-Muñoz, P.; Meneses-Salas, E.; Grewal, T.; Enrich, C. Annexins—Coordinators of Cholesterol Homeostasis in Endocytic Pathways. *Int. J. Mol. Sci.* **2018**, *19*, 1444. [CrossRef]

10

Annexin A1 Regulates NLRP3 Inflammasome Activation and Modifies Lipid Release Profile in Isolated Peritoneal Macrophages

José Marcos Sanches [1,2], Laura Migliari Branco [3], Gustavo Henrique Bueno Duarte [4], Sonia Maria Oliani [5], Karina Ramalho Bortoluci [3], Vanessa Moreira [6] and Cristiane Damas Gil [1,5,*]

1 Departamento de Morfologia e Genética, Universidade Federal de São Paulo, São Paulo 04023-900, Brazil; ms.marcossanches@gmail.com

2 Faculdade de Medicina, Universidade do Oeste Paulista, Guarujá, São Paulo 11410-980, Brazil

3 Departamento de Ciências Biológicas e Centro de Terapia Celular e Molecular, Universidade Federal de São Paulo, São Paulo 04044-010, Brazil; laurambranco@gmail.com (L.M.B.); karina.bortolucci@unifesp.br (K.R.B.)

4 Instituto de Química, Universidade Estadual de Campinas, Campinas, São Paulo 13083-862, Brazil; gustavo_duarte95@hotmail.com

5 Programa de Pós-Graduação em Biociências, Instituto de Biociências, Letras e Ciências Exatas (IBILCE), Universidade Estadual Paulista, São José do Rio Preto, São Paulo 15054-000, Brazil; sonia.oliani@unesp.br

6 Departamento de Farmacologia, Universidade Federal de São Paulo, São Paulo 04044-020, Brazil; vmoreira@unifesp.br

* Correspondence: cristiane.gil@unifesp.br

Abstract: Annexin A1 (AnxA1) is a potent anti-inflammatory protein that downregulates proinflammatory cytokine release. This study evaluated the role of AnxA1 in the regulation of NLRP3 inflammasome activation and lipid release by starch-elicited murine peritoneal macrophages. C57bl/6 wild-type (WT) and AnxA1-null (AnxA1$^{-/-}$) mice received an intraperitoneal injection of 1.5% starch solution for macrophage recruitment. NLRP3 was activated by priming cells with lipopolysaccharide for 3 h, followed by nigericin (1 h) or ATP (30 min) incubation. As expected, nigericin and ATP administration decreased elicited peritoneal macrophage viability and induced IL-1β release, more pronounced in the AnxA1$^{-/-}$ cells than in the control peritoneal macrophages. In addition, nigericin-activated AnxA1$^{-/-}$ macrophages showed increased levels of NLRP3, while points of co-localization of the AnxA1 protein and NLRP3 inflammasome were detected in WT cells, as demonstrated by ultrastructural analysis. The lipidomic analysis showed a pronounced release of prostaglandins in nigericin-stimulated WT peritoneal macrophages, while ceramides were detected in AnxA1$^{-/-}$ cell supernatants. Different eicosanoid profiles were detected for both genotypes, and our results suggest that endogenous AnxA1 regulates the NLRP3-derived IL-1β and lipid mediator release in macrophages.

Keywords: inflammation; nigericin; pyroptosis; mass spectrometry; lipidomics

---

1. Introduction

Annexin A1 (AnxA1) is a 37-kDa protein that can mimic the anti-inflammatory action of glucocorticoids by inhibiting eicosanoid and phospholipase A2 synthesis, affecting components of inflammatory reaction and arachidonic acid release [1,2]. The ability of AnxA1 to down-modulate cellular and molecular processes of inflammation contributes to tissue homeostasis and reprogramming of macrophages [3]. The development of the AnxA1-null mice (AnxA1$^{-/-}$) strain has allowed for a better understanding of the role of endogenous AnxA1 protein in leukocyte biology and the inflammatory process. In models of inflammation induced by carrageenan or zymosan, AnxA1$^{-/-}$ animals exhibit an exacerbated response characterized by a prominent leukocyte influx and IL-1β release [4]. In addition, macrophages from AnxA1$^{-/-}$ mice demonstrate reduced ability to phagocytose non-opsonized zymosan particles [5], and show higher TLR4 mRNA expression and IL-1β production after lipopolysaccharide (LPS) stimulation than wild-type cells [6]. These data demonstrate that AnxA1 exerts a negative inflammatory response through its down-modulation effects on macrophage cells, which are important leukocytes in an innate response.

In addition, the stimulation of a P2 × 7 receptor (P2X7R) in resting and M2 macrophages, but not in M1 cells, provokes the rapid release of AnxA1 through its exposure with phosphatidylserine to the outer plasma membrane leaflet [7]. Also, the release of AnxA1 after P2X7R activation is not affected in inflammasome knockout macrophages, suggesting that its release is independent of caspase-1 activation. Considering that P2X7R activation is necessary to promote the assembly of the NLRP3 inflammasome and cytokine release [8], the release of AnxA1 represents another P2X7R macrophage signaling pathway for the resolution of the inflammation.

Nucleotide-binding oligomerization domain (NOD)-like receptor family pyrin domain containing 3 (NLRP3 or NALP3) is a cytoplasmic sensor that oligomerizes to form a platform known as inflammasome, a protein complex that controls the release of IL-1β and IL-18 by activating caspase-1 [9]. In macrophages, NLRP3 inflammasome activation can be triggered by the pore-forming ionophore nigericin, extracellular ATP, and crystalline substances that induce pyroptosis, a type of cell death [10]. In addition, previous research has shown that some fatty acid-derived lipids, such as the prostaglandin E_2 (PGE_2), regulates NLRP3 and the maturation of IL-1β. PGE_2 can be associated with inhibition of the NLRP3 activation in human macrophages through the EP4 receptor and by the EP2 receptor in murine macrophages, decreasing IL-1β maturation [11,12]. NLRP3 activation driven by damage-associated molecular patterns is also associated with the production and release of several lipidic mediators, such as eicosanoids derived from arachidonic acids and ceramides, which cause cellular and systemic damages to the organism [13]. Eicosanoids, such as prostaglandins and leukotrienes, are lipid mediators that play a crucial role in initiating the acute inflammatory response [14]. Systemic activation of inflammasomes leads to the production of large amounts of eicosanoids in several minutes, contributing to rapid initiation of inflammation characterized by increased vascular permeability, culminating in pathological inflammatory effects [15].

Inflammasome activation is vital for the control of infections and the regulation of metabolic processes and immune responses [16]. However, altered functions of these platforms are implicated in the pathogenesis of several human diseases. Therefore, investigations that highlight novel signaling components that regulate inflammasome activation are crucial to prevent or treat human infections/inflammatory diseases. Considering that AnxA1 is a potent anti-inflammatory protein that down-regulates proinflammatory mediator release, phospholipase A_2 and, consequently, the critical cascade pathways of eicosanoid production such as that of cyclooxygenase 2 (COX-2) [17,18], this study evaluated its role in regulating the NLRP3 inflammasome and lipid release by macrophages.

2. Materials and Methods

2.1. Animals

Male C57BL/6 wild-type (WT) and AnxA1-null (AnxA1$^{-/-}$) mice aged 7–8 weeks and weighing 20–25 g were kept in cages ($n = 4$) in a temperature-controlled environment (22–25 °C) with a 12-h light-dark cycle. They received water and food ad libitum. All animal procedures were approved by the Ethics Committee in Animal Experimentation of the Federal University of São Paulo-UNIFESP (CEUA agreement number: N° 6493130318) and by the Internal Biosafety Commission (CIBio).

2.2. Cell Culture and Treatments

Lipopolysaccharide (LPS), nigericin, and ATP were obtained from InvivoGen (San Diego, CA, USA). LPS and ATP were reconstituted in endotoxin-free water and nigericin in 100% ethanol. The stock solutions were diluted in endotoxin-free water to prepare intermediate concentration solutions, stored at −20 °C.

WT and AnxA1$^{-/-}$ peritoneal macrophages were obtained by the intraperitoneal injection of a 1.5% starch solution (Sigma Aldrich, St. Louis, MO, USA) in sterile PBS, and after four days, cells were collected by peritoneal wash. Differential cell counts were made on Diff–Quick-stained cell smears prepared by cytocentrifugation. The macrophage population obtained was more than 85% pure and at least 90% viable, as examined by trypan blue exclusion. Additionally, macrophage morphology was confirmed by ultrastructural analysis using transmission electron microscopy. Peritoneal cells (1×10^6 cells/well) were cultured in Opti-MEM (Thermo Fisher Scientific, Waltham, MA, USA) overnight at 37 °C under an atmosphere of 5% CO_2. Experiments were performed in triplicate in 24-well plates. WT and AnxA1$^{-/-}$ cells were primed with LPS (500 ng/mL for 3 h) followed by stimulation with nigericin (10 μM for 1 h) or ATP (5 mM, 30 min) to activate the NLRP3 inflammasome.

2.3. Analysis of Cell Viability and IL-1β Release

Cell viability was determined by a 3-(4,5-dimethylthiazol-2-yl)-2,5-diphenyltetrazolium bromide (MTT) assay. After the treatment process, the supernatants were discarded, and the RPMI medium (Invitrogen, Gibco, Portland, OR, USA) with 10% MTT solution (5 mg/mL) was added to the cells. After incubating the cells for 4 h at 37 °C under a 5% CO_2 atmosphere, 300 μL of dimethyl sulfoxide (DMSO; Sigma Aldrich, St. Louis, MO, USA) was added to each well (24-well plate), and 100 μL triplicates of the same sample were transferred to a 96-well plate. The spectrophotometric absorbance values at 490 nm were determined. The percentage of viable cells was calculated by optical density normalization for LPS-stimulated cells only.

IL-1β levels were tested in culture supernatants by an enzyme-linked immunosorbent assay (ELISA) using a commercially available immunoassay kit (BioLegend, San Diego, CA, USA), according to the manufacturer's instructions. All experiments were conducted in duplicate, and the data were expressed as the mean ± standard error of the mean (SEM) of protein (pg/mL).

2.4. Western Blot Analysis

After nigericin and ATP stimulation, the supernatant was removed, and cells were washed three times with sterile PBS, and to each well, 50 μL of lysis buffer was added for cell lysis and protein extraction. Equal amounts of supernatants and cell extracts were loaded onto a 15% sodium dodecyl sulphate-polyacrylamide gel with appropriate molecular weight markers (Bio-Rad Life Science, Hercules, CA, USA) for electrophoresis and transferred to ECL Hybond nitrocellulose membranes. Reversible protein staining of the membranes with 0.1% Ponceau-S in 5% acetic acid (Santa Cruz Biotechnology, Dallas, TX, USA) was used to verify protein transfer. Membranes were incubated 30 min in 5% milk in Tris-buffered saline (TBS) prior to incubation with the antibodies. Primary antibodies were rabbit polyclonal anti-AnxA1 (Invitrogen-Thermo Fisher Scientific, Waltham, MA, USA; 1:1000), goat polyclonal anti-IL-1β (R&E Systems, MN, USA; 1:500), mouse monoclonal anti-caspase-1 (Santa Cruz Biotechnology Dallas, TX, USA; 1:200), and polyclonal rabbit anti-β-actin (Cell Signaling Technology, Beverly, MA, USA; 1:1000), all diluted in TBS. The membranes were then incubated with the appropriate peroxidase-conjugated secondary antibodies (Millipore Corporation, Burlington, MA, USA; 1:2500). Finally, membranes were washed for 15 min with TBS, and immunoreactive proteins were detected (Clarity™ Western ECL Substrate; Bio-Rad, Hercules, CA, USA) using a GeneGnome5 chemiluminescence detection system (SynGene, Cambridge, UK).

2.5. *Ultrastructural Immunocytochemical Analysis*

WT and AnxA1$^{-/-}$ nigericin-stimulated macrophages were fixed in 4% paraformaldehyde and 0.5% glutaraldehyde, 0.1% sodium cacodylate buffer (pH 7.4) for 24 h at 4 °C. Samples were washed in sodium cacodylate, dehydrated through a graded methanol series, and embedded in LR Gold (Sigma Aldrich Corp., St. Louis, MO, USA).

To detect AnxA1 and NLRP3, ultrathin macrophage sections (~90 nm) were submitted for immunocytochemistry, as previously described [19]. To detect the proteins, the sheep polyclonal antibody anti-AnxA1 (1:200) and rabbit polyclonal antibody anti-NLRP3 (1:200; Cusabio, Houston, TX, USA), following a donkey anti-sheep IgG and goat anti-rabbit IgG antibody (1:50) conjugated to 10-nm and 20-nm colloidal gold (British Biocell, Cardiff, UK), respectively, were used. Ultrathin sections were stained with uranyl acetate and lead citrate and examined using a ZEISS EM900 electron microscope (Carl Zeiss, Jena, Germany). Randomly photographed sections of macrophages were analyzed using Axiovision software. The density of immunogold (number of gold particles/μm^2) was calculated and reported as the mean ± SEM of 20–40 cells per experimental condition.

2.6. *Lipidomic Analysis*

After treatments, WT and AnxA1$^{-/-}$ cell supernatants (1 × 10^6 cells/well) were collected and stored at −80 °C until sample processing. For lipid extraction, each sample was randomized and resuspended in 1 mL of 1:2 $CHCl_3$: MeOH solution (Sigma Aldrich, Basel, Switzerland), followed by the addition of 0.33 mL $CHCl_3$ and 0.33 mL deionized water. The solution was stirred for 5 min, then centrifuged at 13,000 rpm for 5 min. Derived-organic fractions with lipids were collected from the bottom layer of the tubes and transferred to 1.5-mL glass tubes. These fractions were dried in a SpeedVac Savant SPD131DDA concentrator (Thermo Scientific) for 30 min at 30 °C and stored at −80 °C. Mass spectrometric analysis was performed in an ultra-high-performance liquid chromatography (UHPLC) Agilent 1290 Infinity system (Agilent, Santa Clara, CA, USA) and chromatographic elution in a Kinetex C18 column (4.6 mm × 50 mm × 2.6 μm) (Phenomenex, Torrance, CA, USA). All samples were randomized before injection and analyzed by the positive and negative mode in a hybrid mass spectrometer with QTOF 6550 mass analyzer (Agilent, Santa Clara, CA, USA).

The mass spectra were acquired in centroid mode, and the mass range used for the acquisition was 50–1700 Da. The raw data were converted by the MassHunter Qualitative software (Agilent, Santa Clara, California, USA) and then imported to XCMS online software (Version 3.7.1, Scripps Center for Metabolomics, La Jolla, CA, USA). For the final statistical analysis, the Metaboanalyst 3.0 platform was used (McGill University, Montreal, Quebec, Canada), as well as the potential lipid biomarkers annotation by the measurement of their exact mass, retention time, and elution profile in METLIN (Scripps Center for Metabolomics, La Jolla, CA, USA), Human Metabolome Database (HMDB) (http://www.hmdb.ca/metabolites), and Lipid Maps databases (http://www.lipidmaps.org/).

2.7. *Statistical Analyses*

The data were analyzed using GraphPad Prism 5.0 software. Results were confirmed to follow a normal distribution using the Kolmogorov–Smirnov test of normality with Dallal–Wilkinson–Lillie for corrected *P*-value. Data that passed the normality assumption were analyzed using analysis of variance (ANOVA) with a Bonferroni post hoc test. Data failing the normality assumption were analyzed using the non-parametric Kruskal–Wallis test followed by Dunn's post-test, and differences were considered statistically significant at a value of $p < 0.05$.

3. Results

3.1. The Lack of Endogenous AnxA1 Exacerbates the IL-1β Release and Increases NLRP3 Levels after Inflammasome Activation

First, we verified the endogenous effect of AnxA1 on the activation and regulation of NLRP3 inflammasome in macrophages. As expected, the administration of nigericin caused a significant reduction in cell viability (Figure 1A) without a difference between the two genotypes. Both nigericin and ATP induced IL-1β release by macrophages, which was more pronounced in nigericin-stimulated AnxA1$^{-/-}$ cells than in their respective controls (Figure 1B,C). This latter result was corroborated by the presence of mature IL-1β observed in AnxA1$^{-/-}$ cells (Figure 1C), and pro caspase 1 (Figure 1D) in the same experimental condition. ATP-stimulated WT macrophages presented decreased levels of AnxA1 compared with nigericin-stimulated cells (Figure 1D). In addition, pro caspase 1 levels were similar between only primed and ATP-stimulated AnxA1$^{-/-}$ cells (Figure 1D).

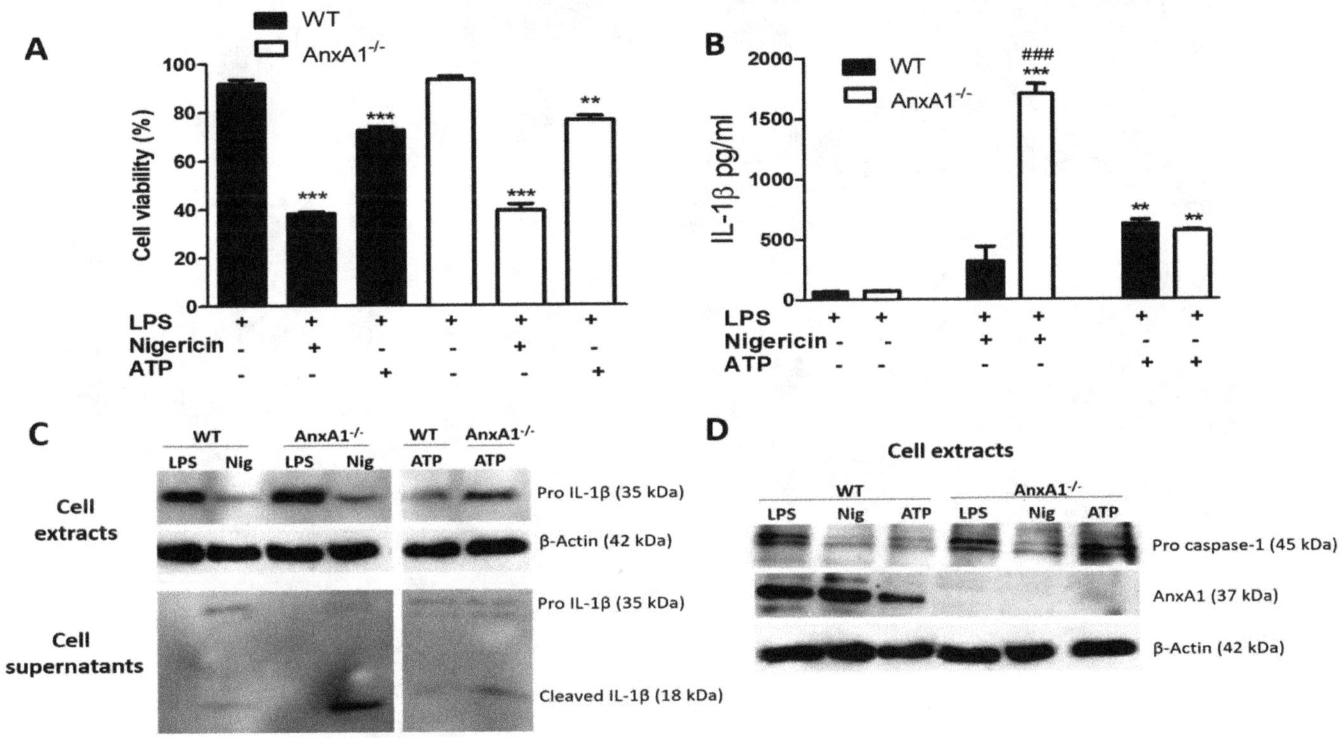

Figure 1. Lack of endogenous Annexin A1 (AnxA1) produced a marked release of IL-1β in macrophages. (**A**) 3-(4,5-dimethylthiazol-2-yl)-2,5-diphenyltetrazolium bromide (MTT) assay. Macrophages of both genotypes showed a marked decrease in cell viability under nigericin and ATP exposure. The percentage of viable cells was calculated by optical density normalization for only lipopolysaccharide (LPS)-stimulated cells. Data are shown as mean ± S.E.M. of cell ratio (%). ** $p < 0.01$; *** $p < 0.001$ vs. LPS-stimulated cells of corresponding genotype (ANOVA, Bonferroni post-test). (**B**,**C**) IL-1β levels. Treatment with nigericin produced a marked release of IL-1β in AnxA1-null (AnxA1$^{-/-}$) macrophages compared with the other groups. Increased levels of pro-IL-1β were detected in the LPS-stimulated wild-type (WT) and AnxA1$^{-/-}$ cell extracts. Values are expressed as mean ± SEM of IL-1β levels (pg/mL). *** $p < 0.001$, ** $p < 0.01$ vs. LPS-stimulated cells of respective genotype; ### $p < 0.001$ vs. WT nigericin-treated cells (ANOVA, Bonferroni post-test). (**D**) Pro caspase 1 and AnxA1 levels in the cell extracts under different experimental conditions. β-actin was used as an endogenous control (representative image of three experiments performed).

After detecting that nigericin-stimulated AnxA1$^{-/-}$ cells exhibited exacerbated IL-1β production, the AnxA1 and NLRP3 levels were analyzed using ultrastructural immunocytochemistry. The lack of endogenous AnxA1 was associated with increased levels of NLRP3 in nigericin-stimulated macrophages compared with the primed AnxA1$^{-/-}$ cells and WT cells (Figure 2). In contrast, WT nigericin-stimulated macrophages showed a marked increase of AnxA1 levels compared with the only primed cells (LPS) (Figure 3A,B,E). In addition, points of co-localization between AnxA1 and NLRP3 were detected in the cytoplasm of nigericin-stimulated cells (Figure 3C). No immunogold labelling was detected in the negative control of the reaction (Figure 3D).

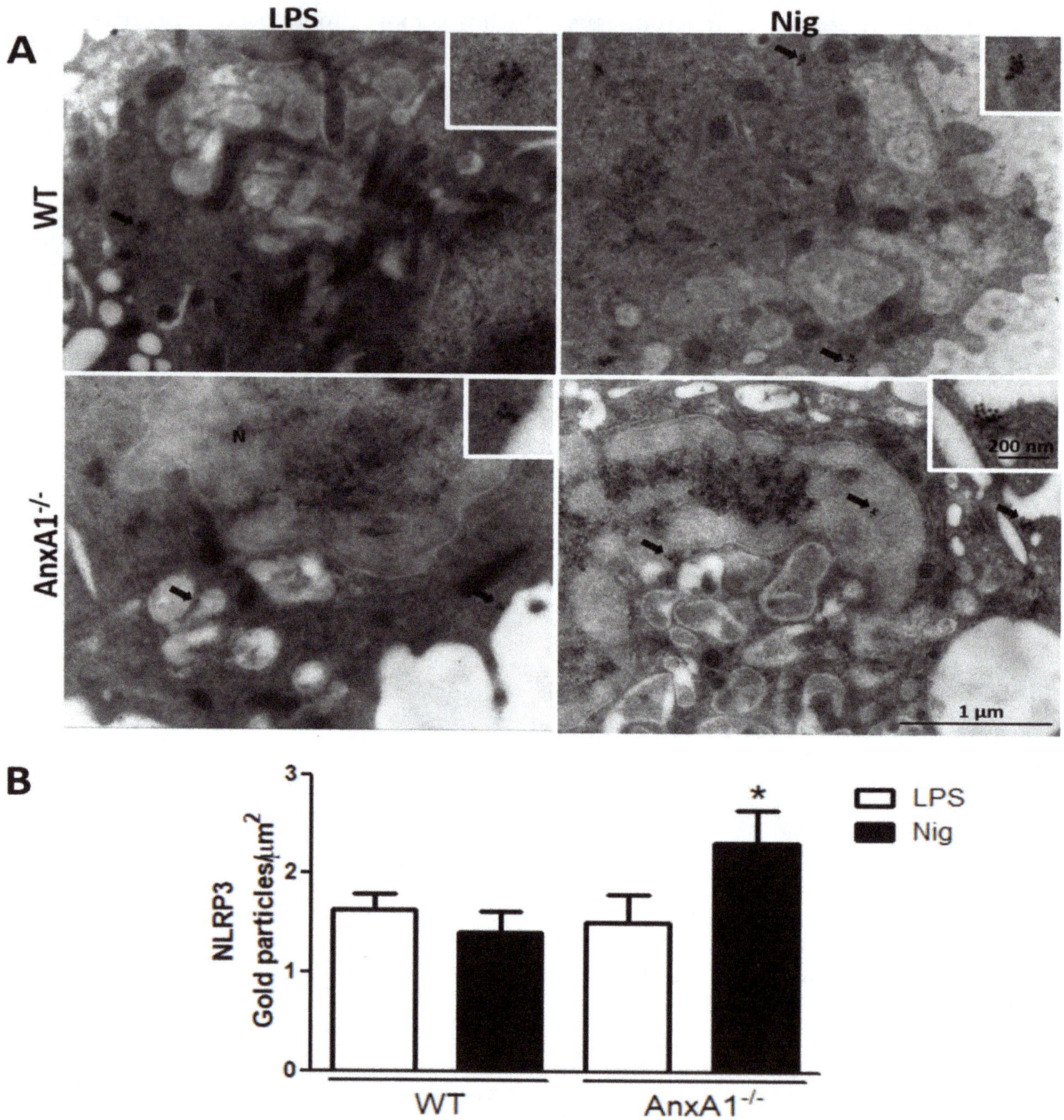

Figure 2. Lack of endogenous AnxA1 increased NLRP3 levels in nigericin-stimulated macrophages. (**A**) NLRP3 expression detected in the cytoplasm (arrows) of cells. Insets: details of cytoplasmic gold labelling of NLRP3. (**B**) Density of NLRP3 immunogold particles in macrophages. Data are mean ± SEM of distinct cells analyzed for each condition. * $p < 0.05$ vs. LPS of corresponding genotype (ANOVA, Bonferroni post-test).

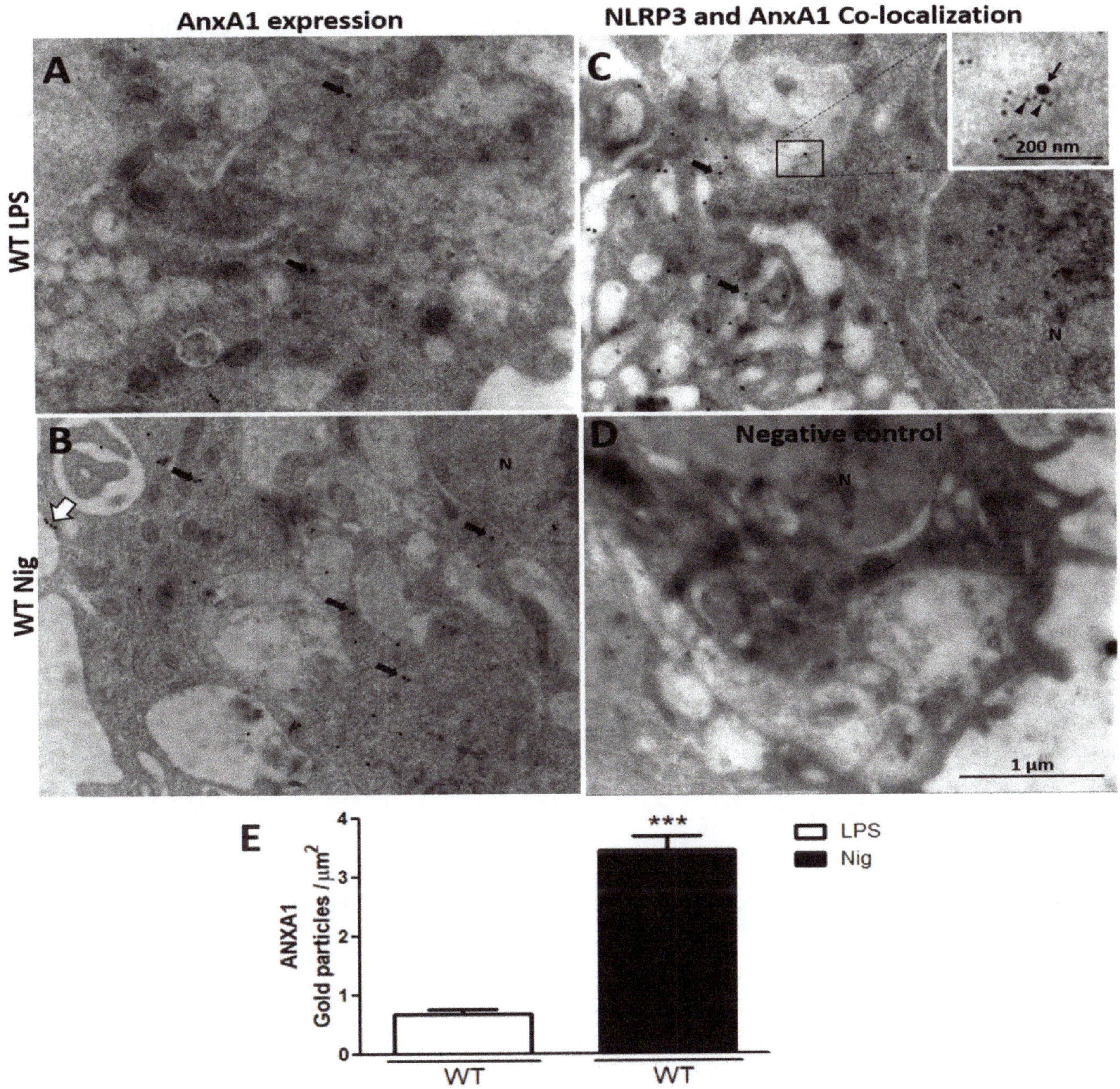

Figure 3. Nigericin-stimulated WT macrophages increase AnxA1 endogenous levels. (**A**,**B**) AnxA1 expression detected in the cytoplasm (arrows) and plasma membrane (white arrow) of cells. (**C**) Points of co-localization of AnxA1 (arrowheads) and NLRP3 (arrow) were detected in the cytoplasm of nigericin-stimulated cells. (**D**) Negative control. (**E**) Density of AnxA1 immunogold particles in macrophages. Data are mean ± SEM of distinct cells analyzed for each condition (*t*-test). *** $p < 0.001$ vs. LPS.

3.2. NLRP3 Activation Induces Different Lipid Release by WT and AnxA1$^{-/-}$ Macrophages

To investigate whether the lack of endogenous AnxA1 also alters lipidomic profiling of macrophages under NLRP3 activation, lipidomic analysis of cell supernatants was performed. Figure 4 shows the score plots of principal component analysis (PCA) and partial least squares discriminant analysis (PLS-DA) obtained in positive and negative ion modes. As expected, the lipid profiles from the WT and AnxA1$^{-/-}$ control groups (LPS-stimulated cells) were different from each other, as shown by the separated dark blue and red ellipses in PCA and PLS-DA, either in the positive and negative ion modes. In the positive mode of the PCA analysis, the first two components of the score plot described 63.6% of explained variation in which the WT LPS separation from the WT treated with nigericin is more pronounced (Figure 4A) in PCA, whereas PLS-DA shows overlapped ellipses (Figure 4B). The negative mode in PCA and PLS-DA analysis showed similar results between groups, as evidenced by the overlapping of ellipses (Figure 4C,D).

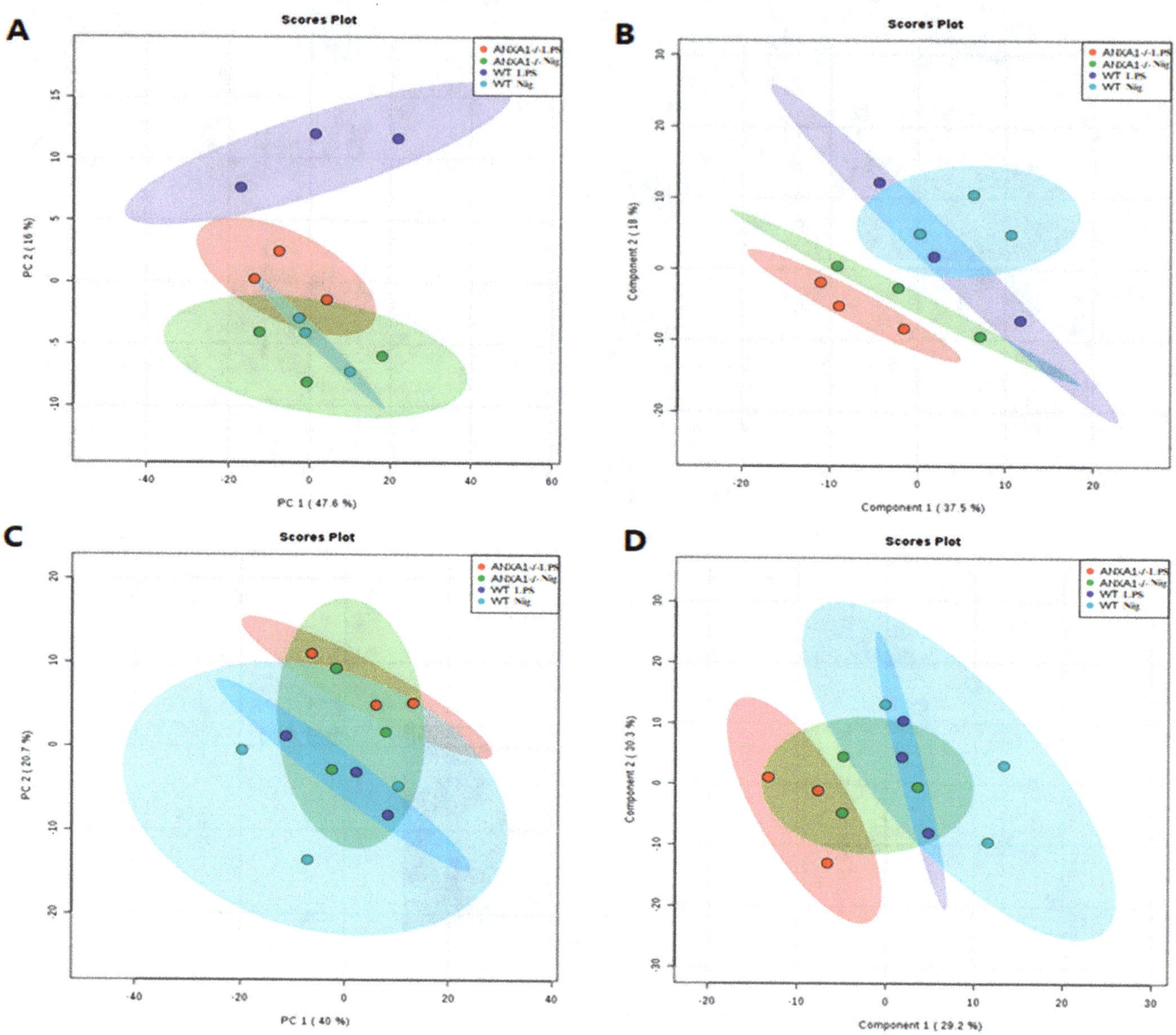

Figure 4. Principal component analysis (PCA) and partial least squares discriminant analysis (PLS-DA) score plots of the lipid fraction. Each spot represents one supernatant sample from control (LPS-stimulated cells - WT: dark blue; AnxA1$^{-/-}$: red) and nigericin-stimulated cells (WT: light blue; AnxA1$^{-/-}$: green). The ellipses show the differences and similarities between groups. (**A**,**C**), PCA in positive and negative ion mode, respectively. (**B**,**D**), PLS-DA in positive and negative ion mode, respectively.

Heatmaps and dendrograms highlight the normalized concentrations of different lipids in the supernatants of WT and AnxA1$^{-/-}$ macrophages. In the positive mode (Figure 5A), the majority of potential lipid biomarkers were produced by nigericin-stimulated AnxA1$^{-/-}$ cells, if compared with the other experimental conditions. These cells showed increased production of sphingolipids, especially ceramides Cer(t18:0/18:O(2OH)), Cer(d14:1/26:0), PE-Cer(d14:2(4E,6E)/16:0), and PE-Cer(d14:2(4E,6E)/19:0) (Figure 5A). In addition, supernatants from nigericin-stimulated AnxA1$^{-/-}$ exhibited a higher concentration of the eicosanoid 11S-15S-dihydroxy-14R-(S-glutathionyl)-5Z,8Z,12E-eicosatrienoic acid and the neutral lipid PI(P-18:0/00). In contrast, heptanoic acid was identified as a potential lipid biomarker in the supernatants of nigericin-stimulated WT cells, with a higher concentration than control WT cells (LPS), while arachidonoyl ethanolamide and the plasma membrane compound phosphatidylglycerol PG (16:1(9Z)/17:2(9Z,12Z)) were lower. PS (15:1(9Z)/16:1(9Z)) is a phosphatidylserine which only appeared in the supernatants of AnxA1$^{-/-}$ control cells and those treated with nigericin (Figure 5A).

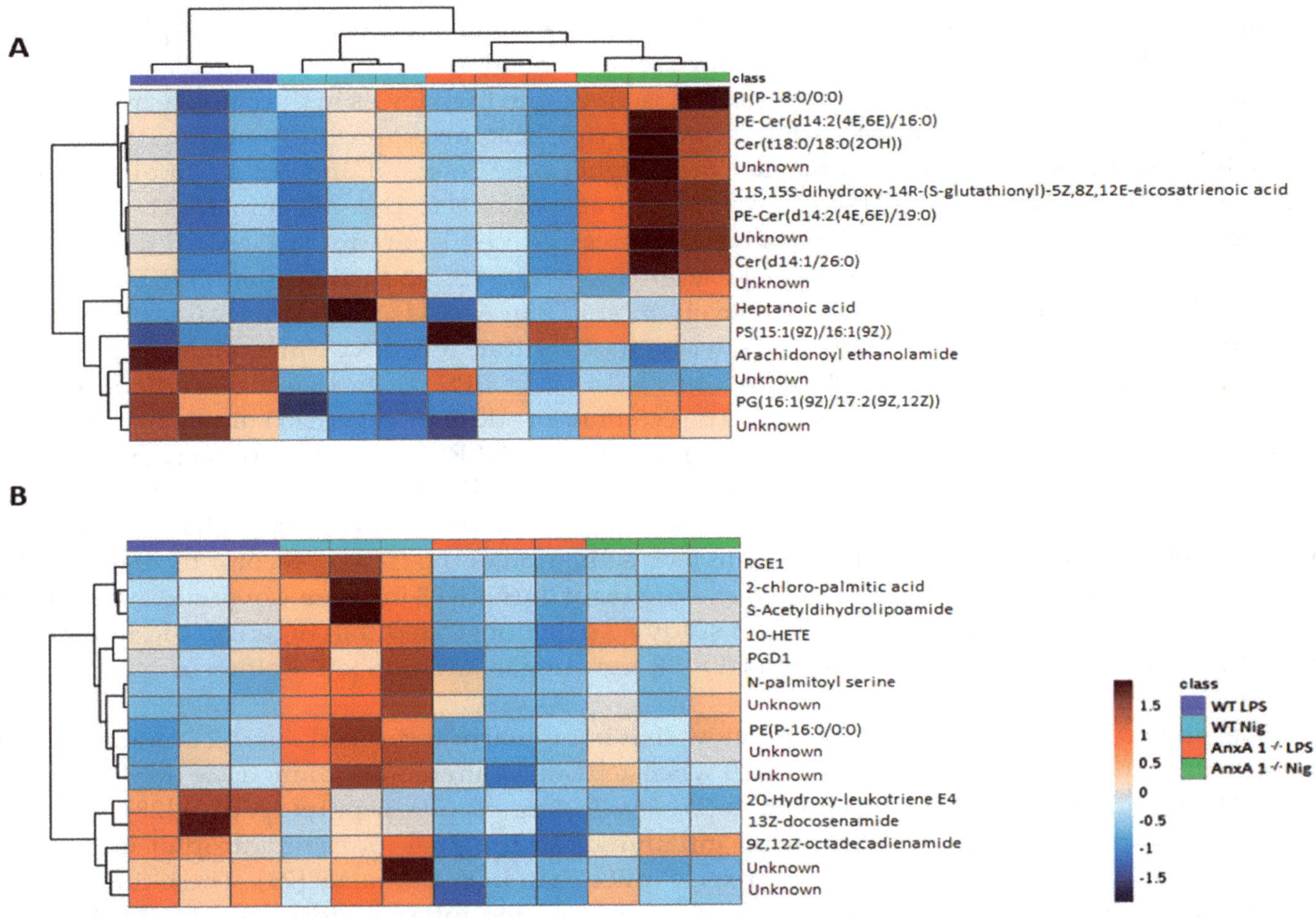

Figure 5. Lipidomic analysis of WT and AnxA1$^{/}$ macrophage supernatants. Heatmaps and dendrograms show the hierarchical clustering of potential lipid biomarkers of the control (WT: dark blue; AnxA1$^{-/-}$: red) and nigericin-stimulated cells (WT: light blue; AnxA1$^{-/-}$: green). (**A**) Positive mode. (**B**) Negative mode. Lipidomic analysis demonstrated a completely different lipid profile between WT and AnxA1$^{-/-}$ supernatant cells. In WT cells, nigericin induced a pronounced release of eicosanoids and prostaglandins, while AnxA1$^{-/-}$ cells showed precursors of prostaglandin and some ceramides. The right bar in Figure B represents the blue–red code (−1.5 to 1.5) of the lipid concentrations. Unknown: noncharacterized lipids.

Figure 5B shows the potential lipid biomarkers in the negative mode. In the supernatants of nigericin-stimulated WT cells, the more concentrated lipids were associated with the arachidonic acid metabolism, such as the 10-hydroxyeicosatetraenoic acid (10-HETE), prostaglandins D1 and E1 (PGD1, PGE1), as well as palmitic acid, S-acetyldihydrolipoamide, N-palmitoyl serine, and PE(P-16:0/0:0). Curiously, there are different lipid profiles in the supernatants of WT and AnxA1$^{-/-}$ control cells, as characterized by a higher concentration of the eicosanoid 20-hydroxy-leukotriene E4 and fatty acid amides (13Z-docosenamide and 9,12Z-octadecadienamide) in the WT cells.

4. Discussion

Monocytes and macrophages express NOD-like receptors, which are very important for the immune response during inflammation [20]. By NLRP3 inflammasome activation in these cells, pro-IL-1β and pro-IL-18 are cleaved and released, increasing the pro-inflammatory response [21]. Once macrophages are activated by the NLRP3 inflammasome agonists, such as nigericin and ATP, there is an IL-1β production and release, and the K^+ concentration is reduced [22]. IL-1β is a proinflammatory cytokine, and once it is released by macrophages, IL-6, TNF, nitric oxide, and prostaglandin E_2 (PGE_2) are produced [23]. Our data provide previously unknown details regarding the interplay between AnxA1 and NLRP3-derived IL-1β in macrophages and its relationship with lipid-mediator release.

Under normal conditions, the AnxA1 protein is present in high levels in the cytoplasm of human and rodent leukocytes, such as neutrophils, monocytes, and macrophages, and once these cells are activated, the AnxA1 moves to the cell membrane to be released and act as an autocrine or paracrine mediator [24,25]. The ATP-binding cassette transporter is responsible for AnxA1 secretion in macrophages [26], and under conditions of cellular stress, AnxA1 is rapidly released [27].

Our results show that the lack of AnxA1 exacerbates the IL-1β production after NLRP3 activation. These findings were supported by increased levels of NLRP3 in AnxA1$^{-/-}$ macrophages, as observed by ultrastructural immunocytochemistry. Considering that peritoneal AnxA1$^{-/-}$ macrophages presented increased levels of TLR4 [6], the lack of AnxA1 could favor LPS "over-priming" and consequent increase in NLRP3 inflammasome-derived IL-1β secretion. In addition, nigericin stimulation increased endogenous AnxA1 that presents points of co-localization with NLRP3 in WT macrophages, supporting a role of AnxA1 in the activation and regulation of the NLRP3 inflammasome. Regarding ATP stimulation, western blotting detected more decreased levels of AnxA1 in WT macrophages than in the priming LPS and nigericin-stimulated cells. In fact, activation of the P2 × 7 receptor by extracellular ATP in macrophages has been widely studied as a trigger of the NLRP3 inflammasome and is associated with the release of AnxA1 [7]. However, this study also demonstrated that nigericin did not induce AnxA1 release, suggesting pathways within P2 × 7R signaling in addition to K+ efflux in macrophages for the release of this protein.

Despite our findings, recent analyses show that bone-marrow-derived AnxA1$^{-/-}$ macrophages produced significantly lower secretion of IL-1β when activated with monosodium urate (MSU) crystals and ATP, but not NLRC4 or AIM2 activators (*Legionella pneumophila* or poly(dA:dT)) [28]. Notably, during the setting of MSU crystal-induced inflammation, the peak of the neutrophil influx was greater, and the resolution was slower in AnxA1$^{-/-}$ mice than in WT animals [29]. These findings are consistent with many studies that have shown an exacerbated inflammatory response in AnxA1$^{-/-}$ mice characterized by a marked leukocyte influx and production of proinflammatory mediators, such as IL-1β and IL-6 [4,6,30–32]. Additionally, the administration of the AnxA1 peptide (Ac2–26) 1 h before and 12 h after challenge with MSU crystals induced decreased levels of IL-1β in periarticular tissue, showing the important anti-inflammatory and proresolving activity of this protein on the course of MSU crystal-induced inflammation in mice [29]. Thus, the opposite effects described for the role of AnxA1 on NLRP3 inflammasome-derived IL-1β secretion could be a result of testing different macrophage populations, bone-marrow versus peritoneal-derived, and different strains, BALB/c versus C57BL/6, an important factor in the mouse immunology response [33,34].

The NLRP3 inflammasome is a cytosolic platform formed by a multi-protein complex containing a nucleotide-binding oligomerization domain-like receptor and the adaptor apoptosis-associated spec-like protein (ASC) containing an amino-terminal caspase-recruitment domain (CARD) [9]. The interaction of the NLRP3 with ASC promotes the recruitment of the procaspase-1 and its autoproteolysis, driving IL-1β and IL-18 maturation, membrane pore formation by the gasdermin D action, and then cytokine secretion and pyroptosis [35,36]. Besides the cleavage and release of the IL-1β, the activation of the NLRP3 inflammasome is directly related to the production of lipid mediators, including eicosanoids and ceramides. These lipids can be involved in metabolic and immunological pathways, and the deregulation of NLRP3 activation causes an increase of the lipidic mediators released after pyroptosis, and it is possible to conduct metabolic damage on a systemic level [13].

The current lipidomics approach is a great tool to understand biological systems and many diseases. The major lipid classes can be categorized as fatty acyls, glycerolipids, glycerophospholipids, sphingolipids, sterol lipids, prenol lipids, saccharolipids, and polyketides [37]. In addition, cells can synthesize lipids, such as the eicosanoids, which are derived by the arachidonic acid oxidation and represented by prostaglandins, leukotrienes, thromboxanes, lipoxins, and epoxyeicosatrienoic acids [38]. By providing the exact mass, retention time, and elution profile, our study presents new data about the lipidomic profile of the supernatant of macrophages after NLRP3 activation by nigericin.

In our study, we showed the ceramides PE-Cer(d14:2(4E,6E)/16:0), Cer(t18:0/18:O(2OH)), PE-Cer(d14:2(4E,6E)/19:0), and Cer(d14:1/26:0) as potential lipid biomarkers released by AnxA1$^{-/-}$ macrophages after nigericin stimulation. Ceramides are bioactive sphingolipids present in the plasma membrane and they mediate cell signaling, with a close relationship to many pathophysiological processes associated with inflammation [39]. Previous studies showed that the exposure of macrophages to ceramides causes activation of caspase-1, and this effect is prevented by the absence of NLRP3 [40]. Ceramides have also been related to the activation of TLR4, augmenting LPS-induced pro-inflammatory response [41]. In this regard, it is reasonable to infer that increased levels of TLR4 in AnxA1$^{-/-}$ macrophages [6] contribute to the more pronounced ceramide concentration in their supernatants and also in NLRP3 activation.

Supernatants from nigericin-stimulated AnxA1$^{-/-}$ macrophages also exhibited a higher concentration of the 11S-15S-dihydroxy-14R-(S-glutathionyl)-5Z,8Z,12E-eicosatrienoic acid, a type of eicosanoid, and PI(P-18:0/0:0). Some studies have shown that 12-hydroxyeicosatrienoic acid (12-HETE) plays an important role as a paracrine mediator of inflammation, as well as in the regulation of neutrophil infiltration in damaged tissue [42,43]. PI(P-18:0/0:0) is an important phosphatidylinositol in cell membranes and for metabolic processes, such as being the primary source of arachidonic acid metabolism for eicosanoid synthesis and intracellular signals in animal tissues [44]. The lack of AnxA1 in macrophages increased the arachidonic acid metabolism and eicosanoid production, as evidenced by the high concentration of 11S-15S-dihydroxy-14R-(S-glutathionyl)-5Z,8Z,12E-eicosatrienoic acid and PI(P-18:0/0:0) after NLRP3 activation and pyroptosis induction. The 11S-15S-dihydroxy-14R-(S-glutathionyl)-5Z,8Z,12E-eicosatrienoic acid is present in leukocytes and red blood cells, and earlier studies have demonstrated that this fatty acid can be converted by arachidonic acid by the lipoxygenase pathway [45,46], which is basically synthesized during the inflammatory process [47].

In the supernatants of nigericin-stimulated WT macrophages, more concentrated lipids are also associated with the arachidonic acid metabolism, such as the 10-HETE, PGD1, PGE1, as well as palmitic acid, S-acetyldihydrolipoamide, N-palmitoyl serine, PE(P-16:0/0:0), and heptanoic acid, indicating a completely different lipid profile of AnxA1$^{-/-}$ supernatants; 10-HETE is a hydroxyeicosatrienoic acid with proinflammatory action, increasing TNF-α and IL-6 production in macrophages [48]. In contrast, PGE1 and PGD1 are anti-inflammatory lipid mediators that inhibit leukocyte migration and adhesion and mast cell activation [49–51]. Palmitic acid (PA), also called hexadecanoic acid, is one of the most common saturated fatty acids in animals. PA acts as a lipid mediator in inflammation and its derived-metabolic products accumulate in the endoplasmic reticulum (ER) and increase reactive oxygen species (ROS) generation, leading to cell death [52]. Additionally, the inflammatory response caused by ER stress and ROS generation from high concentrations of PA drives NF-κB and NLRP3 activation and, consequently, proinflammatory cytokine release by monocytes/macrophages [53–56]. S-Acetyldihydrolipoamide, N-palmitoyl serine and PE(P-16:0/00) (2-Hexadecanoyl-1-(1Z-hexadecenyl)-sn-glycero-3-phosphoethanolamine) are associated with cell metabolism [57], membrane receptor [58], and cell membrane compounds [59], respectively. However, there is little information about heptanoic acid in biological systems. Metabolomic studies have shown a high concentration of heptanoic acid in the faeces of autistic children [60], while in patients with

Crohn's disease, ulcerative colitis, and pouchitis, lower levels of this lipid were found [61]. Altogether, our data show that macrophages can release potential lipid biomarkers after NLRP3 activation that can regulate the inflammatory responses in the damaged tissue.

This study also detected a different lipid profile in supernatants from WT and AnxA1$^{-/-}$ LPS-stimulated macrophages. Higher concentrations of the eicosanoid 20-hydroxy-leukotriene E4 and fatty acid amides (13Z-docosenamide and 9,12Z-octadecadienamide) were found in the WT samples, while PS(15:1(9Z)/16:1(9Z)) was found in AnxA1$^{-/-}$ samples. In addition, 20-hydroxy-leukotriene E4 is an eicosanoid metabolite originating from the lipid oxidation of leukotriene E4 that plays an essential role in cell proliferation, differentiation, and immunoregulation [62]. Moreover, the leukotriene E4 contributes to prolonged intracellular signaling, increasing intracellular $Ca2^+$ and ERK phosphorylation [63], confirming signal pathway triggering by LPS on WT macrophages. Biological functions for both fatty acid amides detected in this study still need to be addressed.

Finally, PS(15:1(9Z)/16:1(9Z)) is a phosphatidylserine (PS) involved in cell signaling, including an important role in cell death, either by apoptosis, necroptosis, or pyroptosis, by its exposure on the outer plasma membrane layer [64]. The detection of this potential lipid biomarker in both AnxA1$^{-/-}$ supernatants (LPS and nigerin) can be related to the release of extracellular vesicles by macrophages. Macrophages can release extracellular vesicles after *Mycobacterium tuberculosis* infection, or spontaneously with a high concentration of phosphatidylserine [65].

5. Conclusions

The lack of AnxA1 favors LPS "over-priming" and the release of lipid mediators (e.g., ceramides) that produce exacerbated NLRP3 activation under nigericin stimulation. Although more detailed investigations are warranted, this study identifies AnxA1 as a novel signaling component of inflammasome activation and a potential therapeutic target to treat inflammatory diseases.

Author Contributions: Conceptualization, J.M.S. and C.D.G.; Methodology, Validation, J.M.S., L.M.B., G.H.B.D., K.R.B., V.M. and C.D.G.; Formal analysis, J.M.S., G.H.B.D., V.M. and C.D.G.; Investigation, J.M.S. and C.D.G.; Resources, S.M.O., K.R.B., V.M. and C.D.G.; Writing—original draft preparation, J.M.S. and C.D.G.; Writing—review and editing, J.M.S., G.H.B.D., S.M.O., K.R.B., V.M. and C.D.G.; Funding acquisition, C.D.G. and S.M.O. All authors have read and agreed to the published version of the manuscript.

References

1. Flower, R.J. Eleventh Gaddum memorial lecture. Lipocortin and the mechanism of action of the glucocorticoids. *Br. J. Pharmacol.* **1988**, *94*, 987–1015. [CrossRef]
2. Sheikh, M.H.; Solito, E. Annexin A1: Uncovering the Many Talents of an Old Protein. *Int. J. Mol. Sci.* **2018**, *19*, 1045. [CrossRef]
3. Sugimoto, M.A.; Vago, J.P.; Teixeira, M.M.; Sousa, L.P. Annexin A1 and the Resolution of Inflammation: Modulation of Neutrophil Recruitment, Apoptosis, and Clearance. *J. Immunol. Res.* **2016**, *2016*, 8239258. [CrossRef]
4. Hannon, R.; Croxtall, J.D.; Getting, S.J.; Roviezzo, F.; Yona, S.; Paul-Clark, M.J.; Gavins, F.N.; Perretti, M.; Morris, J.F.; Buckingham, J.C.; et al. Aberrant inflammation and resistance to glucocorticoids in annexin 1-/- mouse. *FASEB J.* **2003**, *17*, 253–255. [CrossRef]
5. Yona, S.; Buckingham, J.C.; Perretti, M.; Flower, R.J. Stimulus-specific defect in the phagocytic pathways of annexin 1 null macrophages. *Br. J. Pharmacol.* **2004**, *142*, 890–898. [CrossRef]
6. Damazo, A.S.; Yona, S.; D'Acquisto, F.; Flower, R.J.; Oliani, S.M.; Perretti, M. Critical protective role for annexin 1 gene expression in the endotoxemic murine microcirculation. *Am. J. Pathol.* **2005**, *166*, 1607–1617. [CrossRef]
7. de Torre-Minguela, C.; Barberà-Cremades, M.; Gómez, A.I.; Martín-Sánchez, F.; Pelegrín, P. Macrophage activation and polarization modify P2 × 7 receptor secretome influencing the inflammatory process. *Sci. Rep.* **2016**, *6*, 22586. [CrossRef]

8. Idzko, M.; Ferrari, D.; Eltzschig, H.K. Nucleotide signalling during inflammation. *Nature* **2014**, *509*, 310–317. [CrossRef]
9. He, Y.; Hara, H.; Núñez, G. Mechanism and Regulation of NLRP3 Inflammasome Activation. *Trends Biochem. Sci.* **2016**, *41*, 1012–1021. [CrossRef]
10. Hughes, M.M.; O'Neill, L.A.J. Metabolic regulation of NLRP3. *Immunol. Rev.* **2018**, *281*, 88–98. [CrossRef]
11. Sokolowska, M.; Chen, L.Y.; Liu, Y.; Martinez-Anton, A.; Qi, H.Y.; Logun, C.; Alsaaty, S.; Park, Y.H.; Kastner, D.L.; Chae, J.J.; et al. Prostaglandin E2 Inhibits NLRP3 Inflammasome Activation through EP4 Receptor and Intracellular Cyclic AMP in Human Macrophages. *J. Immunol.* **2015**, *194*, 5472–5487. [CrossRef] [PubMed]
12. Zasłona, Z.; Pålsson-McDermott, E.M.; Menon, D.; Haneklaus, M.; Flis, E.; Prendeville, H.; Corcoran, S.E.; Peters-Golden, M.; O'Neill, L.A.J. The Induction of Pro-IL-1β by Lipopolysaccharide Requires Endogenous Prostaglandin E. *J. Immunol.* **2017**, *198*, 3558–3564. [CrossRef] [PubMed]
13. Dennis, E.A.; Norris, P.C. Eicosanoid storm in infection and inflammation. *Nat. Rev. Immunol.* **2015**, *15*, 511–523. [CrossRef]
14. Serhan, C.N.; Savill, J. Resolution of inflammation: The beginning programs the end. *Nat. Immunol.* **2005**, *6*, 1191–1197. [CrossRef] [PubMed]
15. von Moltke, J.; Trinidad, N.J.; Moayeri, M.; Kintzer, A.F.; Wang, S.B.; van Rooijen, N.; Brown, C.R.; Krantz, B.A.; Leppla, S.H.; Gronert, K.; et al. Rapid induction of inflammatory lipid mediators by the inflammasome in vivo. *Nature* **2012**, *490*, 107–111. [CrossRef] [PubMed]
16. Strowig, T.; Henao-Mejia, J.; Elinav, E.; Flavell, R. Inflammasomes in health and disease. *Nature* **2012**, *481*, 278–286. [CrossRef] [PubMed]
17. Jorgensen, I.; Lopez, J.P.; Laufer, S.A.; Miao, E.A. IL-1β, IL-18, and eicosanoids promote neutrophil recruitment to pore-induced intracellular traps following pyroptosis. *Eur. J. Immunol.* **2016**, *46*, 2761–2766. [CrossRef]
18. Seidel, S.; Neymeyer, H.; Kahl, T.; Röschel, T.; Mutig, K.; Flower, R.; Schnermann, J.; Bachmann, S.; Paliege, A. Annexin A1 modulates macula densa function by inhibiting cyclooxygenase 2. *Am. J. Physiol. Renal Physiol.* **2012**, *303*, F845–F854. [CrossRef]
19. Oliani, S.M.; Paul-Clark, M.J.; Christian, H.C.; Flower, R.J.; Perretti, M. Neutrophil interaction with inflamed postcapillary venule endothelium alters annexin 1 expression. *Am. J. Pathol.* **2001**, *158*, 603–615. [CrossRef]
20. Bortolotti, P.; Faure, E.; Kipnis, E. Inflammasomes in Tissue Damages and Immune Disorders After Trauma. *Front. Immunol.* **2018**, *9*, 1900. [CrossRef]
21. Awad, F.; Assrawi, E.; Jumeau, C.; Georgin-Lavialle, S.; Cobret, L.; Duquesnoy, P.; Piterboth, W.; Thomas, L.; Stankovic-Stojanovic, K.; Louvrier, C.; et al. Impact of human monocyte and macrophage polarization on NLR expression and NLRP3 inflammasome activation. *PLoS ONE* **2017**, *12*, e0175336. [CrossRef] [PubMed]
22. Kanneganti, T.D.; Lamkanfi, M. K^+ drops tilt the NLRP3 inflammasome. *Immunity* **2013**, *38*, 1085–1088. [CrossRef] [PubMed]
23. Schett, G.; Dayer, J.M.; Manger, B. Interleukin-1 function and role in rheumatic disease. *Nat. Rev. Rheumatol.* **2016**, *12*, 14–24. [CrossRef] [PubMed]
24. Perretti, M.; Christian, H.; Wheller, S.K.; Aiello, I.; Mugridge, K.G.; Morris, J.F.; Flower, R.J.; Goulding, N.J. Annexin I is stored within gelatinase granules of human neutrophil and mobilized on the cell surface upon adhesion but not phagocytosis. *Cell Biol. Int.* **2000**, *24*, 163–174. [CrossRef]
25. Kamal, A.M.; Flower, R.J.; Perretti, M. An overview of the effects of annexin 1 on cells involved in the inflammatory process. *Mem. Inst. Oswaldo Cruz* **2005**, *100* (Suppl. 1), 39–47. [CrossRef]
26. Wein, S.; Fauroux, M.; Laffitte, J.; de Nadaï, P.; Guaïni, C.; Pons, F.; Coméra, C. Mediation of annexin 1 secretion by a probenecid-sensitive ABC-transporter in rat inflamed mucosa. *Biochem. Pharmacol.* **2004**, *67*, 1195–1202. [CrossRef]
27. de Jong, R.; Leoni, G.; Drechsler, M.; Soehnlein, O. The advantageous role of annexin A1 in cardiovascular disease. *Cell Adhes. Migr.* **2017**, *11*, 261–274. [CrossRef]
28. Galvão, I.; de Carvalho, R.V.H.; Vago, J.P.; Silva, A.L.N.; Carvalho, T.G.; Antunes, M.M.; Ribeiro, F.M.; Menezes, G.B.; Zamboni, D.S.; Sousa, L.P.; et al. The role of Annexin A1 in the modulation of the NLRP3 inflammasome. *Immunology* **2020**. [CrossRef]
29. Galvão, I.; Vago, J.P.; Barroso, L.C.; Tavares, L.P.; Queiroz-Junior, C.M.; Costa, V.V.; Carneiro, F.S.; Ferreira, T.P.; Silva, P.M.; Amaral, F.A.; et al. Annexin A1 promotes timely resolution of inflammation in murine gout. *Eur. J. Immunol.* **2017**, *47*, 585–596. [CrossRef]

30. Roviezzo, F.; Getting, S.J.; Paul-Clark, M.J.; Yona, S.; Gavins, F.N.; Perretti, M.; Hannon, R.; Croxtall, J.D.; Buckingham, J.C.; Flower, R.J. The annexin-1 knockout mouse: What it tells us about the inflammatory response. *J. Physiol. Pharmacol.* **2002**, *53*, 541–553.
31. Gimenes, A.D.; Andrade, T.R.; Mello, C.B.; Ramos, L.; Gil, C.D.; Oliani, S.M. Beneficial effect of annexin A1 in a model of experimental allergic conjunctivitis. *Exp. Eye Res.* **2015**, *134*, 24–32. [CrossRef] [PubMed]
32. Parisi, J.D.S.; Corrêa, M.P.; Gil, C.D. Lack of Endogenous Annexin A1 Increases Mast Cell Activation and Exacerbates Experimental Atopic Dermatitis. *Cells* **2019**, *8*, 51. [CrossRef] [PubMed]
33. Garcia-Pelayo, M.C.; Bachy, V.S.; Kaveh, D.A.; Hogarth, P.J. BALB/c mice display more enhanced BCG vaccine induced Th1 and Th17 response than C57BL/6 mice but have equivalent protection. *Tuberculosis (Edinb)* **2015**, *95*, 48–53. [CrossRef] [PubMed]
34. Miralles, G.D.; Stoeckle, M.Y.; McDermott, D.F.; Finkelman, F.D.; Murray, H.W. Th1 and Th2 cell-associated cytokines in experimental visceral leishmaniasis. *Infect. Immun.* **1994**, *62*, 1058–1063. [CrossRef] [PubMed]
35. Broz, P.; von Moltke, J.; Jones, J.W.; Vance, R.E.; Monack, D.M. Differential requirement for Caspase-1 autoproteolysis in pathogen-induced cell death and cytokine processing. *Cell Host Microbe* **2010**, *8*, 471–483. [CrossRef]
36. Shi, J.; Zhao, Y.; Wang, K.; Shi, X.; Wang, Y.; Huang, H.; Zhuang, Y.; Cai, T.; Wang, F.; Shao, F. Cleavage of GSDMD by inflammatory caspases determines pyroptotic cell death. *Nature* **2015**, *526*, 660–665. [CrossRef]
37. Fahy, E.; Subramaniam, S.; Murphy, R.C.; Nishijima, M.; Raetz, C.R.; Shimizu, T.; Spener, F.; van Meer, G.; Wakelam, M.J.; Dennis, E.A. Update of the LIPID MAPS comprehensive classification system for lipids. *J. Lipid Res.* **2009**, *50*, S9–S14. [CrossRef]
38. Esser-von Bieren, J. Eicosanoids in tissue repair. *Immunol. Cell Biol.* **2019**, *97*, 279–288. [CrossRef]
39. Albeituni, S.; Stiban, J. Roles of Ceramides and Other Sphingolipids in Immune Cell Function and Inflammation. *Adv. Exp. Med. Biol.* **2019**, *1161*, 169–191. [CrossRef]
40. Vandanmagsar, B.; Youm, Y.H.; Ravussin, A.; Galgani, J.E.; Stadler, K.; Mynatt, R.L.; Ravussin, E.; Stephens, J.M.; Dixit, V.D. The NLRP3 inflammasome instigates obesity-induced inflammation and insulin resistance. *Nat. Med.* **2011**, *17*, 179–188. [CrossRef]
41. Płóciennikowska, A.; Hromada-Judycka, A.; Borzęcka, K.; Kwiatkowska, K. Co-operation of TLR4 and raft proteins in LPS-induced pro-inflammatory signaling. *Cell Mol. Life Sci.* **2015**, *72*, 557–581. [CrossRef] [PubMed]
42. Mieyal, P.A.; Dunn, M.W.; Schwartzman, M.L. Detection of endogenous 12-hydroxyeicosatrienoic acid in human tear film. *Investig. Ophthalmol. Vis. Sci.* **2001**, *42*, 328–332. [PubMed]
43. Wainwright, S.; Falck, J.R.; Yadagiri, P.; Powell, W.S. Metabolism of 12(S)-hydroxy-5,8,10,14-eicosatetraenoic acid and other hydroxylated fatty acids by the reductase pathway in porcine polymorphonuclear leukocytes. *Biochemistry* **1990**, *29*, 10126–10135. [CrossRef] [PubMed]
44. Divecha, N.; Irvine, R.F. Phospholipid signaling. *Cell* **1995**, *80*, 269–278. [CrossRef]
45. Pfister, S.L.; Spitzbarth, N.; Zeldin, D.C.; Lafite, P.; Mansuy, D.; Campbell, W.B. Rabbit aorta converts 15-HPETE to trihydroxyeicosatrienoic acids: Potential role of cytochrome P450. *Arch. Biochem. Biophys.* **2003**, *420*, 142–152. [CrossRef] [PubMed]
46. Bryant, R.W.; Schewe, T.; Rapoport, S.M.; Bailey, J.M. Leukotriene formation by a purified reticulocyte lipoxygenase enzyme. Conversion of arachidonic acid and 15-hydroperoxyeicosatetraenoic acid to 14, 15-leukotriene A4. *J. Biol. Chem.* **1985**, *260*, 3548–3555. [PubMed]
47. Brunnström, A.; Hamberg, M.; Griffiths, W.J.; Mannervik, B.; Claesson, H.E. Biosynthesis of 14,15-hepoxilins in human l1236 Hodgkin lymphoma cells and eosinophils. *Lipids* **2011**, *46*, 69–79. [CrossRef] [PubMed]
48. Zhang, Q.; Wang, X.; Yan, G.; Lei, J.; Zhou, Y.; Wu, L.; Wang, T.; Zhang, X.; Ye, D.; Li, Y. Anti- Versus Pro-Inflammatory Metabololipidome Upon Cupping Treatment. *Cell Physiol. Biochem.* **2018**, *45*, 1377–1389. [CrossRef]
49. Itoh, Y.; Yasui, T.; Kakizawa, H.; Makino, M.; Fujiwara, K.; Kato, T.; Imamura, S.; Yamamoto, K.; Hishida, H.; Nakai, A.; et al. The therapeutic effect of lipo PGE1 on diabetic neuropathy-changes in endothelin and various angiopathic factors. *Prostaglandins Other Lipid Mediat.* **2001**, *66*, 221–234. [CrossRef]
50. Kawamura, T.; Nara, N.; Kadosaki, M.; Inada, K.; Endo, S. Prostaglandin E1 reduces myocardial reperfusion injury by inhibiting proinflammatory cytokines production during cardiac surgery. *Crit. Care Med.* **2000**, *28*, 2201–2208. [CrossRef]

51. Amagai, Y.; Oida, K.; Matsuda, A.; Jung, K.; Kakutani, S.; Tanaka, T.; Matsuda, K.; Jang, H.; Ahn, G.; Xia, Y.; et al. Dihomo-γ-linolenic acid prevents the development of atopic dermatitis through prostaglandin D1 production in NC/Tnd mice. *J. Dermatol. Sci.* **2015**, *79*, 30–37. [CrossRef] [PubMed]
52. Borradaile, N.M.; Han, X.; Harp, J.D.; Gale, S.E.; Ory, D.S.; Schaffer, J.E. Disruption of endoplasmic reticulum structure and integrity in lipotoxic cell death. *J. Lipid Res.* **2006**, *47*, 2726–2737. [CrossRef] [PubMed]
53. Korbecki, J.; Bajdak-Rusinek, K. The effect of palmitic acid on inflammatory response in macrophages: An overview of molecular mechanisms. *Inflamm. Res.* **2019**, *68*, 915–932. [CrossRef] [PubMed]
54. Nguyen, M.T.; Favelyukis, S.; Nguyen, A.K.; Reichart, D.; Scott, P.A.; Jenn, A.; Liu-Bryan, R.; Glass, C.K.; Neels, J.G.; Olefsky, J.M. A subpopulation of macrophages infiltrates hypertrophic adipose tissue and is activated by free fatty acids via Toll-like receptors 2 and 4 and JNK-dependent pathways. *J. Biol. Chem.* **2007**, *282*, 35279–35292. [CrossRef]
55. Suganami, T.; Tanimoto-Koyama, K.; Nishida, J.; Itoh, M.; Yuan, X.; Mizuarai, S.; Kotani, H.; Yamaoka, S.; Miyake, K.; Aoe, S.; et al. Role of the Toll-like receptor 4/NF-kappaB pathway in saturated fatty acid-induced inflammatory changes in the interaction between adipocytes and macrophages. *Arterioscler. Thromb. Vasc. Biol.* **2007**, *27*, 84–91. [CrossRef]
56. Huang, M.T.; Taxman, D.J.; Holley-Guthrie, E.A.; Moore, C.B.; Willingham, S.B.; Madden, V.; Parsons, R.K.; Featherstone, G.L.; Arnold, R.R.; O'Connor, B.P.; et al. Critical role of apoptotic speck protein containing a caspase recruitment domain (ASC) and NLRP3 in causing necrosis and ASC speck formation induced by Porphyromonas gingivalis in human cells. *J. Immunol.* **2009**, *182*, 2395–2404. [CrossRef]
57. O'Connor, T.P.; Roche, T.E.; Paukstelis, J.V. 13C nuclear magnetic resonance study of the pyruvate dehydrogenase-catalyzed acetylation of dihydrolipoamide. *J. Biol. Chem.* **1982**, *257*, 3110–3112.
58. Liliom, K.; Bittman, R.; Swords, B.; Tigyi, G. N-palmitoyl-serine and N-palmitoyl-tyrosine phosphoric acids are selective competitive antagonists of the lysophosphatidic acid receptors. *Mol. Pharmacol.* **1996**, *50*, 616–623.
59. Simons, K.; Toomre, D. Lipid rafts and signal transduction. *Nat. Rev. Mol. Cell Biol.* **2000**, *1*, 31–39. [CrossRef]
60. De Angelis, M.; Francavilla, R.; Piccolo, M.; De Giacomo, A.; Gobbetti, M. Autism spectrum disorders and intestinal microbiota. *Gut Microbes* **2015**, *6*, 207–213. [CrossRef]
61. De Preter, V.; Machiels, K.; Joossens, M.; Arijs, I.; Matthys, C.; Vermeire, S.; Rutgeerts, P.; Verbeke, K. Faecal metabolite profiling identifies medium-chain fatty acids as discriminating compounds in IBD. *Gut* **2015**, *64*, 447–458. [CrossRef] [PubMed]
62. Singh, R.K.; Tandon, R.; Dastidar, S.G.; Ray, A. A review on leukotrienes and their receptors with reference to asthma. *J. Asthma* **2013**, *50*, 922–931. [CrossRef] [PubMed]
63. Foster, H.R.; Fuerst, E.; Branchett, W.; Lee, T.H.; Cousins, D.J.; Woszczek, G. Leukotriene E4 is a full functional agonist for human cysteinyl leukotriene type 1 receptor-dependent gene expression. *Sci. Rep.* **2016**, *6*, 20461. [CrossRef] [PubMed]
64. Wang, Q.; Imamura, R.; Motani, K.; Kushiyama, H.; Nagata, S.; Suda, T. Pyroptotic cells externalize eat-me and release find-me signals and are efficiently engulfed by macrophages. *Int. Immunol.* **2013**, *25*, 363–372. [CrossRef]
65. García-Martínez, M.; Vázquez-Flores, L.; Álvarez-Jiménez, V.D.; Castañeda-Casimiro, J.; Ibáñez-Hernández, M.; Sánchez-Torres, L.E.; Barrios-Payán, J.; Mata-Espinosa, D.; Estrada-Parra, S.; Chacón-Salinas, R.; et al. Extracellular vesicles released by J774A.1 macrophages reduce the bacterial load in macrophages and in an experimental mouse model of tuberculosis. *Int. J. Nanomed.* **2019**, *14*, 6707–6719. [CrossRef]

11

Membrane Binding Promotes Annexin A2 Oligomerization

Anna Lívia Linard Matos [1,†], Sergej Kudruk [1,†], Johanna Moratz [2], Milena Heflik [1], David Grill [1], Bart Jan Ravoo [2] and Volker Gerke [1,*]

1 Institute of Medical Biochemistry-Center for Molecular Biology of Inflammation, University of Münster, Von-Esmarch-Str. 56, D-48149 Münster, Germany; linardma@uni-muenster.de (A.L.L.M.); sergej.kudruk@uni-muenster.de (S.K.); milena.heflik@web.de (M.H.); david.grill@uni-muenster.de (D.G.)

2 Organic Chemistry Institute, University of Münster, Corrensstrasse 40, D-48149 Münster, Germany; j_mora01@uni-muenster.de (J.M.); b.j.ravoo@uni-muenster.de (B.J.R.)

* Correspondence: gerke@uni-muenster.de

† These authors contributed equally to this work.

Abstract: Annexin A2 (AnxA2) is a cytosolic Ca^{2+} regulated membrane binding protein that can induce lipid domain formation and plays a role in exocytosis and endocytosis. To better understand the mode of annexin-membrane interaction, we analyzed membrane-bound AnxA2 assemblies by employing a novel 3-armed chemical crosslinker and specific AnxA2 mutant proteins. Our data show that AnxA2 forms crosslinkable oligomers upon binding to membranes containing negatively charged phospholipids. AnxA2 mutants with amino acid substitutions in residues predicted to be involved in lateral protein–protein interaction show compromised oligomer formation, albeit still being capable of binding to negatively charged membranes in the presence of Ca^{2+}. These results suggest that lateral protein–protein interactions are involved in the formation of AnxA2 clusters on a biological membrane.

Keywords: annexin A2; microdomain; cross-linker; quartz crystal microbalance with dissipation monitoring (QCM-D)

1. Introduction

Biological membranes can segregate into microdomains of defined lipid and protein composition that serve diverse but yet very specific tasks. One category of these microdomains, often referred to as lipid rafts, is enriched in cholesterol as well as sphingomyelin species present in the extracellular leaflet and in certain cases phosphatidylinositol (4,5)-bisphosphate [PI(4,5)P_2] present in the cytoplasmic leaflet. Rafts are particularly well studied in the plasma membrane of eukaryotic cells, where they serve as assembly and transmission platforms in outside-in as well as inside-out signaling and also regulate membrane trafficking events to and from the plasma membrane (for review see [1–3]). To function in these processes, rafts have to be highly dynamic, both with respect to lipid and protein composition as well as size, thus they assemble into larger structures and also disassemble again on a rapid time scale. Dynamic assembly/disassembly is driven by intrinsic properties of the raft lipids and proteins and is affected, among other things, by cholesterol content and degree of fatty acid saturation in the incorporated phospho- and sphingolipids. Importantly, raft dynamics are also controlled by membrane-associated proteins that function by binding to raft lipids or proteins and affect their properties and distribution. These associated proteins are often regulated in their membrane association and thus raft-controlling properties enable cells to rapidly respond to certain stimuli by altered membrane microdomain assembly. One important regulatory event is a change in intracellular Ca^{2+} concentration and a number of raft-associated proteins whose membrane interaction is regulated by Ca^{2+} (for review see [4]). Annexins are a family of such Ca^{2+} regulated proteins that bind to acidic phospholipids in the cytoplasmic leaflets of cellular membranes in a peripheral and reversible manner (for review see [5–7]).

Annexin A2 (AnxA2) is a member of the annexin family that has been shown to associate with raft-like microdomains in certain cells and certain physiological scenarios (for review see [8]). As other annexins, it directly binds to headgroups of negatively charged phospholipids and requires the presence of these lipids in membranes (and raft domains) for high-affinity association. One such lipid is $PI(4,5)P_2$, and AnxA2 has been shown to interact with this phosphoinositide in a specific manner [9,10]. In addition to binding to $PI(4,5)P_2$ and other acidic phospholipids, in particular phosphatidylserine (PS), in a Ca^{2+} regulated manner, the protein can also form two-dimensional assemblies on $PI(4,5)P_2$ or PS containing model membranes and can cluster these lipids into domains [11–14]. Most likely, this lipid segregating property of AnxA2 is responsible for a function of the protein in regulating membrane-cytoskeleton contacts, cell polarity, and exocytotic granule docking and fusion [15–19]. However, the molecular basis underlying the phospholipid segregating properties of AnxA2, in particular the role of potential protein-protein interactions in this process, is not known.

Structurally, AnxA2 has a fold similar to all other annexins. It comprises a conserved core domain that is built of four repeats, each with five α helices, and a unique N-terminal domain mediating interactions with protein ligands. The core forms a slightly bent structure, where type-II Ca^{2+} binding sites as well as the membrane binding site are located on the convex side. A high resolution crystal structure of AnxA2 has been obtained revealing anti-parallel dimers of AnxA2 in the crystallized unit [20] and amino acid side chains residing at this dimer interface could provide lateral contacts in two-dimensional AnxA2 assemblies.

Here we have investigated the oligomeric state of membrane-bound AnxA2 to assess whether protein–protein interactions could represent the molecular mechanism underlying AnxA2-driven $PI(4,5)P_2$ and PS segregation. Therefore, we developed a novel chemical crosslinker that revealed the existence of AnxA2 oligomers. These oligomers only form in the presence of Ca^{2+} and require membrane binding, thus could represent the molecular structures driving a segregation of AnxA2-bound phospholipids. Moreover, mutating residues present at the dimer interface identified positions that reside in close proximity in membrane-bound AnxA2 oligomers and could participate in oligomer formation.

2. Materials and Methods

2.1. Lipids, Chemicals

Cholesterol (Chol), 1,2-dioleoyl-sn-glycero-3-phosphocholine (DOPC), 1-palmitoyl-2-oleoyl-sn-glycero-3-phoshocholine (POPC), 1-palmitoyl-2-oleoylsn-glycero-3-phosho-L-serine (sodium salt) (POPS), and 1,2-dioleoyl-sn-glycero-3 [phosphoinositol-4,5-bisphosphate](triammonium salt) ($PI(4,5)P_2$) were purchased from Avanti Polar Lipids Inc. (Alabaster, AL, USA). Lipids and cholesterol were dissolved in chloroform/methanol (1:1, v/v), except for $PI(4,5)P_2$ that was dissolved in chloroform/methanol/water (20:9:1 v/v). Other chemicals were purchased from Applichem (Darmstadt, Germany), Merck KGaA (Darmstadt, Germany), Carl Roth GmbH (Karlsruhe, Germany) and Sigma-Aldrich (Munich, Germany). Water was purified and deionized with a cartridge system from Millipore (18.2 MΩ). For all the SDS-PAGE and Western blots the marker PageRuler Plus Prestained Protein Ladder from Thermo Scientific was used (Waltham, MA, USA).

2.2. Crosslinker Synthesis

Biotinyl *N*-Tris((2-(2.5-dioxopyrrolidin-1-yl propionate triethyleneglycolamido) ethoxy) methyl) methylamide, herein referred to as $Biotin_{3xNHS}$X-Linker, was synthesized as described in detail in the Supplementary Information (SI). Briefly, spacer synthesis started with 2- (2- (2-chloroethoxy) ethoxy) ethanol, which was converted into the ester (**1**) in a MICHAEL reaction with *t*-butyl acrylate and sodium in tetrahydrofuran (THF) (Figure 1). The chloride (**1**) was then mixed with sodium azide (NaN_3) and the azide obtained was reduced to the amine-terminated spacer with triphenylphosphine. In the convergent procedure, Tris (hydroxymethyl) aminomethane (THAM) was converted with *t*-butyl acrylate and sodium hydroxide (NaOH) in THF to the triple-functionalized amine (**3**), which was then protected with benzyl chloroformate (**4**). This protective group shows stable behavior in the case of tri-fluoro acetic acid (TFA) initiated acidic hydrolysis. This was followed by selective TFA deprotection of the ester-protected hydroxyl group with subsequent peptide coupling of the molecule (**4**) and the spacer (**2**). In the next step, the *N*-Cbz protective group was removed under reductive conditions using hydrogen and palladium on activated carbon (Pd/C). The amine (**5**) obtained was further processed under peptide coupling conditions with D(+)biotin, diisopropylethylamine (DIPEA), and benzotriazol-1-yl-oxytripyrrolidinophosphonium hexafluorophosphate (PyBOP) to yield the biotin-functionalized product (**7**). The activated crosslinker ($Biotin_{3xNHS}$X-linker) was obtained after deprotection of the ester units and direct functionalization of the carboxylic acids with N-hydroxysuccinimide (NHS) and N, N′-dicyclohexylcarbodiimide (DCC) in THF.

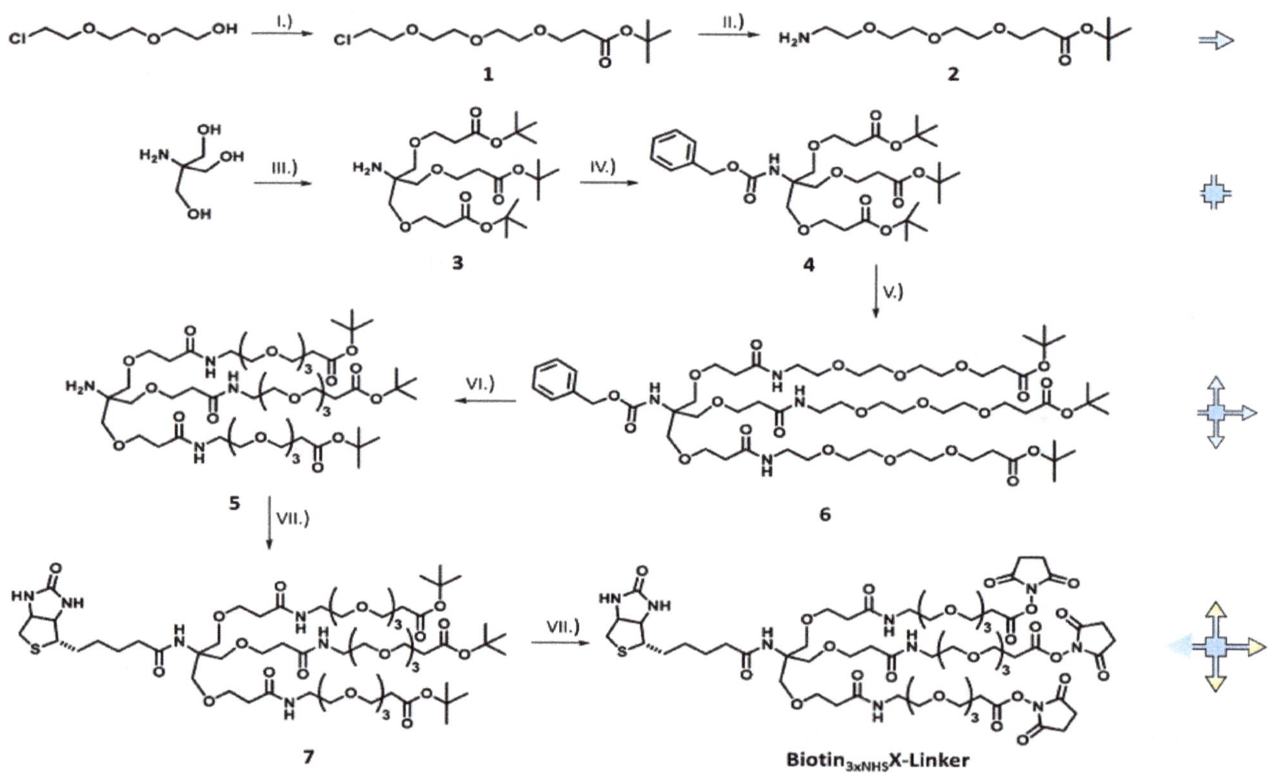

Figure 1. Synthesis of the $Biotin_{3xNHS}$X-Linker: (**I.**) *t*-butylacrylate, sodium, THF, 16h, 13%, (**II.**) 1. NaN_3, DMF, 70 °C, 2d, 2. PPh_3, H_2O, THF, 4d, quant., (**III.**) *t*-butylacrylate, NaOH, DMSO, 15 °C and then warm up to rt, 20h, 23%, (**IV.**) benzyl chloroformate, Na_2CO_3 (aq.), DCM, 4d, 94%, (**V.**) 1. TFA, DCM, 2h, 2. (**2**), DIPEA, PyBoP, DMF, 24 h, 69%, (**VI.**) H_2, Pd/C, MeOH, 3d, 88%, (**VII.**) D(+)Biotin, DIPEA, PyBoP, DMF, 18h, 38%, (**VII.**) 1. TFA, DCM, Toluol, 2h, 2. *N*-hydroxysuccinimide, DCC, THF, 3d, quant. See SI and Figures S1 and S2 for details.

2.3. Liposome Preparation and Co-Sedimentation

Lipids dissolved in chloroform/methanol were mixed at the desired molar ratio and composition. Chloroform was evaporated under a stream of nitrogen and traces of solvent were removed in vacuum for 4 h. Lipid films were stored at 4 °C until use. Liposomes were formed by hydration of the lipid film in a PBS −/− buffer. Small unilamellar vesicles (SUVs, 50 nm) or large unilamellar vesicles (LUVs, 100 or 200 nm) were obtained by extrusion through polycarbonate membranes (Avanti Polar Lipids

Inc.). SUVs were employed in quartz crystal microbalance with dissipation monitoring (QCM-D) experiments to facilitate vesicle rupture that occurs following vesicle coalescence on the sensor surface and results in the formation of a stable bilayer [21]. LUVs were used in liposome co-sedimentation and crosslinking experiments to ensure efficient co-pelleting and prevent rupture more often observed with high curvature vesicles.

Co-sedimentation experiments employed liposomes composed of POPC:Chol:POPS (60:20:20) with a defined size of 200 nm at a final concentration of 1 mg/mL. Liposomes were incubated for 1h at 4 °C with the desired AnxA2 derivative at a liposome/protein ratio of 10/1 (μL/μg) in PBS −/− buffer containing 1 mM $CaCl_2$. Ultracentrifugation (UC) was performed to pellet the liposomes (96600 *g*, 4 °C for 20 min), the supernatant was collected and the pellet resuspended in 500 μL of PBS with 1 mM $CaCl_2$, followed by 20 min incubation at 4 °C. After a second UC, the supernatant was collected and the pellet was resuspended in 500 μL of PBS with 5 mM EGTA and incubated for 30 min at 4 °C. A third UC yielded a supernatant (EGTA eluate) and a pellet that was resuspended in 500 μL of PBS with 5 mM EGTA. All fractions were analyzed via SDS-PAGE and immunoblotting with AnxA2-specific antibodies [22].

2.4. QCM-D Measurements

Quartz Crystal Microbalance with Dissipation (QCM-D) analysis was performed as described before [21,23] using a Q-Sense E4 QCM-D (Q-Sense, Gothenburg, Sweden) equipped with four temperature controlled flow cells in a parallel configuration connected to a peristaltic pump (Ismatec IPC, Glattbrugg, Switzerland), at a flow rate of 80.4 μL/min. A bilayer was established by fusion of SUVs composed of POPC/DOPC/Chol/POPS/PI(4,5)P_2 (37:20:20:20:3). Binding measurements were performed at 20 °C in HBS buffer supplemented with 250 μM Ca^{2+} and AnxA2 constructs at 50 nM. 250 μM Ca^{2+} and a relatively complex lipid mixture were chosen in the QCM-D experiments to directly compare the results to our previous data obtained by QCM-D analysis of AnxA2 and other AnxA2 mutants [21,24]. Frequency and dissipation shifts of the 7th overtone resonance frequency of the quartz sensor (QSX 303, 50 nm SiO2, 4.95 MHz) were recorded. OriginPro v. 9.1 (OriginLab Corp.) was used for data analysis.

2.5. Crosslinking of AnxA2

LUVs (100 nm) composed of POPC:DOPC:Chol:POPS (40:20:20:20) were used at 3.33 μg/μL in HBS pH 7,4. Control #1 contained only the AnxA2 derivative (60 μg) in 1 mM $CaCl_2$, i.e., a reaction in the absence of membranes, while control #2 contained a mixture of AnxA2 (60 μg) with 100 μg of LUVs, 5 mM EGTA, and 0.3 mM $Biotin_{3xNHS}$X-Linker, i.e., a reaction in the absence of Ca^{2+}. The actual Ca^{2+}/crosslink sample consisted of LUVs (100 μg), AnxA2 (60 μg), 1 mM $CaCl_2$, and 0.3 mM $Biotin_{3xNHS}$X-Linker. A Ca^{2+} concentration of 1 mM was used in these experiments to ensure efficient phospholipid binding of the protein. All components were mixed with exception of the $Biotin_{3xNHS}$X-Linker, which was added after 30 min, and incubation was then continued for another 30 min while shaking. The reaction was stopped with 5× PAGE sample buffer without β-mercaptoethanol. For a better separation in the gel, 3 μL of a 100 mM EGTA solution was added to each sample. Samples were kept for 15 min at RT before analysis by 10% SDS-PAGE. Gels were stained with Coomassie Brilliant Blue. Quantification of crosslinked oligomer bands was achieved by gating the area in the stained gel lane above the position of AnxA2 dimers in all samples and relating its intensity to that of the monomer band in the respective sample. In this quantification, the actual dimer band was excluded because AnxA2 species migrating at the dimer position in SDS-PAGE, which do not reflect the physical state of the protein in solution and most likely form during SDS sample preparation, have been observed before [22]. They would mask an association not caused by the crosslink. Image Studio Lite (LI-COR Corporate Offices, NE, USA) and Graphpad Prism 4 (GraphPad Software, San Diego, CA, USA) were used for quantification.

2.6. *Mutagenesis*

The human AnxA2 cDNA carrying a substitution at amino acid 66 (glutamate-for-alanine) to establish a monoclonal antibody epitope was cloned into the pSE420 expression vector as described [11] to yield pSE420-AnxA2A66E.

AnxA2 6x (pSE420-AnxA2A66E_6x) was generated by mutating 6 amino acids in the template pSE420-AnxA2A66E using site-directed mutagenesis as described [23]. Mutations were introduced at amino acid positions: 81 (K to A), 189 (E to K), 196 (R to S), 206 (K to A), 212 (K to S) and 219 (E to K). The following primers were employed in the mutagenesis reactions: K81A_For 5′-CCAGAGAAGGACCAAAGCGGAACTTGCATCAGCAC-3′ and K81A_Rev 5′-GTGCTGATGCAAGTTCCGCTTTGGTCCTTCTCTGG-3′. E189K_For 5′-GGCTCTGTCATTGATTATAAACTGATTGACCAAGATGCTC-3′ E189K_Rev 5′-GAGCATCTTGGTCAATCAGTTTATAATCAATGACAGAGCC-3′, K206A_For 5′-CGCTGGAGTGAAGAGGGCAGGAACTGATGTTCCC-3′ K206A_Rev 5′-GGGAACATCAGTTCCTGCCCTCTTCACTCCAGCG-3′, R196S_For 5′-CTGATTGACCAAGATGCTAGTGATCTCTATGACGCTGGAG-3′ R196S_Rev 5′-CTCCAGCGTCATAGAGATCACTAGCATCTTGGTCAATCAG-3′, K212S_For 5′-CAGGAACTGATGTTCCCTCGTGGATCAGCATCATG-3′ K212S_Rev 5′-CATGATGCTGATCCACGAGGGAACATCAGTTCCTG-3′, E219K_For 5′-ATCAGCATCATGACCAAGCGGAGCGTGCCC-3′ E219K_Rev 5′-GGGCACGCTCCGCTTGGTCATGATGCTGAT-3′, (Biomers, Ulm, Germany):

AnxA2 10x (pSE420-AnxA2A66E_10x) was generated using pSE420-AnxA2A66E_6x as template by introducing 4 additional amino acid substitutions at amino acid positions: 36 (R to S), 53 (V to A), 54 (T to A) and 328 (K to A). The following primers were used: R36A_For 5′-CCTATACTAACTTTGATGCTGAGAGCGATGCTTTGAACATTG-3′ R36A_Rev 5′-GTTTCAATGTTCAAAGCATCGCTCTCAGCATCAAAG-3′, V53A_T54A_For 5′-CAAAGGTGTGGATGAGGCCGCCATTGTCAACATTTTG-3′ V53A_T54A_Rev 5′-CAAAATGTTGACAATGGCGGCCTCATCCACACCTTTG-3′, K328A_For: 5′-TA AGGGCGACTACCAGGCAGCGCTGCTGTACCTG-3′ K328A_Rev 5′-AGGTACAGCAGCGCTGCCTGGTAGTCGCCCTTAG-3′ (Biomers, Ulm, Germany).

2.7. *Protein Expression and Purification*

For protein expression, E. coli cells transformed with the respective pSE420 plasmid were grown at 37 °C in LB medium supplemented with ampicillin to an optical density of 0.6 at 600 nm (OD600). Protein expression was then induced by addition of isopropyl β-D-1-thiogalactopyranoside to a final concentration of 1 mM. After expression for 4 h, cells were harvested by centrifugation at 5000× *g* for 10 min at 4 °C. Protein purification was performed by diethylamioethyl- and carboxymethyl-cellulose ion exchange chromatography and proteins were alkylated specifically at Cys-8 by 2-iodoacetamide treatment to prevent disulfide mediated protein crosslink as previously described [23].

3. Results

3.1. *Protein-Protein Interaction in Membrane-Bound AnxA2*

AnxA2 has been shown by atomic force microscopy to form two-dimensional assemblies on model membranes containing negatively charged phospholipids [13]. To analyze whether these assemblies are characterized by homotypic protein–protein interactions, we performed chemical crosslinking studies of membrane-bound versus soluble AnxA2. Therefore, we first developed a novel crosslinker, herein referred to as $Biotin_{3xNHS}$X-linker, that due to its trifunctional nature should efficiently link proximal amino groups in proteins (Figure 1). $Biotin_{3xNHS}$X-linker also contained a biotin group enabling streptavidin-mediated detection and enrichment.

$Biotin_{3xNHS}$X-linker was then used to study the nature of AnxA2 assemblies on model membranes. Purified AnxA2 (Figure S3) was treated with $Biotin_{3xNHS}$X-linker either in the absence of membranes or following Ca^{2+}-dependent binding to liposomes containing the negatively charged AnxA2-binding lipid phosphatidylserine (PS). Figure 2 shows the results of these crosslinking experiments. While a very small amount of higher molecular mass species was observed in the control reactions, i.e., AnxA2 samples in the absence of membranes or Ca^{2+}, significant crosslink products indicative of oligomeric AnxA2 assemblies were generated when AnxA2 bound to PS-containing liposomes was subjected to the crosslinking reaction. Thus, our crosslink approach involving the Biotin3xNHSX-Linker indicates that membrane binding triggers the formation of AnxA2 oligomers, in which the proteins engage in lateral contacts spatially close enough to allow an effective covalent linkage by the trifunctional crosslinker.

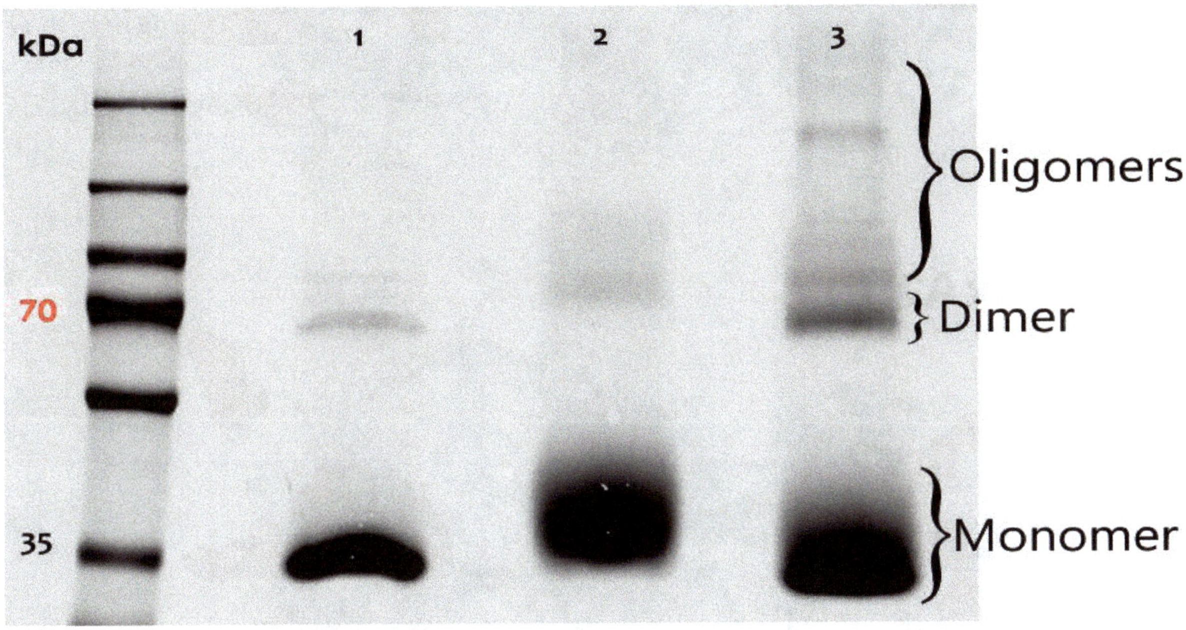

Figure 2. SDS-PAGE of crosslinking reactions involving alkylated AnxA2 wild-type (WT). Lane 1: Control #1 (AnxA2 WT + Ca^{2+}); lane 2: Control #2 (AnxA2 WT + LUVs + EGTA + Biotin3xNHSX-Linker); lane 3: Ca^{2+}/membrane sample (AnxA2 WT + LUVs + Ca^{2+} + Biotin3xNHSX-Linker). Brackets on the right indicate the positions of AnxA2 monomers, dimers, and oligomers. Dimer formation most likely occurs during sample preparation, whereas the oligomers likely present AnxA2 assemblies that form following membrane interaction and are then stabilized by the crosslinker. A representative result of n = 5 independently performed experiments is shown.

3.2. Annexin A2 Oligomers on Model Membranes are Stabilized by Lateral Protein-Protein Interactions

To address the nature of the homotypic AnxA2 interaction, which occurs following membrane binding and can be stabilized by $Biotin_{3xNHS}$X-linker, we generated two AnxA2 derivatives, in which residues predicted to participate in lateral protein-protein interactions in the crystal structure of an anti-parallel AnxA2 dimer [20] (see also pdb entry of the crystal structure of this AnxA2 dimer at 1XJL) were mutated to side chains of opposite charge or to alanine or serine (Figure 3). Importantly, the residues mutated are not part of the known type-II or type-III Ca^{2+}-binding sites of AnxA2 [25] and so far have not been identified as sites of posttranslational modification. Moreover, the residues selected are characterized by polar or charged side chains and thus could engage in salt bridges and/or other ionic interactions that would favor oligomer formation. Provided that the two-dimensional AnxA2 assemblies on membranes involve these residues located on the lateral surface of the folded AnxA2 molecule, the mutants, herein named AnxA2 6x and AnxA2 10x, should show a compromised oligomer formation and thus $Biotin_{3xNHS}$X-linker mediated crosslink. Moreover, as the mutations do not involve residues of the Ca^{2+}/membrane binding sites, AnxA2 6x and AnxA2 10x are expected to retain the capability to bind to membranes containing acidic phospholipids.

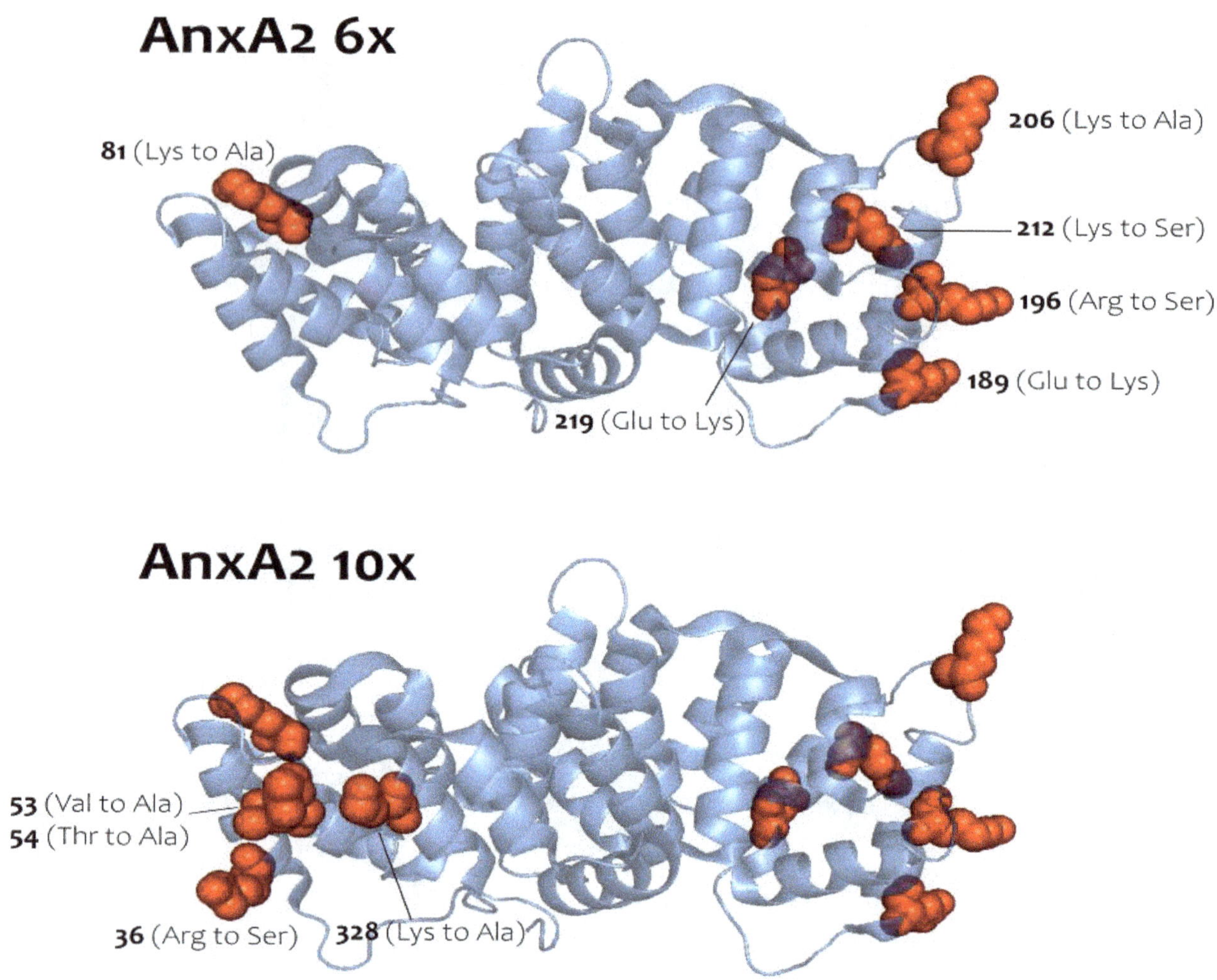

Figure 3. AnxA2 crystal structure highlighting mutations introduced in the AnxA2 6x and 10x constructs. AnxA2 6x top [81 (Lys to Ala), 189 (Glu to Lys), 196 (Arg to Ser), 206 (Lys to Ala), 212 (Lys to Ser) and 219 (Glu to Lys)] and AnxA2 10x bottom [36 (Arg to Ser), 53 (Val to Ala), 54 (Thr to Ala) and 328 (Lys to Ala)]. Illustrations were created using the AnxA2 crystal structure (PDB code: 1XJL).

AnxA2 6x and 10x were purified following the protocol developed for the wild-type protein (Figure S1). Importantly, this also involved alkylation of the exposed cysteine-8 as disulfide bridge formation involving this cysteine residue is observed under oxidative conditions [23]. The mutants were first characterized with respect to their ability to bind to membranes containing acidic phospholipids in a Ca^{2+}-dependent manner by employing liposome co-pelleting and solid-supported membrane-binding assays. Figure 4 shows that AnxA2 6x and 10x effectively bind to PS-containing liposomes in the presence of Ca^{2+}. As observed for the wild-type protein, this binding is fully reversible, i.e., the bound protein is released when the liposome-protein mixtures are treated with the Ca^{2+} chelator EGTA. Analysis of protein binding to solid supported membrane bilayers was carried out using a quartz crystal microbalance with dissipation (QCM-D). QCM-D is a well-established tool to evaluate and quantify protein–lipid interactions [21]. By applying SUVs, a bilayer can be formed on a sensor chip connected to a quartz microbalance and the ability of proteins to interact with this lipid bilayer can be measured via decrease in the resonance frequency of the quartz crystal that occurs as a result of mass adsorption. Importantly, due to the unique Ca^{2+}-dependent and fully reversible interaction of AnxA2 with membranes, AnxA2 bound to the solid-supported membrane can be readily released from the bilayer by Ca^{2+} chelation with EGTA [24]. QCM-D recordings performed with the different AnxA2 derivatives revealed that AnxA2 wild-type (WT) and the AnxA2 6x and 10x mutants show similar binding kinetics and resonance frequency shifts ($\Delta\Delta F$) of 19 Hz for AnxA2 WT, 18.5 Hz for

AnxA2 6x, and 16.7 Hz for AnxA2 10x (Figure 4). Moreover, in each case the binding is fully reversible upon addition of EGTA and the dissipation increase is rather minimal indicating that a relatively rigid protein layer is formed on the solid-supported bilayer.

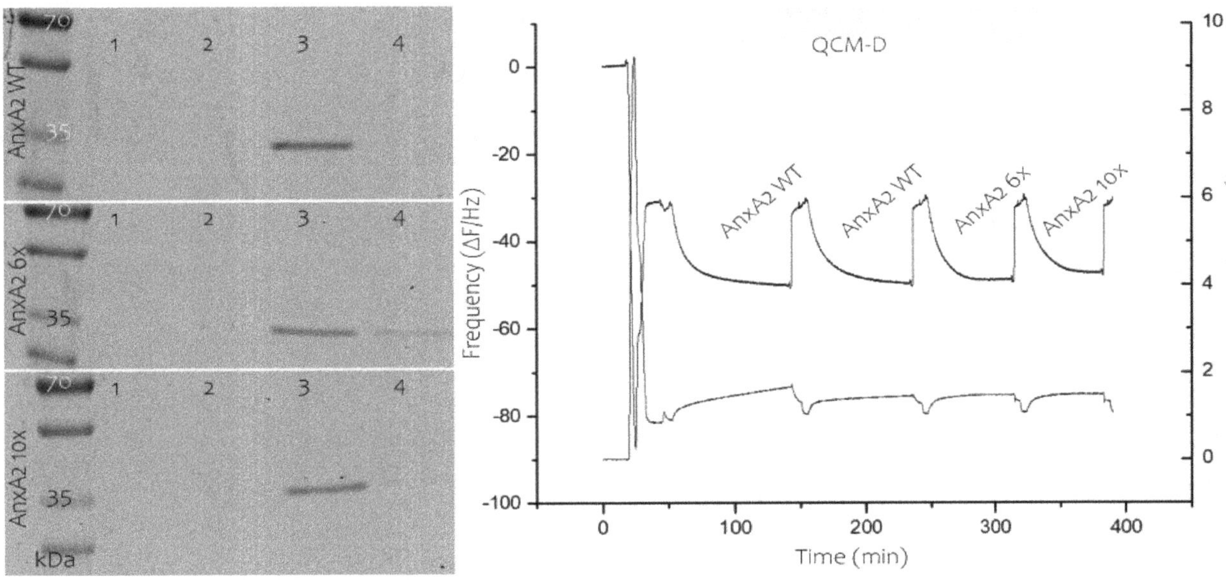

Figure 4. Membrane binding of AnxA2 constructs. Left, Liposome co-pelleting assay analyzed by SDS-PAGE of the different fractions. AnxA2 wild-type (WT), AnxA2 6x, or AnxA2 10x were mixed with PS-containing liposomes in the presence of 1 mM Ca^{2+}. Liposomes were pelleted and the supernatant, i.e., non-bound material, was collected (lane 1). Liposomes were then washed in Ca^{2+}-containing buffer, yielding a second supernatant (Ca^{2+} wash, lane 2). Subsequently, the pelleted liposomes were washed with EGTA-containing buffer resulting in release of the Ca^{2+}-dependently bound material (EGTA eluate, lane 3). The final liposome pellet containing non-released protein is shown in lane 4. The gel shows a representative result of $n = 5$ independently performed experiments. Right, QCM-D measurements, frequency (as deviation from resonance frequency, ΔF) is shown in the upper recordings and dissipation (ΔD) in the lower. Following formation of a solid-supported bilayer (at a ΔF of around −30 Hz in these settings), AnxA2 WT was added in the presence of Ca^{2+}, resulting in a drop in resonance frequency to approximately −49 Hz. Addition of EGTA removed all bound protein with resonance frequency returning to its initial bilayer value (−30 Hz). Recording was continued with subsequent additions (in Ca^{2+} containing buffer) and release (in EGTA containing buffer) of AnxA2 WT (to show reversibility of the reaction), AnxA2 6x, and AnxA2 10x. The QCM-D recordings were performed at least three times each for the different, independently purified AnxA2 derivatives.

Next, the lateral side chain mutants, AnxA2 6x and AnxA2 10x, were characterized with respect to their ability to form crosslinkable oligomers following membrane binding. Mutant proteins were bound to PS-containing liposomes and proteins residing in close proximity were covalently linked employing the $Biotin_{3xNHS}$X-linker. Figure 5 shows that the capability of forming crosslinked high molecular mass products was significantly compromised in both mutants when compared to the wild-type protein (Figure 2). A quantification of the oligomeric products revealed that higher molecular mass products representing crosslinked AnxA2 oligomers are reduced to approximately 48% and 33% for AnxA2 6x and AnxA2 10x, respectively, when compared to the wild-type protein. Thus, side chains identified in AnxA2 crystals as potential protein–protein contact sites appear to reside in close proximity in membrane-bound AnxA2, suggesting that lateral protein–protein interactions accompany the membrane association of AnxA2

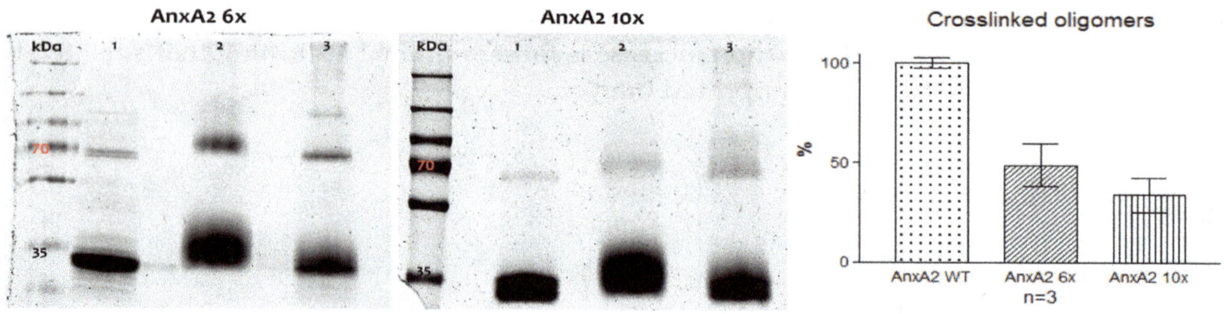

Figure 5. Chemical crosslinking of alkylated AnxA2 mutant proteins. Left, SDS-PAGE of crosslinking reactions involving AnxA2 6x and 10x. Each AnxA2 derivative was subjected to chemical crosslinking in the presence or absence of Ca^{2+} and LUVs. Lanes 1: Controls #1 (AnxA2 + Ca^{2+}); lanes 2: Controls #2 (AnxA2 + LUVs + EGTA + $Biotin_{3xNHS}$X-Linker); lanes 3: Ca^{2+}/membrane sample (AnxA2 + LUVs + Ca^{2+} + $Biotin_{3xNHS}$X-Linker). Right, quantification of crosslinked oligomer bands for each AnxA2 derivative (calculated in relation to the respective monomer band, see Materials and Methods). Given is the relative percentage of these oligomer bands compared to those obtained for the wild-type protein analyzed in a parallel reaction. Three independent crosslinking reactions were analyzed for each protein species and the standard error of means is indicated.

4. Discussion

In addition to binding to membranes containing acidic phospholipids in the presence of elevated Ca^{2+} concentrations, at least some annexins can also form two-dimensional clusters or assemblies on model membranes. Such assemblies have been extensively studied in the case of AnxA5, which forms highly ordered 2D crystals on the membrane [26–28]. In the case of AnxA5, the building blocks in these crystalline arrays are trimers, and residues mediating and stabilizing trimer formation have been mapped [29]. In contrast to AnxA5, AnxA2 does not form such ordered crystalline arrays on model membranes, but rather appears to associate into more irregular two-dimensional assemblies of amorphous shapes [13]. Interestingly, these assemblies have been shown to segregate the AnxA2-binding phospholipids [PS, PI(4,5)P_2] into domains underneath the bound proteins, and this activity has been suggested to underlie the function of AnxA2 in Ca^{2+}-regulated exocytosis [11,12,14, 16,30]. Thus, understanding the molecular basis of the AnxA2-mediated lipid segregation is of high relevance for understanding the cellular properties of the protein.

Here, we have used a chemical cross-linking approach employing a novel trifunctional compound and combined this with the characterization of AnxA2 mutant derivatives to address the nature of the 2D assemblies formed by this annexin on model membranes. Our analysis revealed that these assemblies are characterized by close protein proximity most likely involving protein–protein interactions that can be stabilized by chemical crosslinking. Moreover, we identified amino acid residues residing on the lateral surface of the folded protein that are either directly involved in the crosslink or mediate interactions between bound AnxA2 moieties that are lost in the 6x and 10x mutants. Thus, it appears likely that the assemblies containing AnxA2 and its interacting lipids [PS, PI(4,5)P_2] are at least partly stabilized by lateral protein–protein interactions.

Our analysis benefitted from the introduction of a novel trifunctional crosslinker. Typically, chemical crosslinkers are capable of connecting spatially proximate amino acids of proteins. When selecting the crosslinker, the chemical selectivity and activity of the functional groups towards the amino acids to be linked should be taken into account. In addition, the "length/size" of the crosslinker is crucial, since it defines the maximum spatial distance of the linked amino acids. The $Biotin_{3xNHS}$X-linker synthesized here is a homo-trifunctionalized NHS ester crosslinker, whose N-hydroxy-succinimide group can react with primary amines or alcohol groups, thus yielding high crosslink efficiency that allowed the detection of AnxA2 protein proximity in the membrane-bound form.

Although AnxA2 has been shown to cluster the lipids bound by the protein, the composition of the membrane patch underneath bound AnxA2 is not known. Most likely it also contains cholesterol as it has been shown that membrane cholesterol renders the AnxA2 binding cooperative, suggesting a cooperative nature of the assembly formation. As the protein assemblies formed by AnxA2 on the membrane are accompanied by close protein apposition, it appears plausible that the cooperation is at least in part mediated by a conformational change in the membrane-bound AnxA2, which renders it capable of undergoing protein–protein interactions. Future experiments involving streptavidin-mediated enrichment of membrane patches bound to $Biotin_{3xNHS}$X-linker-crosslinked AnxA2 should shed some light on the lipid composition in the associated membrane. The Ca^{2+}-regulated and reversible manner underlying the cluster formation of AnxA2 on cholesterol and PS/PI(4,5)P_2 containing membranes is most likely highly relevant for its cellular function as it would support the dynamic formation of lipid platforms that could function as sites for exocytosis, and are also involved in epithelial cell polarization shown to depend on AnxA2 [15].

In summary, we introduce a novel water-soluble trifunctional NHS crosslinker, which was suitable to covalently link AnxA2 protein moieties bound to model membranes containing negatively charged phospholipids. Moreover, we could identify amino acid side chains residing on the lateral surface of folded AnxA2 that are required for efficient crosslink, and thus likely support lateral protein–protein interactions of membrane-bound AnxA2, possibly by providing salt bridges and/or other ionic interactions.

Author Contributions: Conceptualization, B.J.R., V.G.; resources, A.L.L.M., S.K., J.M., M.H., D.G.; data curation, A.L.L.M., S.K., J.M., M.H., D.G., B.J.R., V.G.; writing—original draft preparation, V.G., A.L.L.M., S.K.; writing—review and editing, B.J.R., V.G., A.L.L.M., S.K.; visualization, A.L.L.M., S.K., J.M., M.H., D.G.; supervision, B.J.R., V.G..; project administration, B.J.R., V.G.; funding acquisition, B.J.R., V.G. All authors have read and agreed to the published version of the manuscript.

Acknowledgments: We would like to thank Ursula Rescher for helpful discussions and Abigail Cornwell and Michael Hülskamp for protein purification.

References

1. Brown, M.F. Soft Matter in Lipid–Protein Interactions. *Annu. Rev. Biophys.* **2017**, *46*, 379–410. [CrossRef] [PubMed]
2. Lingwood, D.; Simons, K. Lipid Rafts As a Membrane-Organizing Principle. *Sci.* **2009**, *327*, 46–50. [CrossRef] [PubMed]
3. Sezgin, E.; Levental, I.; Mayor, S.; Eggeling, C. The mystery of membrane organization: Composition, regulation and roles of lipid rafts. *Nat. Rev. Mol. Cell Boil.* **2017**, *18*, 361–374. [CrossRef] [PubMed]
4. Lemmon, M.A. Membrane recognition by phospholipid-binding domains. *Nat. Rev. Mol. Cell Boil.* **2008**, *9*, 99–111. [CrossRef] [PubMed]
5. Gerke, V.; Moss, S.E. Annexins: From Structure to Function. *Physiol. Rev.* **2002**, *82*, 331–371. [CrossRef] [PubMed]
6. Monastyrskaya, K.; Babiychuk, E.B.; Draeger, A. The annexins: Spatial and temporal coordination of signaling events during cellular stress. *Cell. Mol. Life Sci.* **2009**, *66*, 2623–2642. [CrossRef]
7. Rescher, U.; Gerke, V. Annexins - unique membrane binding proteins with diverse functions. *J. Cell Sci.* **2004**, *117*, 2631–2639. [CrossRef]
8. Gerke, V.; Creutz, C.E.; Moss, S.E. Annexins: Linking Ca2+ signalling to membrane dynamics. *Nat. Rev. Mol. Cell Boil.* **2005**, *6*, 449–461. [CrossRef]
9. Hayes, M.J.; Merrifield, C.J.; Shao, N.; Ayala-Sanmartin, J.; Schorey, C.D.; Levine, T.P.; Proust, J.; Curran, J.; Bailly, M.; Moss, S.E. Annexin 2 binding to phosphatidylinositol 4,5-bisphosphate on endocytic vesicles is regulated by the stress response pathway. *J. Boil. Chem.* **2004**, *279*, 14157–14164. [CrossRef]
10. Rescher, U.; Richardson, G.D.; Robson, C.; Lang, S.H.; Neal, D.E.; Maitland, N.J.; Collins, A.T. Annexin 2 is a phosphatidylinositol (4,5)-bisphosphate binding protein recruited to actin assembly sites at cellular membranes. *J. Cell Sci.* **2004**, *117*, 3473–3480. [CrossRef]
11. Drücker, P.; Pejic, M.; Galla, H.-J.; Gerke, V. Lipid Segregation and Membrane Budding Induced by the Peripheral Membrane Binding Protein Annexin A2*. *J. Boil. Chem.* **2013**, *288*, 24764–24776. [CrossRef] [PubMed]

12. Gokhale, N.A.; Abraham, A.; A Digman, M.; Gratton, E.; Cho, W. Phosphoinositide Specificity of and Mechanism of Lipid Domain Formation by Annexin A2-p11 Heterotetramer. *J. Boil. Chem.* **2005**, *280*, 42831–42840. [CrossRef] [PubMed]
13. Menke, M.; Ross, M.; Gerke, V.; Steinem, C. The Molecular Arrangement of Membrane-Bound Annexin A2-S100A10 Tetramer as Revealed by Scanning Force Microscopy. *ChemBioChem* **2004**, *5*, 1003–1006. [CrossRef] [PubMed]
14. Menke, M.; Gerke, V.; Steinem, C. Phosphatidylserine Membrane Domain Clustering Induced by Annexin A2/S100A10 Heterotetramer †. *Biochemistry* **2005**, *44*, 15296–15303. [CrossRef]
15. Martin-Belmonte, F.; Gassama, A.; Datta, A.; Yu, W.; Rescher, U.; Gerke, V.; Mostov, K. PTEN-Mediated Apical Segregation of Phosphoinositides Controls Epithelial Morphogenesis through Cdc42. *Cell* **2007**, *128*, 383–397. [CrossRef]
16. Brandherm, I.; Disse, J.; Zeuschner, D.; Gerke, V. cAMP-induced secretion of endothelial von Willebrand factor is regulated by a phosphorylation/dephosphorylation switch in annexin A2. *Blood* **2013**, *122*, 1042–1051. [CrossRef]
17. Gabel, M.; Delavoie, F.; Demais, V.; Royer, C.; Bailly, Y.; Vitale, N.; Bader, M.-F.; Chasserot-Golaz, S. Annexin A2-dependent actin bundling promotes secretory granule docking to the plasma membrane and exocytosis. *J. Gen. Physiol.* **2015**, *146*, 1463. [CrossRef]
18. Hayes, M.A.; Shao, D.-M.; Grieve, A.; Levine, T.P.; Bailly, M.; Moss, S.E. Annexin A2 at the interface between F-actin and membranes enriched in phosphatidylinositol 4,5,-bisphosphate. *Biochim. et Biophys. Acta (BBA) - Bioenerg.* **2009**, *1793*, 1086–1095. [CrossRef]
19. Zobiack, N.; Rescher, U.; Laarmann, S.; Michgehl, S.; Schmidt, M.A.; Gerke, V. Cell-surface attachment of pedestal-forming enteropathogenic E. coli induces a clustering of raft components and a recruitment of annexin 2. *J. Cell Sci.* **2002**, *115*, 91–98.
20. Rosengarth, A.; Luecke, H. Annexin A2: Does it induce membrane aggregation by a new multimeric state of the protein. *Annexins* **2004**, *1*, 129–136.
21. Matos, A.L.L.; Grill, D.; Kudruk, S.; Heitzig, N.; Galla, H.-J.; Gerke, V.; Rescher, U. Dissipative Microgravimetry to Study the Binding Dynamics of the Phospholipid Binding Protein Annexin A2 to Solid-supported Lipid Bilayers Using a Quartz Resonator. *J. Vis. Exp.* **2018**, *141*, e58224. [CrossRef] [PubMed]
22. Thiel, C.; Osborn, M.; Gerke, V. The tight association of the tyrosine kinase substrate annexin II with the submembranous cytoskeleton depends on intact p11- and Ca(2+)-binding sites. *J. Cell Sci.* **1992**, *103*, 103.
23. Grill, D.; Matos, A.L.L.; De Vries, W.C.; Kudruk, S.; Heflik, M.; Dörner, W.; Mootz, H.D.; Ravoo, B.J.; Galla, H.-J.; Gerke, V. Bridging of membrane surfaces by annexin A2. *Sci. Rep.* **2018**, *8*, 14662. [CrossRef] [PubMed]
24. Drücker, P.; Pejic, M.; Grill, D.; Galla, H.-J.; Gerke, V. Cooperative Binding of Annexin A2 to Cholesterol- and Phosphatidylinositol-4,5-Bisphosphate-Containing Bilayers. *Biophys. J.* **2014**, *107*, 2070–2081. [CrossRef] [PubMed]
25. Jost, M.; Weber, K.; Gerke, V. Annexin II contains two types of Ca2+-binding sites. *Biochem. J.* **1994**, *298*, 553–559. [CrossRef] [PubMed]
26. Pigault, C.; Follenius-Wund, A.; Schmutz, M.; Freyssinet, J.-M.; Brisson, A. Formation of Two-dimensional Arrays of Annexin V on Phosphatidylserine-containing Liposomes. *J. Mol. Boil.* **1994**, *236*, 199–208. [CrossRef]
27. Reviakine, I.; Bergsma-Schutter, W.; Brisson, A. Growth of Protein 2-D Crystals on Supported Planar Lipid Bilayers Imagedin Situby AFM. *J. Struct. Boil.* **1998**, *121*, 356–362. [CrossRef]
28. Oling, F.; Bergsma-Schutter, W.; Brisson, A. Trimers, Dimers of Trimers, and Trimers of Trimers Are Common Building Blocks of Annexin A5 Two-Dimensional Crystals. *J. Struct. Boil.* **2001**, *133*, 55–63. [CrossRef]
29. Bouter, A.; Gounou, C.; Bérat, R.; Tan, S.; Gallois, B.; Granier, T.; D'Estaintot, B.L.; Pöschl, E.; Brachvogel, B.; Brisson, A.R. Annexin-A5 assembled into two-dimensional arrays promotes cell membrane repair. *Nat. Commun.* **2011**, *2*, 270. [CrossRef]
30. Gabel, M.; Chasserot-Golaz, S. Annexin A2, an essential partner of the exocytotic process in chromaffin cells. *J. Neurochem.* **2016**, *137*, 890–896. [CrossRef]

12

Targeting AnxA1/Formyl Peptide Receptor 2 Pathway Affords Protection against Pathological Thrombo-Inflammation

Shantel A. Vital [1], Elena Y. Senchenkova [1], Junaid Ansari [1,2] and Felicity N. E. Gavins [1,2,3,*]

1 Department of Molecular and Cellular Physiology, Louisiana State University Health Sciences Center-Shreveport, Shreveport, LA 71130, USA; SVital@lsuhsc.edu (S.A.V.); ESench@lsuhsc.edu (E.Y.S.); junaid.em42@gmail.com (J.A.)

2 Department of Neurology, Louisiana State University Health Sciences Center-Shreveport, Shreveport, LA 71130, USA

3 Department of Life Sciences, Centre for Inflammation Research and Translational Medicine (CIRTM), Brunel University London, Uxbridge, Middlesex UB8 3PH, UK

* Correspondence: felicity.gavins@brunel.ac.uk

Abstract: Stroke is a leading cause of death and disability globally and is associated with a number of co-morbidities including sepsis and sickle cell disease (SCD). Despite thrombo-inflammation underlying these co-morbidities, its pathogenesis remains complicated and drug discovery programs aimed at reducing and resolving the detrimental effects remain a major therapeutic challenge. The objective of this study was to assess whether the anti-inflammatory pro-resolving protein Annexin A1 (AnxA1) was able to reduce inflammation-induced thrombosis and suppress platelet activation and thrombus formation in the cerebral microvasculature. Using two distinct models of pathological thrombo-inflammation (lipopolysaccharide (LPS) and sickle transgenic mice (STM)), thrombosis was induced in the murine brain using photoactivation (light/dye) coupled with intravital microscopy. The heightened inflammation-induced microvascular thrombosis present in these two distinct thrombo-inflammatory models was inhibited significantly by the administration of AnxA1 mimetic peptide $AnxA1_{Ac2\text{-}26}$ (an effect more pronounced in the SCD model vs. the endotoxin model) and mediated by the key resolution receptor, Fpr2/ALX. Furthermore, $AnxA1_{Ac2\text{-}26}$ treatment was able to hamper platelet aggregation by reducing platelet stimulation and aggregation (by moderating $\alpha_{IIb}\beta_3$ and P-selectin). These findings suggest that targeting the AnxA1/Fpr2/ALX pathway represents an attractive novel treatment strategy for resolving thrombo-inflammation, counteracting e.g., stroke in high-risk patient cohorts.

Keywords: thrombosis; inflammation; Annexin A1; formyl peptide receptors; sickle cell disease; sepsis

---

1. Introduction

Stroke is a leading cause of death and disability, with functional impairments producing significant losses in quality of life and accompanying financial burden [1–3]. Although the exact mechanisms responsible for post-ischaemic cerebral damage in stroke remain undefined, the intertwined processes of thrombosis and inflammation play crucial roles in the pathophysiology [4–6]. Unregulated thrombo-inflammation, which involves a complex relationship between inflammatory leukocytes (e.g., neutrophils), platelets, and the vascular endothelium, is also associated with a number of comorbidities such as sickle cell disease ("SCD" [7,8]) and infections (e.g., sepsis [9,10]), which predispose individuals to ischaemic stroke. In the case of SCD, ~11% of SCD patients have a stroke before the age of 20, increasing to 24% by the age of 45 [11]. Furthermore, stroke patients are not only more susceptible to bacterial infections, but infection itself is an independent risk factor for stroke and a major contributor to worse outcome post stroke, increasing recurrent stroke risk [4]. Therefore, reducing and resolving the impact and detrimental effects of microvascular thrombosis and inflammation associated with underlying co-morbidities, such as those already discussed, represent a major therapeutic challenge.

It is now widely accepted that endogenous pro-resolving mediators released during an inflammatory response play a critical role in effective recovery from inflammation and repair [12,13]. Resolution of inflammation is a tightly orchestrated process, involving specific endogenous mediators such as Annexin A1 ("AnxA1") and its biologically active N-terminal domain, "Ac2-26" (Ac-AMVSEFLKQAWFIENEEQEYVQTVK); Lipoxins e.g., "Lipoxin A_4" (5S,6R,15S-trihydroxy-7E,9E,11Z,13E-eicosatetraenoic acid) and aspirin-triggered lipoxin A_4 (15(R)-epi-LXA_4, "ATL"); resolvins ("Rv") e.g., "RvD1" and "RvD2"; protectins e.g., protectin D1 (10R,17S-dihydroxy-4Z,7Z,11E,13E,15Z,19Z-docosahexaenoic acid); maresins e.g., maresin 1 (7R,14S-dihydroxy-4Z,8E,10E,12Z,16Z,19Z-docosahexaenoic, "MaR1") [12,14–16]. These specific endogenous mediators moderate and resolve inflammation through protective pro-resolution pathways such as the formyl peptide receptor 2 (also termed "FPR2/ALX", Fpr2, or ALX receptor i.e., the receptor for LXA_4) pathway, which is a key resolution pathway [17–22].

The human FPR family consists of three specific receptors termed "FPR1", FPR2/ALX, and "FPR3", all of which are well conserved G-protein-coupled receptors expressed in a host of different cells and tissues, all having pluripotent and diverse roles in the initiation, propagation, and resolution of inflammation [17,19]. All three receptors are clustered together on chromosome 19q13.3 and share significant sequence homology, with FPR1 sharing ~69% and 56% sequence homology with FPR2/ALX and FPR3, respectively, and FPR2/ALX sharing 83% sequence homology with FPR3 [23].

AnxA1 is a 37 kDa protein that belongs to the annexin superfamily of Ca^{2+}-dependent phospholipid binding proteins. It is an anti-inflammatory, pro-resolving protein that is up-regulated by glucocorticoids. Annexins share a common structure formed of a core region and a unique N-terminal domain, which acts as the fingerprint for each annexin of the superfamily. A number of investigations have focussed on the anti-inflammatory effects of AnxA1 and its mimetic peptides (including the blockade of leukocyte recruitment, inhibition of cytokine release, promotion of apoptosis, stimulation of phagocytosis, and decreasing vascular permeability) [17] in a variety of clinically related disease models [17,20,24–28], often via an engagement with FPR1 or FPR2/ALX (murine orthologues Fpr1 and Fpr2/ALX, respectively).

In the context of an ischaemic stroke setting, we have previously shown that AnxA1 (via Fpr2/ALX) is able to act both as a therapeutic and a prophylactic drug, reducing infarct volume and improving stroke outcome in a mouse model of acute experimental stroke [17,29] without increasing the risk of intracerebral haemorrhage [3,29,30]. Moreover, we demonstrated that this anti-inflammatory/pro-resolving compound was able to act as an anti-thrombotic agent suppressing thrombin-induced inside-out signalling events such as Akt activation, intracellular calcium release, and Ras-associated protein 1 ("Rap1") [31], thereby exerting protection by altering platelet phenotypes from pro-pathogenic to regulatory [29]. These findings provided novel physio-pharmacological properties of AnxA1 as a powerful pro-resolving mediator of thrombo-inflammation and opened new avenues for an attractive therapeutic treatment strategy for patients with stroke. However, the anti-thrombotic effect of AnxA1 in patients with co-morbidities susceptible to stroke is less well characterized. Therefore, the objective of this current study was to use pharmacological and genetic approaches together with a photoactivation thrombosis (light/dye) model to test the hypothesis that targeting the AnxA1/FPR-pathway protects the cerebral microvascular system against inflammation-induced microvascular thrombosis in co-morbidities susceptible to stroke.

2. Materials and Methods

2.1. Animals

All studies were done blinded and performed on adult male mice weighing 25–30 g. C57BL/6 mice were purchased from Jackson Laboratory (Bar Harbor, ME, USA). The sickle cell transgenic mice (STM) (also termed β^s mice, heterozygous BERK, trait BERK, or sickle cell trait model) were homozygous for knockout of murine α-globin and heterozygous for knockout of murine β-globin and had one copy of the linked transgenes for human α- and β^s-globins [32]. STM used in this study were a gift from Professor Robert Hebbel (University of Minnesota). STM were bred onsite and showed no obvious phenotype and were fertile. Mice were maintained on a 12 h (h) light–dark cycle, during which room temperature was maintained at 21–23 °C. Mice had access to a standard chow pellet diet and tap water *ad libitum*. All animal experiments were approved by the Louisiana State University Health Sciences Center—Shreveport (LSUHSC-S) Institutional Animal Care and Use Committee (IACUC), were in accordance with the guidelines of the American Physiological Society and complied with ARRIVE (Animal Research: Reporting In Vivo Experiments) guidelines.

2.2. Endotoxin (LPS) Administration

Mice were injected intraperitoneally (i.p.) with "LPS" (Escherichia coli serotype 0111:B4, purified initially by phenol extraction and purified further by ion exchange chromatography. L3024. Sigma-Aldrich, St Louis, MO, USA), two hours before the experiment at a dose of 0.4 mg/kg in 100 μL vehicle sterile saline [33].

2.3. Photoactivation Thrombosis Model. (Light/Dye Method)

Intravital microscopy ("IVM") was performed as previously described [29]. Briefly, mice were anaesthetized via i.p. injection of ketamine (150 mg/kg) and xylazine (7.5 mg/kg), and the femoral vein was cannulated for dye (fluorescein isothiocyanate (FITC)-dextran) administration. The head of each mouse was fixed in a frame (sphinx position) and the parietal bone was exposed by a midline skin incision, followed by a craniectomy [29]. A total of 10 mg/kg of 5% FITC dextran (150,000 mol wt, Sigma-Aldrich) was injected i.v. and allowed to circulate for ten minutes (mins) prior to photoactivation. Photoactivation was initiated (excitation, 495 nm; emission, 519 nm) by exposing 100 μm of vessel length to epi-illumination with a 175-W xenon lamp (Lambda LS, Sutter, Novato, CA, USA) and a fluorescein filter cube (HQ-FITC, Chroma, San Francisco, CA, USA) [29]. The excitation power density was measured daily (ILT 1700 Radiometer, SED033 detector; International Light, Peabody, MA, USA) and maintained within 1% of 0.77 W/cm^2. Epi-illumination was applied continuously, and onset and blood flow cessation times were quantified in cerebral vessels (diameter range: 30–70 μm) until blood flow had ceased in the vessel under study [29]. In some cases, 20 min prior to onset of thrombosis, mice were treated with vehicle (saline), AnxA1$_{\text{Ac2-26}}$ (Ac-AMVSEFLKQAWFIENEEQEYVQTVK, 4 mg/kg. Cambridge Research Biochemicals, Cambridge, UK), pan FPR antagonist Boc2 (N-tert-butoxycarbonyl-L-Phe-D-Leu-L-Phe-D-Leu-L-Phe, 0.4 mg/kg; MP Biomedicals, Cambridge, UK) or FPR2/ALX antagonist WRW4 (2.2 mg/kg. Tocris, Bristol, UK) administered (100 μL) i.v. [20,29]. Doses/concentrations used in the study are congruent with binding affinities for the Fprs and were chosen based on our previous findings [17,20,29]. All compounds were made in vehicle sterile saline.

2.4. Bleeding Time

Bleeding times were quantified in mice treated with vehicle saline, LPS, or AnxA1$_{\text{Ac2-26}}$, as previously described. Briefly, mice were anaesthetized with i.p. injection of ketamine (150 mg/kg) and xylazine (7.5 mg/kg). A small tail segment (3 mm) was cut cleanly with a scalpel blade, and bleeding was monitored at 15-s intervals by absorbing the bead of blood with filter paper without contacting the wound site [29]. When no blood was observed on the paper, bleeding was determined to have ceased [34].

2.5. Platelet Cell Counts

Platelets from peripheral blood were stained with 1% buffered ammonium oxalate and counted using a haemocytometer.

2.6. Platelet Flow Cytometry

Whole blood was collected (0.9 mL) via carotid artery into a syringe containing 0.1 mL anticoagulant citrate dextrose ("ACD") buffer, transferred into an Eppendorf tube, and centrifuged at 1200 rpm for 8 min. The platelet-rich plasma ("PRP") layer was transferred to a new Eppendorf tube and centrifuged at 3000 rpm for 10 min. The supernatant was removed and the pellet was resuspended in Tyrodes buffer (containing 1 mM Ca^{2+}) along with the appropriate antibody and allowed to incubate for 15 min at 37 °C. The reaction was stopped by the addition of 450 μL of 1% paraformaldehyde. P-selectin and JON/A-PE FITC antibodies (Emfret Analytics, Eibelstadt, Germany) were used to measure platelet activation and surface P-selectin exposure in murine platelets using flow cytometry, as previously described [35,36]. IgG isotype antibodies were used as controls. Briefly, the platelets were treated with Fc block (15 min, room temperature) to inhibit non-specific binding according to the manufacturer's instructions (eBioscience, San Diego, CA, USA). This was followed by treatment with $AnxA1_{Ac2\text{-}26}$ (30 μM) or vehicle for 30 min and then, the addition of antibodies (1:8 dilution). Next, the glycoprotein VI (GPVI) collagen receptor agonist convulxin (CVX) (1.7 ng/mL. Cayman Chemical Company, MI, USA) was used to stimulate the platelets for 15 min. In some cases, prior to CVX, platelets were treated with LPS (7.5 μg/mL). The activation was stopped by the addition of 450 μL of 1% paraformaldehyde. Platelets were identified by their light scattering using an LSRII flow cytometer (Becton Dickinson, Franklin Lakes, NJ, USA) and Diva8 software by assessing at least 10,000 events per sample, as previously described [34].

2.7. Platelet Aggregation Assay

Arterial blood was freshly collected from the carotid artery of mice, as described above. Platelets 8–10 × 10^6/mL were used to monitor platelet aggregation velocity after agonist exposure in a cuvette loaded with 6 mL of platelet media (140 mM NaCl, 10 mM HEPES, 10 mM $NaHCO_3$, 2 mM KCl, 1 mM $MgCl_2$, 2 mM $CaCl_2$, 5.5 mM D-glucose, pH 7.4) using a laser particle analyser (LasCa-1C, Lumex Ltd., St. Petersburg, Russia) by a low angle light scattering method, as previously described [29]. Platelets were administered either vehicle (saline) or LPS (7.5 μg/mL) and the normalized velocity of aggregation was calculated using original software LasCa_32, as previously described [29].

2.8. Statistical Analysis

All data were analysed using GraphPad Prism8 software. Data are shown as mean values ± standard error of the mean (SEM), with n values given in the respective figures. Results from thrombosis experiments were confirmed to follow a normal distribution using a Kolmogorov–Smirnov test of normality with Dallal–Wilkinson–Lillie for the corrected p value. Data that passed the normality assumption were analysed using Student's t-test (two groups) or ANOVA with Bonferroni post hoc tests (more than two groups). Data that failed the normality assumption were analysed using the non-parametric Mann–Whitney U test (two groups) or Kruskal–Wallis with Dunn's test (more than two groups). Differences were considered statistically significant at a value of $p < 0.05$.

3. Results

3.1. *AnxA1$_{Ac2\text{-}26}$ Inhibits LPS-Induced Thrombus Formation in Cerebral Microcirculation of C57/BL6 Mice*

The light/dye model coupled with intravital microscopy was used to quantify thrombus formation in cerebral arterioles and venules, as determined by onset time (i.e., the time to onset of visible aggregates) and blood flow cessation time (Figure 1A). Figure 1B,C shows that onset time was significantly faster in both arterioles and venules in LPS-treated mice vs. saline (vehicle)-treated mice (with the exception of LPS+AnxA1$_{Ac2\text{-}26}$+Boc2). We found that administration of AnxA1$_{Ac2\text{-}26}$ delayed the time of onset in the venules of mice treated with LPS vs. LPS alone ($p < 0.05$), suggesting a propensity for the AnxA1 peptide to initiate an early anti-thrombo-inflammatory response against inflammation-induced thrombosis.

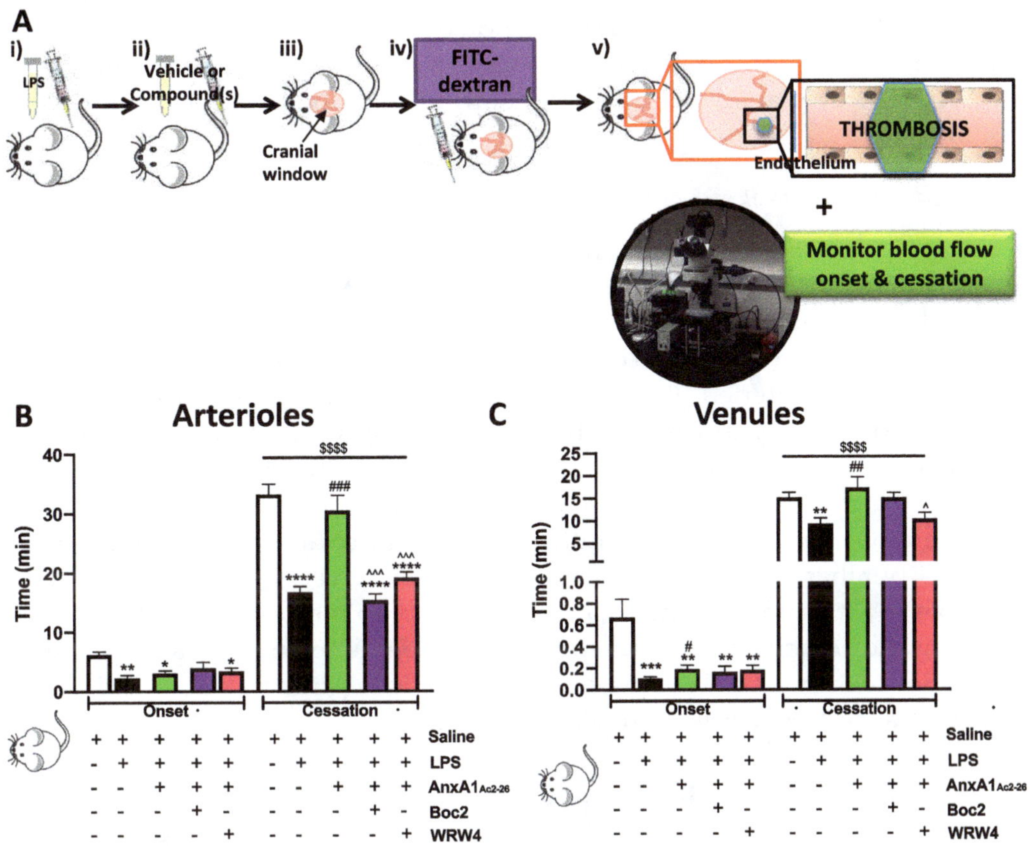

Figure 1. AnxA1$_{Ac2\text{-}26}$ protects against the effects of endotoxin (LPS) in light/dye-induced thrombosis responses in the cerebral microcirculation. (**A**) Schematic showing the experimental protocol: (i) Mice (C57BL/6) were subjected to vehicle (saline) or LPS (0.4 mg/kg) for 2 h. (ii) Mice were treated with vehicle (saline) or compound(s): AnxA1$_{Ac2\text{-}26}$ (4 mg/kg) with/without pan FPR antagonist Boc2 (0.4 mg/kg) or FPR2/ALX antagonist WRW4 (2.2 mg/kg) 20 min prior to thrombosis. (iii) A cranial window was performed and (iv) fluorescein isothiocyanate (FITC)-dextran was injected (10 mg/kg of 5%). (v) Mice were subjected to intravital microscopy and light/dye-induced thrombus formation, with onset and blood flow cessation times recorded for cerebral (**B**) arterioles and (**C**) venules. Data are means ± SEM of 5–6 mice/group. * $p < 0.05$, ** $p < 0.01$, *** $p < 0.001$, **** $p < 0.0001$ vs. corresponding vehicle (saline). # $p < 0.05$, ## $p < 0.01$, ### $p < 0.001$ vs. corresponding LPS group. $$$$ $p < 0.0001$ vs. same group for onset time. ^ $p < 0.05$, ^^^ $p < 0.001$ vs. corresponding LPS+AnxA1$_{Ac2\text{-}26}$ group.

As previously reported in C57BL/6 mice [34], LPS treatment augmented thrombus formation in the cerebral microcirculation (as noted by the decreased blood flow cessation times in LPS-treated mice vs. their saline-treated counterparts). This pro-thrombotic environment was mitigated by the treatment of AnxA1$_{Ac2\text{-}26}$, which was effective in both vessel types analysed (arterioles: 16.7 ± 1.0 min vs. 30.5 ± 2.7 min; venules: 9.3 ± 1.3 min vs. 17.3 ± 2.4 min; LPS vs. LPS+ AnxA1$_{Ac2\text{-}26}$, respectively).

Having established the anti-thrombotic effect of AnxA1$_{Ac2\text{-}26}$ against inflammation-induced thrombosis, we next ascertained whether these effects were mediated by the classic AnxA1/Fpr pathway. Figure 1B,C shows the FPR pan-antagonist Boc2 inhibited the actions of the AnxA1 mimetic peptide ($p < 0.0001$) in cerebral arterioles, but not in venules, suggesting a vessel-specific mechanism of action via the FPR family. (No effect was observed in either onset or blood flow cessation times for Boc2 administration alone vs. LPS; see Figure S1).

To further expand on these findings and tease out through which member(s) of the FPR family AnxA1$_{Ac2\text{-}26}$ was exerting its protective effects, mice were co-administered AnxA1$_{Ac2\text{-}26}$ with WRW4 (specific FPR2/ALX antagonist). We found that WRW4 blocked AnxA1$_{Ac2\text{-}26}$ afforded protection in both cerebral arterioles ($p < 0.001$) and venules ($p < 0.05$), thereby confirming an FPR2/ALX mechanism (Figure 1B,C). (Figure S1 shows that no effect was noted in either onset or blood flow cessation times for administration of WRW4 alone vs. LPS). These data further support the growing body of evidence that the FPR2/ALX pathway plays a crucial role in both thrombotic and inflammatory pathways in the brain microvasculature.

3.2. *AnxA1$_{Ac2\text{-}26}$ Reduces the Effect of Endotoxin-Induced Platelet Activation*

Since AnxA1$_{Ac2\text{-}26}$ was able to reduce thrombosis, we next sought to determine whether exogenous administration of AnxA1$_{Ac2\text{-}26}$ had any effect on the haemostatic activity of platelets. Figure 2A shows that AnxA1$_{Ac2\text{-}26}$-treated LPS mice exhibited decreased bleeding times ($p < 0.001$), and these observations were not related to changes in circulating platelet cell counts as AnxA1$_{Ac2\text{-}26}$ administration had no effect on the LPS-heightened platelet cell counts (Figure 2B). The interaction of platelets with neutrophils is involved in many thrombo-inflammatory responses. Furthermore, LPS is a potential mediator of neutrophil extracellular traps (NETosis) and induces platelet–neutrophil aggregate formation, and neutrophils have been shown to be essential for platelet recruitment in endotoxaemic models [37,38]. Here, we found that circulating neutrophil counts were heightened in those mice challenged with LPS, but these levels were not modified by AnxA1$_{Ac2\text{-}26}$ (Figure S3). We also found that the peptide was able to act as an anti-aggregant in its ability to moderate platelet aggregation induced by LPS (by reducing the velocity of platelet–platelet aggregate formation; see Figure 2C). These data provide further evidence that not only is the AnxA1 mimetic peptide a known anti-inflammatory drug, but it also has potential as an effective anti-platelet drug.

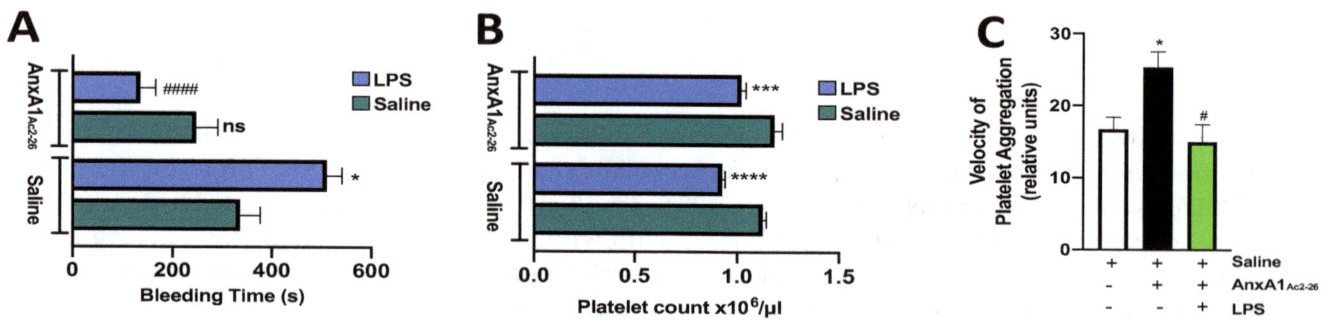

Figure 2. AnxA1$_{Ac2\text{-}26}$ treatment inhibits endotoxin-induced activation of the platelet activation. (**A**) tail bleeding times and (**B**) peripheral blood platelet counts were assessed following saline (vehicle) or AnxA1$_{Ac2\text{-}26}$ (4 mg/kg) administration for 20 min following 2 h saline (vehicle) or LPS (0.4 mg/kg) administration. (**C**) Isolated platelets were treated with saline or LPS (7.5 µg/mL) with or without AnxA1$_{Ac2\text{-}26}$ (1 µg/mL) and the velocity of aggregate formation was measured using a low-angle light-scattering technique. Data are means ± SEM of 5–6 mice/group. * $p < 0.05$, *** $p < 0.001$, **** $p < 0.0001$ vs. corresponding vehicle (saline). # $p < 0.05$, #### $p < 0.0001$ vs. corresponding LPS group. ns—non-significant vs. corresponding saline group.

3.3. *AnxA1$_{Ac2\text{-}26}$ Affords Protection against Cerebral Thrombo-Inflammation*

Having assessed the anti-thrombotic effect of AnxA1$_{Ac2\text{-}26}$ on endotoxin (LPS)-enhanced thrombosis, a model which is known to favour venular thrombosis [39], we next wanted to determine whether the protective effect of the peptide was model-specific. SCD assumes a pro-inflammatory and pro-thrombotic phenotype throughout the microvasculature. As such, we performed the light/dye thrombosis model in STM, which are known to share these clinical features [32,40,41]. Figure 3 shows that without stimulation, STM assume a pro-thrombotic phenotype, as evidenced by much quicker (~50% quicker) blood flow cessation times than non-STM (Figure S2) in both sides of the vascular tree (arterioles: 33.2 ± 1.8 min vs. 15.1 ± 1.2 min ($p < 0.05$); venules: 15.1 ± 1.2 vs. 6.5 ± 0.4 ($p < 0.05$); non-STM vs. STM). Moreover, in comparison to saline-treated STM, administration of AnxA1$_{Ac2\text{-}26}$ significantly protected ($p < 0.01$) against thrombotic events by causing a significant increase in blood flow in arterioles by 47.8% and in venules by 63.5%, respectively, an effect that was again mitigated by the co-administration of either Boc2 or WRW4 (Figure 3B,C). (No differences were observed in blood flow onset times when either of the two antagonists were administered alone; see Figure S4). Interestingly, AnxA1$_{Ac2\text{-}26}$ also delayed the onset time in venules (Figure 3B), an effect that concurred with the effect on cerebral venules following LPS-induced thrombosis (Figure S2).

3.4. *AnxA1$_{Ac2\text{-}26}$ Primes Platelet Activation via GPVI Pathway Regulation in SCD-Associated Thrombo-Inflammation*

Having found that AnxA1$_{Ac2\text{-}26}$ was able to modify the thrombo-inflammatory environment by prolonging blood flow cessation in the cerebral microcirculation in both arterioles and venules, we next sought to investigate the influence of this AnxA1 mimetic peptide on a key platelet receptor which is known to play a central role in thrombosis, i.e., GPVI. This receptor is able to stimulate platelet adhesion by its ability to enhance the affinity of other integrins such as $\alpha_{IIb}\beta_3$ via inside-out signalling mechanisms [42]. Figure 3D,E show that stimulation of platelets with the GPVI collagen receptor agonist CVX potentiated both $\alpha_{IIb}\beta_3$ (Figure 3C) and P-selectin in STM (Figure 3D). Direct treatment of platelets with AnxA1$_{Ac2\text{-}26}$ (30 μM) was found to suppress these levels ($p < 0.05$); although the effects on P-selectin expression were significant, they were less dramatic as those observed with $\alpha_{IIb}\beta_3$. Taken together, these findings suggest that AnxA1$_{Ac2\text{-}26}$ is able to reduce the susceptibility of platelets to interact with collagen receptors, thereby reducing the propensity for platelets to aggregate and cause thrombosis in a thrombo-inflammatory environment.

3.5. *Exploiting the AnxA1/FPR2/ALX Pathway as a Therapeutic Strategy to Alleviate Thrombo-Inflammation*

A pro-inflammatory disease state exists in SCD, which predisposes patients to vaso-occlusion (VOC) in response to triggering factors such as infection [43], thus heightening the risk for stroke [11], acute chest syndrome [44], and early death [45]. As such, in the final part of the study, we validated the therapeutic potential of AnxA1$_{Ac2\text{-}26}$ as a treatment therapy against thrombo-inflammation in STM following LPS injection and photoactivation to induce a VOC (Figure 4A). Markedly, STM mice treated with LPS did not display an additive thrombo-inflammatory response (Figure 4B,C) above that observed in the absence of LPS (Figure 3B,C), suggesting an already exhaustive SCD phenotype. Nonetheless, Figure 4B shows that STM mice treated with AnxA1$_{Ac2\text{-}26}$ caused a prolongation of blood flow cessation responses in arterioles, which was abrogated in the presence of Boc2. This protective effect afforded by the peptide was not mirrored in cerebral venules, despite there being a trend towards an increase (Figure 4C).

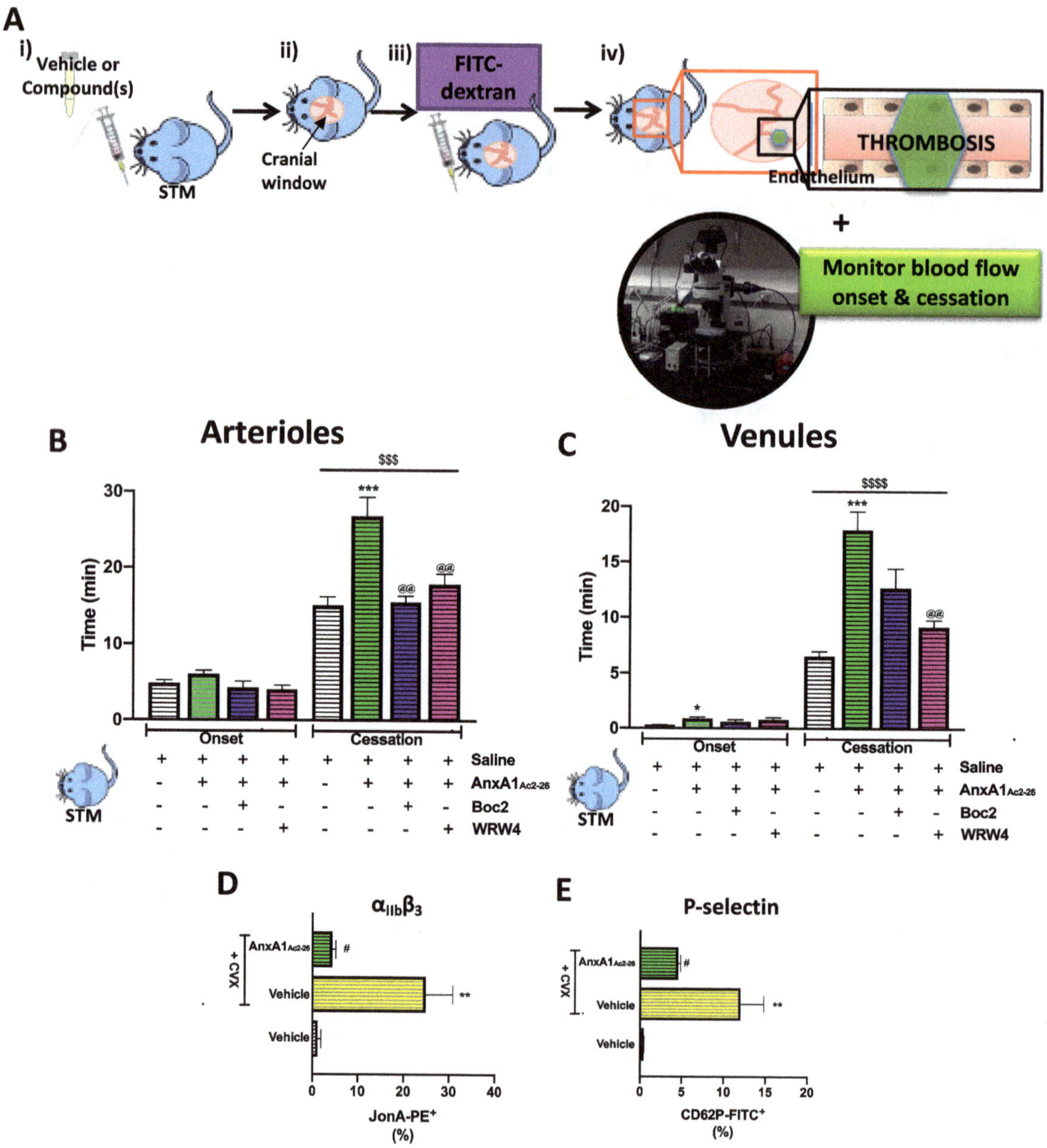

Figure 3. $AnxA1_{Ac2\text{-}26}$ affords protection against cerebral thrombo-inflammation in STM and modifies cell surface expression of platelet receptors. (**A**) Schematic showing the experimental protocol: (i) sickle cell transgenic mice (STM) were treated with vehicle (saline) or compound(s): $AnxA1_{Ac2\text{-}26}$ (4 mg/kg) with/without pan FPR antagonist Boc2 (0.4 mg/kg) or FPR2/ALX antagonist WRW4 (2.2 mg/kg) 20 min (mins) prior to thrombosis. (ii) A cranial window was performed and (iii) FITC-dextran was injected (10 mg/kg of 5%) and allowed to circulate for 10 min before STMs were subjected to (iv) intravital microscopy and light/dye-induced thrombus formation. Onset time and blood flow cessation times were recorded for cerebral. (**B**) arterioles and (**C**) venules. (**D+E**) Activated integrin $\alpha_{IIb}\beta_3$ and P-selectin were quantified using flow cytometry in platelets isolated from STM following 15 min stimulation with GPVI collagen receptor agonist, convulxin (CVX, 1.7 ng/mL). Vehicle (saline) or $AnxA1_{Ac2\text{-}26}$ (30 μM) was given 30 min prior to activation, followed by vehicle (saline) or LPS (7.5 μg/mL for 5 min). (**D**) % of activated integrin $\alpha_{IIb}\beta_3$ (JON-A) and (**E**) % of activated P-selectin expression (CD62P). Data are means ± SEM of 5–6 mice/group. * $p < 0.05$, ** $p < 0.01$, *** $p < 0.001$ vs. corresponding vehicle (saline). @@ $p < 0.01$ vs. corresponding Ac2-26 group. $$$ $p < 0.001$, $$$$ $p < 0.0001$ vs. same group for onset time. # $p < 0.05$ vs. CVX+vehicle group.

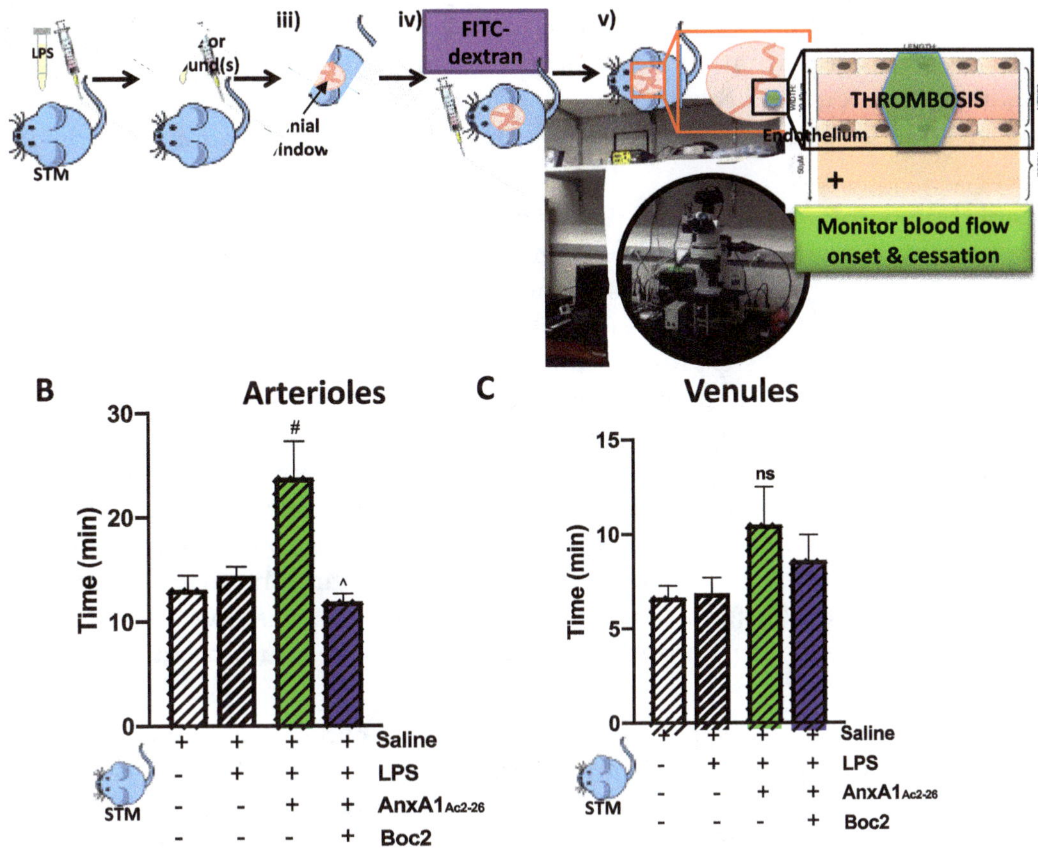

Figure 4. AnxA1$_{Ac2-26}$ affords arteriolar protection against cerebral inflammation-induced thrombosis in sickle cell transgenic mice (STM). (**A**) schematic showing the experimental protocol: (i) STM were subjected to vehicle (saline) or LPS (0.4 mg/kg) for 2 h and (ii) treated with vehicle (saline) or compound(s): AnxA1$_{Ac2-26}$ (4 mg/kg) with/without pan FPR antagonist Boc2 (0.4 mg/kg) 20 min prior to thrombosis. (iii) A cranial window was performed and (iv) FITC-dextran was injected (10 mg/kg of 5%). (v) Mice were subjected to intravital microscopy and light/dye-induced thrombus formation, with onset and blood flow cessation times recorded for cerebral (**B**) arterioles and (**C**) venules. Data are means ± SEM of 5–6 mice/group. $^{\#}$ $p < 0.05$ vs. LPS group. $^{\wedge}$ $p < 0.05$, vs. LPS+AnxA1$_{Ac2-26}$ group. ns—non-significant vs. corresponding LPS group.

4. Discussion

The interplay between thrombosis and inflammation (thrombo-inflammation) occurs in a broad range of human disorders (including stroke, sepsis, and SCD), with ensuing complications transpiring to be more hazardous in the microvasculature of injured tissues and organs [46]. Given the devastating impact of pathological thrombo-inflammation, there is an unmet clinical need to understand the complex pathophysiology for therapeutic development of drugs that are more efficacious, have fewer side-effects, and are devoid of bleeding complications that would ultimately undermine the clinical benefit. Here, using pharmacological and genetic approaches together with a photoactivation thrombosis (light/dye) model coupled with intravital microscopy, we provide further evidence of the multifaceted role of AnxA1 N-terminal mimetic peptide AnxA1$_{Ac2-26}$ as an anti-thrombotic and anti-coagulant agent. Our data demonstrate that AnxA1$_{Ac2-26}$ affected not only the haemostatic action of platelets (e.g., reduced bleeding times), but moderated platelet aggregation (e.g., diminished the propensity for platelets to form platelet–platelet aggregates) and lowered thrombogenesis (i.e., reduced the time of onset of platelet deposition/aggregation within cerebral vessels), thereby reducing LPS-induced thrombosis. Furthermore, the AnxA1 peptide was able to decrease $\alpha_{IIb}\beta_3$ activation and reduce P-selectin expression elicited by GPVI, inhibiting platelet activation and thrombosis in

STM. Taken together, our data reveal that $AnxA1_{Ac2\text{-}26}$ affords protection by altering the haemostatic action of platelets, modulating platelet cell surface molecules (e.g., $\alpha_{IIb}\beta_3$ and P-selectin), and reducing platelet heterotypic aggregation. In so doing, $AnxA1_{Ac2\text{-}26}$ reduces platelet activation, adhesion, and aggregation and tempers thrombosis. Collectively, these results highlight the potential for $AnxA1_{Ac2\text{-}26}$ as a viable therapy for the management of thrombo-inflammatory disorders (as shown in Figure 5).

The resolution of inflammation is a tightly orchestrated process that is controlled by endogenous biosynthetic mediators such as AnxA1 and its mimetic peptide ($AnxA1_{Ac2\text{-}26}$). Both these compounds have been shown to act at various points in the inflammation-resolution pathway [47]. The known anti-inflammatory actions of $AnxA1_{Ac2\text{-}26}$ (e.g., inhibiting neutrophil recruitment, decreasing neutrophil endothelium interactions, and suppressing inflammatory cytokine production [47]) have been widely studied and these actions eventually contribute to inflammation resolution by enabling apoptosis [14] and phagocytosis [29]. Previously, it was thought that inflammation and thrombosis were two independent processes. However, the innate and coagulation systems are so intertwined that these two processes are now considered to be in part the same, in that microvascular thrombosis is accompanied by inflammation and vice versa, an association referred to as thrombo-inflammation [17]. However, a greater understanding of the links between inflammation and thrombosis is needed in order to develop new therapeutic opportunities [29]. Our laboratory is actively pursuing therapeutic drug discovery programs focused on the concept of thrombo-inflammation resolution, with the AnxA1/FPR2/ALX pathway being of significance.

It is estimated that >80% of sepsis patients have either clinical or subclinical hypercoagulopathy, increasing the risks for both thrombosis and haemorrhage [48,49]. These increased thrombotic risks can function not only as acute triggers for stroke but are associated with poor long-term prognosis after stroke [48]. Endotoxin (LPS) is a component of Gram-negative bacteria cell walls, producing many clinical manifestations of Gram-negative sepsis [50]. Using an animal model of endotoxaemia, we found $AnxA1_{Ac2\text{-}26}$ treatment counteracted the enhanced microvascular thrombus formation, an effect which was found to be mediated through Fpr2/ALX. These protective effects are in line with the defensive actions of $AnxA1_{Ac2\text{-}26}$ on the inflammatory cascade that have been observed previously in the microcirculation post-LPS challenge. In these studies, $AnxA1_{Ac2\text{-}26}$ was able to abrogate leukocyte adhesion and plasma protein extravasation in the brain [33] and the mesentery. Furthermore, AnxA1 and $AnxA1_{Ac2\text{-}26}$ both cause leukocyte detachment, reducing the inflammatory environment and driving resolution [51]. Within this study, we have found that although $AnxA1_{Ac2\text{-}26}$ treatment prolonged blood flow cessation time within a thrombotic LPS environment, it also lowered thrombogenesis (i.e., reduced the time of onset of platelet deposition/aggregation within cerebral vessels). Previously, we have shown that whole protein AnxA1 treatment reduces platelet adhesion to the cerebral endothelium following cerebral I/RI [17]. However, although thrombosis and inflammation are interlinked processes, the lack of effect on platelet adhesion observed here may relate to the type of model (thrombotic vs. inflammatory) and/or the possibility that, as with the parent compound in cerebral I/RI, $AnxA1_{Ac2\text{-}26}$ is orchestrating a complex change in the platelet phenotype from pro-pathogenic to regulatory (which concurs in cerebral I/RI [29]) and in doing so, enhances blood flow cessation times.

In the clinic, the pro-thrombogenic phenotype observed in sepsis patients is often accompanied by thrombocytopenia (with prolonged time span of thrombocytopenia being correlated with increased mortality in intensive care patients [52]), reactivity of platelets and by an imbalance between procoagulant and anticoagulant mechanisms [49]. Thus, platelet count is a useful marker of adverse disease activity, although it remains unclear as to whether the reduction in the number of platelets is mechanistically linked to heightened platelet activity and increased thrombosis. Here, LPS treatment caused thrombocytopenia which concurred with previous findings, not only by our group [53], but also by others e.g., George et al. showed LPS to induce severe thrombocytopenia and inflammation resulting in spontaneous intra-alveolar haemorrhage [54] and Hillgruber et al. concluded that thrombocytopenia in the skin and lungs can be limited by targeting neutrophil diapedesis through the endothelial barrier [55], demonstrating the importance of the cross-talk between neutrophils and platelets. In our study, although $AnxA1_{Ac2-26}$ did not affect thrombocytopenia induced by LPS (which may relate to the mechanisms that regulate thrombosis and haemostasis), it did reduce bleeding time in an in vivo assay that is widely used to assess the haemostatic action of platelets [56]. These findings are of clinical significance because the efficacy of anti-platelet agents is often limited by bleeding complications. Genetic *AnxA1* deletion is not associated with spontaneous thrombosis or bleeding but leads to exacerbated cerebral inflammatory responses following experimental ischaemic stroke [29]. Furthermore, decreased circulating levels of AnxA1 are present in thrombo-inflammatory conditions including ischaemic stroke, SCD, sepsis, Crohn's disease, and obesity [29]. Taken together, these results demonstrate $AnxA1_{Ac2-26}$ attenuates LPS-induced peripheral bleeding and intravascular thrombosis, affording a therapeutic strategy for thrombotic complications.

Defining molecular mechanisms regulating thrombo-inflammation in specific disease states is of major clinical importance. Having characterised the effects of $AnxA1_{Ac2-26}$ on inflammation-induced thrombosis instigated by LPS, we next wanted to ascertain whether the protective effects elicited by the AnxA1 peptide were specific to the LPS experimental model of thrombo-inflammation. As such, we focused on SCD, an inherited autosomal recessive disorder (resulting from a single amino acid substitution in the haemoglobin β chain) whose pathophysiology is characterized by relentless thrombo-inflammation, enabling heightened propensity for I/RI events such as stroke [7,8].

As observed with mice stimulated with LPS, STM had comparable cerebral responses in both arterioles and venules when exposed to light/dye thrombosis, supporting findings demonstrating that a pro-inflammatory phenotype exists within their cerebral microvasculature (as quantified by enhanced leukocyte–endothelial cell adhesion and increased ROS production [32,57]). $AnxA1_{Ac2-26}$ treatment reduced cerebral thrombosis and the use of Fpr antagonists Boc2 and WRW4 confirmed that these protective effects were caused by receptor engagement of peptide $AnxA1_{Ac2-26}$ to Fpr2/ALX. Furthermore, research from our laboratory (Ansari et al., 2020, under review) has also shown $AnxA1_{Ac2-26}$ to reduce citrullinated histone-3 (H3Cit+)-rich neutrophil extracellular trap (NET) production in SCD. These findings (along with those from other groups [58,59]) are of clinical significance as NETs have been recognized as critical components for venous and arterial thrombosis and inhibition of pathological NET formation may be beneficial for thrombo-inflammatory events and disorders such as SCD [59]. Taken together, these results demonstrate the versatility and innovative approach of $AnxA1_{Ac2-26}$ as a therapeutic compound for the resolution of thrombo-inflammation.

We uncovered that although the Fpr-pan antagonist Boc2 abrogated the effects of the peptide in arterioles, no statistically significant effect was observed in venules irrespective of the thrombo-inflammatory environment, although there was a trend towards abrogation. The same could not be said for the Fpr2/ALX-specific antagonist WRW4, which was effective in annulling the peptide's responses irrespective of vessel type or thrombo-inflammatory model. This dichotomy of behaviour of the peptide is less likely to be due to antagonist doses (as these were chosen based on dose–response curves) but could be due the fact that FPRs have a large number of diverse unrelated ligands that are able to bind to the receptor family to elicit pro- and anti-inflammatory effects that are specific to ligand and cell type. More recently, the concept of biased agonism has been coined to describe "the ability of a ligand to selectively activate subsets of downstream signalling pathways coupled to a receptor while inhibiting others" [60], which could help to explain differences observed between $AnxA1_{Ac2-26}$ and co-administration with either the pan-antagonist Boc2 or the Fpr2-specific antagonist WRW4. Other factors that could be accountable for the variations in peptide and antagonist responses could be the known physiological differences between arterioles and venules e.g., shear rates (being predominantly higher in arterioles vs. venules) and differences in the thrombi formed in these respective vessels e.g., arteriolar thrombi being platelet-rich, but venular thrombi also containing leukocytes (typically near the surface of the microvascular thrombi) [61]. Despite these disparities, our data demonstrate the potent activity and versatility of $AnxA1_{Ac2-26}$ to mitigate both LPS- and

SCD-associated cerebral thrombosis (in both arterioles and venules) in an Fpr2/ALX-dependent manner. Correspondingly, whole protein AnxA1 has also been shown to afford protection against subsequent thrombotic events post stroke, demonstrating the diversity of AnxA1 and its mimetic peptide [29].

Platelets play a pivotal role in normal haemostasis and thrombus formation, and more recently have been found to also play a role in maintaining barrier function [6]. Once activated, platelets undergo a shape change to enhance adhesiveness, a process mediated by exposed glycoprotein receptors on the platelet surface [6]. GPVI is a unique platelet membrane glycoprotein whose binding with collagen (which is exposed on the extracellular matrix following stroke [6]) results in platelet activation and adhesion, and ultimately, thrombus formation. Platelets primarily rely on signalling through GPVI and C-type lectin-like type II transmembrane receptor (CLEC-2) to prevent bleeding [62]. GPVI, via inside-out signalling, enhances the affinity of integrins such as αIIbβ3 (the most abundant platelet receptor (80,000–100,000 copies per platelet) and necessary for aggregation), leading to platelet adhesion [53]. As such, given the important role that GPVI plays in thrombosis, we used flow cytometry to investigate whether the anti-thrombotic effects of AnxA1$_{Ac2\text{-}26}$ involved GPVI signalling. By directly activating the GPVI pathway using CVX, we were able to discern that the AnxA1 peptide was able to decrease $\alpha_{IIb}\beta_3$ activation and the surface expression of P-selectin, which in turn inhibits platelet activation. Interestingly, we have previously found that the parent protein does not reduce P-selectin expression in thrombin-stimulated platelets [29]. One might speculate that these discrepancies may lie in the different platelet stimuli and/or the downstream responses elicited by the binding of either the whole protein or the peptide to Fpr2/ALX and subsequent conformational changes of the receptor [26]. Equally, these effects may simply relate to the varied regulatory functions afforded by the peptide during the host defence response. Further experiments will help to tease out these mechanisms.

Other regulatory mechanisms (such as those involving cyclooxygenase 2 ('COX2')/ hydroxyeicosatetraenoic acid-5 (HETE-5)/Lipooxygenase) may also be involved in the role that the AnxA1/Fpr2/ALX pathway plays in thrombo-inflammation. AnxA1 (and its mimetic peptide AnxA1$_{Ac2\text{-}26}$), Lipoxin A_4, and ATL (produced after aspirin acetylation of inducible COX-2) all exert their protective effects through FPR2/ALX at different phases of the inflammatory response. Phospholipase (e.g., calcium-dependent cytosolic phospholipase A_2 ("cPLA_2")) activity releases arachidonic acid ("AA") from phospholipids in the outer nuclear membrane. Once released, the free fatty acid can be metabolized *via* enzymatic pathways including the (COX) and lipoxygenase (LOX) pathways, generating 2-series prostaglandins (PGs) and thromboxanes (Txs) (COX pathway) or 4-series leukotrienes (LTs) and hydroxyeicosatetraenoic acids (HETEs) (LOX pathway) [63]. Transcellular production of lipoxins and leukotrienes ensues when contact is made between e.g., neutrophils and platelets. Previous research from our laboratory has shown that when aspirin is administered to mice with experimental stroke, they can produce ATL, which triggers pro-resolving responses through the engagement of the AnxA1/Fpr2/ALX pathway [17]. Furthermore, ASA administered to healthy volunteers is also capable of producing bioactive levels of ATL [64]. More recently, Sanches et al. showed that macrophages lacking AnxA1 have increased AA metabolism and eicosanoid production. This lack of AnxA1 favours LPS "over-priming" and exacerbated NLR Family Pyrin Domain Containing 3 ("NLRP3") activation, demonstrating an important role for AnxA1 in inflammasome activation [63]. Other studies have also shown AnxA1 to exert its effects via the FPR2/ALX/p38 mitogen-activated protein kinase ("MAPK")/COX-2 pathway in an experimental model of intracerebral haemorrhage, although other MAPKs may also be involved [65].

Patients with thrombo-inflammation (including stroke, sepsis, and SCD) all present with abnormal circulating platelet–leukocyte aggregates (with platelets binding to neutrophils in a P-selectin-dependent mechanism, an interaction that primes the leukocyte and promotes integrin activation), increasing their propensity to develop disseminated intravascular coagulation [49]. Studies have indicated that ischaemic stroke is not simply mediated by platelet aggregation, but also by other intravascular cells including neutrophils, although it remains unclear whether the direct interaction between platelets and neutrophils is critical for the pathogenesis of ischaemic stroke [59].

Von Brühl et al. demonstrated leukocytes (predominantly neutrophils) crawl and adhere to the endothelium, initiating and propagating venous thrombosis [66]. Additionally, neutrophils have now been shown to be involved in arterial thrombosis as well [67], although platelet activation and aggregation is the main driving factor. Furthermore, several distinct mechanisms have been postulated for neutrophil involvement in thrombosis including: the transfer of tissue factor from neutrophils to platelets, inducing thrombosis [68]; the release of specific mediators that affect thrombosis, including cathepsin G and neutrophil elastase (which inactivate anticoagulant systems such as tissue factor pathway inhibitor, thrombomodulin, and antithrombin), neutrophil oxidants (which e.g., inactivate thrombomodulin and ADAMTS13), and micro-RNAs [59]; and the generation of NETs which can trap and activate platelets via histone production [69]. Of interest, more recently, COVID-19 patients have been shown to have increased serum markers of NETs including myeloperoxidase-DNA (MPO-DNA) and H3Cit. Given our recent findings (Ansari et al., 2020, under review) that $AnxA1_{Ac2\text{-}26}$ is able to reduce $H3Cit^{+}$-rich NET production, transforming neutrophil phenotype from pro-NETotic to pro-apoptotic (thereby driving thrombo-inflammation resolution in SCD), we are currently investigating whether these findings can be exploited to provide therapeutic value for COVID-19 patients [70].

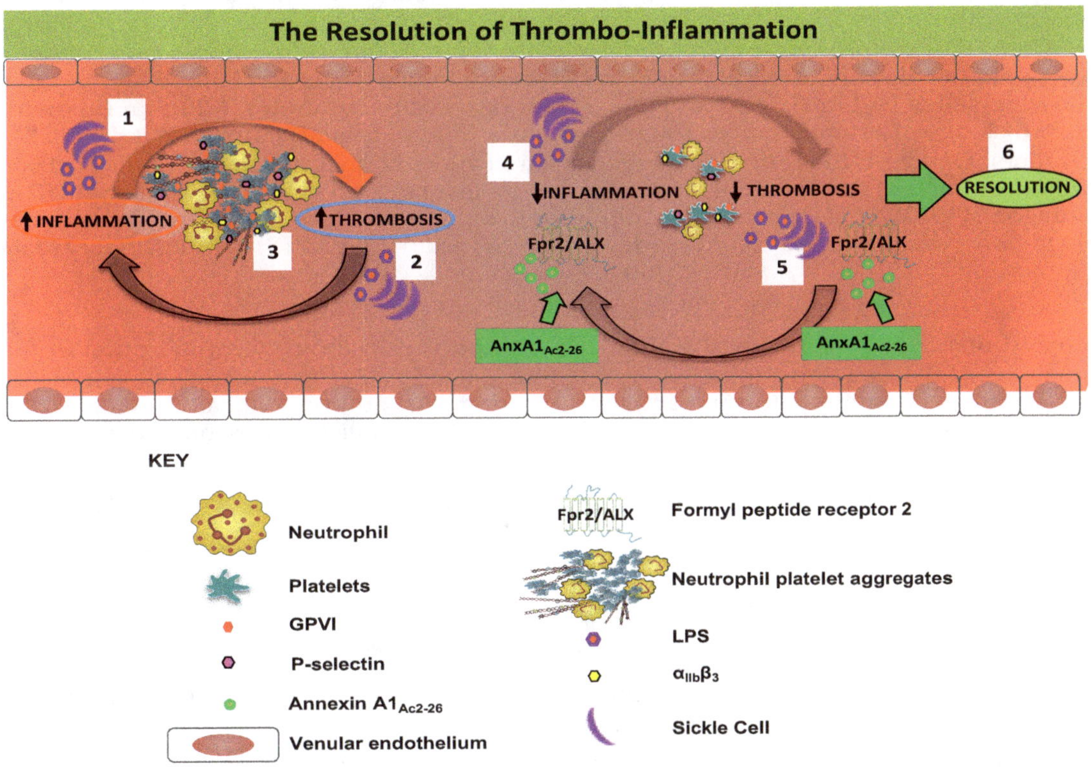

Figure 5. Anti-inflammatory and pro-resolving effects of targeting the AnxA1/Fpr2/ALX pathway against thrombo-inflammation. Thrombo-inflammatory conditions such as sepsis [35] and sickle cell disease (SCD) [71] induce inflammation (Box 1) as well as thrombus formation (Box 2) in the cerebral microcirculation enabling thrombo-inflammation (as indicated by e.g., heightened platelet GPVI, $\alpha_{IIb}\beta_3$ and P-selectin expression, and increased platelet–platelet (homotypic) aggregates and platelet-neutrophil (heterotypic) aggregates). Neutrophils also produce neutrophil extracellular traps (NETs) which can exacerbate thrombo-inflammation (Box 3). $AnxA1_{Ac2\text{-}26}$ inhibits lipopolysaccharide (LPS) and SCD-induced cerebral inflammation (Box 4) and cerebral thrombosis (Box 5) via Fpr2/ALX. The AnxA1 peptide affords protection by altering the haemostatic action of platelets, modulating platelet cell surface molecules elicited by GPVI (e.g., $\alpha_{IIb}\beta_3$ and P-selectin), and reducing platelet (homotypic and heterotypic) aggregation. Thus, $AnxA1_{Ac2\text{-}26}$ promotes the resolution of thrombo-inflammation by reducing platelet activation, adhesion, and aggregation which cause and promote thrombosis (Box 6).

5. Conclusions

In conclusion, we provide strong supporting evidence that AnxA1 mimetic peptide AnxA1$_{Ac2-26}$ possesses an arsenal of immune responses extending beyond that of an anti-inflammatory (e.g., attenuation of leukocyte-platelet responses post stroke, reduction of lipopolysaccharide-induced leukocyte adhesion and migration) and pro-resolution mediator to also include a role as an anti-coagulant and anti-thrombotic agent. These combined effects make AnxA1$_{Ac2-26}$ a promising therapeutic candidate for promoting resolution in the context of thrombo-inflammatory diseases/conditions.

Supplementary Materials
Figure S1: Effect of FPR antagonists. Mice (C57BL/6) were subjected to vehicle LPS (10 µg/mouse) for 2 h and treated with vehicle (saline), pan FPR antagonist Boc2 (10 µg/mouse) or FPR2/ALX antagonist WRW4 (55 µg/mouse) 20 min prior to light/dye-induced thrombus formation, with time of onset and blood flow cessation times recorded for cerebral (A) arterioles and (B) venules. Data are means ± SEM of 5–6 mice/group. $^{\$\$\$\$}$ $p < 0.0001$ vs. same group for onset time, Figure S2: Onset and blood flow cessation times in mice with/without endotoxaemia. C57BL/6 mice or sickle cell transgenic mice (STM) were subjected to vehicle (saline) or LPS (0.4 mg/kg) for 2 h. A cranial window was performed, and FITC-dextran injected (10 mg/kg of 5%).Mice were then subjected to intravital microscopy and light/dye-induced thrombus formation, with time of onset and blood flow cessation times recorded for cerebral (A) arterioles and (B) venules. Data are means ± SEM of 5–6 mice/group. * $p < 0.05$, ** $p < 0.01$ vs. C57BL/6 saline group. ## $p < 0.01$ vs. STM saline group. $^{\$}$ $p < 0.05$, $^{\$\$}$ $p < 0.0001$ vs. same group for onset time, Figure S3: AnxA1$_{Ac2-26}$ did not affect neutrophil counts in LPS-treated mice. Peripheral blood neutrophil counts were assessed following saline (vehicle) or AnxA1$_{Ac2-26}$ (4 mg/kg) administration for 20 min following 2 h saline (vehicle) or LPS (0.4 mg/kg) administration. Data are means ± SEM of 5–6 mice/group. ** $p < 0.01$, *** $p < 0.001$ vs. LPS vehicle (saline), Figure S4: Effect of FPR antagonists in sickle cell transgenic mice (STM). STM were subjected to vehicle LPS (10 µg/mouse) for 2 h and treated with vehicle (saline), pan FPR antagonist Boc2 (10 µg/mouse) or FPR2/ALX antagonist WRW4 (55 µg/mouse) 20 min prior to light/dye-induced thrombus formation, with time of onset and blood flow cessation times recorded for cerebral (A) arterioles and (B) venules. Data are means ± SEM of 5–6 mice/group. $^{\$\$\$\$}$ $p < 0.0001$ vs. same group for onset time.

Author Contributions: S.A.V., E.Y.S., and F.N.E.G. were involved in the conceptualisation of the study and collected and analysed the data. All authors interpreted the data. S.A.V., J.A. and F.N.E.G. wrote the manuscript. F.N.E.G. was the project administrator and acquired funding. All authors critically reviewed the paper and approved its final version. All authors have read and agreed to the published version of the manuscript.

Acknowledgments: The authors would like to thank Professor Robert Hebbel (University of Minnesota) for the STM.

References

1. Virani, S.S.; Alonso, A.; Benjamin, E.J.; Bittencourt, M.S.; Callaway, C.W.; Carson, A.P.; Chamberlain, A.M.; Chang, A.R.; Cheng, S.; Delling, F.N.; et al. Heart Disease and Stroke Statistics-2020 Update: A Report From the American Heart Association. *Circulation* **2020**, *141*, e139–e596. [CrossRef]
2. Luengo-Fernandez, R.; Gray, A.M.; Bull, L.; Welch, S.; Cuthbertson, F.; Rothwell, P.M. Quality of life after TIA and stroke: Ten-year results of the Oxford Vascular Study. *Neurology* **2013**, *81*, 1588–1595. [CrossRef]
3. Nieswandt, B.; Kleinschnitz, C.; Stoll, G. Ischaemic stroke: A thrombo-inflammatory disease? *J. Physiol.* **2011**, *589*, 4115–4123. [CrossRef]
4. Ruhnau, J.; Schulze, J.; Dressel, A.; Vogelgesang, A. Thrombosis, Neuroinflammation, and Poststroke Infection: The Multifaceted Role of Neutrophils in Stroke. *J. Immunol. Res.* **2017**, *2017*, 7. [CrossRef]
5. Jin, R.; Yang, G.; Li, G. Inflammatory mechanisms in ischemic stroke: Role of inflammatory cells. *J. Leukoc. Biol.* **2010**, *87*, 779–789. [CrossRef]
6. Gupta, S.; Konradt, C.; Corken, A.; Ware, J.; Nieswandt, B.; Di Paola, J.; Yu, M.; Wang, D.; Nieman, M.T.; Whiteheart, S.W.; et al. Hemostasis vs. homeostasis: Platelets are essential for preserving vascular barrier function in the absence of injury or inflammation. *Proc. Natl. Acad. Sci. USA* **2020**, *117*, 24316–24325. [CrossRef]

7. Hebbel, R.P.; Belcher, J.D.; Vercellotti, G.M. The multifaceted role of ischemia/reperfusion in sickle cell anemia. *J. Clin. Investig.* **2020**, *130*, 1062–1072. [CrossRef] [PubMed]
8. Ansari, J.; Gavins, F.N.E. Ischemia-Reperfusion Injury in Sickle Cell Disease: From Basics to Therapeutics. *Am. J. Pathol.* **2019**, *189*, 706–718. [CrossRef] [PubMed]
9. Miller, E.C.; Elkind, M.S. Infection and Stroke: An Update on Recent Progress. *Curr. Neurol. Neurosci. Rep.* **2016**, *16*, 2. [CrossRef] [PubMed]
10. Miller, C.M.; Behrouz, R. Impact of Infection on Stroke Morbidity and Outcomes. *Curr. Neurol. Neurosci. Rep.* **2016**, *16*, 83. [CrossRef]
11. Ohene-Frempong, K.; Weiner, S.J.; Sleeper, L.A.; Miller, S.T.; Embury, S.; Moohr, J.W.; Wethers, D.L.; Pegelow, C.H.; Gill, F.M. Cerebrovascular accidents in sickle cell disease: Rates and risk factors. *Blood* **1998**, *91*, 288–294.
12. Serhan, C.N. Pro-resolving lipid mediators are leads for resolution physiology. *Nature* **2014**, *510*, 92–101. [CrossRef]
13. Serhan, C.N.; Levy, B.D. Resolvins in inflammation: Emergence of the pro-resolving superfamily of mediators. *J. Clin. Investig.* **2018**, *128*, 2657–2669. [CrossRef]
14. Sugimoto, M.A.; Vago, J.P.; Teixeira, M.M.; Sousa, L.P. Annexin A1 and the Resolution of Inflammation: Modulation of Neutrophil Recruitment, Apoptosis, and Clearance. *J. Immunol. Res.* **2016**, *2016*, 13. [CrossRef]
15. Perretti, M.; Dalli, J. Exploiting the Annexin A1 pathway for the development of novel anti-inflammatory therapeutics. *Br. J. Pharmacol.* **2009**, *158*, 936–946. [CrossRef]
16. Serhan, C.N.; Dalli, J.; Colas, R.A.; Winkler, J.W.; Chiang, N. Protectins and maresins: New pro-resolving families of mediators in acute inflammation and resolution bioactive metabolome. *Biochim. Biophys. Acta* **2015**, *1851*, 397–413. [CrossRef]
17. Vital, S.A.; Becker, F.; Holloway, P.M.; Russell, J.; Perretti, M.; Granger, D.N.; Gavins, F.N. Formyl-Peptide Receptor 2/3/Lipoxin A4 Receptor Regulates Neutrophil-Platelet Aggregation and Attenuates Cerebral Inflammation: Impact for Therapy in Cardiovascular Disease. *Circulation* **2016**, *133*, 2169–2179. [CrossRef]
18. Yazid, S.; Norling, L.V.; Flower, R.J. Anti-inflammatory drugs, eicosanoids and the annexin A1/FPR2 anti-inflammatory system. *Prostaglandins Other Lipid Mediat.* **2012**, *98*, 94–100. [CrossRef]
19. Ye, R.D.; Boulay, F.; Wang, J.M.; Dahlgren, C.; Gerard, C.; Parmentier, M.; Serhan, C.N.; Murphy, P.M. International Union of Basic and Clinical Pharmacology. LXXIII. Nomenclature for the formyl peptide receptor (FPR) family. *Pharmacol. Rev.* **2009**, *61*, 119–161. [CrossRef]
20. Smith, H.K.; Gil, C.D.; Oliani, S.M.; Gavins, F.N. Targeting formyl peptide receptor 2 reduces leukocyte-endothelial interactions in a murine model of stroke. *FASEB J.* **2015**, *29*, 2161–2171. [CrossRef]
21. Zhuang, Y.; Liu, H.; Edward Zhou, X.; Kumar Verma, R.; de Waal, P.W.; Jang, W.; Xu, T.H.; Wang, L.; Meng, X.; Zhao, G.; et al. Structure of formylpeptide receptor 2-Gi complex reveals insights into ligand recognition and signaling. *Nat. Commun.* **2020**, *11*, 885. [CrossRef]
22. Perretti, M.; Godson, C. Formyl peptide receptor type 2 agonists to kick-start resolution pharmacology. *Br. J. Pharmacol.* **2020**, *177*, 4595–4600. [CrossRef]
23. Dorward, D.A.; Lucas, C.D.; Chapman, G.B.; Haslett, C.; Dhaliwal, K.; Rossi, A.G. The role of formylated peptides and formyl peptide receptor 1 in governing neutrophil function during acute inflammation. *Am. J. Pathol.* **2015**, *185*, 1172–1184. [CrossRef]
24. Nonaka, M.; Suzuki-Anekoji, M.; Nakayama, J.; Mabashi-Asazuma, H.; Jarvis, D.L.; Yeh, J.C.; Yamasaki, K.; Akama, T.O.; Huang, C.T.; Campos, A.R.; et al. Overcoming the blood-brain barrier by Annexin A1-binding peptide to target brain tumours. *Br. J. Cancer* **2020**, 1–11. [CrossRef]
25. Yap, G.L.R.; Sachaphibulkij, K.; Foo, S.L.; Cui, J.; Fairhurst, A.M.; Lim, L.H.K. Annexin-A1 promotes RIG-I-dependent signaling and apoptosis via regulation of the IRF3-IFNAR-STAT1-IFIT1 pathway in A549 lung epithelial cells. *Cell Death Dis.* **2020**, *11*, 463. [CrossRef]
26. Bena, S.; Brancaleone, V.; Wang, J.M.; Perretti, M.; Flower, R.J. Annexin A1 interaction with the FPR2/ALX receptor: Identification of distinct domains and downstream associated signaling. *J. Biol. Chem.* **2012**, *287*, 24690–24697. [CrossRef]
27. Yu, C.; Chen, H.; Qi, X.; Chen, P.; Di, G. Annexin A1 mimetic peptide Ac2-26 attenuates mechanical injury induced corneal scarring and inflammation. *Biochem. Biophys. Res. Commun.* **2019**, *519*, 396–401. [CrossRef]

28. Hughes, E.L.; Becker, F.; Flower, R.J.; Buckingham, J.C.; Gavins, F.N.E. Mast cells mediate early neutrophil recruitment and exhibit anti-inflammatory properties via the formyl peptide receptor 2/lipoxin A4 receptor. *Br. J. Pharmacol.* **2017**, *174*, 2393–2408. [CrossRef]
29. Senchenkova, E.Y.; Ansari, J.; Becker, F.; Vital, S.A.; Al-Yafeai, Z.; Sparkenbaugh, E.M.; Pawlinski, R.; Stokes, K.Y.; Carroll, J.L.; Dragoi, A.M.; et al. Novel Role for the AnxA1-Fpr2/ALX Signaling Axis as a Key Regulator of Platelet Function to Promote Resolution of Inflammation. *Circulation* **2019**, *140*, 319–335. [CrossRef]
30. Hankey, G.J.; Eikelboom, J.W. Antithrombotic drugs for patients with ischaemic stroke and transient ischaemic attack to prevent recurrent major vascular events. *Lancet Neurol.* **2010**, *9*, 273–284. [CrossRef]
31. Bergmeier, W.; Schulte, V.; Brockhoff, G.; Bier, U.; Zirngibl, H.; Nieswandt, B. Flow cytometric detection of activated mouse integrin alphaIIbbeta3 with a novel monoclonal antibody. *Cytometry* **2002**, *48*, 80–86. [CrossRef]
32. Gavins, F.N.; Russell, J.; Senchenkova, E.L.; De Almeida Paula, L.; Damazo, A.S.; Esmon, C.T.; Kirchhofer, D.; Hebbel, R.P.; Granger, D.N. Mechanisms of enhanced thrombus formation in cerebral microvessels of mice expressing hemoglobin-S. *Blood* **2011**, *117*, 4125–4133. [CrossRef]
33. Gavins, F.N.; Hughes, E.L.; Buss, N.A.; Holloway, P.M.; Getting, S.J.; Buckingham, J.C. Leukocyte recruitment in the brain in sepsis: Involvement of the annexin 1-FPR2/ALX anti-inflammatory system. *FASEB J.* **2012**, *26*, 4977–4989. [CrossRef]
34. Gavins, F.N.; Li, G.; Russell, J.; Perretti, M.; Granger, D.N. Microvascular thrombosis and CD40/CD40L signaling. *J. Thromb. Haemost.* **2011**, *9*, 574–581. [CrossRef]
35. Yan, S.L.; Russell, J.; Harris, N.R.; Senchenkova, E.Y.; Yildirim, A.; Granger, D.N. Platelet abnormalities during colonic inflammation. *Inflamm. Bowel Dis.* **2013**, *19*, 1245–1253. [CrossRef]
36. Senchenkova, E.Y.; Russell, J.; Yildirim, A.; Granger, D.N.; Gavins, F.N.E. Novel Role of T Cells and IL-6 (Interleukin-6) in Angiotensin II-Induced Microvascular Dysfunction. *Hypertension* **2019**, *73*, 829–838. [CrossRef]
37. Ma, A.C.; Kubes, P. Platelets, neutrophils, and neutrophil extracellular traps (NETs) in sepsis. *J. Thromb. Haemost.* **2008**, *6*, 415–420. [CrossRef]
38. Schattner, M. Platelet TLR4 at the crossroads of thrombosis and the innate immune response. *J. Leukoc. Biol.* **2019**, *105*, 873–880. [CrossRef]
39. Broeders, M.A.; Tangelder, G.J.; Slaaf, D.W.; Reneman, R.S.; oude Egbrink, M.G. Hypercholesterolemia enhances thromboembolism in arterioles but not venules: Complete reversal by L-arginine. *Arterioscler. Thromb. Vasc. Biol.* **2002**, *22*, 680–685. [CrossRef]
40. Chantrathammachart, P.; Mackman, N.; Sparkenbaugh, E.; Wang, J.G.; Parise, L.V.; Kirchhofer, D.; Key, N.S.; Pawlinski, R. Tissue factor promotes activation of coagulation and inflammation in a mouse model of sickle cell disease. *Blood* **2012**, *120*, 636–646. [CrossRef]
41. Solovey, A.A.; Solovey, A.N.; Harkness, J.; Hebbel, R.P. Modulation of endothelial cell activation in sickle cell disease: A pilot study. *Blood* **2001**, *97*, 1937–1941. [CrossRef]
42. Huang, J.; Li, X.; Shi, X.; Zhu, M.; Wang, J.; Huang, S.; Huang, X.; Wang, H.; Li, L.; Deng, H.; et al. Platelet integrin alphaIIbbeta3: Signal transduction, regulation, and its therapeutic targeting. *J. Hematol. Oncol.* **2019**, *12*, 26. [CrossRef]
43. Vichinsky, E.P.; Neumayr, L.D.; Earles, A.N.; Williams, R.; Lennette, E.T.; Dean, D.; Nickerson, B.; Orringer, E.; McKie, V.; Bellevue, R.; et al. Causes and outcomes of the acute chest syndrome in sickle cell disease. National Acute Chest Syndrome Study Group. *N. Engl. J. Med.* **2000**, *342*, 1855–1865. [CrossRef]
44. Castro, O.; Brambilla, D.J.; Thorington, B.; Reindorf, C.A.; Scott, R.B.; Gillette, P.; Vera, J.C.; Levy, P.S. The acute chest syndrome in sickle cell disease: Incidence and risk factors. The Cooperative Study of Sickle Cell Disease. *Blood* **1994**, *84*, 643–649. [CrossRef]
45. Platt, O.S.; Brambilla, D.J.; Rosse, W.F.; Milner, P.F.; Castro, O.; Steinberg, M.H.; Klug, P.P. Mortality in sickle cell disease. Life expectancy and risk factors for early death. *N. Engl. J. Med.* **1994**, *330*, 1639–1644. [CrossRef]
46. Jackson, S.P.; Darbousset, R.; Schoenwaelder, S.M. Thromboinflammation: Challenges of therapeutically targeting coagulation and other host defense mechanisms. *Blood* **2019**, *133*, 906–918. [CrossRef]
47. Ansari, J.; Kaur, G.; Gavins, F.N.E. Therapeutic Potential of Annexin A1 in Ischemia Reperfusion Injury. *Int. J. Mol. Sci.* **2018**, *19*, 1211. [CrossRef]

48. Shao, I.Y.; Elkind, M.S.V.; Boehme, A.K. Risk Factors for Stroke in Patients With Sepsis and Bloodstream Infections. *Stroke* **2019**, *50*, 1046–1051. [CrossRef]
49. Ishikura, H.; Nishida, T.; Murai, A.; Nakamura, Y.; Irie, Y.; Tanaka, J.; Umemura, T. New diagnostic strategy for sepsis-induced disseminated intravascular coagulation: A prospective single-center observational study. *Crit. Care* **2014**, *18*, R19. [CrossRef]
50. Branchford, B.R.; Carpenter, S.L. The Role of Inflammation in Venous Thromboembolism. *Front. Pediatr.* **2018**, *6*, 142. [CrossRef]
51. Gavins, F.N.; Yona, S.; Kamal, A.M.; Flower, R.J.; Perretti, M. Leukocyte antiadhesive actions of annexin 1: ALXR- and FPR-related anti-inflammatory mechanisms. *Blood* **2003**, *101*, 4140–4147. [CrossRef]
52. Akca, S.; Haji-Michael, P.; de Mendonça, A.; Suter, P.; Levi, M.; Vincent, J.L. Time course of platelet counts in critically ill patients. *Crit. Care Med.* **2002**, *30*, 753–756. [CrossRef]
53. Gillespie, S.; Holloway, P.M.; Becker, F.; Rauzi, F.; Vital, S.A.; Taylor, K.A.; Stokes, K.Y.; Emerson, M.; Gavins, F.N.E. The isothiocyanate sulforaphane modulates platelet function and protects against cerebral thrombotic dysfunction. *Br. J. Pharmacol.* **2018**, *175*, 3333–3346. [CrossRef]
54. Goerge, T.; Ho-Tin-Noe, B.; Carbo, C.; Benarafa, C.; Remold-O'Donnell, E.; Zhao, B.Q.; Cifuni, S.M.; Wagner, D.D. Inflammation induces hemorrhage in thrombocytopenia. *Blood* **2008**, *111*, 4958–4964. [CrossRef]
55. Hillgruber, C.; Pöppelmann, B.; Weishaupt, C.; Steingräber, A.K.; Wessel, F.; Berdel, W.E.; Gessner, J.E.; Ho-Tin-Noé, B.; Vestweber, D.; Goerge, T. Blocking neutrophil diapedesis prevents hemorrhage during thrombocytopenia. *J. Exp. Med.* **2015**, *212*, 1255–1266. [CrossRef]
56. Zhang, G.; Han, J.; Welch, E.J.; Ye, R.D.; Voyno-Yasenetskaya, T.A.; Malik, A.B.; Du, X.; Li, Z. Lipopolysaccharide stimulates platelet secretion and potentiates platelet aggregation via TLR4/MyD88 and the cGMP-dependent protein kinase pathway. *J. Immunol.* **2009**, *182*, 7997–8004. [CrossRef]
57. Wood, K.C.; Granger, D.N. Sickle cell disease: Role of reactive oxygen and nitrogen metabolites. *Clin. Exp. Pharmacol. Physiol.* **2007**, *34*, 926–932. [CrossRef]
58. Chen, G.; Zhang, D.; Fuchs, T.A.; Manwani, D.; Wagner, D.D.; Frenette, P.S. Heme-induced neutrophil extracellular traps contribute to the pathogenesis of sickle cell disease. *Blood* **2014**, *123*, 3818–3827. [CrossRef]
59. Li, J.; Kim, K.; Barazia, A.; Tseng, A.; Cho, J. Platelet-neutrophil interactions under thromboinflammatory conditions. *Cell. Mol. Life Sci.* **2015**, *72*, 2627–2643. [CrossRef]
60. Raabe, C.A.; Groper, J.; Rescher, U. Biased perspectives on formyl peptide receptors. *Biochim. Biophys. Acta Mol. Cell Res.* **2019**, *1866*, 305–316. [CrossRef]
61. Rumbaut, R.E.; Slaff, D.W.; Burns, A.R. Microvascular thrombosis models in venules and arterioles in vivo. *Microcirculation* **2005**, *12*, 259–274. [CrossRef] [PubMed]
62. Boulaftali, Y.; Hess, P.R.; Getz, T.M.; Cholka, A.; Stolla, M.; Mackman, N.; Owens, A.P., 3rd; Ware, J.; Kahn, M.L. Platelet ITAM signaling is critical for vascular integrity in inflammation. *J. Clin. Investig.* **2013**, *123*, 908–916. [CrossRef] [PubMed]
63. Sanches, J.M.; Branco, L.M.; Duarte, G.H.B.; Oliani, S.M.; Bortoluci, K.R.; Moreira, V.; Gil, C.D. Annexin A1 Regulates NLRP3 Inflammasome Activation and Modifies Lipid Release Profile in Isolated Peritoneal Macrophages. *Cells* **2020**, *9*, 926. [CrossRef] [PubMed]
64. Brancaleone, V.; Gobbetti, T.; Cenac, N.; le Faouder, P.; Colom, B.; Flower, R.J.; Vergnolle, N.; Nourshargh, S.; Perretti, M. A vasculo-protective circuit centered on lipoxin A4 and aspirin-triggered 15-epi-lipoxin A4 operative in murine microcirculation. *Blood* **2013**, *122*, 608–617. [CrossRef] [PubMed]
65. Ding, Y.; Flores, J.; Klebe, D.; Li, P.; McBride, D.W.; Tang, J.; Zhang, J.H. Annexin A1 attenuates neuroinflammation through FPR2/p38/COX-2 pathway after intracerebral hemorrhage in male mice. *J. Neurosci. Res.* **2020**, *98*, 168–178. [CrossRef]
66. von Brühl, M.L.; Stark, K.; Steinhart, A.; Chandraratne, S.; Konrad, I.; Lorenz, M.; Khandoga, A.; Tirniceriu, A.; Coletti, R.; Köllnberger, M.; et al. Monocytes, neutrophils, and platelets cooperate to initiate and propagate venous thrombosis in mice in vivo. *J. Exp. Med.* **2012**, *209*, 819–835. [CrossRef]
67. Darbousset, R.; Thomas, G.M.; Mezouar, S.; Frère, C.; Bonier, R.; Mackman, N.; Renné, T.; Dignat-George, F.; Dubois, C.; Panicot-Dubois, L. Tissue factor-positive neutrophils bind to injured endothelial wall and initiate thrombus formation. *Blood* **2012**, *120*, 2133–2143. [CrossRef]
68. Giesen, P.L.; Rauch, U.; Bohrmann, B.; Kling, D.; Roqué, M.; Fallon, J.T.; Badimon, J.J.; Himber, J.; Riederer, M.A.; Nemerson, Y. Blood-borne tissue factor: Another view of thrombosis. *Proc. Natl. Acad. Sci. USA* **1999**, *96*, 2311–2315. [CrossRef]

69. Lisman, T. Platelet-neutrophil interactions as drivers of inflammatory and thrombotic disease. *Cell Tissue Res.* **2018**, *371*, 567–576. [CrossRef]
70. Zuo, Y.; Yalavarthi, S.; Shi, H.; Gockman, K.; Zuo, M.; Madison, J.A.; Blair, C.; Weber, A.; Barnes, B.J.; Egeblad, M.; et al. Neutrophil extracellular traps in COVID-19. *JCI Insight* **2020**, *5*, e138999. [CrossRef]
71. Ansari, J.; Moufarrej, Y.E.; Pawlinski, R.; Gavins, F.N.E. Sickle cell disease: A malady beyond a hemoglobin defect in cerebrovascular disease. *Expert Rev. Hematol.* **2018**, *11*, 45–55. [CrossRef] [PubMed]

13

Annexin-A6 in Membrane Repair of Human Skeletal Muscle Cell: A Role in the Cap Subdomain

Coralie Croissant, Céline Gounou, Flora Bouvet, Sisareuth Tan and Anthony Bouter *

Institute of Chemistry and Biology of Membranes and Nano-objects, UMR 5248, CNRS, University of Bordeaux, IPB, F-33600 Pessac, France; coralie33400@free.fr (C.C.); celine.gounou@u-bordeaux.fr (C.G.); flora.bouvet@u-bordeaux.fr (F.B.); s.tan@cbmn.u-bordeaux.fr (S.T.)

* Correspondence: a.bouter@cbmn.u-bordeaux.fr

Abstract: Defects in membrane repair contribute to the development of some muscular dystrophies, highlighting the importance to decipher the membrane repair mechanisms in human skeletal muscle. In murine myofibers, the formation of a cap subdomain composed notably by annexins (Anx) is critical for membrane repair. We applied membrane damage by laser ablation to human skeletal muscle cells and assessed the behavior of annexin-A6 (AnxA6) tagged with GFP by correlative light and electron microscopy (CLEM). We show that AnxA6 was recruited to the site of membrane injury within a few seconds after membrane injury. In addition, we show that the deficiency in AnxA6 compromises human sarcolemma repair, demonstrating the crucial role played by AnxA6 in this process. An AnxA6-containing cap-subdomain was formed in damaged human myotubes in about one minute. Through transmission electron microscopy (TEM), we observed that extension of the sarcolemma occurred during membrane resealing, which participated in forming a dense lipid structure in order to plug the hole. By properties of membrane folding and curvature, AnxA6 helped in the formation of this tight structure. The compaction of intracellular membranes—which are used for membrane resealing and engulfed in extensions of the sarcolemma—may also facilitate elimination of the excess of lipid and protein material once cell membrane has been repaired. These data reinforce the role played by AnxA6 and the cap subdomain in membrane repair of skeletal muscle cells.

Keywords: annexin-A6; membrane repair; human skeletal muscle; cap subdomain; fluorescence; electron microscopy; CLEM

---

1. Introduction

Plasma membrane of numerous cell types, such as cardiac or skeletal muscle cells during contraction [1,2], epithelial cells during the passage of food bolus [3] or endothelial cells submitted to blood flux [4], endure mechanical stress that can induce membrane damages. For example, it has been observed that up to 20% of skeletal muscle cells can be damaged during physical exercise [1]. Currently, the absence or failure of membrane resealing leads to cell death and may contribute to the development of degenerative diseases, such as muscular dystrophies. Among these, limb-girdle muscular dystrophy type R2 (LGMDR2, formerly 2B) or Miyoshi myopathy and LGMDR12 (formerly 2 L) are, respectively characterized by mutations in the dysferlin and anoctamin-5 genes, leading to several dysfunctions including a failure in cell membrane repair process [5,6].

Following plasma membrane damage, extracellular mM Ca^{2+} concentration enters the cell. The increase of intracellular Ca^{2+} concentration triggers the repair machinery in eukaryotic cells ensuring a rapid resealing of large plasma membrane ruptures [7–9]. Several proteins have been identified as belonging to the membrane repair machinery such as AHNAK, acid sphingomyelinase, ESCRT complexes, MG53, S100 proteins, SNAREs, synaptotagmins, calpains, caveolins, dysferlin or Anx [10]. The Anx family consists of twelve soluble proteins in mammals, named AnxA1 to A13 (the number 12 is non-assigned) [11]. Anx share the property of binding to membranes exposing negatively charged phospholipids in a Ca^{2+}-dependent manner [11,12], with little specificity for anionic lipid head-groups, yet the Ca^{2+} concentration required for binding varies considerably between Anx [13–15]. Anx present a common carboxy-terminal membrane-binding domain formed by the cyclic arrangement of conserved 70 amino-acid repeats [11,16]. This membrane-binding domain has the shape of a slightly curved rhomboid with a convex membrane-binding face where Ca^{2+}-binding loops are exposed and a concave face from which protrudes the amino-terminal domain. The amino-terminal part is very variable in size and is assumed to be responsible for the functional specificity of Anx [11,12,16,17]. The presence of different Anx in skeletal muscle [18–21]—together with the participation of several members of the Anx family in membrane repair processes [22,23]—raises the question of a collective role of these proteins in the protection and repair of sarcolemma injuries.

AnxA6, which is the largest Anx with a molecular weight of 68 kDa, exhibits eight 70 amino-acid conserved sequences forming two modules of four repeats [12]. It has been reported that at high Ca^{2+} concentration (about 2 mM), AnxA6 is able to bind two distinct phosphatidylserine-containing membranes, probably by the two distinct modules [24]. The use of this property for promoting membrane repair remains elusive so far. Implication of AnxA6 in the sarcolemma repair machinery has been mainly characterized in animal models. Roostalu and Strähle have shown that AnxA6 knock-down in muscle cells leads to a form of myopathy due to a defective membrane repair in zebrafish [19]. They have observed that dysferlin and AnxA6 accumulate at the disruption site independently one to each other and are immediately followed by AnxA1 and A2, building a crucial scaffold for membrane repair. The absence of AnxA6 inhibits the accumulation of AnxA1 and A2, indicating that the sequential recruitment of Anx is therefore crucial for membrane repair [19]. The formation of a similar scaffold has been observed in murine myofibers [25]. After membrane injury, AnxA1, A2, A5 and A6 aggregate into a tight structure positioned on the exterior surface of the myofiber that the authors termed the "repair cap" subdomain (hereafter termed cap subdomain). Adjacent to the cap subdomain is a region the authors referred to as the "shoulder" subdomain"—where phosphatidylinositol-4,5-bisphosphate, phosphatidylserine, dysferlin, MG53, BIN1 and EHD2 accumulate [25]. Regarding human cells, the rapid recruitment of Anx in a tight structure, which may look like a cap subdomain, has been observed for AnxA5 in skeletal muscle cells and AnxA4 and AnxA6 in cancer epithelial cells [26,27]. In these cancer epithelial cells, Nylandsted and collaborators have shown that AnxA6 induces membrane curvature, which may facilitate membrane resealing [27,28].

Hitherto, no muscular dystrophy has been linked to AnxA6 deficiency, but the etiology of numerous muscular dystrophies remains unknown. Nevertheless, AnxA6 has been identified as being a genetic modifier of muscular dystrophy [21,29]. Indeed, a truncated form of AnxA6, called AnxA6N32, prevents the translocation of wild-type AnxA6 to the disruption site in mice muscle fibers, leading to a reduced or inexistent cap subdomain and inhibition of membrane repair in these cells. The importance of AnxA6 in membrane repair has been recently reinforced by the demonstration that the treatment with recombinant AnxA6 protects against acute muscle injury in wild-type mice and reduced the level of serum creatinine kinase, a biomarker of disease, in a model of muscular dystrophy [30]. A dysfunction of AnxA6 in human muscle cells may therefore greatly impact the severity of the disease.

Here, we analyzed the response of myotubes established from a human skeletal muscle cell line to membrane damage by laser ablation. We observed that myotubes rendered deficient for AnxA6 suffered from a defect of membrane repair. In damaged myotubes, the behavior of AnxA6 coupled to GFP was analyzed and compared to AnxA5 and dysferlin. We observed that AnxA6 belongs to the cap subdomain built during membrane resealing. By means of CLEM, we characterized the cap subdomain, which is notably composed of an extension of the cell membrane. This extension of the cell membrane may promote membrane resealing and also participate in the compaction of lipid and protein material that has to be eliminated after membrane repair.

2. Materials and Methods

2.1. Culture of Human Skeletal Muscle Cells

The healthy LHCN-M2 (referred to hereafter as LHCN) cell line was provided by the platform for immortalization of human cells from the Center of Research in Myology (Paris, France). This cell line was established from satellite cells of the pectoralis major muscle of a 41-year-old subject [31]. Myoblasts were cultured in a skeletal muscle medium composed by one volume of Medium 199 with glutaMAX™ (Gibco® by by Thermo Fisher Scientific, Waltham, MA, USA), four volumes of Dubelcco's modified Eagle Medium (DMEM) with high-glucose and glutaMAX™ and without pyruvate (Gibco® by Thermo Fisher Scientific, Waltham, MA, USA) supplemented with 20% fetal bovine serum, 50-µg/mL gentamycin and a commercial mix of skeletal muscle cell growth medium supplements (ref. C-39365, Promocell, Heidelberg, Germany), which included 12.5-µg/mL fetuin, 2.5-ng/mL human recombinant epidermal growth factor, 0.25-ng/mL basic fibroblast growth factor, 2.5-µg/mL insulin, 0.1-µg/mL dexamethasone. Myotubes were obtained by cultivating 90%-confluence myoblasts for three days in a differentiation medium, composed of the skeletal muscle medium supplemented only with 10-µg/mL insulin.

The MDA-MB-231 cancer cell line was cultured in DMEM containing high-glucose, pyruvate and glutaMAX™ (Gibco® by Thermo Fisher Scientific) supplemented with 10% fetal bovine serum.

2.2. Western Blot

Myotubes were trypsinized, pelleted and resuspended in 300 µL of Dulbecco's phosphate buffer saline (D-PBS) depleted in Ca^{2+} and supplemented with 1-mM EGTA. Sonication of ice-cold cell suspension was performed with a Branson digital sonifier (amplitude 20%, duration 2 min, interval 5 s and pulse 5 s) and two successive centrifugations at 13,000 g for 1 min allowed to remove cell debris. 10 µg protein extracts were separated on 10% SDS-PAGE. Tank electrophoretic transfer (Bio-Rad, Hercules, CA, USA) onto PVDF membrane was performed for 1 h at 100 V. AnxA5 (35 kDa) and AnxA6 (68 kDa) were, respectively detected with mouse anti-AnxA5 (Sigma, Saint-Louis, MI, USA) and anti-A6 (Santa Cruz Biotechnology, Dallas, TX, USA) monoclonal antibodies and GAPDH (loading control) was detected with a rabbit anti-GAPDH polyclonal antibody (Santa Cruz Biotechnology). Primary antibodies were diluted 1:1000 in saturation solution composed by Tris buffer saline (10-mM Tris, 150-mM NaCl, pH 8.0) supplemented with 0.1% Tween20 and 5% non-fat dry milk. Revelation was performed using secondary antibodies conjugated to horse-radish peroxidase (GE-Healthcare, Chicago, IL, USA) diluted 1:2000 in saturation solution and Opti-4CN™ colorimetric kit (Bio-Rad, Hercules, CA, USA). ImageJ software was used to measure the relative intensity of protein bands.

2.3. Membrane Rupture and Repair Assay

Membrane repair was assayed as recently reported [32], according to a protocol that differed slightly from our previous studies [26,33]. Myotubes were cultured in complete growth medium on 18*18-mm glass coverslips (Nunc). A solution of 5-µg/mL FM1-43 (Thermo Fisher Scientific, Waltham, MA, USA) in D-PBS containing 2-mM Ca^{2+} was maintained over ice and subsequently added in a homemade coverslip cell chamber, where the coverslip was mounted. To induce membrane damage, cells were irradiated at 820 nm with a tunable pulsed depletion laser Mai Tai HP (Spectra-Physics, Irvine, USA) of an upright two-photon confocal scanning microscope (TCS SP5, Leica, Wetzlar, Germany) equipped with an HCX APO L U–V-I 63.0 × 0.90 water lens. Irradiation consisted of 1 scan (1.3 s) of a 1 µm × 1 µm area with a power of 110 (±5) mW. 512 × 512 images were acquired at 1.3 s intervals with pinhole set to 1 Airy unit. Membrane rupture and repair processes were monitored by measuring variations in fluorescence intensity of FM1-43 as previously described [26,33]. The FM1-43 was excited by the 488-nm laser line (intensity set at 30% of maximal power) and fluorescence emission was measured between 520 nm and 650 nm. For quantitative analysis, the fluorescence intensity was integrated over the whole cell surface and corrected for the fluorescence value recorded before irradiation, using ImageJ software.

2.4. Transduction of AnxA6-Targetting shRNA Lentiviral Particles in LHCN Myotubes

The following shRNA sequences, cloned into the pLKO.1 puro-vector (MISSION® shRNA plasmids, Sigma, Saint-Louis, MI, USA), were used: AnxA6-targetting shRNAA6-1 (TRCN0000011461, Sigma): $^{5'}$CCGGCGG GCACTTCTGCCAAGAAATCTCGAGATTTCTTGGCAGAAGTGCCCGTT TTT$^{3'}$; shRNAA6-2 (TRC N0000008686, Sigma): $^{5'}$CCGGCGGTTGGTGTTCGATGAGTATCTCGAGATA CTCATCGAACACCAACCGTTTTT$^{3'}$; control shRNA: $^{5'}$CCTAAGGTTAAGTCGCCCTCGCTCGAGCG AGGGCGACTTAACCTTAGG$^{3'}$. The two AnxA6-targetting shRNAs were selected from a commercial library according to the "mean knockdown level" given by the supplier (0.9 and 0.87, respectively). Lentiviral-based particles containing shRNAs were produced by Bordeaux University lentiviral vectorology platform (US005, Bordeaux, France) by transient transfection of 293T cells. MDA-MB-231 cells and LHCN myoblasts were cultured in their respective growth medium for 24 h. Transduction was carried out by adding concentrated lentiviral particles to the cells at a multiplicity of infection (MOI) of 100 in Opti-MEM® medium for 24 h. Cells were subsequently incubated with growth medium for 24 h and then 2-μg/mL puromycin (Sigma, Saint-Louis, MI, USA) were added in order to select transduced cells. 72 h after antibiotic treatment, western-blot analysis or immunocytofluorescence experiments were performed as described in Sections 2.2 and 2.6, respectively. To establish Anx-deficient myotubes, myoblasts were incubated in differentiation medium for 7 h 30 and then transduced with shRNA lentiviral particles at a MOI of 100 in Opti-MEM® medium for 24 h. Subsequently, cells were incubated in differentiation medium for 24 h and 2-μg/mL puromycin were added for 72 h in order to select transduced myotubes.

2.5. Subcellular Traffic of AnxA6-GFP in Damaged Myotubes

The pA6-GFP plasmid was constructed by cloning AnxA6 cDNA into the pEGFP-N1 (Takara Bio USA, Mountain View, CA, USA) plasmid and was a gift of Volker Gerke (University of Munster, Germany). The pA5-mCh plasmid, which contained recombinant AnxA5-mCherry, was constructed from pA6-GFP using the In-Fusion® HD Cloning kit (Takara Bio USA, Mountain View, CA, USA). Anx-GFP or Anx-mCherry fusion proteins harbored the fluorescent tag at the C-terminus. Myotubes at 65 h of differentiation were transfected using Lipofectamine 2000 reagent (Thermo Fisher Scientific, Waltham, MA, USA). 0.4 μg plasmid was mixed with 0.52 μL Lipofectamine 2000 for 20 min at room temperature and the mixture was incubated with cells cultured on 18*18-mm glass coverslip for 4 h at 37 °C in 200 μL Opti-MEM® (Thermo Fisher Scientific, Waltham, MA, USA). Transfection was renewed once and myotubes were then incubated for at least 24 h in differentiation medium before analysis. For myotubes co-expressing pA6-GFP together with pA5-mCh, fluorescent signal of GFP and mCherry is, respectively presented in green and magenta within the manuscript, for a better visualization of AnxA5-mCherry. Magenta is the color that the human eye best distinguishes from green [34].

2.6. Immunocytofluorescence

Cells were fixed and permeabilized in 100% methanol solution (5 min, at −20 °C). All subsequent steps (antibody incubation and washes) were performed using 2% BSA in D-PBS solution. Primary antibody at 1:100 and secondary antibody at 1:1000 were successively incubated with cells for 1 h at 37 °C. Antibodies were mouse monoclonal anti-AnxA6 antibody (Santa Cruz Biotechnology, sc-271859), mouse monoclonal anti-Dysferlin antibody (Leica, NCL-Hamlet), Alexa Fluor 488-coupled anti-mouse goat antibody and Alexa Fluor 546-coupled anti-rabbit donkey antibody (Thermo Fisher Scientific, Waltham, MA, USA). Finally, cells were washed in D-PBS and nuclear counterstaining was performed with DAPI (Sigma, Saint-Louis, MI, USA).

2.7. *Immuno-TEM*

For subcellular localization of endogenous Anx and CLEM in damaged cells, myotubes were formed and cultured in a 35-mm glass bottom dish equipped with a square-patterned coverslip (MatTek, Ashland, MA, USA). Cell membrane rupture was performed by laser ablation according to the protocol described above, but in the absence of FM1-43 to avoid fluorescence crosstalk. Myotubes were fixed in 1% glutaraldehyde solution for 20 min at room temperature, as described previously [35]. They were subsequently incubated in 25-mM ammonium chloride for 15 min and permeabilization and saturation of nonspecific sites were performed in a mixture of 2% BSA and 0.1% TritonX100 in D-PBS for 30 min. All subsequent steps (antibody incubation and washes) were performed using 2% BSA in D-PBS solution. Mouse anti-AnxA6 monoclonal antibody at 1:100 and secondary Alexa Fluor 488- and gold nanoparticles-conjugated anti-mouse goat antibody (FluoroNanogold, Nanoprobes, NY, USA) at 1:100 were successively incubated with myotubes for 1 h at 37 °C. After three rinses, myotubes were observed by fluorescence microscopy. Cells were then postfixed overnight at 4 °C in a mixture of 4% paraformaldehyde and 2% glutaraldehyde in 0.1-M cacodylate buffer (pH 7.4). Signal amplification was performed using the HQ silver kit (Nanoprobes, NY, USA) according to the manufacturer's instructions. Cells were treated with 1% osmium tetroxide in 0.1-M cacodylate buffer for 1 h at room temperature and then dehydrated with ethanol and finally embedded in Epon-Araldite. Thin sections (65 nm) were collected using EM UC7 ultramicrotome (Leica, Wetzlar, Germany) and stained successively with 5% uranyl acetate and 1% lead citrate. TEM observation was performed with a FEI CM120 operated at 120 kV. Images were recorded with a USC1000 slow scan CCD camera (Gatan, Pleasanton, CA, USA).

3. Results

3.1. *Expression and Distribution of AnxA6 in Human Skeletal Muscle Cells*

In order to investigate the relative expression of AnxA6 in human skeletal muscle cells, western-blot analysis was performed from the healthy human myogenic cell line LHCN, which was established from satellite cells isolated from pectoralis major adult muscle [31]. We observed that myotubes showed a 3-fold increase in AnxA6 expression compared to myoblasts (Figure 1A,B), whereas AnxA5 expression is unchanged between myoblasts and myotubes as previously reported [26]. By immunocytofluorescence, we observed that endogenous AnxA6 is exclusively cytoplasmic in LHCN myoblasts and myotubes (Figure 1C). Homogenous distribution of the fluorescence suggested that AnxA6 is mainly cytosolic. When LHCN myoblasts and myotubes were transfected by an AnxA6-GFP expression plasmid, we observed that the fusion protein also exhibited a cytosolic localization (Figure 1D).

We concluded therefore that AnxA6 localizes exclusively in the cytoplasm of human myoblasts and myotubes, as previously described for other cell types or species [19,36]. AnxA6 expression is enhanced in myotubes, which suggests it is important for physiological processes in differentiated muscle cells.

3.2. *AnxA6 is Mandatory for Membrane Repair in Human Skeletal Muscle Cells*

In order to study the involvement of AnxA6 in membrane repair, we had to generate AnxA6-deficient myotubes. As with other post-mitotic cells, transfection or transduction of differentiated muscle cells remains a challenging task. Here, we applied a shRNA strategy as previously performed for the knock-down of AnxA5 in LHCN myotubes [26]. Screening experiments were performed using the MDA-MB-231 cell line, which is easier to be transduced, with two candidate shRNAs selected from a commercial library. Western-blot analysis showed that both shRNAs are able to specifically reduce AnxA6 expression of more than 90% systematically (n = 3, Figure S1). Using experimental conditions established with MDA-MB-231 cells, LHCN myoblasts were then transduced with both shRNAs. AnxA6 expression in transduced LHCN myoblasts was about 40% lower than in control cells, in a fairly similar way whatever the shRNA sequence (Figure S2). No synergy effect was observed when both shRNA sequences were mixed. As expected, the level of knock-down was lower in myoblasts compared to MDA-MB-231 cells. LHCN myotubes, which are multinucleated cells measuring several hundred μm long, are normally established from fusion of myoblasts cultured three days in the differentiation medium [26] (Figure S3A). However, as previously reported for AnxA5-deficient LHCN myoblasts [26], we observed that LHCN myoblasts rendered deficient in AnxA6 were unable to form myotubes (Figure S3B), suggesting that AnxA6 is involved in the process of cell differentiation and/or fusion. This result also implied that shRNA transduction had to be carried out during the formation of myotubes. We determined that transduction had to be performed 8 h after starting incubation of myoblasts in differentiation medium. Since western-blot analysis requires a large number of cells and prevents to distinguish between myotubes and myoblasts remaining in culture, we quantified the effect of shRNAA6 transduction specifically in myotubes by immunocytofluorescence. Preliminary experiments showed that quantification by immunocytofluorescence gave similar results to western-blot analysis regarding the relative expression of AnxA6 in shRNA-transduced or control LHCN myoblasts (Figure S4). Whatever the shRNA sequence used; we observed a decrease of about 60% of the expression of AnxA6 in shRNA-transduced LHCN myotubes (Figure S5). Subsequent experiments were performed using the LHCN myotubes transduced with the shRNAA6-2 sequence, which are hereafter named AnxA6-deficient myotubes.

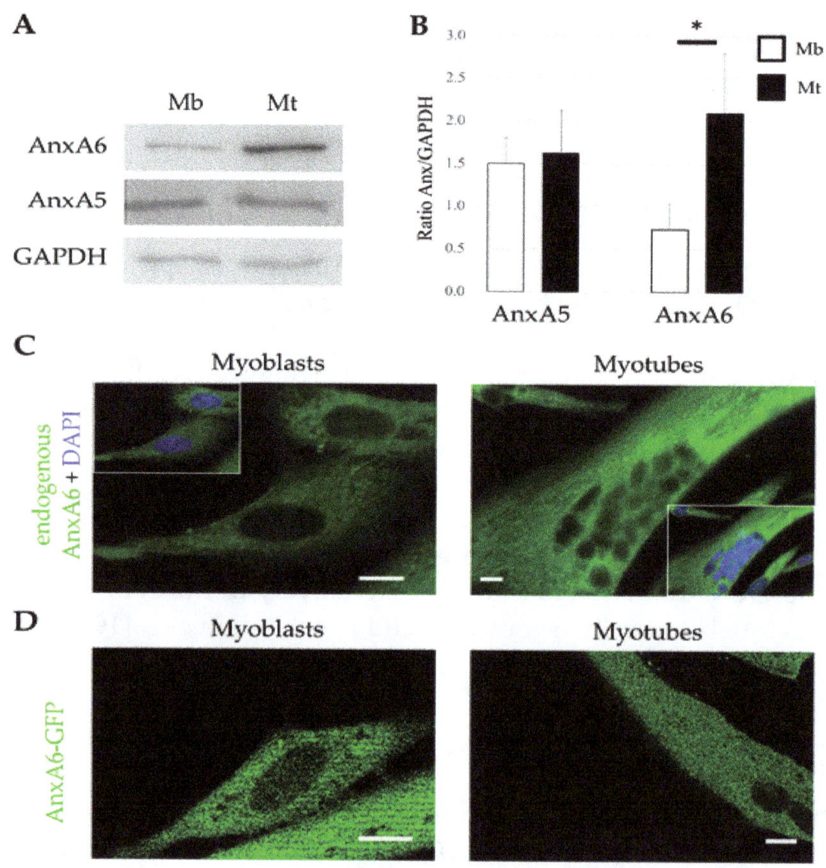

Figure 1. Expression and subcellular distribution of AnxA6 in human myoblasts and myotubes. (**A**,**B**) Cellular content in endogenous AnxA5 and AnxA6 in healthy LHCN-M2 (LHCN) myoblasts (Mb) and myotubes (Mt) was quantified through western-blot analysis; (B) data are mean ± SEM from five experiments. Wilcoxon test, * $p < 0.05$; (**C**) subcellular localization of endogenous AnxA6 (green) in LHCN myoblasts and myotubes by immunocytofluorescence. In the insets, nuclear counterstaining with DAPI is displayed (blue). Scale bars: 10 μm; (**D**) subcellular localization of AnxA6-GFP in living LHCN myoblasts and myotubes by fluorescence microscopy. Scale bars: 10 μm.

Sarcolemma repair assay was performed by laser ablation in the presence of Ca^{2+} and FM1-43, as previously described [32,33]. By analyzing changes in the FM1-43 intracellular fluorescence intensity, we first confirmed that LHCN myotubes are able to reseal a µm-size membrane damage in about 80 s (Figure 2A,C) as previously reported [26]. When AnxA6-deficient LHCN myotubes were submitted to the same irradiation conditions, different types of response were observed. Some myotubes exhibited an increase in fluorescence intensity limited to an area close to the disruption site (Figure 2B, repaired). The kinetics of fluorescence intensity changes were characterized by an increase for about 70–90 s and then the presence of a plateau (Figure 2D, repaired), which indicated a rapid resealing of cell membrane. Some other myotubes exhibited a strong and large increase of the intracellular fluorescence intensity (Figure 2B,D, unrepaired), which meant that they were unable to reseal cell membrane damage. Finally, for a significant part (about 50%) of them we were unable to determine whether cell membrane resealing occurred or not (Table S1). Indeed, kinetics associated with these myotubes exhibited whether a weak, but continuous raise of the intracellular fluorescence intensity or the presence of a plateau at a very high level of fluorescence intensity (Figure 2B,D undetermined). These complex and unclear responses of AnxA6-deficient myotubes to laser membrane damage result probably from a variable AnxA6 expression due to an uncomplete shRNA knock-down. We had therefore to slightly modify the way to analyze membrane repair. Control and AnxA6-deficient myotubes were cultured on glass bottom dishes equipped with a square-patterned coverslip that enabled cell tracking during different stages of the experiment (Figure 3). A field identified by means of the alphanumeric code was imaged and myotubes present in the field were successively damaged by laser ablation exactly as in the standard membrane repair assay. In the case of control LHCN myotubes, which are able to rapidly reseal sarcolemma damage, we observed that a large majority (about 90%, Table S2A) of damaged myotubes were still adhered on the coverslip 1 h after laser ablation (Figure 3A). A similar result was observed when LHCN myotubes were transduced with a scrambled shRNA (data not shown). However, when AnxA6-deficient LHCN myotubes were submitted to the same experimental conditions, we observed that only 30% of myotubes were still present on the coverslip 1 h after laser ablation (Figure 3B and Table S2B). It is likely that the loss of 70% of myotubes is due to the absence of membrane repair, which leads to the death of cells that are released from the coverslip. We therefore concluded that the deficiency in AnxA6 compromises sarcolemma repair, suggesting that AnxA6 is a crucial component of the membrane repair machinery in human skeletal muscle cells.

3.3. *Recruitment of AnxA6 at the Membrane Disruption Site and Formation of the Cap Subdomain*

All proteins composing the repair machinery, notably dysferlin [5], Anx [18,19,37–39] and MG-53 [40], are rapidly recruited to the membrane disruption site. In order to investigate the behavior of AnxA6 during human sarcolemma injury, we analyzed the subcellular trafficking of AnxA6 fused to GFP in laser-damaged LHCN myotubes. For this purpose, laser ablation was performed at two different positions on cells in order to study the kinetics of recruitment from two different perspectives. Membrane damage was carried out either at the lateral edge—which is the conventional way to perform membrane damage by laser ablation—or on the surface of the myotube (Figure S6). Irradiation at the edge of the cell allowed to visualize the wounded area sideways, whereas the irradiation on the surface allowed a view from the top.

We observed that AnxA6-GFP was recruited to the disruption site immediately after membrane damage and appeared to be primarily associated with the plasma membrane (Figure 4A, +1.3 s and Video S1). Fluorescence intensity increased mainly at the plasma membrane around the rupture site over a large area (+13.0 s). From about 20 s after membrane injury, AnxA6-GFP that was interacting with the plasma membrane seemed to be released from the inner leaflet and went back to the cytoplasm (Video S1). This result suggested that intracellular Ca^{2+} concentration rapidly decreased, probably while plasma membrane was resealed. Then, we observed that AnxA6-GFP concentrated in a structure (+26.0 s) that persisted beyond 65 s (Figure 4B). We checked by immunocytofluorescence that endogenous AnxA6 accumulated similarly in damaged control LHCN myotubes (Figure 4C). This tight structure that was positioned on the exterior surface of the myotube is identical to the cap subdomain previously described in murine myofibers [25]. In these cells, adjacent to the cap subdomain is the shoulder subdomain, which is composed notably by dysferlin, MG53, BIN1 and EHD2 [25]. To confirm the presence of both subdomains in repaired LHCN myotubes, we damaged AnxA6-GFP-expressing myotubes by laser ablation and performed immunostaining of dysferlin. If we observed a region where AnxA6 and dysferlin overlapped, we were able also to distinguish two specific regions with either AnxA6-GFP or dysferlin corresponding, respectively to the cap and shoulder subdomains (Figure 4D). When laser ablation was performed on the surface of the myotube, we observed a slower recruitment of AnxA6-GFP to the disruption site (Figure 4E and Video S2). For instance, no increase of fluorescence intensity was observed around the disruption site within the first 20 s after membrane damage. This slower dynamic resulted probably from a stronger membrane damage, as the laser penetrated deeply in the cytoplasm. AnxA6-GFP was finally recruited about 20 s after membrane injury in a wave-like motion (Video S2). Such a behavior, which was not identified when the wounded area was visualized sideways, has been previously reported for S100A11 during its recruitment to the laser-injured membrane site in HeLa cells [41]. Molecular mechanisms responsible for this specific motion remain to be elucidated. After about 30 s, we observed that AnxA6-GFP accumulated at the disruption site, plugging the hole created by the membrane injury (Figure 4E). A strictly different behavior was observed in damaged MCF-7 cancer cells, in which AnxA6-GFP is recruited to wound edges where it promotes the closure of the hole by triggering contraction of the membrane edges [27]. The role played by AnxA6 in the processes of membrane repair may therefore be different in human skeletal muscle and cancer epithelial cells.

We previously reported that AnxA5 participates in the membrane repair of human skeletal myotubes [26] and AnxA5 has also been observed as composing the cap subdomain in murine myofibers [25]. In order to compare the traffic of AnxA5 and AnxA6 in human skeletal muscle cells during sarcolemma repair, we imaged AnxA5-mCherry and AnxA6-GFP in LHCN myotubes damaged by laser ablation. We observed that AnxA5-mCherry and AnxA6-GFP were recruited nearly immediately and simultaneously to the damaged site of the plasma membrane (Figure 5A and Video S3). Nevertheless, a major difference distinguished AnxA5 from AnxA6: AnxA5 was present in a limited area at the disruption site, while AnxA6 spread more widely on the plasma membrane around the disruption site. AnxA5 and AnxA6 were present in the cap subdomain, AnxA5 composing the core of the subdomain (Figure 5B). This result is consistent with previous studies showing the presence of AnxA5 in the cap subdomain of murine myofibers [25] and human myotubes [26]. We previously showed that AnxA5 forms two-dimensional arrays around the disruption site to prevent the expansion of the tear and promote membrane repair [26,37]. The observation of a small cluster of AnxA5-mCherry in the cap subdomain could correspond to AnxA5 arrays that interacted with the edges of the torn membrane and were then packaged in the cap after membrane resealing. All together, these results indicated that membrane repair in human skeletal muscle cells is based on the formation of a cap subdomain, composed notably by AnxA6.

3.4. *Visualization of AnxA6 and the Cap Subdomain by TEM*

To further characterize the cap subdomain and the role played by AnxA6 in this structure, we imaged by CLEM damaged AnxA6-GFP-expressing myotubes, as previously described [26]. Briefly, we observed AnxA6-GFP in a single damaged myotube successively through fluorescence microscopy and TEM. Immunostaining of AnxA6 using a secondary antibody coupled to gold nanoparticles enabled to analyze the distribution of the protein in the damaged myotube at high resolution. We first carried out membrane damage at the edge of the myotube (Figure 6A,B) and focused our attention on the characterization of the ultrastructure of the cap subdomain. Indeed, little information was known about the origin, structure and function of the cap subdomain. It has been proposed that the cap subdomain may correspond to an ordered protein scaffold required for membrane repair, which is composed notably by Anx [25]. Even if the cap subdomain can vary considerably in size, TEM images revealed that it is characterized by an accumulation of disorganized membrane material on the exterior surface of the myotube, exactly where laser ablation was performed (Figure 6A and Figure S7). In most cases, membrane material that was accumulated in the cap subdomain appeared as a tangle of filaments, suggesting it is of cell membrane origin. This excess of plasma membrane may arise from lysosome exocytosis, which would increase the membrane surface and thus reduce the tension exerted by the cortical cytoskeleton, as previously proposed [42]. AnxA6-GFP appeared to be accumulated mainly outside the cell in the cap subdomain, interacting primarily with membrane filaments that may correspond to extensions of the sarcolemma (Figure 6A). Endogenous AnxA6 exhibited a similar distribution (Figure S7).

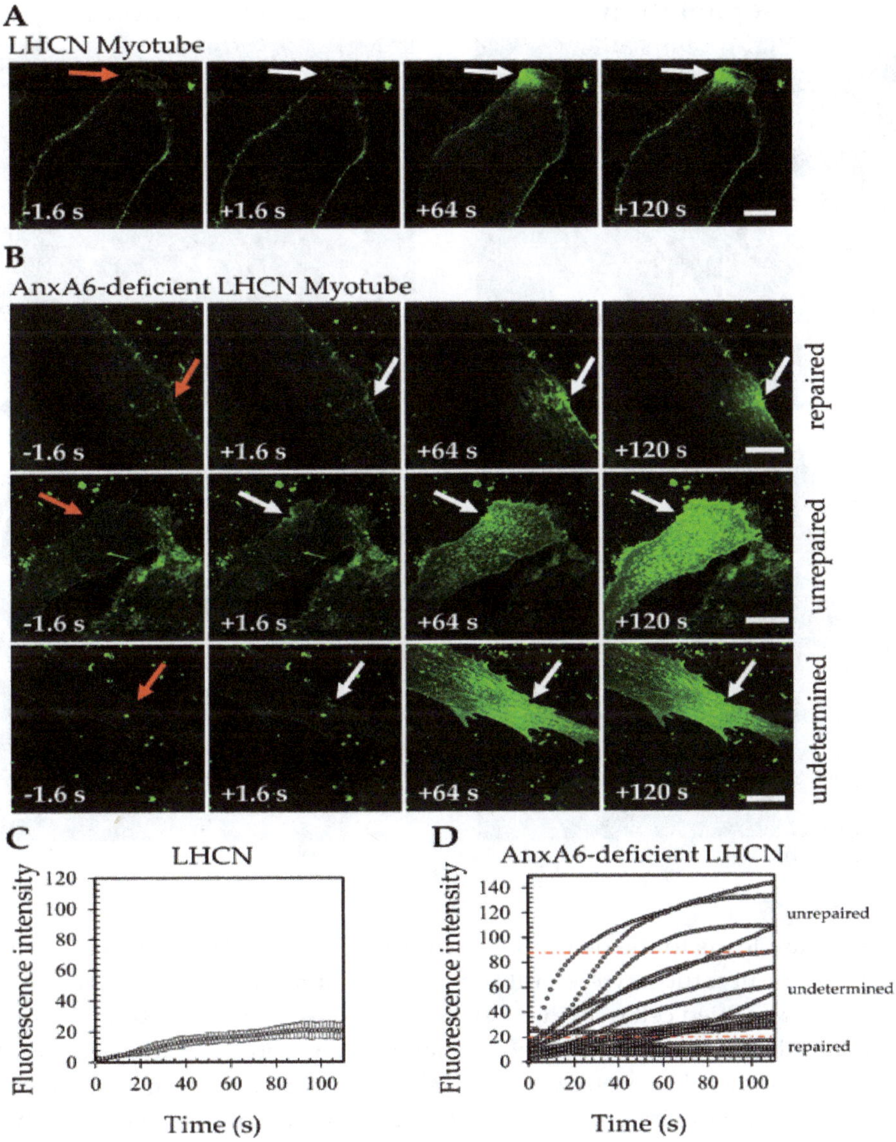

Figure 2. Responses of AnxA6-deficient LHCN myotubes to laser ablation. (A-B) Sequences of images showing the response of a wild-type (**A**) and three different AnxA6-deficient (**B**) LHCN myotubes to 110-mW laser irradiation in the presence of 2-mM Ca^{2+} and FM1-43 (green). In these panels, the area of membrane irradiation is marked with a red arrow before irradiation and a white arrow after irradiation. Time at which the image was recorded before (-) or after (+) irradiation is indicated. Scale bars = 20 μm; (**C**,**D**) Data represent the FM1-43 fluorescence intensity integrated over whole cell section, which is plotted versus time. For LHCN myotubes (mean ± SEM of 20 cells within 2 independent experiments), the fluorescence intensity reached a plateau after ~80 s (**C**). For AnxA6-deficient myotubes, very different

kinetics were observed from one myotube to another. As an example, one typical experiment with 20 cells is presented in (**D**). When the presence of a plateau and a maximal fluorescence intensity lower than 20 a.u. were observed, the cell was classified as "repaired cell" (lower part of the graph, as indicated). When the kinetics of fluorescence changes showed a continuous increase with values higher than 90 a.u., the cell was classified as "unrepaired cell" (upper part). Otherwise, cells were classified as "undetermined cells". Data analysis is presented in Table S1. Red dashed lines divide the three types of response.

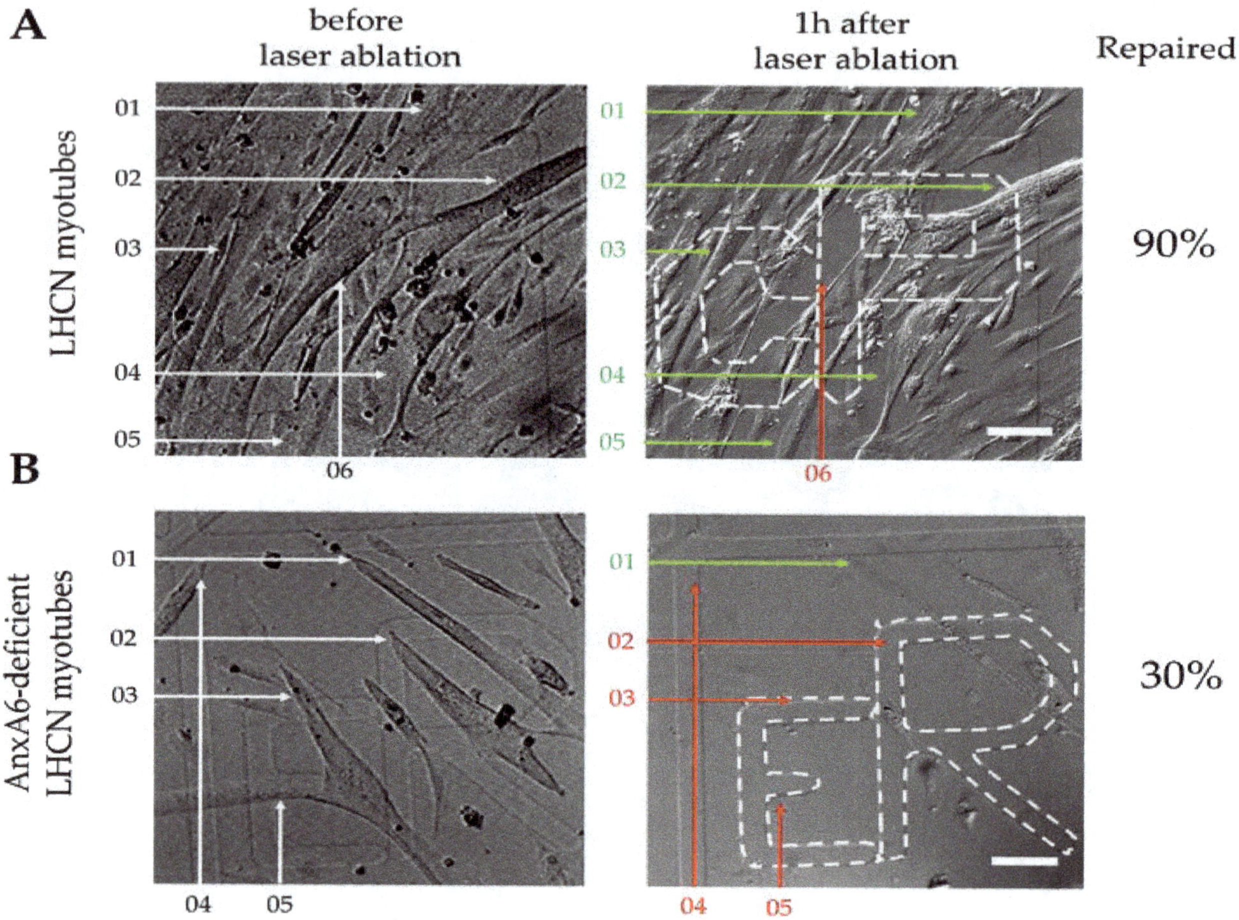

Figure 3. Laser ablation leads to the death of AnxA6-deficient LHCN myotubes. (**A**) Control or (**B**) AnxA6-deficient LHCN myotubes were cultivated on a square-patterned coverslip displaying an alphanumeric code that enabled cell tracking. Left-hand images exhibit myotubes before laser ablation. Each arrow indicates the area to be damaged, which is identified by a number. On the right-hand images, the alphanumeric code has been drawn for sake of clarity. Green arrows indicate myotubes that are still present 1 h after laser ablation. In contrast, red arrows point out myotubes that disappeared. About 100 cells over 5 independent experiments were analyzed for each condition. Data analysis is presented in Table S2. Scale bars = 100 μm.

When LHCN myotubes were damaged on their surface, accumulation of membrane material at the disruption site was visible on several sections (Figure 6E). A tangle of membranes was observed in sections located above the plane of the damaged sarcolemma, which may correspond to the cap subdomain that is localized outside the cell (Figure 6E, Section 1). Deeply in the cytoplasm we observed that an accumulation of membrane material filled the entire area damaged by laser ablation. We can hypothesize that the laser penetrated deeply in the cytoplasm and created a significant membrane damage, which required greater mobilization of intracellular material to plug the tear (Figure 6E, Sections 2 and 3). AnxA6-GFP was found mainly associated with membranes present above the plane of the initial position of the sarcolemma (Section 1) and rarely in cytoplasmic sections (Sections 2 and 3). AnxA6 seems therefore accumulated essentially in the cap subdomain outside the damaged area, in agreement with the fluorescence images of AnxA6-GFP (Figure 4A).

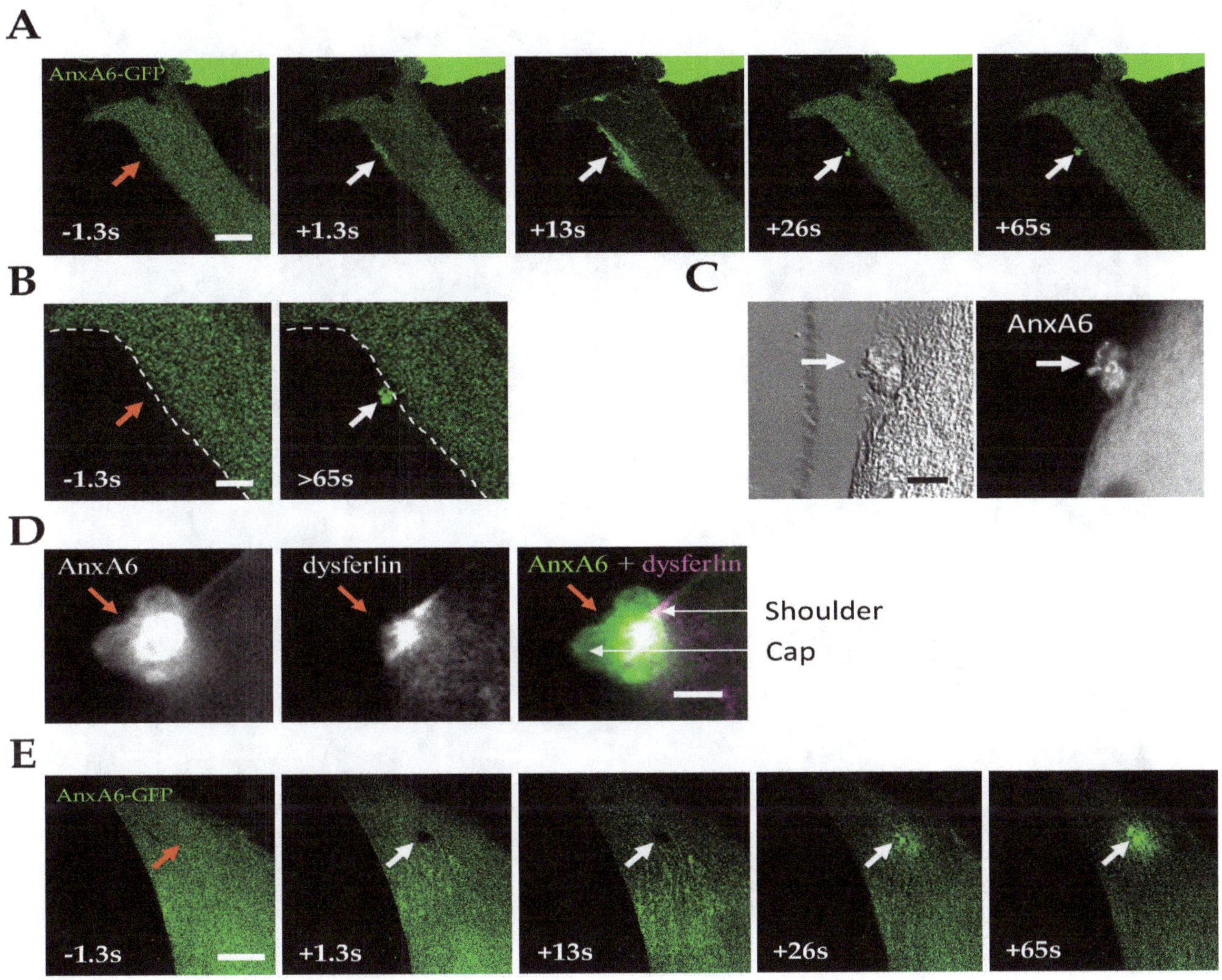

Figure 4. Recruitment of AnxA6 to the damaged membrane site and formation of the cap subdomain in LHCN myotube. (**A**) LHCN myotubes were transfected with the plasmid pA6-GFP and membrane damage experiments were performed by laser ablation at the lateral edge of the myotube. Red arrow, area before irradiation; white arrow, area after irradiation. Scale bar: 10 μm; (**B**) Magnified images of the disruption site before (left) and more than 65 s (right) after laser ablation. The dashed white line corresponds to the position of the plasma membrane before irradiation. Scale bar: 5 μm; (**C**) Distribution of endogenous AnxA6 in a damaged LHCN myotube. The plasma membrane of LHCN myotubes was damaged by laser irradiation. Myotubes were immunostained for AnxA6 and observed in bright-field (left) and fluorescence microscopy (right). White arrow, irradiated area. Scale bar: 5 μm; (**D**) Distribution of AnxA6 and dysferlin between cap and shoulder subdomains. A LHCN myotube expressing AnxA6-GFP (green) was damaged at the lateral edge and then fixed and immunostained for dysferlin (magenta). Red arrow, irradiated area (**E**) LHCN myotubes were transfected with the plasmid pA6-GFP and membrane damage experiments were performed by laser ablation on the surface of the myotube. Red arrow, area before irradiation; white arrow, area after irradiation. Scale bar: 20 μm.

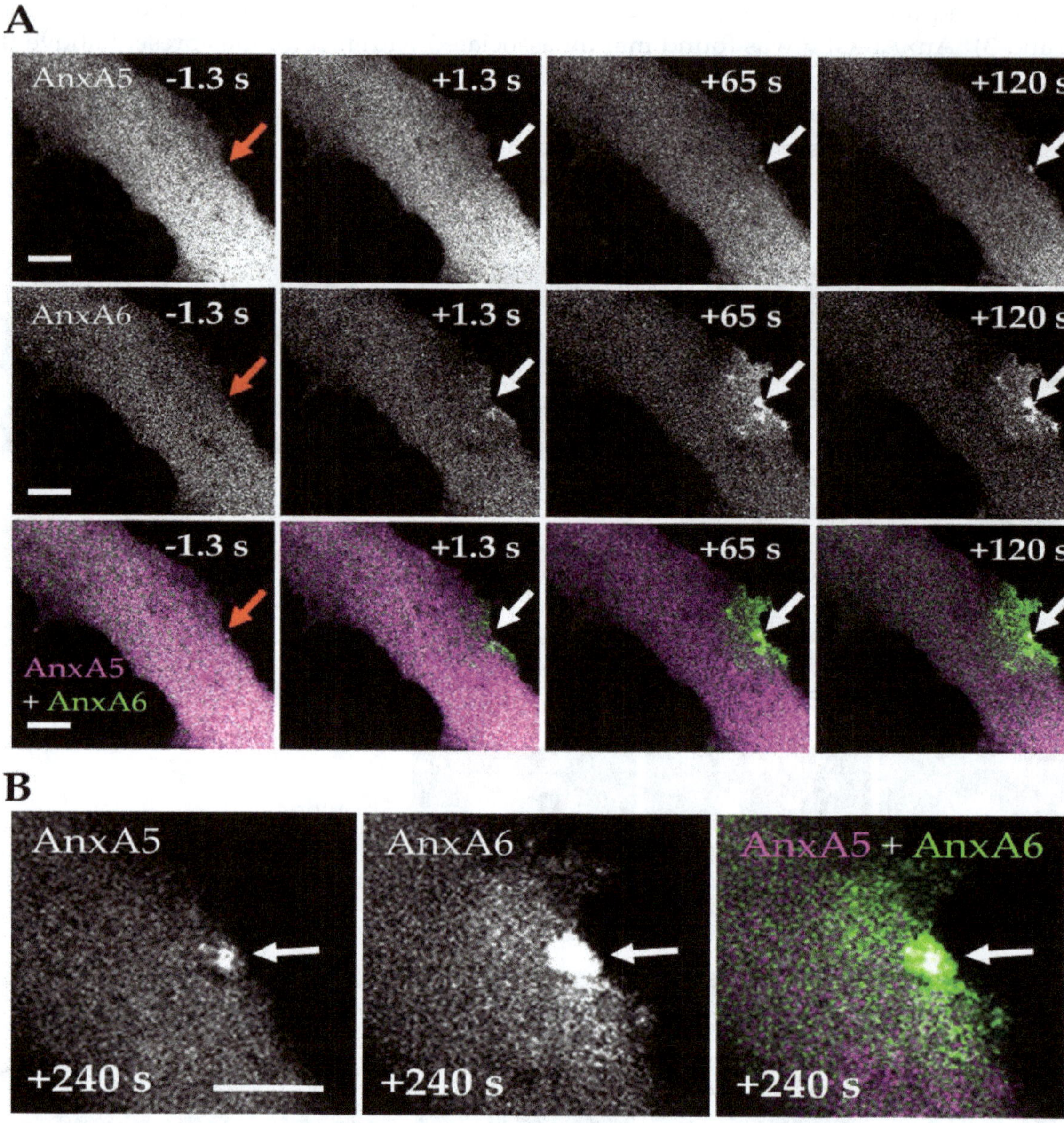

Figure 5. Distribution of AnxA5 and AnxA6 within the cap subdomain. LHCN myotubes were transfected with the plasmids pA6-GFP and pA5-mCherry to express AnxA6-GFP (green) and AnxA5-mCherry (magenta) and membrane damage experiments were performed by laser ablation at the edge of the myotube. (**A**) Follow-up of the recruitment of AnxA6 and AnxA5 within 120 s after membrane damage. Red arrow, area before irradiation; white arrow, area after irradiation; (**B**) Enlarged images of the damaged area 240 s after membrane damage. Scale bars: 10 μm.

To get further insight in the formation of the cap subdomain, we sought to fix myotubes rapidly after membrane damage. We had the opportunity to observe the membrane disruption site during the formation of the cap subdomain (Figure 7). This event was characterized by the fact that AnxA6 was still interacting with a large area around the disruption site (Figure 7B).

The typical image displayed in Figure 7C shows the complex structure of the disruption site during membrane resealing (Figure 7C). Multiple intracellular vesicles were present at the base of the membrane material accumulated at the disruption site, supporting the hypothesis of the formation of a lipid patch. We also observed a tangle of membrane filaments that formed the main part of the protrusion. AnxA6 was mainly interacting with these membrane filaments, which obviously looked like extensions of the sarcolemma.

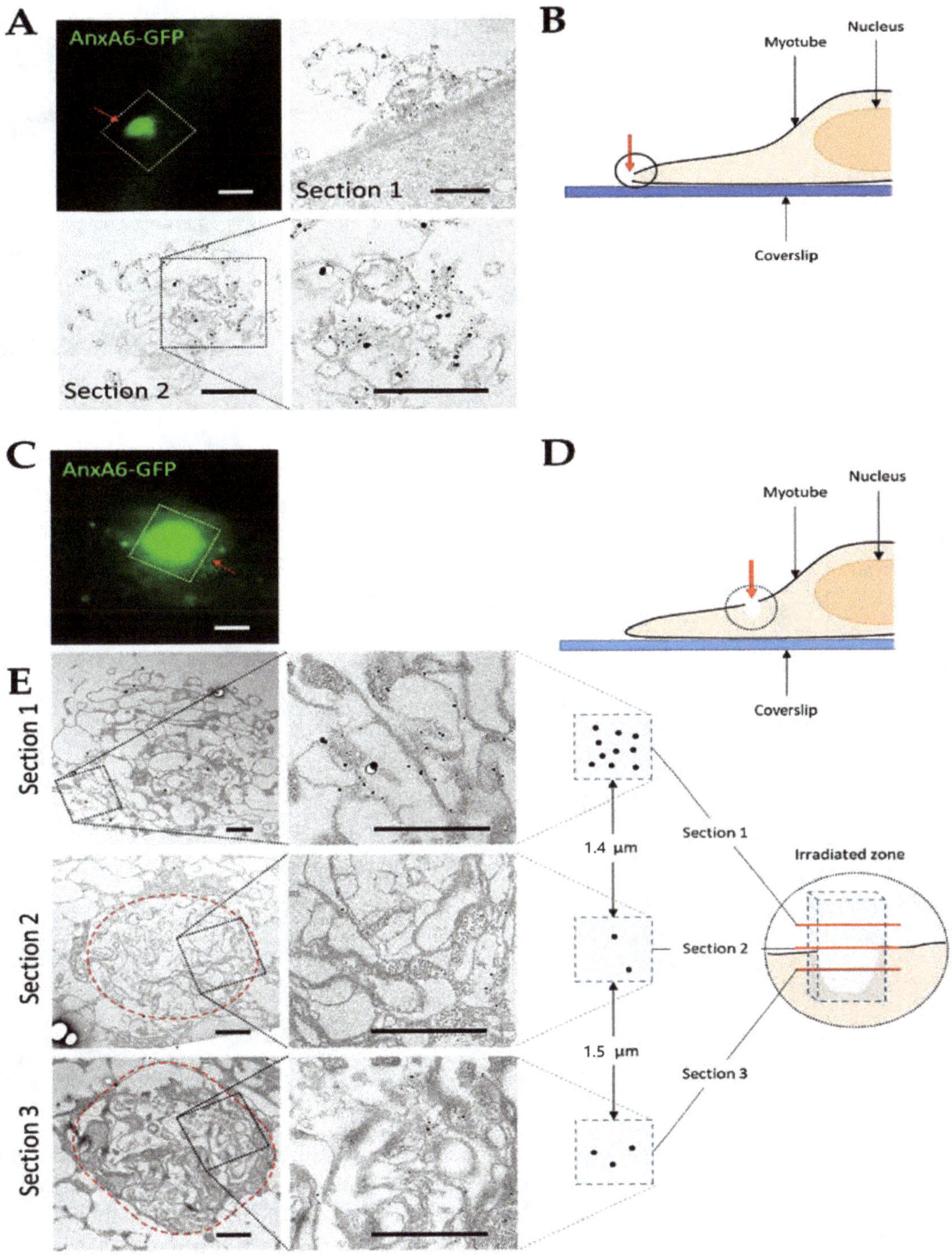

Figure 6. Correlative light and electron microscopy (CLEM) imaging of damaged AnxA6-GFP-expressing LHCN myotubes. LHCN myotubes were transfected with pA6-GFP plasmid and membrane damage was performed on the edge (**A**) or on the surface (**C**, **E**) of the myotubes as featured in (**B**) and (**D**), respectively. After membrane injury, AnxA6-GFP (green) was observed in fluorescence microscopy for about two minutes and chemically fixed. Myotubes were then immunostained for AnxA6 with a secondary antibody coupled with a gold nanoparticle (black particles in **A** and **E**) and analyzed by TEM (bright images in **A** and **E**). Fluorescence images in (**A**) and (**C**) correspond to the localization of AnxA6-GFP about two minutes after membrane injury. In (**E**), the diagram on the right represents the irradiated area and the location of the three sections observed. Red arrow, irradiated area; red dotted lines, boundaries of the damaged area. Scale bar in fluorescence microscopy: 5 µm. Scale bar in TEM: 1 µm.

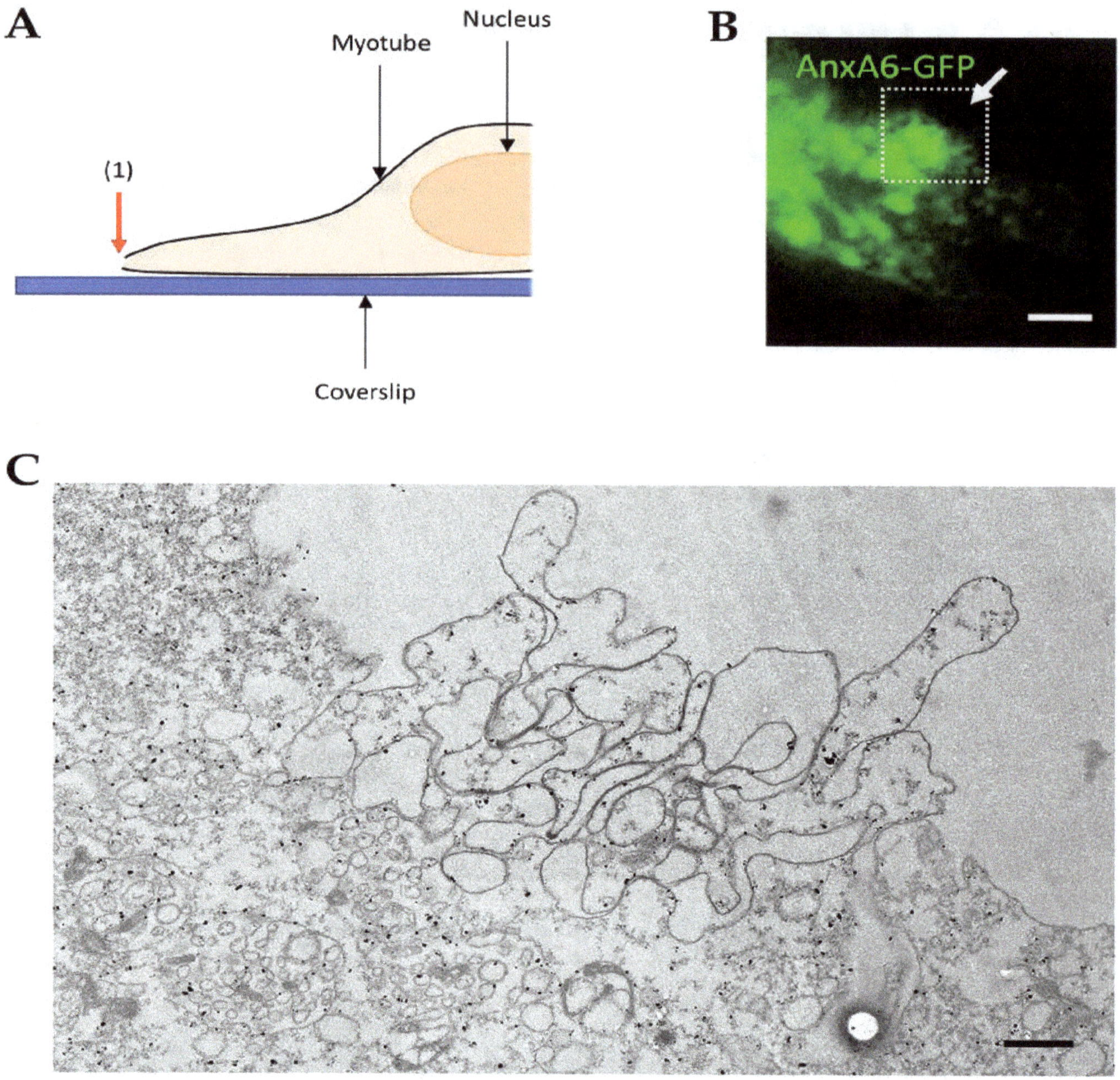

Figure 7. CLEM imaging of AnxA6-GFP-expressing LHCN myotubes during the formation of the cap subdomain. (**A**) LHCN myotubes were transfected with pA6-GFP plasmid and membrane damage was performed at the edge of the myotubes, which were then rapidly fixed. (**B**) Recruitment of AnxA6-GFP was checked by fluorescence microscopy as described in the legend of Figure 6. (**C**) Myotubes were then immunostained for AnxA6 with a secondary antibody coupled with a gold nanoparticle and observed in TEM. Red and white arrows, irradiated area. Scale bar in fluorescence microscopy: 5 μm. Scale bar in TEM: 1 μm.

4. Discussion

Here we show that the deficiency of AnxA6 in human myotubes compromises sarcolemma repair. This result confirms previous studies that have shown the crucial role played by AnxA6 in membrane repair [19,25,27,30]. We observed that AnxA6 expression increases during differentiation of myoblasts into myotubes, which correlates with the fact that myotubes are more prone to membrane damage and must be well-equipped to cope with more frequent mechanical stress due to contraction. In laser damaged human myotubes, AnxA6 is recruited to the disruption site within seconds after sarcolemma rupture, where it mainly interacts with cell membrane. About one minute after injury, AnxA6 is found in a very tight structure outside the myotube, which is identical to the cap subdomain previously described by McNally and collaborators [25]. The presence of dysferlin at the base of the cap subdomain indicates the existence of shoulder subdomains in human resealing sarcolemma, as described in murine myofibers [25].

Structure, origin and specific function of the cap subdomain remained unclear. It has been reported that it may correspond to a protein scaffold mainly constituted of Anx and essential for membrane repair [25]. Currently, our TEM images reveal the presence of a dense and disorganized membrane structure outside the myotube several minutes after membrane resealing, which corresponds to the cap subdomain. The cap subdomain would therefore not be a protein scaffold that participates in membrane repair, but rather the excess of lipid and protein material to be eliminated after membrane repair. Our TEM images also reveal that during the process of membrane resealing a large part of the membrane material is constituted of "filaments" that seem to correspond to extension of the cell membrane. Expansion of cell membrane triggered by lysosome exocytosis has been already proposed in order to reduce tension exerted by cortical cytoskeleton and thus promote membrane repair [42]. Our finding has resolved a longstanding mystery about "filaments" stained by FM1-43 and randomly observed by fluorescence microscopy during our membrane repair assays. This applied in particular to murine perivascular cells damaged by laser ablation, when the protocol used a powerful laser (160 mW) irradiating a large area (9 μm^2) (Video S4) [37].

In collaboration with the cluster of intracellular vesicles, the extension of the cell membrane may participate in the resealing process by forming a dense lipid structure plugging the hole. Compaction of the membrane material in this tangle of cell membrane may also facilitate its elimination once cell membrane was resealed. In zebrafish myofibers, it was observed that this task is performed by macrophages [43]. This would explain why the cap subdomain persists for several tens of minutes after membrane injury in our experiments, where only muscle cells are present.

Numerous studies question the existence of the lipid "patch" for membrane-resealing. We hypothesize that the formation of a lipid "patch" [8] and the cap subdomain are not mutually exclusive and may coexist. They could actually correspond to two successive stages of the membrane repair process, the formation of the cap subdomain resulting from the excess of membranes produced for membrane resealing and intended to be released.

Which is the role of AnxA6 in these processes? During membrane resealing, we observed that AnxA6 interacts essentially with plasma membrane. With AnxA5, AnxA6 is part of less Ca^{2+}-sensitive Anx for membrane binding [44]. The Ca^{2+} gradient created at the disruption site after membrane injury supports the binding of AnxA6 to cell membrane rather than to intracellular vesicles present deeply in the cytoplasm. It has been proposed that AnxA6 induces a constriction force closing the tear after AnxA4-induced plasma-membrane invagination in damaged MCF-7 cells [27]. Unlike MCF-7 cells, no inward invagination was observed at the disruption site in damaged human skeletal muscle cells. In addition, when membrane damage was performed on the surface of the myotube, neither fluorescence video-microscopy nor TEM images revealed the presence of AnxA6 surrounding the hole created by membrane disruption. AnxA6 may therefore act differently in cancer epithelial MCF-7 and muscle LHCN cells. It is important to note that AnxA4 is highly expressed in cancer cells [45,46] and that MCF-7 cells were exposed to strong injuries by laser irradiation inducing a large wound diameter (up to 3 μm) in the study cited above [27].

As the membrane repair mechanism may vary depending on the cell type, the extent of the damage and the spatial position of the injury [23,27,47], different experimental conditions may explain different behaviors observed for AnxA6.

The Anx family has been shown to be able to modify membrane morphology by inducing curvature and folding [27,28]. Together with their ability to aggregate adjacent membranes [14,24], Anx could allow the compaction of extended cell membrane present at the disruption site in order to promote membrane resealing and then to facilitate the release of the membrane excess.

We have shown that AnxA6 is rapidly recruited to the disruption site, by mainly interacting with the plasma membrane. At the disruption site, the excess of plasma membrane covered by AnxA6 could initiate the formation of the cap subdomain. By inducing folding and curvature of the extensions of cell membrane, AnxA6 may allow to form a tight structure plugging the hole. Once sarcolemma was resealed, AnxA6 is exclusively found in the cap subdomain interacting with accumulated membranes. It may therefore help at condensing membranes to be eliminated, this elimination being probably performed by macrophages [43].

The presence of two Anx core domains gives the ability to AnxA6 to bridge two adjacent membranes [24], such as two regions of the cell membrane and/or vesicle membranes. In addition, phylogenetic analysis of Anx has revealed that the core domain present at the N-terminus was similar to AnxA3 whereas at the C-terminus it was similar to AnxA5 [48]. Anx induce different and specific morphologies upon interaction with membranes [28]. For instance, AnxA3 rolls the membrane in a fragmented manner from free edges, producing multiple thin rolls, whereas AnxA5 induces cooperative roll-up of the membrane. AnxA6 may therefore induce multiple membrane rearrangements at the disruption site, which promote membrane repair and the formation of the cap subdomain.

Taken together our results enable to propose a model of the implication of AnxA6 in membrane repair of human skeletal muscle cells (Figure 8). The entry of Ca^{2+} induces the binding of AnxA6 to membranes surrounding the disruption site (Figure 8A,A'). At this stage, depolymerization of cortical actin and lysosome exocytosis induced by the massive increase of intracellular Ca^{2+} concentration may lead to a reduction in the tension exerted by the cortical cytoskeleton, as proposed by Togo et al. [42]. Increase in membrane surface by lysosome exocytosis may lead in the formation of an extension of sarcolemma at the disruption site, on which AnxA6 is attached (Figure 8B,B'). At this time, AnxA5 forms 2D arrays at the edges of the torn sarcolemma to prevent the expansion of the tear, as previously reported [26].

The aggregation of intracellular vesicles recruited to the disruption site may then form a lipid patch in order to plug the hole and interrupt exchanges between the intra- and extracellular environments (Figure 8C,C'). Anx and particularly AnxA6, may induce folding and curvature [28] of the excess of sarcolemma, thus initiating the formation of a tight membrane structure. Membrane excess due to lengthened sarcolemma and input of intracellular vesicles may form a structure where Anx accumulate (Figure 8D,D'). The ability of Anx to induce membrane aggregation and curvature [28] allows the compaction of this membrane excess to form the cap subdomain. The newformed membrane and the elimination of the cap subdomain by other cell types such as macrophages [43], may allow the membrane integrity of the cell to be regained (Figure 8E,E').

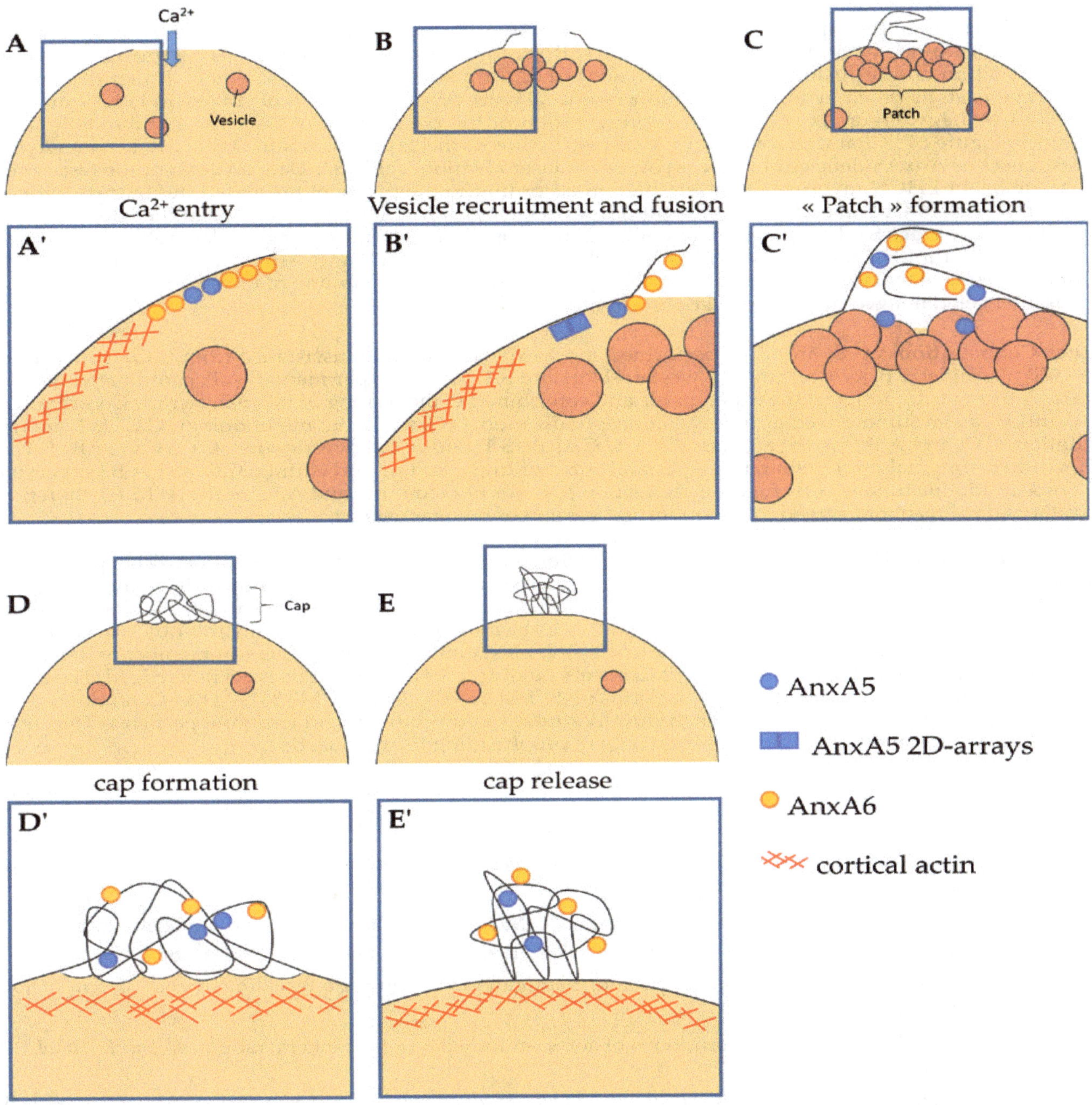

Figure 8. Model of membrane repair in human skeletal muscle cells. (**A,A′**) Entry of Ca^{2+} induces the recruitment of Anx to the plasma membrane, notably AnxA5 and AnxA6. (**B,B′**) The membrane tension is reduced by depolymerization of actin and exocytosis of lysosomes [42]. The increase in sarcolemma surface leads to excess membrane at the disruption site on which AnxA6 is associated. AnxA5 forms 2D arrays that strengthen the sarcolemma and limit the expansion of the tear, as previously reported [26]. Intracellular vesicles are recruited to the disruption site. (**C,C′**) Aggregation of intracellular vesicles forms a "patch" that plugs the rupture. AnxA6 and maybe AnxA5, induce the folding of the extensions of sarcolemma in order to form a tight structure. (**D,D′**) Accumulation of Anx leads to the folding and curvature [28] of membranes and the formation of the cap subdomain. (**E,E′**) The integrity of the sarcolemma is restored by the newformed plasma membrane and the elimination of the cap subdomain by macrophages (not represented) [43].

Supplementary Materials
Figure S1: Decrease in the expression of endogenous AnxA6 in MDA-MB-231 cells by shRNA strategy, Figure S2: Decrease in the expression of endogenous AnxA6 in shRNA-transduced LHCN myoblasts, Figure S3: The absence of differentiation and/or fusion of AnxA6-deficient LHCN myoblasts, Figure S4: Knock-down of AnxA6 in LHCN myoblasts assessed by immunocytofluorescence, Figure S5: Knock-down of AnxA6 in LHCN myotubes by shRNA strategy, Figure S6: Schematic representation of the positioning of laser for membrane injury on myotubes, Figure S7: Characterization of the ultrastructure of the cap subdomain, Table S1: Data analysis of the responses of AnxA6-deficient LHCN myotubes to laser ablation, Table S2: Data analysis of the responses of AnxA6-deficient LHCN myotubes to laser ablation according to the protocol presented in Figure 3, Video S1: Behavior of AnxA6-GFP in response to membrane damage by laser ablation performed at the lateral edge of a LHCN myotube, Video S2: Behavior of AnxA6-GFP in response to membrane damage by laser ablation performed on the surface of a LHCN myotube, Video S3: Behavior of AnxA6-GFP and AnxA5-mCherry in response to membrane damage by laser ablation in a LHCN myotube, Video S4: Membrane filaments emerge from murine perivascular cell damaged by powerful laser irradiation.

Author Contributions: C.C. and A.B. performed most experiments with assistance from C.G. (generation of Anx-GFP plasmids), F.B. (cell culture and western-blot) and S.T. (CLEM experiments). A.B. coordinated the entire project. A.B. and C.C. designed the experiments and contributed to the writing of the manuscript. Contribution of each author can be summarized as follows: conceptualization, C.C. and A.B.; methodology, C.C., S.T. and A.B.; validation, C.C. and A.B.; formal analysis, C.C., C.G., F.B., S.T. and A.B.; investigation, C.C., C.G., F.B., S.T. and A.B.; writing—original draft preparation, C.C. and A.B.; writing—review and editing, C.C. and A.B.; supervision, A.B.; project administration, A.B.; funding acquisition, C.C. for PhD funding and A.B. for the AFM-telethon grants. All authors have read and agreed to the published version of the manuscript.

Acknowledgments: The authors thank the AFM-Telethon for their financial support. The platform for immortalization of human cells of the Center of research in Myology (Paris, France) is acknowledged for access to human immortalized LHCN myoblasts. The authors thank Volker Gerke and Sophia Koerdt from the University of Muenster for the gift of pAnxA6-GFP-N1 plasmid. The fluorescence microscopy was done in the Bordeaux Imaging Center, a service unit of the CNRS-INSERM and Bordeaux University, member of the national infrastructure France BioImaging. The help of Christel Poujol and Sébastien Marais is acknowledged. The authors thank the staff of Vect'UB, the vectorology platforms (INSERM US 005- CNRS UMS 3427- TBM-Core, Université de Bordeaux, France) for technical assistance in shRNA strategy and production of lentiviral particles. The authors thank Melissa Sarps and Rachel Ngai for their help in proofreading the manuscript.

References

1. McNeil, P.L.; Khakee, R. Disruptions of muscle fiber plasma membranes. Role in exercise-induced damage. *Am. J. Pathol.* **1992**, *140*, 1097–1109. [PubMed]
2. Clarke, M.S.; Caldwell, R.W.; Chiao, H.; Miyake, K.; McNeil, P.L. Contraction-induced cell wounding and release of fibroblast growth factor in heart. *Circ. Res.* **1995**, *76*, 927–934. [CrossRef] [PubMed]
3. McNeil, P.L.; Ito, S. Gastrointestinal cell plasma membrane wounding and resealing in vivo. *Gastroenterology* **1989**, *96*, 1238–1248. [CrossRef]
4. Yu, Q.C.; McNeil, P.L. Transient disruptions of aortic endothelial cell plasma membranes. *Am. J. Pathol.* **1992**, *141*, 1349–1360. [PubMed]
5. Bansal, D.; Miyake, K.; Vogel, S.S.; Groh, S.; Chen, C.-C.; Williamson, R.; McNeil, P.L.; Campbell, K.P. Defective membrane repair in dysferlin-deficient muscular dystrophy. *Nature* **2003**, *423*, 168–172. [CrossRef] [PubMed]
6. Griffin, D.A.; Johnson, R.W.; Whitlock, J.M.; Pozsgai, E.R.; Heller, K.N.; Grose, W.E.; Arnold, W.D.; Sahenk, Z.; Hartzell, H.C.; Rodino-Klapac, L.R. Defective membrane fusion and repair in Anoctamin5-deficient muscular dystrophy. *Hum. Mol. Genet.* **2016**, *25*, 1900–1911. [CrossRef]
7. Steinhardt, R.A.; Bi, G.; Alderton, J.M. Cell membrane resealing by a vesicular mechanism similar to neurotransmitter release. *Science* **1994**, *263*, 390–393. [CrossRef]
8. McNeil, P.L.; Steinhardt, R.A. Plasma membrane disruption: Repair, Prevention, Adaptation. *Annu. Rev. Cell Dev. Biol* **2003**, *19*, 697–731. [CrossRef]
9. Steinhardt, R.A. The mechanisms of cell membrane repair: A tutorial guide to key experiments. *Ann. N. Y. Acad. Sci.* **2005**, *1066*, 152–165. [CrossRef]
10. Blazek, A.D.; Paleo, B.J.; Weisleder, N. Plasma membrane repair: A central process for maintaining cellular homeostasis. *Physiology* **2015**, *30*, 438–448. [CrossRef]
11. Moss, S.E.; Morgan, R.O. The annexins. *Genome Biol.* **2004**, *5*, 219. [CrossRef]

12. Gerke, V.; Creutz, C.E.; Moss, S.E. Annexins: Linking Ca^{2+} signalling to membrane dynamics. *Nat. Rev. Mol. Cell Biol.* **2005**, *6*, 449–461. [CrossRef] [PubMed]
13. Andree, H.A.; Reutelingsperger, C.P.; Hauptmann, R.; Hemker, H.C.; Hermens, W.T.; Willems, G.M. Binding of vascular anticoagulant alpha (VAC alpha) to planar phospholipid bilayers. *J. Biol. Chem.* **1990**, *265*, 4923–4928.
14. Blackwood, R.A.; Ernst, J.D. Characterization of Ca^{2+}-dependent phospholipid binding, vesicle aggregation and membrane fusion by annexins. *Biochem. J.* **1990**, *266*, 195–200. [CrossRef]
15. Monastyrskaya, K.; Babiychuk, E.B.; Hostettler, A.; Rescher, U.; Draeger, A. Annexins as intracellular calcium sensors. *Cell Calcium* **2007**, *41*, 207–219. [CrossRef] [PubMed]
16. Huber, R.; Römisch, J.; Paques, E.P. The crystal and molecular structure of human annexin V, an anticoagulant protein that binds to calcium and membranes. *EMBO J.* **1990**, *9*, 3867–3874. [CrossRef] [PubMed]
17. Swairjo, M.A.; Roberts, M.F.; Campos, M.B.; Dedman, J.R.; Seaton, B.A. Annexin V binding to the outer leaflet of small unilamellar vesicles leads to altered inner-leaflet properties: 31P- and 1H-NMR studies. *Biochemistry* **1994**, *33*, 10944–10950. [CrossRef] [PubMed]
18. Lennon, N.J.; Kho, A.; Bacskai, B.J.; Perlmutter, S.L.; Hyman, B.T.; Brown, R.H. Dysferlin interacts with annexins A1 and A2 and mediates sarcolemmal wound-healing. *J. Biol. Chem.* **2003**, *278*, 50466–50473. [CrossRef] [PubMed]
19. Roostalu, U.; Strähle, U. In vivo imaging of molecular interactions at damaged sarcolemma. *Dev. Cell* **2012**, *22*, 515–529. [CrossRef] [PubMed]
20. Selbert, S.; Fischer, P.; Menke, A.; Jockusch, H.; Pongratz, D.; Noegel, A.A. Annexin VII Relocalization as a Result of Dystrophin Deficiency. *Exp. Cell Res.* **1996**, *222*, 199–208. [CrossRef]
21. Swaggart, K.A.; Demonbreun, A.R.; Vo, A.H.; Swanson, K.E.; Kim, E.Y.; Fahrenbach, J.P.; Holley-Cuthrell, J.; Eskin, A.; Chen, Z.; Squire, K.; et al. Annexin A6 modifies muscular dystrophy by mediating sarcolemmal repair. *Proc. Natl. Acad. Sci. USA* **2014**, *111*, 6004–6009. [CrossRef] [PubMed]
22. Draeger, A.; Schoenauer, R.; Atanassoff, A.P.; Wolfmeier, H.; Babiychuk, E.B. Dealing with damage: Plasma membrane repair mechanisms. *Biochimie* **2014**, *107 Pt A*, 66–72. [CrossRef]
23. Boye, T.L.; Nylandsted, J. Annexins in plasma membrane repair. *Biol. Chem.* **2016**, *397*, 961–969. [CrossRef]
24. Buzhynskyy, N.; Golczak, M.; Lai-Kee-Him, J.; Lambert, O.; Tessier, B.; Gounou, C.; Bérat, R.; Simon, A.; Granier, T.; Chevalier, J.M.; et al. Annexin-A6 presents two modes of association with phospholipid membranes. A combined QCM-D, AFM and cryo-TEM study. *J. Struct. Biol.* **2009**, *168*, 107–116. [CrossRef] [PubMed]
25. Demonbreun, A.R.; Quattrocelli, M.; Barefield, D.Y.; Allen, M.V.; Swanson, K.E.; McNally, E.M. An actin-dependent annexin complex mediates plasma membrane repair in muscle. *J. Cell Biol.* **2016**, *213*, 705–718. [CrossRef] [PubMed]
26. Carmeille, R.; Bouvet, F.; Tan, S.; Croissant, C.; Gounou, C.; Mamchaoui, K.; Mouly, V.; Brisson, A.R.; Bouter, A. Membrane repair of human skeletal muscle cells requires Annexin-A5. *Biochim. Biophys. Acta Mol. Cell Res.* **2016**, *1863*, 2267–2279. [CrossRef]
27. Boye, T.L.; Maeda, K.; Pezeshkian, W.; Sønder, S.L.; Haeger, S.C.; Gerke, V.; Simonsen, A.C.; Nylandsted, J. Annexin A4 and A6 induce membrane curvature and constriction during cell membrane repair. *Nat. Commun.* **2017**, *8*, 1623. [CrossRef]
28. Boye, T.L.; Jeppesen, J.C.; Maeda, K.; Pezeshkian, W.; Solovyeva, V.; Nylandsted, J.; Simonsen, A.C. Annexins induce curvature on free-edge membranes displaying distinct morphologies. *Sci. Rep.* **2018**, *8*, 10309. [CrossRef]
29. Demonbreun, A.R.; Allen, M.V.; Warner, J.L.; Barefield, D.Y.; Krishnan, S.; Swanson, K.E.; Earley, J.U.; McNally, E.M. Enhanced Muscular Dystrophy from Loss of Dysferlin Is Accompanied by Impaired Annexin A6 Translocation after Sarcolemmal Disruption. *Am. J. Pathol.* **2016**, *186*, 1610–1622. [CrossRef]
30. Demonbreun, A.R.; Fallon, K.S.; Oosterbaan, C.C.; Bogdanovic, E.; Warner, J.L.; Sell, J.J.; Page, P.G.; Quattrocelli, M.; Barefield, D.Y.; McNally, E.M. Recombinant annexin A6 promotes membrane repair and protects against muscle injury. *J. Clin. Investig.* **2019**, *129*, 4657–4670. [CrossRef]
31. Zhu, C.-H.; Mouly, V.; Cooper, R.N.; Mamchaoui, K.; Bigot, A.; Shay, J.W.; Di Santo, J.P.; Butler-Browne, G.S.; Wright, W.E. Cellular senescence in human myoblasts is overcome by human telomerase reverse transcriptase and cyclin-dependent kinase 4: Consequences in aging muscle and therapeutic strategies for muscular dystrophies. *Aging Cell* **2007**, *6*, 515–523. [CrossRef] [PubMed]

32. Bouvet, F.; Ros, M.; Bonedeau, E.; Croissant, C.; Frelin, L.; Saltel, F.; Moreau, V.; Bouter, A. Defective membrane repair machinery impairs survival of invasive cancer cells. *bioRxiv* **2020**. [CrossRef]
33. Carmeille, R.; Croissant, C.; Bouvet, F.; Bouter, A. Membrane repair assay for human skeletal muscle cells. *Methods Mol. Biol.* **2017**, *1668*, 195–207. [PubMed]
34. Geissbuehler, M.; Lasser, T. How to display data by color schemes compatible with red-green color perception deficiencies. *Opt. Express* **2013**, *21*, 9862. [CrossRef]
35. Croissant, C.; Bouvet, F.; Tan, S.; Bouter, A. Imaging Membrane Repair in Single Cells Using Correlative Light and Electron Microscopy. *Curr. Protoc. Cell Biol.* **2018**, e55. [CrossRef]
36. Vilá de Muga, S.; Timpson, P.; Cubells, L.; Evans, R.; Hayes, T.E.; Rentero, C.; Hegemann, A.; Reverter, M.; Leschner, J.; Pol, A.; et al. Annexin A6 inhibits Ras signalling in breast cancer cells. *Oncogene* **2009**, *28*, 363–377. [CrossRef]
37. Bouter, A.; Gounou, C.; Bérat, R.; Tan, S.; Gallois, B.; Granier, T.; d'Estaintot, B.L.; Pöschl, E.; Brachvogel, B.; Brisson, A.R. Annexin-A5 assembled into two-dimensional arrays promotes cell membrane repair. *Nat. Commun.* **2011**, *2*, 270. [CrossRef]
38. Carmeille, R.; Degrelle, S.A.; Plawinski, L.; Bouvet, F.; Gounou, C.; Evain-Brion, D.; Brisson, A.R.; Bouter, A. Annexin-A5 promotes membrane resealing in human trophoblasts. *Biochim. Biophys. Acta Mol. Cell Res.* **2015**, *1853*, 2033–2044. [CrossRef] [PubMed]
39. McNeil, A.K.; Rescher, U.; Gerke, V.; McNeil, P.L. Requirement for annexin A1 in plasma membrane repair. *J. Biol. Chem.* **2006**, *281*, 35202–35207. [CrossRef]
40. Cai, C.; Masumiya, H.; Weisleder, N.; Matsuda, N.; Nishi, M.; Hwang, M.; Ko, J.-K.; Lin, P.; Thornton, A.; Zhao, X.; et al. MG53 nucleates assembly of cell membrane repair machinery. *Nat. Cell Biol.* **2009**, *11*, 56–64. [CrossRef]
41. Jaiswal, J.K.; Lauritzen, S.P.; Scheffer, L.; Sakaguchi, M.; Bunkenborg, J.; Simon, S.M.; Kallunki, T.; Jäättelä, M.; Nylandsted, J. S100A11 is required for efficient plasma membrane repair and survival of invasive cancer cells. *Nat. Commun.* **2014**, *5*, 3795. [CrossRef] [PubMed]
42. Togo, T.; Krasieva, T.B.; Steinhardt, R.A. A decrease in membrane tension precedes successful cell-membrane repair. *Mol. Biol. Cell* **2000**, *11*, 4339–4346. [CrossRef] [PubMed]
43. Middel, V.; Zhou, L.; Takamiya, M.; Beil, T.; Shahid, M.; Roostalu, U.; Grabher, C.; Rastegar, S.; Reischl, M.; Nienhaus, G.U.; et al. Dysferlin-mediated phosphatidylserine sorting engages macrophages in sarcolemma repair. *Nat. Commun.* **2016**, *7*. [CrossRef] [PubMed]
44. Skrahina, T.; Piljić, A.; Schultz, C. Heterogeneity and timing of translocation and membrane-mediated assembly of different annexins. *Exp. Cell Res.* **2008**, *314*, 1039–1047. [CrossRef] [PubMed]
45. Choi, C.H.; Chung, J.Y.; Chung, E.J.; Sears, J.D.; Lee, J.W.; Bae, D.S.; Hewitt, S.M. Prognostic significance of annexin A2 and annexin A4 expression in patients with cervical cancer. *BMC Cancer* **2016**, *16*, 1–11. [CrossRef] [PubMed]
46. Deng, S.; Wang, J.; Hou, L.; Li, J.; Chen, G.; Jing, B.; Zhang, X.; Yang, Z. Annexin A1, A2, A4 and A5 play important roles in breast cancer, pancreatic cancer and laryngeal carcinoma, alone and/or synergistically. *Oncol. Lett.* **2013**, *5*, 107–112. [CrossRef]
47. Jimenez, A.J.; Perez, F. Plasma membrane repair: The adaptable cell life-insurance. *Curr. Opin. Cell Biol.* **2017**, *47*, 99–107. [CrossRef]
48. Morgan, R.O.; Fernández, M.P. Molecular phylogeny of annexins and identification of a primitive homologue in Giardia lamblia. *Mol. Biol. Evol.* **1995**, *12*, 967–979. [CrossRef]

14

Selective Degradation Permits a Feedback Loop Controlling Annexin A6 and Cholesterol Levels in Endolysosomes of NPC1 Mutant Cells

Elsa Meneses-Salas [1,2], Ana García-Melero [1,2], Patricia Blanco-Muñoz [1,2], Jaimy Jose [3], Marie-Sophie Brenner [3], Albert Lu [4], Francesc Tebar [1,2], Thomas Grewal [3,*], Carles Rentero [1,2,*] and Carlos Enrich [1,2,*]

1 Departament de Biomedicina, Unitat de Biologia Cel·lular, Facultat de Medicina i Ciències de la Salut, Universitat de Barcelona, 08036 Barcelona, Spain; elsameneses.s@gmail.com (E.M.-S.); anagarciamelero@gmail.com (A.G.-M.); patrii__4@live.com (P.B.-M.); tebar@ub.edu (F.T.)

2 Centre de Recerca Biomèdica CELLEX, Institut d'Investigacions Biomèdiques August Pi i Sunyer (IDIBAPS), 08036-Barcelona, Spain

3 School of Pharmacy, Faculty of Medicine and Health, University of Sydney, Sydney 2006, NSW, Australia; jjos6151@uni.sydney.edu.au (J.J.); marie-sophie.brenner@sydney.edu.au (M.-S.B.)

4 Department of Biochemistry, Stanford University School of Medicine, Stanford, CA 94305, USA; alulopez@stanford.edu

* Correspondence: thomas.grewal@sydney.edu.au (T.G.); carles.rentero@ub.edu (C.R.); enrich@ub.edu (C.E.)

Abstract: We recently identified elevated annexin A6 (AnxA6) protein levels in Niemann–Pick-type C1 (NPC1) mutant cells. In these cells, AnxA6 depletion rescued the cholesterol accumulation associated with NPC1 deficiency. Here, we demonstrate that elevated AnxA6 protein levels in NPC1 mutants or upon pharmacological NPC1 inhibition, using U18666A, were not due to upregulated AnxA6 mRNA expression, but caused by defects in AnxA6 protein degradation. Two KFERQ-motifs are believed to target AnxA6 to lysosomes for chaperone-mediated autophagy (CMA), and we hypothesized that the cholesterol accumulation in endolysosomes (LE/Lys) triggered by the NPC1 inhibition could interfere with the CMA pathway. Therefore, AnxA6 protein amounts and cholesterol levels in the LE/Lys (LE-Chol) compartment were analyzed in NPC1 mutant cells ectopically expressing lysosome-associated membrane protein 2A (Lamp2A), which is well known to induce the CMA pathway. Strikingly, AnxA6 protein amounts were strongly decreased and coincided with significantly reduced LE-Chol levels in NPC1 mutant cells upon Lamp2A overexpression. Therefore, these findings suggest Lamp2A-mediated restoration of CMA in NPC1 mutant cells to lower LE-Chol levels with concomitant lysosomal AnxA6 degradation. Collectively, we propose CMA to permit a feedback loop between AnxA6 and cholesterol levels in LE/Lys, encompassing a novel mechanism for regulating cholesterol homeostasis in NPC1 disease.

Keywords: cholesterol; AnxA6; chaperone-mediated autophagy; endolysosomes; NPC1; Lamp2A

---

1. Introduction

Cholesterol is a fundamental lipid in mammalian cells responsible for the proper organization and function of membranes. The levels of cholesterol in cellular compartments are tightly controlled through *de novo* synthesis in the endoplasmic reticulum (ER), and the uptake of low-density lipoproteins (LDL) by receptor-mediated endocytosis. As excess amounts of cellular unesterified (free) cholesterol are cytotoxic, cells have developed sophisticated circuits to regulate its intracellular sorting, trafficking and storage [1]. Once internalized, LDL-derived cholesterol is targeted to the LE/Lys compartment where cholesterol is first transferred from intraluminal vesicles (ILVs) to the limiting membrane via NPC2, lysobisphosphatidic acid (LBPA), and possibly other transporters [2–5]. In the outer LE/Lys membrane, NPC1 is the major transporter, and together with several other cholesterol-binding proteins [6], is responsible for LE-Chol export and subsequent transfer to other cellular destinations [7], preferentially the plasma membrane and ER, but also mitochondria, peroxisomes, Golgi, or recycling endosomes. In the ER, cholesterol can be re-esterified, permitting cytoplasmic storage of excess cholesterol in lipid droplets. Several pathways regulate the delivery of cholesterol from LE/Lys to other cellular sites. This includes vesicular trafficking via small GTPases (e.g., Rab7, Rab8, and Rab9), non-vesicular transport mediated by lipid transfer proteins, or cholesterol transfer across membrane contact sites (MCS) [8]. In addition, autophagy also contributes to regulate lipid metabolism in the LE/Lys compartment [9–11]. Therefore, it has been suggested that alterations in autophagy may contribute to the pathology of lipid storage disorders. For example, Sarkar et al. (2013) identified defective autophagy in Niemann–Pick type C1 (NPC1) disease models to be associated with cholesterol accumulation [12]. In these studies, failure of the SNAP receptor (SNARE) machinery caused defects in amphisome formation, which impaired the maturation of autophagosomes, while the lysosomal proteolytic function remained unaffected. In this setting, ectopic NPC1 expression rescued the defect in autophagosome formation. Intriguingly, both the inhibition and stimulation of autophagy caused cholesterol accumulation in LE/Lys, suggesting that the regulation of autophagy may be intimately linked to changes in LE-Chol levels [13,14]. To date, the precise way in which autophagy can alter LE-Chol homeostasis still remains elusive.

The complexity of autophagic pathways has been described in detail in recent reviews [15,16]. Calcium (Ca^{2+}) is a well-known regulator of autophagy, yet despite the wide range of lysosomal storage diseases that share defects in both autophagy and Ca^{2+} homeostasis, the intersection between these two pathways is still not well characterized [17]. In fact, a number of Ca^{2+}-binding proteins, including apoptosis-linked gene-2 (ALG-2); calmodulin; several S100 family proteins; ALG-2-interacting protein 1 (AIP1, also called Alix); calcineurin; as well as Ca^{2+} channels in LE/Lys, the ER, or mitochondria [18], have been associated with autophagy.

In addition, three members of the annexin family—AnxA1, A2, and A5—have been associated with autophagic processes [19]. Annexins are a conserved multigene family of proteins that bind to membranes in a Ca^{2+}-dependent manner and are widely expressed [20]. Within the endocytic pathway, they have been associated with a variety of membrane trafficking events, including vesicle transport and fusion, microdomain organization, and LE/Lys positioning, as well as membrane-associated actin cytoskeleton dynamics and cholesterol homeostasis [21–23]. Furthermore, AnxA1 and AnxA6 participate in MCS formation [24,25], regulating the transfer of cholesterol, and possibly other lipids and Ca^{2+}, from LE/Lys to other cellular sites [23].

Despite the accumulating knowledge on the abovementioned annexins and their mode of action in late endocytic circuits, including autophagy, our understanding how these annexins operate in this cellular location is still incomplete. Yet, to exert their various functions, their physical association with the LE/Lys compartment seems essential. The accessibility to membrane lipids that serve as annexin binding sites, in particular, phosphatidylserine and phosphatidic acid, but also cholesterol and phosphatidylinositol (4,5)-bisphosphate (PIP2), is well documented [22]. Therefore, for the association of annexins with LE/Lys membranes in space and time, the lipid composition of LE/Lys, which is likely to undergo dynamic changes not only due to membrane turnover, but also nutrient availability and consequently, the endocytic activity of cells, together with the differential affinity of each annexin for individual lipid species, is most likely critical. In addition, the cytosolic tails of proteins in the limiting LE/Lys membrane may directly or indirectly further add to establish interactions. In support of this, direct protein–protein interactions of annexins with several LE/Lys proteins have been reported, including the two-pore channel (TPC1/2) [26], NPC1 [27], or the vacuolar proton pump (ATPase H^+ transporting VO subunit a2) [28]. This complex interactome of annexins with lipids and proteins could ensure proper LE/Lys functioning and provide opportunity to control the intracellular movement of lipids, in particular, cholesterol, into and out of the LE/Lys compartment. Indeed, we and others demonstrated that AnxA1, A2, A6, and A8 contribute to cholesterol trafficking along endo- and exocytic pathways, suggesting that this subset of annexins, as well as other Ca^{2+}-binding proteins, are good candidates for achieving regulatory tasks not only in regard to cellular cholesterol homeostasis, but also cholesterol-dependent activities in LE/Lys such as autophagy [23].

A recent proteomic analysis of CMA-targeting motifs supported earlier studies that revealed most members of the human annexin family to contain the KFERQ-like motif known to target cytosolic proteins for CMA or endosomal microautophagy (eMi) into lysosomes [15,29–31]. CMA and eMi are initiated by the binding of heat shock cognate 71 kDa protein (hsc70) to the pentapeptide motif in cytosolic proteins, which enables their binding to Lamp2A at the LE/Lys membrane. This initial recruitment step is critical for the subsequent translocation of these selective substrates into lysosomes via CMA. eMi is less well studied and relies on the endosomal sorting complexes required for transport (ESCRT) machinery [32]. Besides, CMA and eMI, macroautophagy can also engulf cytoplasmic proteins and organelles for lysosomal degradation. This process requires the formation of autophagosomes before fusion with lysosomes. Several annexins, including AnxA1 and AnxA5, have been suggested to play a role in autophagosome maturation or further steps of autophagy [33], but the underlying molecular mechanisms still have to be elucidated.

We previously demonstrated that overexpression of AnxA6 was associated with LE-Chol accumulation, mimicking a NPC1 mutant-like phenotype, which was detrimental for the intracellular trafficking along endo- and exocytic routes [27,34]. In addition, we recently documented that NPC1 mutant Chinese hamster ovary (CHO) cells displayed elevated AnxA6 levels. This coincided with an increased and AnxA6-mediated recruitment of a Rab7-GTPase activating protein (Rab7-GAP), TBC1D15, to LE/Lys to inactivate Rab7. AnxA6-dependent Rab7 inactivation in NPC1 mutant cells was associated with a reduction of MCS between LE/Lys and the ER. *Vice versa*, AnxA6 depletion in NPC1 mutant cells caused elevated Rab7 activity, increased MCS formation, and significantly reduced the amount of cholesterol in LE/Lys [25].

In this report, we examined the link between AnxA6 protein levels and the extent of LE-Chol accumulation in LE/Lys of NPC1 mutant cells. Our findings reveal a complex feedback loop involving CMA that controls AnxA6 and LE-Chol levels. In NPC1 deficiency, or other disease settings where lysosomal dysfunction may be associated with LE-Chol accumulation, this may eventually increase the amount of AnxA6 in ILVs of the LE/Lys compartment. Remarkably, these findings could be relevant for intercellular communication, as extracellular vesicles, or exosomes, which deliver biologically active molecules to recipient cells, originate from the LE/Lys compartment, and could explain elevated levels of this annexin in exosomes observed in other pathologies [35,36].

2. Material and Methods

2.1. Cells, Reagents, Cell Culture and Transfections

CHO cells were grown in Ham's F-12 with 10% fetal calf serum (FCS), L-glutamine (2 mM), penicillin (100 units/mL), and streptomycin (100 μg/mL) at 37 °C, 5% CO_2. CHO M12 and CHO 2-2 were kindly provided by Drs. L. Liscum (Tufts University School of Medicine, Boston, MA, USA) and D. Ory (Washington University, St. Louis, MO, USA). The AnxA6-depleted CHO M12-A6ko cell line was generated in our laboratories using CRISPR/Cas9 editing technology as described [25]. Control (GM5659D) and NPC1 mutant human skin fibroblasts (GM03123) were from the Coriell Institute for Medical Research (Camden, NJ, USA) and cultured in DMEM, 10% FCS, L-glutamine (2 mM), penicillin (100 units/mL), and streptomycin (100 μg/mL) at 37 °C, 5% CO_2. Nutrient Mixture Ham's F-12 and DMEM were from Biological Industries (Cromwell, CT, USA). Hank's buffered salt solution (HBSS) was from Gibco (Waltham, MA, USA). Filipin, saponin, U18666A, cycloheximide (CHX), leupeptin, and aprotinin were from Sigma-Aldrich (St. Louis, MO, USA). Paraformaldehyde (PFA) was from Electron Microscopy Sciences (Hatfield, PA, USA), and Mowiol was from Calbiochem (San Diego, CA,

USA). Glutathione S-transferase (GST) (GE Healthcare, Chicago, IL, USA) and GST-fusion protein (Rab Interacting Lysosomal Protein (RILP)-C33-GST) were produced in *E. coli* BL21 cells and purified using Glutathione-Sepharose 4B beads (GE Healthcare, Chicago, IL, USA) as reported previously [37]. For transient transfections, cells were transfected using GenJet (SigmaGen Laboratories, Rockville, MD, USA) following manufacturer's instructions.

2.2. RNA Extraction and Quantitative Real-Time PCR

Total RNA was extracted using the RNeasy Mini Kit (Qiagen, Venlo, Netherlands) in accordance with the manufacturer's protocol. One microgram of RNA was reverse transcribed using High-Capacity cDNA Reverse Transcription Kit (Applied Bioscience, Waltham, MA, USA). In a final volume of 20 µL, real-time PCR Brilliant SYBRGreen QPCR Master Mix (Agilent Technologies, Santa Clara, CA, USA) and 10 µL of 1:20 diluted cDNA was used as a template for PCR analysis using the LightCycler system (Roche Diagnostics, Rotkreuz, Switzerland). Specific hamster primers (*AnxA6*: forward 5′-ccgggaagatgctaggaat-3′, reverse 5′-accctggtgagggtcttctt-3′; *Rpl13*: forward 5′-gccccacttccacaaggatt-3′, reverse 5′-ataccagccaccctgagttc-3′) were added and a standard PCR amplification protocol (10 min at 95 °C, 45 cycles of 30 s. at 95 °C, 15 s. at 60 °C and 30 s. at 72 °C, 10 s. at 95 °C and 60 s. at 65 °C) was performed according to manufacturer's instructions. Values were normalized to the *Rpl13* gene in each sample.

2.3. Immunoblotting

For western blotting, whole cell lysates were prepared in lysis buffer (50 mM Tris, 150 mM NaCl, 5 mM EDTA, pH 7.5), protease inhibitors (1 mM Na_3VO_4, 10 mM NaF, 1 mM phenylmethylsulfonyl fluoride, 10 µg/mL leupeptin, and 10 µg/mL aprotinin), and equal amounts of protein, as determined by Bradford assay, were separated by 10% SDS-PAGE and transferred to a Hybridation Nitrocellulose membrane (Millipore, Burlington, MA, USA). After blocking in 5% skimmed milk, western blots were performed using rabbit anti-AnxA6 serum [38], mouse anti-β-actin (MP Biomedicals, Illkirch-Graffenstaden, France), mouse anti-epidermal growth factor receptor kinase substrate 8 (Eps8; BD Transduction Lab, San Jose, CA, USA), rabbit anti-GFP (Abcam, Cambridge, UK), rabbit anti-Rab7 (Cell Signaling Technology, Danvers, MA, USA) and mouse anti-α-tubulin (Sigma-Aldrich, St. Louis, MO, USA). Appropriate peroxidase-conjugated secondary antibodies (BioRad Laboratories, Hercules, CA, USA) and enhanced detection (EZ-ECL; Enhanced ChemiLuminiscence, Biological Industries, Cromwell, CT, USA) for band visualization were used. The intensity of bands was quantified using ImageJ and results were normalized to actin.

2.4. Rab7-GTP Pull-Down Assays

Cells were solubilized in pull-down buffer (50 mM Tris, 150 mM NaCl, 1% Triton X-100, 0.1 mM $CaCl_2$, pH 7.3) supplemented with protease/phosphatase inhibitor cocktail (see above). Samples were centrifuged at 12,000× *g* for 10 min at 4 °C. Proteins from post-nuclear supernatants (400–700 µg) were incubated with Glutathione Sepharose 4B beads coated with purified recombinant RILP-C33-GST (40–70 µg) fusion protein for 2 h at 4 °C. GST was used as a negative control (data not shown). Samples were washed 3 times, collected in 30 µL of 1× loading buffer, and analyzed by western blotting (see above).

2.5. Fluorescence Microscopy

For fluorescence studies, CHO M12 cells were grown on coverslips, transfected with Lamp2A-GFP as per manufacturer instructions, and after 48 h, fixed with 4% PFA for 20 min, washed and then stained with filipin (0.05 mg/mL) and mounted in Mowiol as described [39]. Samples were observed using a Leica TCS SP5 laser scanning confocal microscope equipped with a DMI6000 inverted microscope, blue diode (405 nm), Argon (458/476/488/496/514), diode pumped solid state (561 nm), HeNe (594/633 nm) lasers, and APO 63× oil immersion objective lenses (Leica Microsystems, Wetzlar, Germany). Image analysis was performed with NIH ImageJ software as described below [40].

The quantification of filipin staining per cell (≥20 cells per group) was performed as described [25] and served to determine the relative amounts of free cholesterol in enlarged, perinuclear cholesterol-enriched and Lamp2A-positive late endosomal vesicles in NPC1 mutant M12 cells.

2.6. Transmission Electron Microscopy (TEM)

CHO-WT and CHO M12 cells were fixed at 4 °C with a mixture of 4% PFA and 0.1% glutaraldehyde in 0.1 M phosphate buffer, then scrapped, pelleted, washed in 2% PFA, then cryofixed by high pressure freezing (EM HPM100; Leica Microsystems, Wetzlar, Germany) and finally freeze-substituted for 80 h at −90 °C in anhydrous acetone containing 0.5% uranyl acetate. Next, samples were embedded in Lowicryl HM20 resin (Electron Microscopy Science, Hatfield, PA, USA). Ultrathin sections were obtained using an Ultracut UC6 ultramicrotome (Leica Microsystems, Wetzlar, Germany) and picked up on Formvar-coated nickel grids and observed using a transmission electron microscope, JEOL JEM-1010 (JEOL USA, Peabody, MA, USA) fitted with a Gatan Orius SC1000 (model 832) digital camera (Gatan Inc, Pleasnaton, CA, USA).

For immunolabeling, sections were incubated at room temperature on drops of 5% bovine serum albumin (BSA) in PBS for 20 min, followed by anti-AnxA6 polyclonal antibody (1:5 dilution) in PBS for 1 h. Then, sections were incubated for 30 min using anti-rabbit 12 nm colloidal gold (Jackson ImmunoResearch, West Grove, PA, USA) using a 1:30 dilution in PBS/1% BSA. This was followed by five washes with drops of PBS and 5 min in 1% glutaraldehyde; then, grids were rinsed with milliQ water and air dried. As a control for non-specific binding of the colloidal gold-conjugated antibody, the primary AnxA6 antibody was omitted. Three independent experiments were analyzed and at least 2 grids were used for each condition per cell line. The minimum number of cells scored for each condition was 25 and the average number of sections (fields) was 40. Data are shown as the mean ± SEM.

2.7. Image Analysis

Image analysis was performed using NIH ImageJ software [40]. When comparing different samples, images were captured and systematically screened using identical microscope settings.

2.8. Bulk Degradation Experiments

For the assessment of bulk degradation, CHO-WT, CHO M12, and CHO M12-A6ko cells were incubated with 10 µg/mL Dye Quenched-Red-BSA (DQ-Red-BSA) (Thermo Fisher Scientific, Waltham, MA, USA) in complete culture medium for 6 h at 37 °C. Then, cells were fixed in 4% PFA, washed extensively, stained with 4′,6-diamidino-2-phenylindole (DAPI) (Sigma-Aldrich, St. Louis, MO, USA) for 20 min, and then mounted with Mowiol. Control experiments with 100 µM Bafilomycin A1 (Sigma-Aldrich, St. Louis, MO, USA), which blocks lysosomal acidification, were performed (data not shown) [41,42]. Images were obtained with a Zeiss LSM880 Confocal Microscope (Zeiss Microscopy, Jena, Germany). Quantification of fluorescence intensity was performed as described [41].

2.9. Statistics

Unless mentioned in the figure legend, group data are presented as mean ± SEM. Comparison between 2 groups was analyzed by Student's *t*-test, comparison between more than 2 groups was analyzed by one-way ANOVA with a Bonferroni *post hoc* test, and comparison between groups was analyzed using Prism software (GraphPad Software, San Diego, CA, USA). Differences were considered statistically significant at * $p < 0.05$, ** $p < 0.01$, *** $p < 0.001$.

3. Results and Discussion

3.1. AnxA6 Expression in NPC1 Mutant Cells

Cholesterol accumulation in LE/Lys is a distinctive feature of cells carrying loss-of-function mutations in the NPC1 gene and we previously showed that AnxA6 overexpression mimics this phenotype [27]. Furthermore, we recently identified that endogenous AnxA6 levels were significantly elevated in CHO cell lines harboring different NPC1 mutations (M12, 2-2) [25]. This prompted us to investigate the relationship between AnxA6 levels and LE-Chol accumulation in the LE/Lys compartment of NPC1 mutant cells. We first aimed to validate these findings by analyzing AnxA6 protein levels in human control (HSF-WT) and NPC1 patient skin fibroblasts (GM03123), or upon incubation of CHO wild type (WT) cells with the pharmacological NPC1 inhibitor U18666A. Indeed, loss of NPC1 function in both settings correlated with increased amounts of AnxA6 (Figure 1A,B). Remarkably, higher AnxA6 protein levels in cells lacking functional NPC1 protein were not associated with a significant increase in AnxA6 mRNA expression, as shown by quantitative RT-PCR of NPC1 mutant CHO cells or after U18666A treatment (Figure 1C). Therefore, we reasoned that alterations in the posttranscriptional regulation could be responsible for elevated AnxA6 protein levels in NPC1 mutant cells, including defects in AnxA6 degradation and/or AnxA6 protein accumulation in the LE/Lys compartment due to lysosomal dysfunction in NPC disease [43].

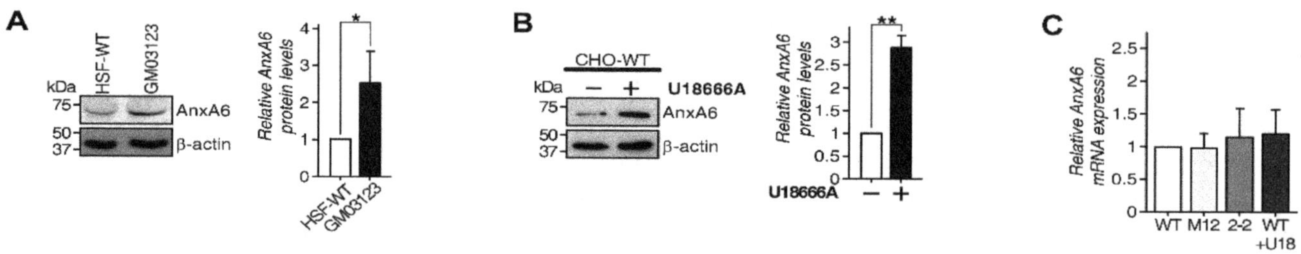

Figure 1. Elevated AnxA6 protein levels in NPC1 mutant cells. Cell lysates from (**A**) human control (HSF) and NPC1 patient (GM03123) skin fibroblasts and (**B**) CHO-WT treated for 16 h with and without U18666A (2 μg/mL) as indicated were analyzed by Western blotting with antibodies against AnxA6 and β-actin as loading control. Western blots representative for three independent experiments are shown (* $p < 0.05$, ** $p < 0.01$). (**C**) Relative mRNA quantification determined by qPCR of *AnxA6* in CHO-WT, CHO M12, CHO 2-2, and CHO-WT cells treated with U18666A (WT + U18) is shown. The housekeeper gene *Rpl13* served as control. Data shown represents the mean ± SEM from three independent experiments with duplicate samples.

Therefore, we examined the bulk degradation using fluorescently-labeled DQ-Red-BSA as cargo [41], which is internalized and transported to lysosomes and then cleaved, resulting in a bright red fluorescent signal. In line with published data [12], CHO M12 cells displayed a 2-3-fold increased capacity of degradative endolysosomes (Supplementary Figure S1A, see quantification in B) compared to CHO-WT cells. Strikingly, AnxA6 depletion in CHO M12 cells (CHO M12-A6ko), which we reported recently to significantly reduce LE-Chol accumulation in these cells [25], also drastically reduced lysosomal DQ-Red-BSA degradation comparable to the CHO-WT control cells. It is tempting to speculate that in addition to the Rab7-mediated restoration of cholesterol export in NPC1 mutant cells upon AnxA6 depletion [25], the reduced bulk degradation in CHO M12-A6ko cells is also connected to elevated Rab7 activity in these cells, as Rab7 is well known to drive the fusion of late endosomes with lysosomes and cargo degradation [44]. Furthermore, as this connects AnxA6 up- or downregulation to the modulation of endolysosomal degradation in NPC1 mutant cells, one could envisage also other degradation routes such as autophagy, to be altered by changes in AnxA6 levels. Yet, as described further below, degradation can be selective and different pathways and mechanisms have been described.

In some models, lipid accumulation in LE/Lys inhibits lysosomal cathepsins, leading to impaired degradation of lysosomal cargo and increased autophagic intermediates [14]. However, in the latter experiments described above (Supplementary Figure S1), DQ-Red-BSA degradation was increased in CHO M12 cells, and we hypothesize that possibly the enlarged LE/Lys compartment in these NPC1 mutant cells [45,46] may be one of the underlying causes for this observation. We have previously reported a reduction in the size of the LE/Lys compartment in AnxA6-depleted NPC1 mutant cells [25], which might reduce their capacity to process DQ-Red-BSA to levels comparable to those observed in CHO-WT cells. Yet, diverse outcomes when studying lysosomal functions in NPC1 mutant cells or other lysosomal storage diseases [47] have been reported. Therefore, additional factors to consider include the differential and cell-specific repertoire of other players driving LE/Lys function, but also the possibility that degradation is selective [12], which is possibly relevant here, could be sensitive to the amount of cholesterol or other lipids in the LE/Lys compartment [48]. Along these lines, several studies have implicated increased LE-Chol as a decisive factor for the regulation of LE functioning such as ILV formation and/or the sorting of LE/multivesicular bodies (MVB)/Lys structures for autophagic pathways such as CMA and eMi or exocytosis [15,49,50]. Moreover, autophagy delivers cholesterol to the lysosome, creating a feedback loop that promotes further lipid storage and lysosomal dysfunction [14,51]. Thus, although accumulating evidence points at LE-Chol accumulation interfering with the proper functioning of a variety of autophagic and lysosomal tasks, more research is needed to identify the factors or environmental changes that create sensitivity towards LE-Chol and/or AnxA6 levels for the various lysosomal activities.

3.2. Selective Degradation of AnxA6 in LE/Lys

As previous reports were in support of AnxA6 being targeted for lysosomal degradation by selective autophagy (CMA/eMi) [30,50], we next examined if LE-Chol accumulation in M12 cells could compromise AnxA6 degradation. For this purpose, CHO-WT and CHO M12 cells were incubated with CHX alone to inhibit protein synthesis, or in combination with leupeptin, which blocks lysosomal degradation. Given the slow protein turnover of AnxA6 [30], cells were incubated with CHX ± leupeptin for 48 h. Then, whole cell lysates were prepared and protein levels of AnxA6 and Eps8, which also contains a pentapeptide-motif and is degraded via CMA, were determined by Western blotting (Figure 2A see quantification in Figure 2B,C). CHX-treated CHO-WT cells displayed only a minor reduction in AnxA6 and Eps8 protein levels, probably reflecting the long half-life of these proteins (>48 h) [30] (compare lanes 1 and 2). Yet, the combined treatment of CHX and leupeptin in CHO-WT cells resulted in AnxA6, as well as Eps8, protein levels that were higher compared to their respective negative controls (compare lane 1 and 3, for quantification see Figure 2B,C), confirming that this assay was suitable to examine lysosomal protein degradation. However, the combined treatment of CHX and leupeptin resulted in AnxA6 and Eps8 protein levels similar to CHO M12 controls (Figure 2A, compare lanes 4 and 6). Likewise and consistent with the results described previously [25] and above (see Figure 1), AnxA6 protein levels in CHO M12 cells were substantially elevated compared to CHO-WT cells (compare lanes 1 and 4; for quantification see Figure 2B,C). Thus, the overall behavior of the CMA marker protein, Eps8, as well as AnxA6 protein pools, in the presence of inhibitors blocking protein synthesis and lysosomal degradation, showed similar trends in both cell lines. Yet, when comparing the response of CHO-WT and CHO M12 cells to leupeptin, CHO-WT showed a much stronger elevation of Eps8 and AnxA6 protein levels compared to CHO M12 cells (compare lanes 1 and 3 with 4 and 6), indicating that in NPC1 mutant cells the degradation route of Eps8, but also AnxA6, most likely via CMA, was inhibited.

To further assess if an inhibition of CMA-mediated AnxA6 degradation could explain upregulated AnxA6 levels in NPC1 mutant cells, we next aimed to activate this pathway through the overexpression of Lamp2A, which acts as a receptor for CMA substrates and thereby activates the CMA pathway [49,50]. Therefore, GFP-tagged Lamp2A was ectopically expressed in CHO-WT and CHO M12 cells and cell lysates were then analyzed for AnxA6 expression levels by Western blotting. As shown in Figure 3A (see quantification in 3B), overexpression of Lamp2A-GFP caused a significant decrease of AnxA6 levels in NPC1 mutant cells, further supporting AnxA6-mediated degradation via the CMA pathway and its reduced efficacy due to LE-Chol enrichment. Moreover, pointing at a link between levels of AnxA6 and LE-Chol, Lamp2A-induced downregulation of AnxA6 was associated with increased Rab7 activity, as judged by elevated Rab7-GTP levels in CHO M12 and CHO 2-2 cells in pulldown assays as described (Figure 3C). This is in agreement with our recent findings that identified AnxA6 depletion to prevent the recruitment of the Rab7-GAP TBC1D15 to LE/Lys [25]. In NPC1 mutant cells, this led to Rab7 activation and was accompanied by the restoration of LE-Chol export in these cells. The latter observation further suggests that the reduced LE-Chol accumulation in cells lacking NPC1 normalizes CMA-mediated AnxA6 degradation.

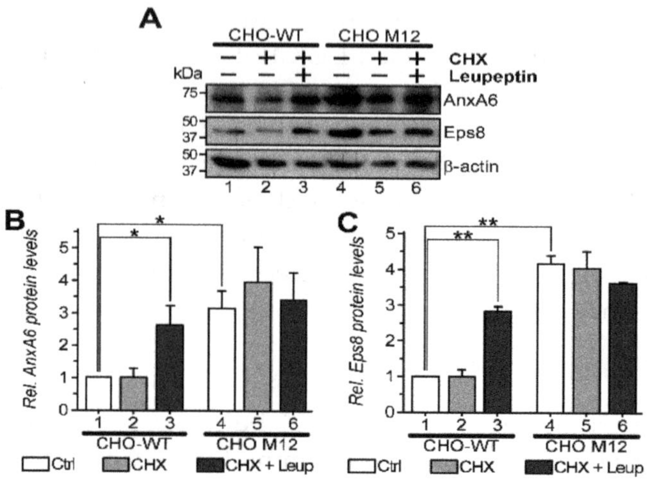

Figure 2. Reduced lysosomal degradation of AnxA6 and Eps8 in NPC1 mutant cells. (**A**) CHO-WT (lane 1–3) and CHO M12 (lane 4–6) cells treated without (control) or with 100 μg/mL cycloheximide (CHX) alone or together with 10 μg/mL leupeptin (CHX + Leup) as indicated for 48 h. Cell lysates were prepared and analyzed by Western blotting for AnxA6, Eps8 and β-actin levels as indicated. (**B**,**C**) The intensity of bands was densitometrically quantified. Data shown represents the mean ± SEM of two independent experiments (* $p < 0.05$, ** $p < 0.01$).

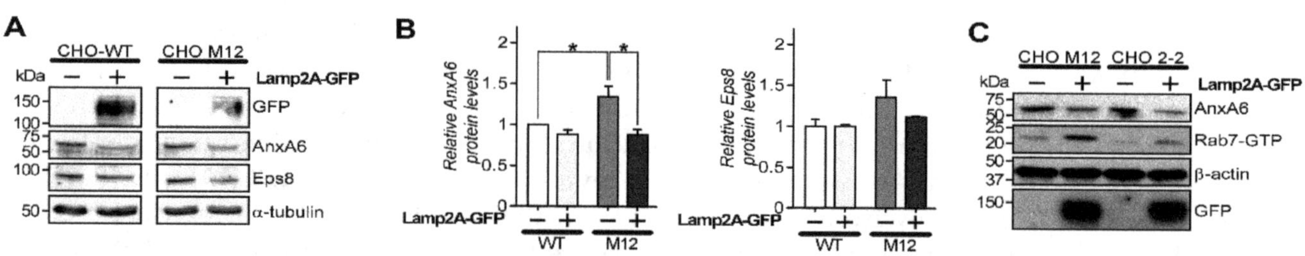

Figure 3. Overexpression of Lamp2A increases AnxA6 protein degradation. (**A**) Cell lysates from CHO-WT and CHO M12 cells ± ectopically expressed Lamp2A-GFP as indicated were analyzed by Western blotting for GFP, AnxA6, Eps8, and α-tubulin. A representative western blot from two independent experiments is shown. (**B**) The intensity of AnxA6 and Eps8 bands was quantified and normalized to α-tubulin. Data shown represents the mean ± SEM (* $p < 0.05$). (**C**) CHO M12 and CHO 2-2 cells were transfected ± Lamp2A-GFP as above. Cell lysates were prepared and subjected to Rab interacting lysosomal protein (RILP)-C33–GST pull-down assays to determine active Rab7 (Rab7-GTP) levels (see Material and Methods for details). Total levels of AnxA6, β-actin, and GFP in cell lysates are shown.

3.3. Lamp2A Regulates Cholesterol Levels in Endolysosomes of NPC1 Mutant Cells

Lamp2A is a lysosomal-associated membrane protein that acts as a protecting barrier for hydrolases and a receptor for CMA substrates, but has also been implicated in cholesterol export from LE/Lys [52–54].

Inversely, cholesterol levels may influence Lamp2A protein conformation or interaction with other lysosomal partners, with consequences for autophagy and cellular metabolism [52]. As mentioned above, ectopic Lamp2A expression in NPC1 mutant cells led to AnxA6 downregulation and elevation of Rab7-GTP levels (Figures 1 and 2), a scenario that we showed was associated with restoration of LE-Chol export in cells lacking NPC1 [25]. This prompted us to study whether Lamp2A-induced CMA activation possibly correlated with diminished LE-Chol accumulation in NPC1 mutant cells. Therefore, Lamp2A-GFP was ectopically expressed in NPC1 mutant CHO M12 cells and to examine LE-Chol accumulation, cells were then fixed and stained with filipin to visualize unesterified (free) cholesterol (Figure 4). Remarkably, quantification of filipin staining per cell, which resembles almost exclusively prominent and enlarged, perinuclear cholesterol-enriched, and Lamp2A-positive late endosomal vesicles in NPC1 mutant M12 cells (see also enlarged inserts in 4C), identified a significant decrease of filipin intensity per cell by 15% in CHO M12 cells (9.474 ± 0.3 vs. 8.059 ± 0.45) ectopically expressing Lamp2A compared to the neighboring non-transfected CHO M12 cells (Figure 4A, see quantification in Figure 4B). Indeed, enlarged images revealed a mixed population of Lamp2A-positive vesicles, many of them only weakly stained with filipin, indicating restoration of LE-Chol export (Figure 4C; see white arrows in the enlarged inset and line profile intensity plot). Taken together, it appears feasible that overexpression of Lamp2A in CHO M12 cells leads to CMA activation that allows AnxA6 protein degradation. Consequently, this instigates the upregulation of Rab7-GTP levels, which triggers restoration of LE-Chol export to eventually reduce the levels of cholesterol in the LE/Lys compartment of NPC1 mutant cells.

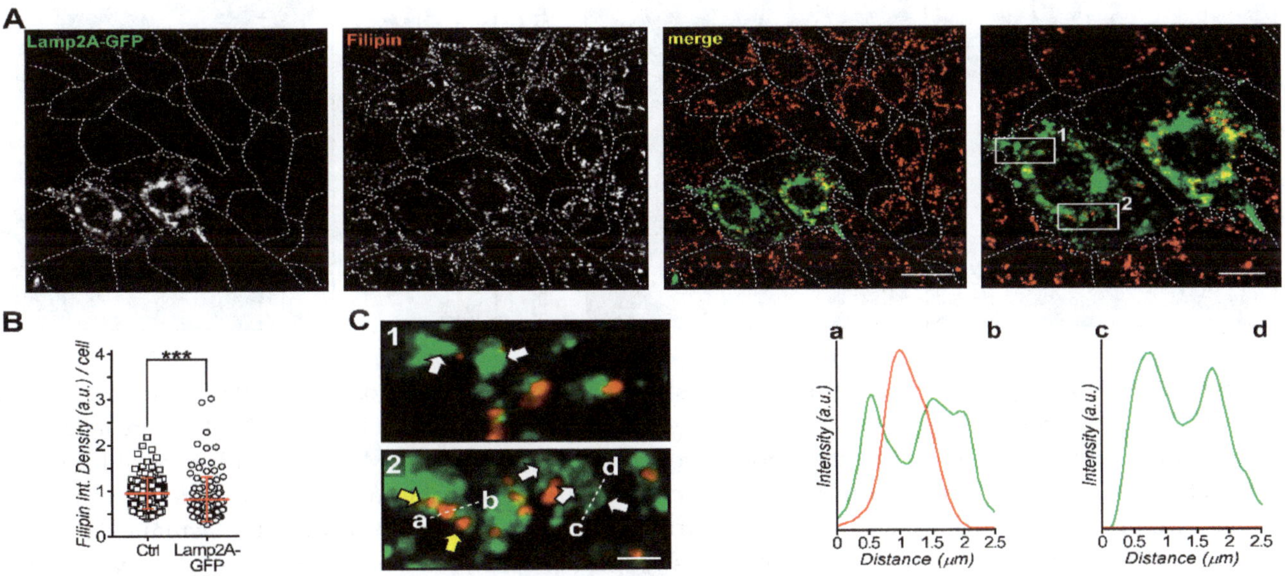

Figure 4. Lamp2A-GFP overexpression reduces late endosomal cholesterol (LE-Chol) accumulation in NPC1 mutant cells. (**A**) CHO M12 cells were transfected with Lamp2A-GFP (green), fixed, and stained with filipin to visualize free cholesterol (red). A representative image, including the merged image, is shown (scale bar, 20 µm; 10 µm in enlarged right panel). The shape of all cells was outlined. (**B**) The fluorescence intensity of filipin staining (integrated density per cell) of at least 20 control (Ctrl) and Lamp2A-GFP transfected cells from three independent experiments was quantified and shown in a dot plot. The mean ± SEM is also given (*** $p < 0.001$). (**C**) Two enlarged regions of interest from A (1 and 2) show details of Lamp2A-GFP/filipin colocalization and vesicle heterogeneity in transfected cells (scale bar, 2 µm). White arrows point at filipin-negative, but Lamp2A-positive vesicles (1,2) and yellow arrows highlight Lamp2A-positive vesicles colocalizing with filipin (2). Line profiles of fluorescence intensities of Lamp2A-GFP (green) and filipin (red) are shown (a–b and c–d, 2.5 µm).

In previous studies we demonstrated that AnxA6 is associated with acidic, endo-, and pre-lysosomal compartments of non-polarized as well as polarized cells [37,55,56]. Conversely, in settings where LE/Lys accumulate cholesterol, such as in NPC1 disease or after pharmacological NPC1 inhibition, increased amounts of AnxA6 proteins were found associated with LE/Lys [57]. The latter findings were based on colocalization studies using confocal microscopy as well as subcellular fractionation techniques, and in order to get additional insights into the AnxA6 location in the complex membrane organization of the LE/Lys compartment of NPC1 mutant cells, we compared the distribution of AnxA6 in CHO-WT and CHO M12 cells at much higher resolution, using electron microscopy. Therefore, ultrathin sections of both cell lines were labeled (for details see Methods) with an anti-AnxA6 polyclonal antibody [57] followed by anti-rabbit 12 nm colloidal gold antibodies (Figure 5A). Quantification of gold particle density (gold particles/μm^2) or vesicle area (μm^2) validated previous findings, identifying a significant increase of AnxA6-gold labeling in LE/Lys structures of CHO M12 compared to CHO-WT cells (Figure 5A,B). Moreover, the complex and heterogeneous inner LE/Lys structures consisting of multilamellar membranes and electrodense material in prototypical endolysosome in CHO M12 cells were enriched (~95%) with anti-AnxA6 labeling (Figure 5C, red arrowheads). Much less prominent AnxA6 labeling was also observed at the perimeter membrane of those LE/Lys structures (white arrowheads). In fact, these findings are supported by others, identifying significant levels of AnxA6 in the lysosomal matrix, compared to the corresponding lysosomal membranes isolated from liver [30].

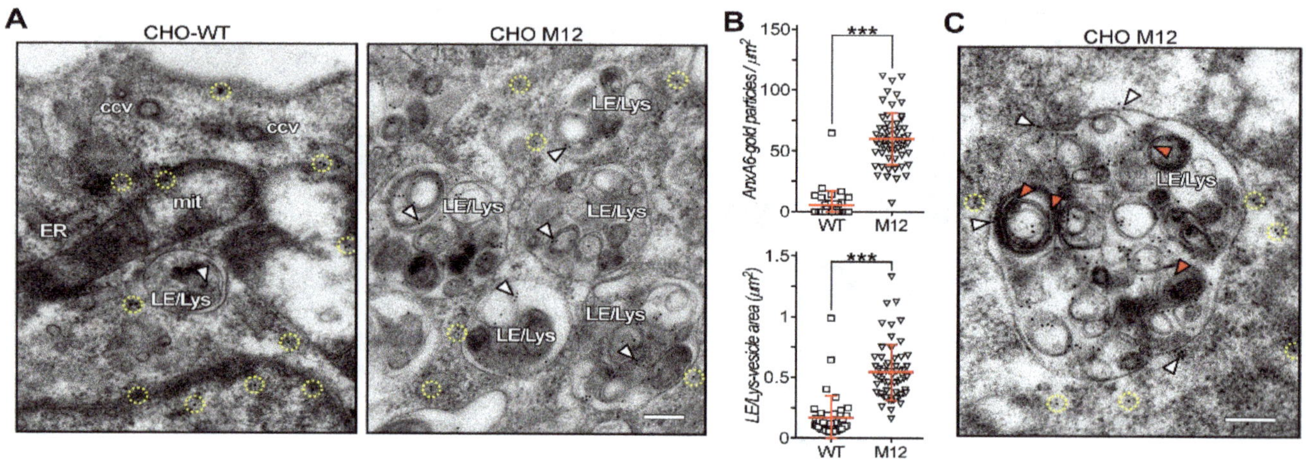

Figure 5. AnxA6 is enriched in inner membranes of the late endosome/lysosome (LE/Lys) compartment in NPC1 mutant cells. Lowicryl HM20 ultrathin sections of CHO-WT and CHO M12 cells were labeled with anti-AnxA6. (**A**) A region of the cytoplasm of CHO-WT and CHO M12 cells with several LE/Lys structures is shown. Note that in CHO M12 cells AnxA6 gold labeling is much more prominent compared to CHO-WT cells. Encircled gold particles (dashed yellow) indicate cytoplasmatic AnxA6 labeling. (**B**) Quantification of gold particle density (gold particle/μm^2) and LE/Lys vesicle area (μm^2). The data is shown as dot plot per cell. The mean ± SEM is also given (*** $p < 0.001$). (**C**) Representative enlarged image of a prototypical endolysosome in CHO M12 cells. AnxA6 labeling is enriched in inner LE/Lys structures (red arrowheads) and less prominent at the LE/Lys perimeter (white arrowheads). mit, mitochondria; ccv, clathrin coated vesicles; ER, endoplasmic reticulum. Scale bar is 200 nm.

Therefore, AnxA6 and LE-Chol seem to be engaged in a feedback loop in the LE/Lys compartment of NPC1 mutant cells (Figure 6). Upregulated AnxA6 levels increase the amount of LE-Chol, and, inversely, increased LE-Chol levels upon loss of NPC1 function raise AnxA6 protein amounts. Given that several studies pointed at LE-Chol accumulation to interfere with CMA and lysosomal function [13,45,48,58–60], we propose that this connectivity of LE-Chol levels and CMA efficacy is the underlying cause for the inhibition of AnxA6 degradation by selective autophagy. In support of this hypothesis, Rodriguez-Navarro et al., 2012 demonstrated that increased lipid intake in mice, either feeding a high-fat or cholesterol-rich diet, dramatically decreased Lamp2A levels due to accelerated degradation [48], causing a substantial decline in CMA activity that was associated with increased amounts of AnxA6.

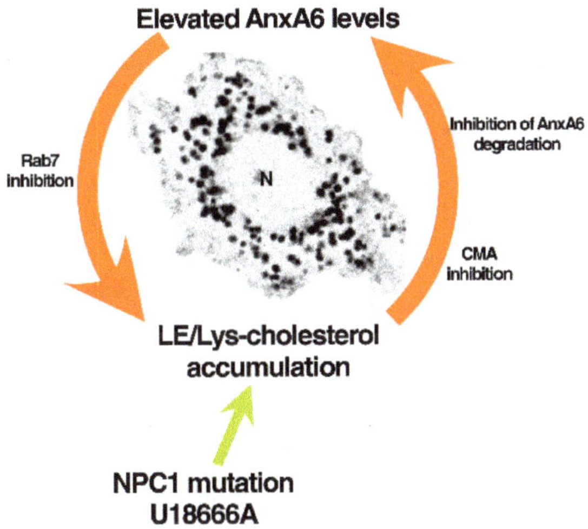

Figure 6. Scheme depicting the feedback loop of AnxA6 and LE-Chol levels in NPC1 mutant cells. NPC1 loss-of-function mutations or pharmacological NPC1 inhibition (U18666A) lead to late endosome/lysosome (LE/Lys)-cholesterol accumulation [1,6,61]. This interferes with Lamp2A-dependent chaperone-mediated autophagy (CMA) [31,48], which leads to reduced AnxA6 protein degradation (this study). Elevated AnxA6 protein levels promote Rab7 inactivation, which further contributes to cholesterol accumulation in LE/Lys (this study and [25]). In fact, diminution of active Rab7 in NPC1 mutant cells could also inhibit CMA [62]. This inhibitory feedback loop can be overcome by Lamp2A overexpression, which activates CMA [49], leading to reduced AnxA6 levels and concomitantly, also induces the diminution of LE/Lys-cholesterol accumulation (this study). Lamp2A-induced restoration of LE-cholesterol export being responsible for reduced AnxA6 protein levels cannot be completely ruled out. Alternatively, AnxA6 depletion in NPC1 mutant cells increased Rab7 activity [25], which restored cholesterol export from LE/Lys. Taken together, this feedback loop connects cholesterol levels in LE/Lys with CMA-mediated AnxA6 protein degradation, with CMA functionality being highly dependent on cholesterol levels in the LE/Lys compartment. N, nucleus.

Although the data presented here may favor Lamp2A-induced CMA activation being responsible for AnxA6 degradation (Figure 6), it should be emphasized that other routes, such as eMi or even specific lysosomes, may also contribute to control AnxA6 protein turnover, and cannot be entirely excluded [30,50]. Importantly, this study clearly identifies a close relationship between AnxA6 protein and LE-Chol levels in NPC1 mutant cells. Future studies, using pharmacological inhibitors to specifically block the abovementioned pathways will be needed to dissect the contribution of CMA or other degradation routes in AnxA6 protein turnover.

The underlying mechanisms that lead to an enrichment of AnxA6 protein pools on inner LE/Lys membranes and its molecular consequences remain to be clarified. AnxA6 is the largest member of the annexin family and instead of the four annexin repeats in all other annexins, has an unusual size and structure with eight annexin domains that are connected by a hinge-like loop. This provides some flexibility in the membrane interactions for AnxA6, which may be sensitive to changes in the lipid microenvironment. Within this context, alterations in the amount of LE-Chol indeed appear to induce conformational changes in proteins residing in the limiting membrane of LE/Lys, with consequences for protein-protein interactions. The LE-Chol-sensitive formation of the ORP1L-VAP-A complex, which participates in cholesterol transfer from LE to the ER, is currently the best example for this mechanism. Moreover, ORP1L-VAP-A interaction in response to alterations in LE-Chol levels impairs Rab7-dependent autophagosome maturation [63,64]. Thus, LE-Chol levels may also promote redistribution and conformational changes that impact on the ability of AnxA6, and possibly another annexin, AnxA1, to participate in the control of cholesterol transfer from LE/Lys to the ER (and vice versa) through MCS, with far reaching implications for cholesterol homeostasis [8,24,25].

Previous studies identified AnxA6 as a cholesterol-interacting protein [65,66]. This coincides with the increased association of AnxA6 with LE-Chol-enriched vesicles [57], which not only contain increased amounts of cholesterol in internal vesicles, but, as shown by several studies, also in the perimeter LE/Lys membrane [52]. Thus, the recruitment of cytosolic pools of AnxA6 proteins with an increased affinity for cholesterol in the limiting membrane of LE/Lys could in part explain these observations. Strikingly, this study identified substantial amounts of AnxA6 associated with inner membranes and ILV of the late endocytic compartment. Although it remains to be determined how AnxA6 is delivered to inner LE/Lys membranes upon LE-Chol accumulation, it is well known that cholesterol can accumulate in ILV of MVB [67].

This observation may be relevant for other biological events that commence in the LE/Lys compartment. For instance, exosomes originate from the ILV of MVB to then fuse with the plasma membrane for subsequent extracellular release. Despite the rapidly increasing knowledge regarding the complex repertoire of bioactive molecules in exosomes, including proteins, RNAs, and lipids, its biogenesis and the loading of these structures with cargo is still not well understood. Interestingly, based on the proteomic databases available on exosomes isolated from a large variety of settings, amongst all kinds of proteins, exosomes are highly enriched in annexins (ExoCarta exosome database: www.exocarta.org, accessed on 29 July 2015).

Several questions arise from these observations: (1) What are the targeting mechanisms that deliver and allow access of annexins into exosomes? Could these mechanisms include an escape from CMA-or eMI-mediated degradation once annexins enter the lumen of MVBs? (2) What is the function of annexins once they are released in exosomes and enter target cells? (3) Finally, and most relevant in relation to NPC disease, do annexins and cholesterol pools deposited in exosomes have a function in cellular cholesterol homeostasis? In fact, recent studies identified drastically increased amounts of cholesterol-rich exosomes released from NPC1 mutants or U18666A-treated cells. Thus, it was proposed that upregulated exosome release of exosomes might serve to ameliorate LE-Chol accumulation [68] and consequently the NPC1 phenotype.

In summary, this study demonstrates that CMA is inhibited in NPC1 mutant cells causing an increase in AnxA6 protein levels. This inhibits Rab7 activity and its ability to overcome accumulation of cholesterol in LE/Lys, which is detrimental for cellular homeostasis. Lamp2A-mediated restoration of CMA promotes AnxA6 protein degradation and is associated with a rescue of LE-Chol accumulation in NPC1 mutant cells. Taken together, AnxA6 protein levels appear to be associated with LE-Chol accumulation and CMA-mediated protein degradation, the functionality of the latter being highly dependent on LE-Chol levels.

Author Contributions: E.M.-S., A.G.-M., P.B.-M., J.J. and M.-S.B. performed experiments and prepared reagents; C.E. conducted electron microscopy studies; C.R. designed automated software for ImageJ quantifications; A.L. and F.T. provided expertise, data analysis and contributed to discussion; C.E., T.G. and C.R. designed the study and wrote, reviewed and edited the manuscript. All authors have read and agreed to the published version of the manuscript.

Acknowledgments: We are thankful to the staff from Centres Científics i Tecnològics (CCiTUB), Universitat de Barcelona, Unitat de Microscòpia Òptica Avançada and Unitat de Microscòpia Electrònica. We are also grateful to Maria Molinos (University of Barcelona) and Lea Strobel (University of Sydney) for technical assistance.

References

1. Ikonen, E. Cellular cholesterol trafficking and compartmentalization. *Nat. Rev. Mol. Cell Biol.* **2008**, *9*, 125–138. [CrossRef] [PubMed]

2. Heybrock, S.; Kanerva, K.; Meng, Y.; Ing, C.; Liang, A.; Xiong, Z.J.; Weng, X.; Ah Kim, Y.; Collins, R.; Trimble, W.; et al. Lysosomal integral membrane protein-2 (limp-2/scarb2) is involved in lysosomal cholesterol export. *Nat. Commun.* **2019**, *10*, 3521. [CrossRef] [PubMed]

3. Infante, R.E.; Wang, M.L.; Radhakrishnan, A.; Kwon, H.J.; Brown, M.S.; Goldstein, J.L. Npc2 facilitates bidirectional transfer of cholesterol between npc1 and lipid bilayers, a step in cholesterol egress from lysosomes. *Proc. Natl. Acad. Sci. USA* **2008**, *105*, 15287–15292. [CrossRef] [PubMed]
4. Chevallier, J.; Chamoun, Z.; Jiang, G.; Prestwich, G.; Sakai, N.; Matile, S.; Parton, R.G.; Gruenberg, J. Lysobisphosphatidic acid controls endosomal cholesterol levels. *J. Biol. Chem.* **2008**, *283*, 27871–27880. [CrossRef] [PubMed]
5. McCauliff, L.A.; Langan, A.; Li, R.; Ilnytska, O.; Bose, D.; Waghalter, M.; Lai, K.; Kahn, P.C.; Storch, J. Intracellular cholesterol trafficking is dependent upon npc2 interaction with lysobisphosphatidic acid. *Elife* **2019**, *8*, e50832. [CrossRef]
6. Ikonen, E. Mechanisms of cellular cholesterol compartmentalization: Recent insights. *Curr. Opin. Cell Biol.* **2018**, *53*, 77–83. [CrossRef]
7. Pfeffer, S.R. Npc intracellular cholesterol transporter 1 (npc1)-mediated cholesterol export from lysosomes. *J. Biol. Chem.* **2019**, *294*, 1706–1709. [CrossRef]
8. Martello, A.; Platt, F.M.; Eden, E.R. Staying in touch with the endocytic network: The importance of contacts for cholesterol transport. *Traffic* **2020**. [CrossRef]
9. Boya, P.; Reggiori, F.; Codogno, P. Emerging regulation and functions of autophagy. *Nat. Cell Biol.* **2013**, *15*, 713–720. [CrossRef]
10. Singh, R.; Kaushik, S.; Wang, Y.; Xiang, Y.; Novak, I.; Komatsu, M.; Tanaka, K.; Cuervo, A.M.; Czaja, M.J. Autophagy regulates lipid metabolism. *Nature* **2009**, *458*, 1131–1135. [CrossRef]
11. Kaushik, S.; Cuervo, A.M. Degradation of lipid droplet-associated proteins by chaperone-mediated autophagy facilitates lipolysis. *Nat. Cell Biol.* **2015**, *17*, 759–770. [CrossRef] [PubMed]
12. Sarkar, S.; Carroll, B.; Buganim, Y.; Maetzel, D.; Ng, A.H.; Cassady, J.P.; Cohen, M.A.; Chakraborty, S.; Wang, H.; Spooner, E.; et al. Impaired autophagy in the lipid-storage disorder niemann-pick type c1 disease. *Cell Rep.* **2013**, *5*, 1302–1315. [CrossRef] [PubMed]
13. Fraldi, A.; Annunziata, F.; Lombardi, A.; Kaiser, H.J.; Medina, D.L.; Spampanato, C.; Fedele, A.O.; Polishchuk, R.; Sorrentino, N.C.; Simons, K.; et al. Lysosomal fusion and snare function are impaired by cholesterol accumulation in lysosomal storage disorders. *EMBO J.* **2010**, *29*, 3607–3620. [CrossRef] [PubMed]
14. Elrick, M.J.; Yu, T.; Chung, C.; Lieberman, A.P. Impaired proteolysis underlies autophagic dysfunction in niemann-pick type c disease. *Hum. Mol Genet.* **2012**, *21*, 4876–4887. [CrossRef]
15. Kaushik, S.; Cuervo, A.M. The coming of age of chaperone-mediated autophagy. *Nat. Rev. Mol. Cell Biol.* **2018**, *19*, 365–381. [CrossRef]
16. Qin, Z.-K. *Autophagy: Biology and Diseases*; Science Press: Beijing, China, 2019; Volume 1206.
17. Medina, D.L.; Ballabio, A. Lysosomal calcium regulates autophagy. *Autophagy* **2015**, *11*, 970–971. [CrossRef]
18. Lieberman, A.P.; Puertollano, R.; Raben, N.; Slaugenhaupt, S.; Walkley, S.U.; Ballabio, A. Autophagy in lysosomal storage disorders. *Autophagy* **2012**, *8*, 719–730. [CrossRef]
19. Ghislat, G.; Aguado, C.; Knecht, E. Annexin a5 stimulates autophagy and inhibits endocytosis. *J. Cell Sci.* **2012**, *125*, 92–107. [CrossRef]
20. Gerke, V.; Creutz, C.E.; Moss, S.E. Annexins: Linking ca2+ signalling to membrane dynamics. *Nat. Rev. Mol. Cell Biol.* **2005**, *6*, 449–461. [CrossRef]
21. Rescher, U.; Gerke, V. Annexins–unique membrane binding proteins with diverse functions. *J. Cell Sci.* **2004**, *117*, 2631–2639. [CrossRef]
22. Gerke, V.; Moss, S.E. Annexins: From structure to function. *Physiol. Rev.* **2002**, *82*, 331–371. [CrossRef] [PubMed]
23. Enrich, C.; Rentero, C.; Meneses-Salas, E.; Tebar, F.; Grewal, T. Annexins: Ca(2+) effectors determining membrane trafficking in the late endocytic compartment. *Adv. Exp. Med. Biol.* **2017**, *981*, 351–385.
24. Eden, E.R.; Sanchez-Heras, E.; Tsapara, A.; Sobota, A.; Levine, T.P.; Futter, C.E. Annexin a1 tethers membrane contact sites that mediate er to endosome cholesterol transport. *Dev. Cell* **2016**, *37*, 473–483. [CrossRef] [PubMed]
25. Meneses-Salas, E.; Garcia-Melero, A.; Kanerva, K.; Blanco-Munoz, P.; Morales-Paytuvi, F.; Bonjoch, J.; Casas, J.; Egert, A.; Beevi, S.S.; Jose, J.; et al. Annexin a6 modulates tbc1d15/rab7/stard3 axis to control endosomal cholesterol export in npc1 cells. *Cell Mol. Life Sci.* **2019**. [CrossRef] [PubMed]
26. Lin-Moshier, Y.; Keebler, M.V.; Hooper, R.; Boulware, M.J.; Liu, X.; Churamani, D.; Abood, M.E.; Walseth, T.F.; Brailoiu, E.; Patel, S.; et al. The two-pore channel (tpc) interactome unmasks isoform-specific roles for tpcs

in endolysosomal morphology and cell pigmentation. *Proc. Natl. Acad. Sci. USA* **2014**, *111*, 13087–13092. [CrossRef]

27. Cubells, L.; Vila de Muga, S.; Tebar, F.; Wood, P.; Evans, R.; Ingelmo-Torres, M.; Calvo, M.; Gaus, K.; Pol, A.; Grewal, T.; et al. Annexin a6-induced alterations in cholesterol transport and caveolin export from the golgi complex. *Traffic* **2007**, *8*, 1568–1589. [CrossRef]
28. Huttlin, E.L.; Ting, L.; Bruckner, R.J.; Gebreab, F.; Gygi, M.P.; Szpyt, J.; Tam, S.; Zarraga, G.; Colby, G.; Baltier, K.; et al. The bioplex network: A systematic exploration of the human interactome. *Cell* **2015**, *162*, 425–440. [CrossRef]
29. Kirchner, P.; Bourdenx, M.; Madrigal-Matute, J.; Tiano, S.; Diaz, A.; Bartholdy, B.A.; Will, B.; Cuervo, A.M. Proteome-wide analysis of chaperone-mediated autophagy targeting motifs. *PLoS Biol.* **2019**, *17*, e3000301. [CrossRef]
30. Cuervo, A.M.; Gomes, A.V.; Barnes, J.A.; Dice, J.F. Selective degradation of annexins by chaperone-mediated autophagy. *J. Biol. Chem.* **2000**, *275*, 33329–33335. [CrossRef]
31. Sahu, R.; Kaushik, S.; Clement, C.C.; Cannizzo, E.S.; Scharf, B.; Follenzi, A.; Potolicchio, I.; Nieves, E.; Cuervo, A.M.; Santambrogio, L. Microautophagy of cytosolic proteins by late endosomes. *Dev. Cell* **2011**, *20*, 131–139. [CrossRef]
32. Lefebvre, C.; Legouis, R.; Culetto, E. Escrt and autophagies: Endosomal functions and beyond. *Semin. Cell Dev. Biol.* **2018**, *74*, 21–28. [CrossRef] [PubMed]
33. Ghislat, G.; Knecht, E. New ca(2+)-dependent regulators of autophagosome maturation. *Commun. Integr. Biol.* **2012**, *5*, 308–311. [CrossRef] [PubMed]
34. Reverter, M.; Rentero, C.; de Muga, S.V.; Alvarez-Guaita, A.; Mulay, V.; Cairns, R.; Wood, P.; Monastyrskaya, K.; Pol, A.; Tebar, F.; et al. Cholesterol transport from late endosomes to the golgi regulates t-snare trafficking, assembly, and function. *Mol. Biol. Cell* **2011**, *22*, 4108–4123. [CrossRef] [PubMed]
35. Leca, J.; Martinez, S.; Lac, S.; Nigri, J.; Secq, V.; Rubis, M.; Bressy, C.; Serge, A.; Lavaut, M.N.; Dusetti, N.; et al. Cancer-associated fibroblast-derived annexin a6+ extracellular vesicles support pancreatic cancer aggressiveness. *J. Clin. Invest.* **2016**, *126*, 4140–4156. [CrossRef] [PubMed]
36. Keklikoglou, I.; Cianciaruso, C.; Guc, E.; Squadrito, M.L.; Spring, L.M.; Tazzyman, S.; Lambein, L.; Poissonnier, A.; Ferraro, G.B.; Baer, C.; et al. Chemotherapy elicits pro-metastatic extracellular vesicles in breast cancer models. *Nat. Cell Biol.* **2019**, *21*, 190–202. [CrossRef] [PubMed]
37. Pons, M.; Ihrke, G.; Koch, S.; Biermer, M.; Pol, A.; Grewal, T.; Jackle, S.; Enrich, C. Late endocytic compartments are major sites of annexin vi localization in nrk fibroblasts and polarized wif-b hepatoma cells. *Exp. Cell Res.* **2000**, *257*, 33–47. [CrossRef]
38. Grewal, T.; Heeren, J.; Mewawala, D.; Schnitgerhans, T.; Wendt, D.; Salomon, G.; Enrich, C.; Beisiegel, U.; Jackle, S. Annexin vi stimulates endocytosis and is involved in the trafficking of low density lipoprotein to the prelysosomal compartment. *J. Biol. Chem.* **2000**, *275*, 33806–33813. [CrossRef]
39. Reverter, M.; Rentero, C.; Garcia-Melero, A.; Hoque, M.; Vila de Muga, S.; Alvarez-Guaita, A.; Conway, J.R.; Wood, P.; Cairns, R.; Lykopoulou, L.; et al. Cholesterol regulates syntaxin 6 trafficking at trans-golgi network endosomal boundaries. *Cell Rep.* **2014**, *7*, 883–897. [CrossRef]
40. Schneider, C.A.; Rasband, W.S.; Eliceiri, K.W. Nih image to imagej: 25 years of image analysis. *Nat. Methods* **2012**, *9*, 671–675. [CrossRef]
41. Marwaha, R.; Sharma, M. Dq-red bsa trafficking assay in cultured cells to assess cargo delivery to lysosomes. *Bio-Protoc.* **2017**, *7*. [CrossRef]
42. Vazquez, C.L.; Colombo, M.I. Assays to assess autophagy induction and fusion of autophagic vacuoles with a degradative compartment, using monodansylcadaverine (mdc) and dq-bsa. *Methods Enzymol.* **2009**, *452*, 85–95. [PubMed]
43. Bi, X.; Liao, G. Autophagic-lysosomal dysfunction and neurodegeneration in niemann-pick type c mice: Lipid starvation or indigestion? *Autophagy* **2007**, *3*, 646–648. [CrossRef] [PubMed]
44. Kummel, D.; Ungermann, C. Principles of membrane tethering and fusion in endosome and lysosome biogenesis. *Curr. Opin. Cell Biol.* **2014**, *29*, 61–66. [CrossRef] [PubMed]
45. Sobo, K.; Le Blanc, I.; Luyet, P.P.; Fivaz, M.; Ferguson, C.; Parton, R.G.; Gruenberg, J.; van der Goot, F.G. Late endosomal cholesterol accumulation leads to impaired intra-endosomal trafficking. *PLoS ONE* **2007**, *2*, e851. [CrossRef] [PubMed]

46. de Araujo, M.E.G.; Liebscher, G.; Hess, M.W.; Huber, L.A. Lysosomal size matters. *Traffic* **2020**, *21*, 60–75. [CrossRef]
47. Ballabio, A.; Bonifacino, J.S. Lysosomes as dynamic regulators of cell and organismal homeostasis. *Nat. Rev. Mol. Cell Biol.* **2020**, *21*, 101–118. [CrossRef]
48. Rodriguez-Navarro, J.A.; Kaushik, S.; Koga, H.; Dall'Armi, C.; Shui, G.; Wenk, M.R.; Di Paolo, G.; Cuervo, A.M. Inhibitory effect of dietary lipids on chaperone-mediated autophagy. *Proc. Natl. Acad. Sci. USA* **2012**, *109*, E705–E714. [CrossRef]
49. Kaushik, S.; Cuervo, A.M. Chaperone-mediated autophagy: A unique way to enter the lysosome world. *Trends Cell Biol.* **2012**, *22*, 407–417. [CrossRef]
50. Tekirdag, K.; Cuervo, A.M. Chaperone-mediated autophagy and endosomal microautophagy: Joint by a chaperone. *J. Biol. Chem.* **2018**, *293*, 5414–5424. [CrossRef]
51. Elrick, M.J.; Lieberman, A.P. Autophagic dysfunction in a lysosomal storage disorder due to impaired proteolysis. *Autophagy* **2013**, *9*, 234–235. [CrossRef]
52. Li, J.; Pfeffer, S.R. Lysosomal membrane glycoproteins bind cholesterol and contribute to lysosomal cholesterol export. *Elife* **2016**, *5*. [CrossRef]
53. Eskelinen, E.L.; Schmidt, C.K.; Neu, S.; Willenborg, M.; Fuertes, G.; Salvador, N.; Tanaka, Y.; Lullmann-Rauch, R.; Hartmann, D.; Heeren, J.; et al. Disturbed cholesterol traffic but normal proteolytic function in lamp-1/lamp-2 double-deficient fibroblasts. *Mol. Biol. Cell* **2004**, *15*, 3132–3145. [CrossRef] [PubMed]
54. Schneede, A.; Schmidt, C.K.; Holtta-Vuori, M.; Heeren, J.; Willenborg, M.; Blanz, J.; Domanskyy, M.; Breiden, B.; Brodesser, S.; Landgrebe, J.; et al. Role for lamp-2 in endosomal cholesterol transport. *J. Cell Mol. Med.* **2011**, *15*, 280–295. [CrossRef] [PubMed]
55. Pons, M.; Grewal, T.; Rius, E.; Schnitgerhans, T.; Jackle, S.; Enrich, C. Evidence for the involvement of annexin 6 in the trafficking between the endocytic compartment and lysosomes. *Exp. Cell Res.* **2001**, *269*, 13–22. [CrossRef] [PubMed]
56. Enrich, C.; Rentero, C.; de Muga, S.V.; Reverter, M.; Mulay, V.; Wood, P.; Koese, M.; Grewal, T. Annexin A6—Linking Ca^{2+} signaling with cholesterol transport. *Biochim. Biophys. Acta-Mol. Cell Res.* **2011**, *1813*, 935–947. [CrossRef] [PubMed]
57. de Diego, I.; Schwartz, F.; Siegfried, H.; Dauterstedt, P.; Heeren, J.; Beisiegel, U.; Enrich, C.; Grewal, T. Cholesterol modulates the membrane binding and intracellular distribution of annexin 6. *J. Biol. Chem.* **2002**, *277*, 32187–32194. [CrossRef]
58. Kaushik, S.; Massey, A.C.; Cuervo, A.M. Lysosome membrane lipid microdomains: Novel regulators of chaperone-mediated autophagy. *EMBO J.* **2006**, *25*, 3921–3933. [CrossRef]
59. Koga, H.; Kaushik, S.; Cuervo, A.M. Altered lipid content inhibits autophagic vesicular fusion. *FASEB J.* **2010**, *24*, 3052–3065. [CrossRef]
60. Napolitano, G.; Johnson, J.L.; He, J.; Rocca, C.J.; Monfregola, J.; Pestonjamasp, K.; Cherqui, S.; Catz, S.D. Impairment of chaperone-mediated autophagy leads to selective lysosomal degradation defects in the lysosomal storage disease cystinosis. *EMBO Mol. Med.* **2015**, *7*, 158–174. [CrossRef]
61. Lu, F.; Liang, Q.; Abi-Mosleh, L.; Das, A.; De Brabander, J.K.; Goldstein, J.L.; Brown, M.S. Identification of npc1 as the target of u18666a, an inhibitor of lysosomal cholesterol export and ebola infection. *Elife* **2015**, *4*. [CrossRef]
62. Zhang, J.; Johnson, J.L.; He, J.; Napolitano, G.; Ramadass, M.; Rocca, C.; Kiosses, W.B.; Bucci, C.; Xin, Q.; Gavathiotis, E.; et al. Cystinosin, the small gtpase rab11, and the rab7 effector rilp regulate intracellular trafficking of the chaperone-mediated autophagy receptor lamp2a. *J. Biol. Chem.* **2017**, *292*, 10328–10346. [CrossRef] [PubMed]
63. Rocha, N.; Kuijl, C.; van der Kant, R.; Janssen, L.; Houben, D.; Janssen, H.; Zwart, W.; Neefjes, J. Cholesterol sensor orp1l contacts the er protein vap to control rab7-rilp-p150 glued and late endosome positioning. *J. Cell Biol.* **2009**, *185*, 1209–1225. [CrossRef] [PubMed]
64. Wilhelm, L.P.; Wendling, C.; Vedie, B.; Kobayashi, T.; Chenard, M.P.; Tomasetto, C.; Drin, G.; Alpy, F. Stard3 mediates endoplasmic reticulum-to-endosome cholesterol transport at membrane contact sites. *EMBO J.* **2017**, *36*, 1412–1433. [CrossRef] [PubMed]
65. Hulce, J.J.; Cognetta, A.B.; Niphakis, M.J.; Tully, S.E.; Cravatt, B.F. Proteome-wide mapping of cholesterol-interacting proteins in mammalian cells. *Nat. Methods* **2013**, *10*, 259–264. [CrossRef]

66. Domon, M.M.; Matar, G.; Strzelecka-Kiliszek, A.; Bandorowicz-Pikula, J.; Pikula, S.; Besson, F. Interaction of annexin a6 with cholesterol rich membranes is ph-dependent and mediated by the sterol oh. *J. Colloid Interface Sci.* **2010**, *346*, 436–441. [CrossRef]
67. Mobius, W.; van Donselaar, E.; Ohno-Iwashita, Y.; Shimada, Y.; Heijnen, H.F.; Slot, J.W.; Geuze, H.J. Recycling compartments and the internal vesicles of multivesicular bodies harbor most of the cholesterol found in the endocytic pathway. *Traffic* **2003**, *4*, 222–231. [CrossRef]
68. Strauss, K.; Goebel, C.; Runz, H.; Mobius, W.; Weiss, S.; Feussner, I.; Simons, M.; Schneider, A. Exosome secretion ameliorates lysosomal storage of cholesterol in niemann-pick type c disease. *J. Biol. Chem.* **2010**, *285*, 26279–26288. [CrossRef]

PERMISSIONS

All chapters in this book were first published in MDPI; hereby published with permission under the Creative Commons Attribution License or equivalent. Every chapter published in this book has been scrutinized by our experts. Their significance has been extensively debated. The topics covered herein carry significant findings which will fuel the growth of the discipline. They may even be implemented as practical applications or may be referred to as a beginning point for another development.

The contributors of this book come from diverse backgrounds, making this book a truly international effort. This book will bring forth new frontiers with its revolutionizing research information and detailed analysis of the nascent developments around the world.

We would like to thank all the contributing authors for lending their expertise to make the book truly unique. They have played a crucial role in the development of this book. Without their invaluable contributions this book wouldn't have been possible. They have made vital efforts to compile up to date information on the varied aspects of this subject to make this book a valuable addition to the collection of many professionals and students.

This book was conceptualized with the vision of imparting up-to-date information and advanced data in this field. To ensure the same, a matchless editorial board was set up. Every individual on the board went through rigorous rounds of assessment to prove their worth. After which they invested a large part of their time researching and compiling the most relevant data for our readers.

The editorial board has been involved in producing this book since its inception. They have spent rigorous hours researching and exploring the diverse topics which have resulted in the successful publishing of this book. They have passed on their knowledge of decades through this book. To expedite this challenging task, the publisher supported the team at every step. A small team of assistant editors was also appointed to further simplify the editing procedure and attain best results for the readers.

Apart from the editorial board, the designing team has also invested a significant amount of their time in understanding the subject and creating the most relevant covers. They scrutinized every image to scout for the most suitable representation of the subject and create an appropriate cover for the book.

The publishing team has been an ardent support to the editorial, designing and production team. Their endless efforts to recruit the best for this project, has resulted in the accomplishment of this book. They are a veteran in the field of academics and their pool of knowledge is as vast as their experience in printing. Their expertise and guidance has proved useful at every step. Their uncompromising quality standards have made this book an exceptional effort. Their encouragement from time to time has been an inspiration for everyone.

The publisher and the editorial board hope that this book will prove to be a valuable piece of knowledge for researchers, students, practitioners and scholars across the globe.

LIST OF CONTRIBUTORS

Peder A. G. Lillebostad, Arne Raasakka, Silje J. Hjellbrekke, Sudarshan Patil, Trude Røstbø, Hanne Hollås, Siri A. Sakya, Peter D. Szigetvari and Anni Vedeler
Department of Biomedicine, University of Bergen, Jonas Lies vei 91, 5009 Bergen, Norway

Petri Kursula
Department of Biomedicine, University of Bergen, Jonas Lies vei 91, 5009 Bergen, Norway
Faculty of Biochemistry and Molecular Medicine & Biocenter Oulu, University of Oulu, Aapistie 7, 90220 Oulu, Finland

Ekeveliny Amabile Veschi, Maytê Bolean, Ana Paula Ramos and Pietro Ciancaglini
Departamento de Química, Faculdade de Filosofia, Ciências e Letras de Ribeirão Preto da Universidade de São Paulo (FFCLRP-USP), Ribeirão Preto, São Paulo 14040-900, Brazil

Agnieszka Strzelecka-Kiliszek, Joanna Bandorowicz-Pikula and Slawomir Pikula
Nencki Institute of Experimental Biology, 3 Pasteur Street, 02-093 Warsaw, Poland

Thierry Granjon, Saida Mebarek, David Magne and Rene Buchet
Universite Lyon 1, UFR Chimie Biochimie, CEDEX, 69 622 Villeurbanne, France
ICBMS UMR 5246 CNRS, CEDEX, 69 622 Villeurbanne, France
INSA, Lyon, CEDEX, 69 622 Villeurbanne, France
CPE, Lyon, CEDEX, 69 622 Villeurbanne, France
Université de Lyon, CEDEX, 69 622 Villeurbanne, France

Nicola Rosato
Department of Experimental Medicine, University of Rome Tor Vergata, 00133 Rome, Italy

José Luis Millán
Sanford Burnham Prebys Medical Discovery Institute, La Jolla, CA 92037, USA

Massimo Bottini
Department of Experimental Medicine, University of Rome Tor Vergata, 00133 Rome, Italy
Sanford Burnham Prebys Medical Discovery Institute, La Jolla, CA 92037, USA

Yu-Shi Wang, He Li, Yang Li, Shiyin Zhang and Ying-Hua Jin
Key Laboratory for Molecular Enzymology and Engineering of the Ministry of Education, School of Life Sciences, Jilin University, Changchun 130012, China

Jieny Gröper, Volker Gerke and Ursula Rescher
Institute of Medical Biochemistry, Center for Molecular Biology of Inflammation, University of Muenster, Von-Esmarch-Str. 56, D-48149 Muenster, Germany
Cells in Motion" Interfaculty Centre, University of Muenster, 48149 Muenster, Germany

Gabriele M. König and Evi Kostenis
Institute for Pharmaceutical Biology, University of Bonn, Nussallee 6, 53115 Bonn, Germany

Carsten A. Raabe
Institute of Medical Biochemistry, Center for Molecular Biology of Inflammation, University of Muenster, Von-Esmarch-Str. 56, D-48149 Muenster, Germany
Institute of Experimental Pathology, Center for Molecular Biology of Inflammation, University of Muenster, Von-Esmarch-Str. 56, D-48149 Muenster, Germany

Cristina B. Hebeda, Silvana Sandri, Cláudia M. Benis, Marina de Paula-Silva, Rodrigo A. Loiola and Sandra H. P. Farsky
Department of Clinical and Toxicological Analyses, School of Pharmaceutical Sciences, University of Sao Paulo, São Paulo CEP 05508-000, Brazil

Chris Reutelingsperger
Faculty of Health, Medicine and Life Sciences, Part of Maastricht University Medical Center, Part of Maastricht University, 6211 LK Maastricht, The Netherlands

Mauro Perretti
The William Harvey Research Institute, Queen Mary University of London, London EC1M 6BQ, UK

Amira F. Mahdi, Beatrice Malacrida, Joanne Nolan and Patrick A. Kiely
Graduate Entry Medical School, University of Limerick, V94 T9PX Limerick, Ireland
Health Research Institute, University of Limerick and Bernal Institute, University of Limerick, V94 T9PX Limerick, Ireland

Mary E. McCumiskey
Stokes Laboratories, Bernal Institute, University of Limerick, V94 T9PX Limerick, Ireland
Department of Surgery, University Hospital Limerick, V94 F858 Limerick, Ireland

Anne B. Merrigan, Ashish Lal and Shona Tormey
Department of Surgery, University Hospital Limerick, V94 F858 Limerick, Ireland

Aoife J. Lowery
Lambe Institute for Translational Research, National University of Ireland Galway, H91 TK33 Galway, Ireland

Kieran McGourty
Health Research Institute, University of Limerick and Bernal Institute, University of Limerick, V94 T9PX Limerick, Ireland
Department of Chemical Sciences, University of Limerick, V94 T9PX Limerick, Ireland

Marion Gabel, Tamou Thahouly, Valérie Calco, Stéphane Gasman, Marie-France Bader and Nicolas Vitale
Centre National de la Recherche Scientifique, Université de Strasbourg, Institut des Neurosciences Cellulaires et Intégratives, F-67000 Strasbourg, France

Cathy Royer
Plateforme Imagerie In Vitro, Neuropôle, Université de Strasbourg, F-67000 Strasbourg, France

Sylvette Chasserot-Golaz
Centre National de la Recherche Scientifique, Université de Strasbourg, Institut des Neurosciences Cellulaires et Intégratives, F-67000 Strasbourg, France
Plateforme Imagerie In Vitro, Neuropôle, Université de Strasbourg, F-67000 Strasbourg, France

Valentina Dallacasagrande and Katherine A. Hajjar
Department of Pediatrics,Weill Cornell Medicine, 1300 York Avenue, New York, NY 10065, USA

Olga Y. Korolkova, Sarrah E. Widatalla, Stephen D.Williams, Diva S. Whalen, Heather K. Beasley, Josiah Ochieng and Amos M. Sakwe
Department of Biochemistry and Cancer Biology, School of Graduate Studies and Research, Meharry Medical College, Nashville, TN 37208, USA

José Marcos Sanches
Departamento de Morfologia e Genética, Universidade Federal de São Paulo, São Paulo 04023-900, Brazil
Faculdade de Medicina, Universidade do Oeste Paulista, Guarujá, São Paulo 11410-980, Brazil

Laura Migliari Branco and Karina Ramalho Bortoluci
Departamento de Ciências Biológicas e Centro de Terapia Celular e Molecular, Universidade Federal de São Paulo, São Paulo 04044-010, Brazil

Gustavo Henrique Bueno Duarte
Instituto de Química, Universidade Estadual de Campinas, Campinas, São Paulo 13083-862, Brazil

Sonia Maria Oliani
Programa de Pós-Graduação em Biociências, Instituto de Biociências, Letras e Ciências Exatas (IBILCE), Universidade Estadual Paulista, São José do Rio Preto, São Paulo 15054-000, Brazil

Cristiane Damas Gil
Departamento de Morfologia e Genética, Universidade Federal de São Paulo, São Paulo 04023-900, Brazil
Programa de Pós-Graduação em Biociências, Instituto de Biociências, Letras e Ciências Exatas (IBILCE), Universidade Estadual Paulista, São José do Rio Preto, São Paulo 15054-000, Brazil

Vanessa Moreira
Departamento de Farmacologia, Universidade Federal de São Paulo, São Paulo 04044-020, Brazil

Anna Lívia Linard Matos, Sergej Kudruk, Milena Heflik and David Grill
Institute of Medical Biochemistry-Center for Molecular Biology of Inflammation, University of Münster, Von-Esmarch-Str. 56, D-48149 Münster, Germany

Johanna Moratz and Bart Jan Ravoo
Organic Chemistry Institute, University of Münster, Corrensstrasse 40, D-48149 Münster, Germany

Shantel A. Vital and Elena Y. Senchenkova
Department of Molecular and Cellular Physiology, Louisiana State University Health Sciences Center-Shreveport, Shreveport, LA 71130, USA

Junaid Ansari
Department of Molecular and Cellular Physiology, Louisiana State University Health Sciences Center-Shreveport, Shreveport, LA 71130, USA
Department of Neurology, Louisiana State University Health Sciences Center-Shreveport, Shreveport, LA 71130, USA

Felicity N. E. Gavins
Department of Molecular and Cellular Physiology, Louisiana State University Health Sciences Center-Shreveport, Shreveport, LA 71130, USA
Department of Neurology, Louisiana State University Health Sciences Center-Shreveport, Shreveport, LA 71130, USA
Department of Life Sciences, Centre for Inflammation Research and Translational Medicine (CIRTM), Brunel University London, Uxbridge, Middlesex UB8 3PH, UK

Coralie Croissant, Céline Gounou, Flora Bouvet, Sisareuth Tan and Anthony Bouter
Institute of Chemistry and Biology of Membranes and Nano-objects, UMR 5248, CNRS, University of Bordeaux, IPB, F-33600 Pessac, France

Elsa Meneses-Salas, Ana García-Melero, Patricia Blanco-Muñoz, Francesc Tebar, Carles Rentero and Carlos Enrich
Departament de Biomedicina, Unitat de Biologia Cel•lular, Facultat de Medicina i Ciències de la Salut, Universitat de Barcelona, 08036 Barcelona, Spain
Centre de Recerca Biomèdica CELLEX, Institut d'Investigacions Biomèdiques August Pi i Sunyer (IDIBAPS), 08036-Barcelona, Spain

Jaimy Jose, Marie-Sophie Brenner and Thomas Grewal
School of Pharmacy, Faculty of Medicine and Health, University of Sydney, Sydney 2006, NSW, Australia

Albert Lu
Department of Biochemistry, Stanford University School of Medicine, Stanford, CA 94305, USA

Index

A

Abbreviations, 144

Acetic Acid, 88, 159, 174

B

Bacterial Infections, 184

Blood-brain Barrier, 83, 126, 197

Blot, 34, 36, 39, 42-43, 92-93, 95-96, 101, 110, 112, 116, 159, 203-206, 218, 228

Blotting, 20, 27, 39, 42-43, 87, 92-93, 100, 166, 224, 226-228

Bone Marrow, 156

Bovine Serum Albumin, 4, 26-28, 69, 225

Breast Cancer, 15, 35, 37, 40, 44-47, 79, 81, 83-89, 91, 93-95, 97-107, 122, 129, 134, 136-138, 141-144, 146-147, 149-150, 152, 154-156, 220, 234

C

Calorimetry, 19, 28

Cardiovascular Disease, 169, 197

Caspase, 47, 118, 126, 158-159, 161, 166-167, 170-171

Cell Adhesion, 37, 64, 104, 121, 124, 139, 142, 193

D

Damage-associated Molecular Patterns, 158

Data Collection, 4-5

E

Enterocytes, 118

Enthalpy, 18, 22, 28-29

F

Fibroblast Growth Factor, 128, 147, 203, 218

Fibroblasts, 31, 33, 108, 118, 142, 150, 223, 226, 234-235

Fibrosis, 34, 78, 82, 124

Flow Cytometer, 52, 70, 186

G

Galectins, 121

Gastroenterology, 218

H

Hippocampus, 119

Histone Deacetylases, 147

Hypoxia, 125, 128, 130, 134, 141, 153

I

Interleukin-6, 80-81, 198

Intraperitoneal Injection, 157, 159

Ion Channels, 31

Ionization Potential, 30

L

Lipopolysaccharide, 157-159, 161, 169, 183, 195-196, 199

Listeria Monocytogenes, 50, 65-66

Lysis, 3, 36, 86-87, 110, 150, 159, 224

Lysosomes, 12, 14, 127, 152, 217, 221, 223, 226, 231, 233-235

M

Macromolecules, 15

Metabolic Syndrome, 126

Metabolism, 91, 141-142, 156, 165, 167, 170, 194, 222, 229, 233

Metabolites, 160, 199

Metastasis, 34-35, 44-47, 83-85, 88-89, 91, 94, 97-106, 123, 129, 134-137, 140, 142-143, 146-148

Metastatic Potential, 134

Methanol, 37, 69-70, 160, 173-174, 204

Methionine, 49-50, 86, 89, 128

Methyltransferase, 112, 147

Microarray, 88, 104, 156

Mitogen-activated Protein Kinase, 50, 65, 68, 144, 194

Mycobacterium Tuberculosis, 168

N

Neuroblastoma, 65

Neuroblastoma Cells, 65

Neurodegeneration, 126, 128, 133, 234

Neurodegenerative Disorders, 127

Neuroendocrine, 108, 117, 119-121

O

Oligomerization, 60, 122, 126, 154, 158, 166, 172

Oligomers, 172-173, 177, 179

Oligonucleotides, 87

Omics, 103

Optical Microscope, 72

Osteoarthritic Cartilage, 127

Osteoarthritis, 127, 133

Overall Survival, 136, 141

P

Paclitaxel, 139, 148, 152

Pancreatic Cancer, 37, 44, 47, 105, 146-147, 151, 154-155, 220, 234
Parkinson's Disease, 127

Q
Quercetin, 82

R
Radiotherapy, 155
Retention Time, 160, 167
Reverse Transcriptase, 2, 219
Reverse Transcription, 36, 51, 224

S
Salmonella Typhimurium, 124-125, 132
Sarcolemma, 153, 201-202, 207-212, 214-217, 219-220
Sarcoplasmic Reticulum, 140, 152

T
Tagging, 84-85, 103
Tamoxifen, 101, 156
Target Cells, 125, 232
Therapeutic Target, 66, 84, 91, 100, 102, 106, 130, 137, 146, 150, 168

U
Ubiquitin, 46, 143-144
Ubiquitination, 143-144, 154
Ulcerative Colitis, 168
Ultracentrifugation, 175

V
Vascular Endothelial Cells, 124, 132
Vascular Endothelial Growth Factor, 67, 70, 125, 128, 134
Vascular Smooth Muscle Cells, 31
Viral Replication, 65

W
Western Blot, 92-93, 95, 101, 112, 116, 159, 203, 228
Western Blot Analysis, 92, 116, 159
Western Blotting, 20, 27, 87, 92-93, 100, 166, 224, 226-228
Wnt Signaling, 142
Wound Healing, 34, 37-40, 42-43, 83-84, 88, 95-96, 99-101, 123, 126, 142

X
X-ray Diffraction, 5
Xenografts, 141
Xenopus, 106

Z
Zebrafish, 202, 215
Zonula Occludens, 82

Printed in the USA
CPSIA information can be obtained
at www.ICGtesting.com
LVHW080045211023
761596LV00010BA/1925

9 781647 403966